ADHESIVES TECHNOLOGY HANDBOOK

ADHESIVES
TECHNOLOGY
HANDBOOK

by

Arthur H. Landrock

Plastics Technical Evaluation Center
U.S. Army Armament Research and Development Center
Dover, New Jersey

np **NOYES PUBLICATIONS**
Park Ridge, New Jersey, U.S.A.

Library of Congress Catalog Card Number: 85-15329
ISBN: 0-8155-1040-3
Printed in the United States

Published in the United States of America by
Noyes Publications
Mill Road, Park Ridge, New Jersey 07656

10 9 8 7 6 5 4 3 2 1

Library of Congress Cataloging-in-Publication Data

Landrock, Arthur H.
    Adhesives technology handbook.

    Bibliography: p.
    Includes index.
    1. Adhesives--Handbooks, manuals, etc. I. Title.
TP968.L36 1985   620.1'99   85-15329
ISBN 0-8155-1040-3

Preface

This book is intended to be useful to anyone working with adhesives and, to a certain extent, adhesives/sealants. The emphasis is on practical rather than theoretical aspects. The book should prove helpful to materials engineers, aerospace engineers, design engineers, chemists, adhesives and sealants manufacturers, and technical sales personnel. It may also find use as a textbook or reference book in materials engineering courses. It is a comprehensive technical treatment of the important field of structural adhesive bonding, and covers information not available in any other single source. Although the emphasis is on structural adhesive bonding, some attention is given to nonstructural bonding aspects.

A brief description of what I have attempted to cover may be helpful, Chapter 1 is a brief introduction covering functions of adhesives, advantages and disadvantages of adhesive bonding over other methods of joining, theories of adhesion, requirements for a good bond, mechanism of bond failure, and statistics on adhesives and sealants usage. Chapter 2 lists definitions of some 263 terms found in the adhesive bonding literature. Chapter 3 covers joint design and design criteria. Chapter 4 is an exhaustive treatment of surface-preparation techniques. Each metal, plastic, elastomer, or other adherend likely to be adhesive-bonded is covered, in some cases with a number of suggested treatments. Chapter 5 discusses the various types of adhesives used, along with their properties and applications. A number of different types of classifications of adhesives are described briefly. The actual listing and discussion of each type is by alphabetical order. Each of the 57 adhesive types covered is discussed in considerable detail.

Chapter 6 covers adhesives recommended for specific adherend types—specific metals, plastics, rubbers and miscellaneous materials, such as wood, glass, and ceramics. In Chapter 7 the entire adhesive bonding process is discussed, from storage of adhesives to adhesive preparation, methods of application, joint assembly methods, and bonding equipment. The weldbonding process is also covered in a completely separate discussion. Chapter 8 covers solvent cementing of thermoplastics. In Chapter 9 attention is given to the durability of adhesive-bonded joints in unfavorable environments, including temperature and humidity extremes, salt water and salt spray, chemicals and solvents, and radiation. Chapter 10 is a brief primer on quality control, including raw material inspection and process control and test methods used in checking quality, including nondestructive methods.

Chapter 11 discusses, in a general way, the various types of testing used on adhesives and adhesive joints. Following this discussion is a tabulation of 54 subject areas listing all relevant ASTM Test Methods and Practices and SAE Aerospace Recommended Practices. Both standard number designations and full titles are given. Chapter 12 is a unique chapter covering 235 published test methods and specifications used in the United States. This lengthy chapter covers ASTM Test Methods, Practices and Specifications, SAE Aerospace Materials Specifications and Aerospace Recommended Practices, Federal Specifications, a Federal Test Method Standard, Commercial Item Descriptions, and Military Specifications. Detailed descriptions are given of all of these documents. Chapter 12, used in conjunction with Chapter 11, should prove helpful to the reader who wishes to learn something about the test methods and specifications available without having to refer to the complete published document, which is frequently not readily available.

In Chapter 13 I have attempted to summarize most of the important applications of adhesive bonding in particular fields, such as automotive, aircraft, space, transportation, electrical, electronic, and building construction. Finally, the Appendix is a brief listing of recommended sources of information on adhesive bonding. The book is well documented, with references given at the end of each chapter. There are two comprehensive indexes, a General Index and an Index to Standards Listed in Chapter 12. A detailed table of contents complements the indexes.

I wish to express my gratitude to the many people who have made this book possible. Robert S. Miller of Franklin Chemical Industries was kind enough to provide a number of excellent line drawings of wood joints from his book (Chapter 3). Tom Kochaba, ASTM Staff Director for ASTM D-14, made a number of figures from ASTM standards available. I have had three excellent typists, each of whom deserves recognition—Mrs. Diane McGreen of Wharton, NJ, Mrs. Carole Nix of Sparta, NJ and Mrs. Laura Socha of Hamburg, NJ. Mr. George Narita, Vice-President and Executive Editor of Noyes Publications, encouraged and supported me throughout this project. Last of all, I must express appreciation to my wife, Rose-Marie, for her patience and understanding in my spending so much time with her chief rival, this book.

Sparta, New Jersey   Arthur H. Landrock
September 1985

ABOUT THE AUTHOR

Arthur H. Landrock is presently on the staff of the DOD Plastics Technical Evaluation Center, in Dover, New Jersey, where he has been a subject specialist in many areas including powder coatings, polymeric materials, cellular plastics, adhesives, specifications and standards, and flammability. He has authored the *Handbook of Plastics Flammability and Combustion Toxicology*, published in 1983. Mr. Landrock is also a member of the Editorial Advisory Board of the Journal of Elastomers and Plastics, and has served as Consulting Editor to the International Plastics Selector, Inc. A member of ASTM, SAMPE, SPE, and the National Fire Protection Association, he is also a member of the USA Technical Advisory Group for ISO TC 61 on Plastics. He serves as Editorial Review Chairman of ASTM Committee D-14 on Adhesives and D-20 on Plastics, and is also a member of the ASTM Standing Committee on Terminology.

NOTICE

This book represents the opinions of the author alone and in no way is to be interpreted as representing the position of the Department of the Army or any other branch of the United States Government.

Contents

1. INTRODUCTION 1
   Functions of Adhesives. 1
   Advantages and Disadvantages of Adhesive Bonding. 2
       Advantages. 2
       Disadvantages 3
   Comments on Adhesive Usage 3
   Theories of Adhesion 4
       Mechanical Theory 5
       Adsorption Theory 5
       Electrostatic Theory 6
       Diffusion Theory. 6
       Weak-Boundary Layer Theory 6
   Requirements for a Good Bond 7
       Cleanliness 7
       Wetting 7
       Solidification 7
       Adhesive Choice 7
       Joint Design 8
   Mechanism of Bond Failure. 8
   Adhesives and Sealants Consumption .10
   References 11

2. DEFINITIONS OF TERMS. 12
   Comments 12
   Definitions. 12
   References. 29

3. JOINT DESIGN AND DESIGN CRITERIA 31
   Basic Principles. 31

Types of Stress..31
Methods of Improving Joint Efficiency......................33
  Joint Design Criteria...................................33
  Typical Joint Designs...................................37
    Butt Joints..37
    Lap Joints...37
  Peeling of Adhesive Joints..............................38
  Stiffening Joints.......................................39
  Cylindrical Joints......................................39
  Angle and Corner Joints.................................41
Joints for Plastics and Elastomers.........................42
  Flexible Materials......................................42
  Rigid Plastics..43
Wood Joints..44
  Glue or Corner Blocks...................................46
  Mortise and Tenon Joints................................46
  Dovetail Joints...47
  Dado Joints...47
  Lap Joints..47
  Miter Joints..47
  Lapped Miter Joints.....................................49
  Corner Joints...50
Metal Joints...50
Different Adherend Types in Joints.........................50
Stress Analysis of Adhesive Joints.........................50
  Theoretical Analysis of Stresses and Strains............51
  Experimental Analyses...................................51
  Failure Analyses..51
  Methods of Stress Analysis..............................52
    Theory of Volkersen..................................52
    Theory of Goland and Reissner........................52
References...53

4. SURFACE PREPARATION OF ADHERENDS.......................54
Introduction...54
Cleaning...56
  General Sequence of Cleaning............................57
    Solvent Cleaning.....................................57
    Intermediate Cleaning................................60
    Chemical Treatment...................................60
Priming..60
Activated-Gas Surface Treatment of Polymers (Plasma Treatment)....61
Methods for Evaluating Effectiveness of Surface Preparation.......63
  Water-Break Test..63
  Contact-Angle Test......................................63
Surface Exposure Time (SET)................................63
Surface Preparation of Metals..............................64
  Aluminum..64

FPL Immersion Etch (Improved Method). 64
FPL Paste Etch. 65
Chromate-Free Etch Process 65
Anodization. 66
Beryllium. 67
Brass. 67
Bronze 68
Cadmium. 68
Copper and Copper Alloys 68
Black Oxide Coating 68
Sodium Dichromate-Sulfuric Acid Process 69
Gold. 69
Magnesium and Magnesium Alloys 69
Alkaline-Detergent Solution (ASTM D 2651, Method A). 69
Hot Chromic Acid (ASTM D 2651, Method B) 70
Anodic and Other Corrosion-Preventive Treatments (ASTM
D 2651, Method D). 70
Conversion Coatings (ASTM D 2651, Method E). 70
Nickel and Nickel Alloys. 70
Abrasive Cleaning 70
Nitric Acid Etch 70
Sulfuric-Nitric Acid Pickle. 71
Manual Cleaning for Plated Parts 71
Nickel-Base Alloy Treatments 71
Platinum 71
Abrasive Cleaning 71
Abrasive Scouring 71
Silver 71
Chromate Conversion Coating 71
Degrease-Abrade-Prime. 72
Steel. 72
Acid Treatment. 73
Potassium Iodide—Phosphoric Acid Method 73
Stainless Steel. 74
Acid Etch (for Types 301 and 302 Stainless Steel). 74
Oxalic-Sulfuric Acid Process for Maximum Heat Resistance. 75
Bromophosphate Treatment 75
ASTM Suggested Methods. 76
Tin. 76
Titanium 76
Stabilized Phosphate-Fluoride Treatment. 77
Alkaline Cleaning 78
Alkaline Etch 78
Pasa Jell Treatment 78
VAST Process. 78
Alkaline-Peroxide Etch (RAE Etch) 79
Recent Studies 80
Tungsten and Alloys 80

Hydrofluoric-Nitric Sulfuric Acid Method 80
Uranium 80
    Abrasive Method 80
    Acetic Acid-Hydrochloric Acid Method 81
    Nitric Acid Bath 81
Zinc and Alloys 81
    Abrasion (for General-Purpose Bonding) 81
    Acid Etch 81
    Sulfuric Acid/Dichromate Etch 82
    Conversion Coatings 82
Weldbonding Metals 82
    Vapor Honing/Pasa Jell 107M Procedure 83
Surface Preparation of Plastics 84
Thermoplastics 84
    Acetal Copolymer 84
    Acetal Homopolymer 85
    Acrylonitrile-Butadiene-Styrene (ABS) 86
    Allyl Diglycol Carbonate 87
    Cellulosics [Cellulose Acetate, Cellulose Acetate Butyrate
       (CAB), Cellulose Nitrate, Cellulose Propionate, Ethyl
       Cellulose] 87
    Ethylene-Chlorotrifluoroethylene Copolymer (E-CTFE) 87
    Ethylene-Tetrafluoroethylene Copolymer (ETFE) 88
    Ethylene-Vinyl Acetate (EVA) 88
    Fluorinated Ethylene-Propylene Copolymer (FEP) 88
    Ionomer 88
    Nylon (Polyamide) 89
    Perfluoroalkoxy Resins (PFA) 89
    Phenylene-Oxide-Based Resins (Polyaryl Ethers) 89
    Polyarylate 90
    Polyaryl Sulfone 90
    Polycarbonate 91
    Polychlorotrifluoroethylene (PCTFE) 92
    Polyester (Saturated) 92
    Polyetheretherketone (PEEK) 93
    Polyethersulfone (PES) 93
    Polyetherimide 94
    Polyethylene (PE) 94
    Polymethylmethacrylate (PMMA) 95
    Polymethylpentene (TPX) 95
    Polyphenylene Sulfide (PPS) 96
    Polypropylene (PP) 96
    Polystyrene (PS) 97
    Polysulfone 98
    Polytetrafluoroethylene (PTFE) 98
    Polyvinyl Chloride (PVC) 98
    Polyvinyl Fluoride (PVF) 99
    Polyvinylidene Fluoride (PVDF) 99

Styrene-Acrylonitrile (SAN) 100
Thermosets 100
  Diallyl Phthalate (DAP) 100
  Epoxies. 101
  Melamine-Formaldehyde (Melamines) 101
  Phenol-Formaldehyde (Phenolics). 101
  Polyester. 102
  Polyimide 102
  Polyurethane 102
  Silicone Resins 103
  Urea-Formaldehyde (U-F). 103
Reinforced Plastics/Composites 104
  Reinforced Plastics/Thermosets 104
  Reinforced Thermoplastics (Glass-Reinforced) 106
Plastic Foams. 106
Surface Preparation of Rubbers 106
Neoprene (Polychloroprene) (CR) 108
  Abrasive Treatment. 108
  Cyclization. 108
  Cyclization (Boeing-Vertol Modification). 109
  Chlorination. 109
  Activated Gas Plasma. 109
Ethylene-Propylene-Diene Terpolymer (EPDM). 110
  Abrasive Treatment 110
  Cyclization. 110
Silicone Rubber (Polydimethylsiloxane) 110
  Solvent Cleaning. 110
  Soap-and-Water Wash. 110
  Primers 110
  Activated-Gas Plasma. 110
Butyl Rubber (IIR) 110
  Abrasive Treatment. 111
  Cyclization. 111
  Chlorination. 111
  Primers 111
  Activated-Gas Plasma. 111
Chlorobutyl Rubber (CIIR). 111
  Abrasive Treatment 111
  Cyclization. 111
  Chlorination. 111
  Other Methods 111
Chlorosulfonated Polyethylene (CSM). 111
  Abrasive Treatment 111
  Primers 112
Nitrile Rubber (Butadiene-Acrylonitrile) (NBR) 112
  Abrasive Treatment 112
  Cyclization. 112
  Chlorination. 112

Primers 112
Activated-Gas Plasma 112
Polyurethane Elastomers 112
Abrasive Treatment 113
Primers 113
Synthetic Natural Rubber (Polyisoprene) (IR) 113
Abrasive Treatment 113
Cyclization 113
Chlorination 113
Styrene-Butadiene Rubber (SBR) (Buna S) 113
Abrasive Treatment 113
Cyclization 113
Chlorination 113
Primers 113
Activated-Gas Plasma 113
Polybutadiene (Butadiene Rubber) (BR) 113
Abrasive Treatment 113
Cyclization 113
Chlorination 114
Solvent Wipe 114
Fluorosilicone Elastomers 114
Fluorocarbon Elastomers 114
Sodium Etch 114
Dry Abrasion 114
Epichlorohydrin Elastomer 114
Polysulfide Rubber (PTR) 115
Abrasive Treatment 115
Chlorination 115
Primers 115
Polypropylene Oxide (Propylene Oxide Rubber) (PO) 115
Polyacrylate (Polyacrylic Rubber) (AMC) (ANM) 115
Dry Abrasion 115
Thermoplastic Rubber (Thermoplastic Elastomer) 115
Surface Preparation of Wood and Wood Products 116
Surface Preparation of Miscellaneous Materials 117
Asbestos (Rigid) 117
Brick and Fired Non-Glazed Building Materials 117
Carbon and Graphite (for General-Purpose Bonding) 117
Glass (Non-Optical) 117
Abrasive Treatment (for General-Purpose Bonding) 117
Acid Etch (for Maximum Strength) 118
Primers 118
Glass (Optical) 118
Ceramics (Unglazed) 118
Ceramics (Glazed) 118
Concrete 119
Portland Cement Type 119
Bituminous Type 119

Painted Surfaces 119
References 120

5. ADHESIVE TYPES AND THEIR PROPERTIES AND
   APPLICATIONS 126
   Introduction 126
   Classification 126
      Classification by Function 126
         Structural Adhesives 126
         Nonstructural Adhesives 126
      Classification by Physical Form 126
         Liquid Adhesives 126
         Paste Adhesives 127
         Tape and Film Adhesives 127
         Powder or Granule Adhesives 127
      Classification by Mode of Application and Setting . 127
      Classification by Specific Adherends or Applications . 128
      Classification by Chemical Composition 128
         Thermosetting Adhesives 128
         Thermoplastic Adhesives 129
         Elastomeric Adhesives 129
         Adhesive Alloys 130
      Natural vs. Synthetic Adhesives 130
         Natural Adhesives 130
         Synthetic Adhesives 130
      SME Classification 130
         Chemically Reactive Types 130
         Evaporative or Diffusion Adhesives 131
         Hot-Melt Adhesives 131
         Delayed-Tack Adhesives 132
         Tape and Film Adhesives 132
         Pressure-Sensitive Adhesives 132
      Classification by Rayner 132
         Thermosetting Resin Adhesives 132
         Thermoplastic Resin Adhesives 132
         Two-Polymer Adhesives (Alloys) 133
      Additional Classification 133
   Adhesive Composition 134
      Adhesive Base or Binder 134
      Hardener (for Thermosetting Adhesives) 134
      Solvents 134
      Diluents 134
      Fillers 134
      Carriers or Reinforcements 134
   Individual Adhesive Types and Their Characteristics . 135
      Acrylics 135
      Allyl Diglycol Carbonate (CR-39) 136
      Alloyed (Two-Polymer) Adhesives 136

Anaerobic Adhesive/Sealants............................137
Aromatic Polymer Adhesives (Polyaromatics)...............138
Asphalt..139
Butyl Rubber Adhesives................................139
Cellulose Ester Adhesives..............................139
Cellulose Ether Adhesives..............................140
Conductive Adhesives..................................140
    Electrically Conductive Adhesives (Chip-Bonding
        Adhesives)....................................140
    Thermally Conductive Adhesives.....................141
Cyanoacrylate Adhesives...............................141
Delayed-Tack Adhesives................................143
Elastomeric Adhesives.................................144
Epoxy Adhesives......................................144
Epoxy-Phenolic Adhesives..............................149
Epoxy-Polysulfide Adhesives............................150
Film and Tape Adhesives...............................150
Furane Adhesives.....................................153
Hot-Melt Adhesives...................................154
    Foamable Hot-Melt Adhesives.......................154
    Ethylene-Vinyl Acetate (EVA).......................155
    Polyamide (Nylon) and Polyester Resins...............155
    Other Hot-Melt Adhesives..........................155
Inorganic Adhesives (Cements)..........................156
    Soluble Silicates (Potassium and Sodium Silicate)........156
    Phosphate Cements...............................157
    Basic Salts (Sorel Cements)........................157
    Litharge Cements.................................157
    Sulfur Cements...................................157
Melamine-Formaldehyde Adhesives (Melamines).............158
Microencapsulated Adhesives...........................158
Natural Glues..158
    Vegetable Glues..................................158
    Glues of Animal Origin............................160
Neoprene (Polychloroprene) Adhesives...................162
Neoprene-Phenolic Adhesives...........................162
Nitrile-Epoxy (Elastomer-Epoxy) Adhesives...............163
Nitrile-Phenolic Adhesives.............................163
Nitrile Rubber Adhesive...............................163
Nylon Adhesives......................................164
Nylon-Epoxy Adhesives................................164
Phenolic Adhesives...................................164
    Acid-Catalyzed Phenolics..........................165
    Hot-Setting Phenolics.............................165
Phenoxy Adhesives....................................166
Polybenzimidazole (PBI) Adhesives......................166
Polyester Adhesives..................................168
Polyimide (PI) Adhesives..............................168

Polyisobutylene Adhesives 169
Polystyrene Adhesives 169
Polysulfides (Thiokols) 170
Polysulfone Adhesives 170
Polyurethane Adhesives 171
Polyvinyl Acetal Adhesives 172
Polyvinyl Acetate Adhesives 173
Polyvinyl Alcohol Adhesives 173
Polyvinyl Butyral Adhesives 174
Premixed Frozen Adhesives 174
Pressure-Sensitive Adhesives (PSAs) 174
Resorcinol-Formaldehyde Adhesives 175
Rubber-Base Adhesives 176
Silicone Adhesives 176
Solvent-Based Systems 178
Thermoplastic Resin Adhesives 179
Thermoplastic Rubber 180
Thermosetting Resin Adhesives 180
Ultraviolet-Curing Adhesives 180
Urea-Formaldehyde Adhesives (Ureas) 181
Vinyl-Epoxy Adhesives 182
Vinyl-Phenolic Adhesives 182
    Polyvinyl Formal-Phenolics 182
    Polyvinyl Butyral-Phenolics 183
Vinyl-Resin Adhesives 183
Water-Based Adhesives 183
References 185

6. ADHESIVES FOR SPECIFIC ADHERENDS 190
Introduction 190
Metals 190
    Aluminum and Alloys 190
    Beryllium 191
    Brass 191
    Bronze 191
    Cadmium (Plated on Steel) 191
    Copper and Copper Alloys 191
    Gold 191
    Lead 191
    Magnesium and Magnesium Alloys 191
    Nickel and Nickel Alloys 192
    Plated Metals 192
    Silver 192
    Steel, Mild, Carbon (Iron) 192
    Stainless Steel 192
    Tin 192
    Titanium and Titanium Alloys 192
    Tungsten and Tungsten Alloys 193

Uranium 193
Zinc and Zinc Alloys 193
Thermoplastics 193
Acetal Copolymer 193
Acetal Homopolymer 194
Acrylonitrile-Butadiene-Styrene (ABS) 194
Cellulosics 194
Ethylene-Chlorotrifluoroethylene (E-CTFE) 194
Fluorinated-Ethylene Propylene (FEP Teflon) 194
Fluoroplastics 194
Ionomer 194
Nylons (Polyamides) 194
Perfluoroalkoxy Resins (PFA) 195
Phenylene-Oxide Based Resins 195
Polyaryl Ether 195
Polyaryl Sulfone 195
Polycarbonate 195
Polychlorotrifluoroethylene (PCTFE) (KEL-F) 195
Polyester (Thermoplastic Polyester) 196
Polyetheretherketone (PEEK) 196
Polyetherimide 196
Polyethersulfone 196
Polyethylene 196
Polymethylmethacrylate (PMMA) 196
Polymethylpentene (TPX) 196
Polyphenylene Sulfide (PPS) 196
Polypropylene 197
Polystyrene 197
Polysulfone 197
Polytetrafluoroethylene (PTFE) (TFE Teflon) 197
Polyvinyl Chloride (PVC) 197
Polyvinyl Fluoride (PVC) 198
Polyvinylidene Fluoride (PVDF) 198
Styrene-Acrylonitrile (SAN) 198
Thermosetting Plastics (Thermosets) 198
Diallyl Phthalate (DAP) 198
Epoxies 198
Melamine-Formaldehyde (Melamines) 198
Phenol-Formaldehyde (Phenolics) 198
Polyester (Thermosetting Polyester) 198
Polyimide 199
Polyurethane 199
Silicone Resins 199
Urea-Formaldehyde 199
Reinforced Plastics/Composites 199
Plastic Foams 199
Rubbers (Elastomers) 201
Wood 201

        Glass and Ceramics 202
    References.. 202

7. ADHESIVE BONDING PROCESS 205
    Introduction...................................... 205
    Adhesive Storage.................................. 205
    Adhesive Preparation 206
        Small-Portion Mixer Dispensers 206
    Methods of Adhesive Application...................... 207
        Liquid Adhesives................................ 207
            Brushing 207
            Flowing...................................... 207
            Spraying 207
            Roll Coating.................................. 208
            Knife Coating................................ 208
            Silk Screening................................ 208
            Oil Can and Squeeze Bottle...................... 208
            Hand Dipping................................ 208
        Pastes 208
            Spatulas, Knives, Trowels 209
        Powders...................................... 209
        Films 209
        Hot Melts.................................... 210
            Melt-Reservoir Systems (Tank-Type Applications).......... 210
            Progressive-Feed Systems 210
    Joint-Assembly Methods............................ 211
        Wet Assembly.................................. 212
        Pressure-Sensitive and Contact Bonding.................. 212
        Solvent Activation.............................. 212
        Heat Activation................................ 213
        Curing.. 213
    Bonding Equipment................................ 214
        Pressure Equipment.............................. 214
        Heating Equipment 215
            Direct Heating Curing.......................... 215
            Radiation Curing.............................. 216
            Electric Resistance Heaters 216
            High-Frequency Dielectric (Radio Frequency) Heating....... 217
            Induction Heating............................ 217
            Low-Voltage Heating.......................... 217
            Ultrasonic Activation.......................... 217
            Adhesive Thickness 218
    Weldbonding 218
        Weldbond Configuration.......................... 220
        Advantages and Limitations........................ 221
        Surface Preparation 221
        Adhesive Choice 221
        Tooling for Weldbonding 222

Weldbonding Techniques 222
Quality Control. 223
References. 223

8. SOLVENT CEMENTING OF PLASTICS 225
   Introduction. 225
   Recommendations for Specific Adherends 227
      Acetal Copolymer 227
      Acetal Homopolymer. 227
      Acrylonitrile-Butadiene-Styrene (ABS) 227
      Cellulosics 228
         Cellulose Acetate. 228
         Cellulose Acetate Butyrate (CAB). 228
         Cellulose Nitrate 229
         Cellulose Propionate (CP) 229
         Ethyl Cellulose 230
      Nylons (Polyamides) 230
      Polycarbonate. 231
      Polystyrene 231
      Styrene-Acrylonitrile (SAN) 232
      Polysulfone 233
      Polybutylene Terephthalate (PBT) 233
      Polymethylmethacrylate (PMMA). 233
      Phenylene-Oxide Based Resins. 233
      Polyvinyl Chloride (PVC) 235
      Chlorinated Polyvinyl Chloride (CPVC). 236
      Polyetherimide 236
   References. 236

9. EFFECTS OF ENVIRONMENT ON DURABILITY OF
   ADHESIVE JOINTS 238
   Introduction. 238
   High Temperature 240
      Epoxies. 241
      Modified Phenolics 241
         Nitrile-Phenolics 241
         Epoxy-Phenolics 241
      Polysulfone 241
      Silicones 242
      Polyaromatics. 242
         Polyimides (PI). 242
         Polybenzimidazoles (PBI) 243
   Low and Cryogenic Temperatures. 243
   Humidity and Water Immersion 246
      Effects of Surface Preparation on Moisture Exposure 246
      Stressed Temperature/Humidity Test 247
      Hot-Water-Soak Test 251
      Fatigue-Life Data 251

Salt Water and Salt Spray 253
    Seacoast Weathering Environment 254
    Salt Water Immersion. 255
    Nitrile-Phenolic Adhesives. 256
    Boeing/Air Force Studies on Salt-Spray Effects. 256
Weathering. 257
    Simulated Weathering/Accelerated Testing. 257
    Outdoor Weathering (Picatinny Arsenal Studies) 258
Chemicals and Solvents. 263
Vacuum. 264
Radiation. 265
Biological. 268
Test Methods 269
References. 270

10. QUALITY CONTROL 273
Introduction. 273
Raw Material Inspection and Process Control 275
    Incoming Material Control 275
        Containers. 276
        Adhesives. 276
        Adhesives—Mechanical Properties. 276
        Adhesives—Miscellaneous Properties (Including Creep) 277
    Surface Preparation Control 277
    Process Control of Bonding. 277
        Prefit 277
        Adhesive Application. 278
        Assembly. 278
        Curing. 278
        Standard Test Specimen 279
    Final Inspection 279
        Nondestructive Tests 279
    Weldbonding 285
    References. 285

11. TEST METHODS AND PRACTICES 287
Introduction. 287
Tensile 287
Shear 288
Peel 289
Cleavage 290
Creep 290
Fatigue 290
Impact 290
Durability 290
Compilation of Test Methods and Practices 291
    Aging 291
    Amylaceous Matter 291

Ash Content. 291
Biodeterioration 291
Blocking Point 292
Bonding Permanency 292
Characterization 292
Chemical Reagents. 292
Cleavage 292
Cleavage/Peel Strength 292
Corrosivity. 292
Creep 292
Cryogenic Temperatures. 292
Density 293
Durability (Including Weathering). 293
Electrical Properties. 293
Electrolytic Corrosion 293
Fatigue 293
Filler Content. 293
Flexibility 293
Flexural Strength 293
Flow Properties. 294
Fracture Strength in Cleavage 294
Gap-Filling Adhesive Bonds. 294
Grit Content. 294
High-Temperature Effects. 294
Hydrogen-Ion Concentration. 294
Impact Strength 294
Light Exposure. 294
Low and Cryogenic Temperatures. 294
Nonvolatile Content. 294
Odor. 295
Peel Strength (Stripping Strength) 295
Penetration 295
pH 295
Radiation Exposure (Including Light) 295
Rubber Cement Tests. 295
Salt Spray (Fog) Testing. 295
Shear Strength (Tensile Shear Strength). 296
Specimen Preparation. 297
Spot-Adhesion Test 297
Spread. 297
Storage Life 297
Strength Development 297
Stress-Cracking Resistance. 297
Stripping Strength. 297
Surface Preparation 297
Tack. 298
Tensile Strength 298
Torque Strength 298

Viscosity 298
Volume Resistivity 298
Water Absorptiveness 299
Weathering 299
Wedge Test 299
Working Life 299
References 299

12. STANDARD TEST METHODS, PRACTICES AND
    SPECIFICATIONS 300
    Introduction 300
    Test Methods and Practices 302
        American Society for Testing and Materials (ASTM) 302
        Society of Automotive Engineers (SAE) 339
            Aerospace Recommended Practices (ARP's) 339
        General Services Administration—Federal Test Methods 340
    Specifications 341
        American Society for Testing and Materials (ASTM) 341
        Society of Automotive Engineers (SAE)—Aerospace Materials
            Specifications (AMS's) 347
        General Services Administration—Federal Specifications and
            Commercial Item Descriptions 353
        Department of Defense (DOD)—Military Specifications 367
    References 392

13. APPLICATIONS OF ADHESIVE BONDING 393
    Introduction 393
    Automotive Applications 393
    Building Construction 395
    Electrical/Electronic Applications 396
        Printed Circuit Boards 397
    Aerospace Applications 397
        Aircraft Applications 397
            Helicopter Rotor Blades 398
            European Airbus 398
        Space Applications 399
    Civil Engineering Applications 399
    Bonded and Coated Abrasives 400
    Recreation Industry 400
    Marine Applications 401
    Packaging 401
    Miscellaneous 401
    References 402

APPENDIX—SOURCES OF INFORMATION 403
    Journals and Other Periodicals 403
    Manufacturers' Bulletins 405
    Technical Conferences 405

Seminars and Workshops. 405
Standardization Activities. 405
    ASTM. 405
    SAE (AMS's and ARP's). 406
Trade and Professional Associations 406
Consultants 407
Information Center 407
Miscellaneous Items. 407

GENERAL INDEX 409

INDEX TO STANDARDS LISTED IN CHAPTER 12. 441

1

Introduction

Adhesive bonding is the process of uniting materials with the aid of an adhesive, a substance capable of holding such materials together by surface attachment.[1] There are two principal types of adhesive bonding, structural and non-structural. *Structural adhesive bonding* is adhesive bonding which stresses the adherend (the object being bonded) to the yield point, thereby taking full advantage of the strength of the adherend. Structural adhesive bonds must be capable of transmitting structural stress without loss of structural integrity within design limits.[2] The emphasis in this book will be on structural adhesive bonding. *Non-structural adhesive bonding*, including such applications as adhesive/sealants and conductive adhesives, will also be discussed to some extent.

FUNCTIONS OF ADHESIVES

The primary function of adhesives is to join parts together. Adhesives do this by transmitting stresses from one member to another in a manner that distributes the stresses much more uniformly than can be achieved with conventional mechanical fasteners. Adhesive bonding often provides structures that are mechanically equivalent to, or stronger than, conventional assemblies at lower cost and weight. In *mechanical fastening* or *spot welding*, the strength of the structure is limited to that of the areas of the members in contact with the fasteners or welds.[3]

Smooth surfaces are an inherent advantage of adhesively joined structures and products. Exposed surfaces are not defaced and contours are not disturbed, as happens with mechanical fastening systems. This feature is important, both in function and appearance. Aerospace structures, including helicopter rotor blades, require such external smoothness to minimize drag and to keep temperatures as low as possible. Lighter-weight materials can often be used with adhesive bonding than with conventional fastening because the uniform stress distribution in the joint permits full utilization of the strength and rigidity of the adherends.[3]

1

Dissimilar materials are readily joined by many adhesives, provided that proper surface treatments are used. Metal adherends that would ordinarily corrode because of their electromotive series relationship can be protected by a layer of nonconductive adhesive that not only joins the adherends but also isolates them. Adhesives can also be formulated to be conductive. In addition to joining metals to metals, adhesives can be used to join metals to plastics, ceramics, cork, rubber, or combinations of materials. Of course, plastics can also be bonded to plastics, to rubber, etc.[3]

Where temperature variations are encountered in the service of an item containing dissimilar materials, adhesives perform another useful·function. Flexible adhesives are able to accommodate differences in the *thermal-expansion coefficients* of the adherends and in this way prevent damage that might occur if a stiff fastening system were used.[3]

Sealing is another important function of adhesive joining. The continuous bond seals out liquids or gases that do not attack the adhesive (or adhesive/sealant). Adhesive/sealants are often used in place of solid or cellular gaskets. Sealing is also carried out by potting and encapsulation compounds in circuit boards, motors, and other electrical and electronic assemblies. These materials are not really adhesives or adhesive/sealants, however. *Mechanical damping* can be imparted to a structure through the use of adhesives formulated for that purpose. A related characteristic, fatigue resistance, can be improved by the ability of such adhesives to withstand cyclic strains and shock loads without cracking. In a properly designed joint the adherends generally fail in fatigue before the adhesive fails. Thin or fragile parts can also be adhesive bonded. Adhesive joints do not usually impose heavy loads on the adherends, as in riveting, or localized heating, as in welding. The adherends will also be relatively free from heat-induced distortion.[3]

ADVANTAGES AND DISADVANTAGES OF ADHESIVE BONDING

The above discussion has considered a number of advantages of adhesive bonding. The following tabulation, however, will cover both advantages and disadvantages, recognizing that some of the points have already been mentioned.

Advantages[4-6]

- Provide uniform distribution of stress and larger stress-bearing area
- Join thin or thick materials of any shape
- Join similar or dissimilar materials
- Minimize or prevent electrochemical (galvanic) corrosion between dissimilar materials
- Resist fatigue and cyclic loads
- Provide joints with smooth contours
- Seal joints against a variety of environments
- Insulate against heat transfer and electrical conductance (in some cases adhesives are designed to provide such conductance)

- The heat required to set the joint is usually too low to reduce the strength of the metal parts
- Damp vibration and absorb shock
- Provide an attractive strength/weight ratio
- Frequently faster or cheaper than mechanical fastening

Disadvantages[4,6,7]

- The bond does not permit visual examination of the bond area (unless the adherends are transparent)
- Careful surface preparation is required to obtain durable bonds, often with corrosive chemicals
- Long cure times may be needed, particularly where high cure temperatures are not used
- Holding fixtures, presses, ovens and autoclaves, not usually required for other fastening methods, are necessities for adhesive bonding
- Upper service temperatures are limited to approximately 350°F (177°C) in most cases, but special adhesives, usually high-cost, are available for limited use to 700°F (371°C)
- Rigid process control, including emphasis on cleanliness, is required for most adhesives
- The useful life of the adhesive joint depends on the environment to which it is exposed
- Natural or vegetable-origin adhesives are subject to attack by bacteria, mold, rodents, or vermin
- Exposure to solvents used in cleaning or solvent cementing may present health problems

COMMENTS ON ADHESIVE USAGE

The most common methods of structural fastening include welding or brazing, useful in heavy-gage metal. These techniques are expensive and require considerable heat. Many lightweight materials, such as aluminum, magnesium, and titanium, are difficult to weld and are distorted by the heat applied. High-strength, reliable joints can be made in these metals with adhesives. Holes needed for rivets or other mechanical fasteners are not required for an adhesive bond, thereby allowing the stresses to be uniformly distributed over the entire bonded areas. The stress-distribution characteristics and inherent toughness of adhesives provide bonds with superior fatigue resistance. In well-designed joints the adherends usually fail in fatigue before the adhesive fails. With proper selection and use of adhesives, shear strengths up to 7,000 psi (48.3 MPa) can be obtained.

In the presence of vibration, a bonded structure usually has a longer life than a riveted one. In addition, because adhesives of difficult moduli (stiffnesses) are available, damaging resonant frequencies can be modified, or even eliminated by careful analysis and selection of the adhesive to be used. In manufacturing aircraft and motor vehicles, adhesive bonding offers advantages over riveting and welding, since (a) riveted structures add weight and drag, (b) rivets and welds are unsightly and difficult to conceal, and (c) holes and welds may bring about corrosion. Film adhesives are especially useful for this type of application since they facilitate close control of weight distribution and total weight of adhesive used.[7]

Some applications, such as bonded honeycomb structures, are completely fabricated with the use of adhesives. One example of this type of structure—thin nonmetallic honeycomb core bonded to thin metal facings—could not be constructed without adhesives. Adhesives may also function as electrical and/or thermal insulators in joints. Their thermal insulating efficiency may be increased by foaming the adhesive in place. On the other hand, the electrical and/or thermal conductivities of adhesives can be raised appreciably by adding metallic fillers. Oxide fillers, such as alumina or beryllia, increase only the thermal conductivity.[4]

Adhesive bonding is not a panacea for assembling all products. The fact that there are so many variations in available adhesive formulations makes selection of the optimum adhesive for a particular application more difficult than selecting a mechanical fastening system. These variations also complicate control procedures on incoming materials, assembly processing, and testing of the finished product. Even though adhesive-bonding operations can be automated, they usually require higher-caliber personnel than other joining methods.[3]

Although adhesives can produce structures that are more reliable than those joined by conventional methods, adhesive-bonded structures must be carefully designed and used under conditions that do not exceed the known operational limitations of the adhesive. Such limitations include types and magnitudes of stresses (whether static or dynamic), and environmental factors such as temperature, humidity, salt environment, or the presence of other vapors or liquids. The effects of salt spray, simulating a marine environment, are particularly corrosive to aluminum-bonded joints.[4]

Probably all organic adhesives provide poor bonds when moisture and stress are present at the same time. Some organic adhesives may sustain constant loads that produce no greater than 10% of their normal ultimate failing stress for only short periods. Joints made with other adhesives may sustain 50% of failing stress for extended periods of time. Reasons for this behavior are not completely known. It is important to know that such behavior exists and to take it into consideration in designing joints. The permanence of an adhesive joint is influenced not only by the properties of the adhesive, but also by the method of preparing the surfaces of the adherends for bonding. If the surface layers are weak and susceptible to moisture, the adhesive joint will also be weak and susceptible to moisture. Thorough surface preparation, sometimes requiring aggressive and messy techniques, is an essential step if reliable bonds are to be obtained.[4]

THEORIES OF ADHESION

The actual mechanism of adhesive attachment is not yet fully understood. Some of the theories suggested are given below.

Mechanical Theory

According to this theory, in order to function properly the adhesive must penetrate cavities on the surface of the substrate and displace the trapped air at the interface. Mechanical anchoring of the adhesive is an important factor in bonding many porous substrates such as plastic foams (cellular plastics). Adhesives also frequently bond better to nonporous abraded surfaces than they do to smooth surfaces. This effect may be due to (1) mechanical interlocking, (2) formation of a clean surface, (3) formation of a highly reactive surface, and (4) increasing surface area. While the surface does become rougher because of abrasion, it is believed that a change in *both* physical and chemical properties of the surface layer produces an increase in adhesive strength.[6]

Adsorption Theory

This theory states that adhesion results from molecular contact between two materials and the surface forces that develop. The process of establishing continuous contact between the adhesive and the adherend is called *wetting*. For an adhesive to wet a solid surface the adhesive should have a lower surface tension (δ) than the critical surface tension of the solid. Figure 1-1 illustrates good and poor wetting of an adhesive spreading over a surface. Good wetting results when the adhesive flows into the valleys and crevices on the substrate surface. Poor wetting results when the adhesive bridges over the valley and results in a reduction of the actual contact area between the adhesive and adherend, resulting in a lower overall joint strength.[6]

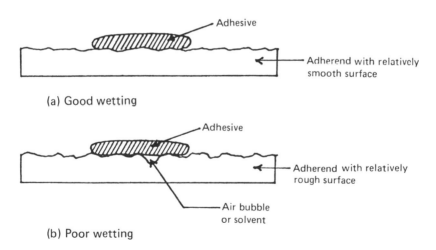

(a) Good wetting

(b) Poor wetting

Figure 1-1: Examples of good and poor wetting by an adhesive spreading across a surface. Modified after Schneberger.[6]

Most organic adhesives readily wet metal adherends. On the other hand, many solid organic substrates have surface tensions lower than those of common

adhesives. The fact that good wetting requires the adhesives to have a lower surface tension than the substrate explains, in part, why organic adhesives such as
epoxies have excellent adhesion to metals, but offer weak adhesion on untreated
polymeric substrates such as polyethylene, polypropylene, and fluoroplastics.[6]

After intimate contact is achieved between the adhesive and adherend
through wetting, it is believed that permanent adhesion results primarily through
forces of molecular attraction. Four general types of chemical bonds are thought
to be involved in adhesion and cohesion. These are:[6]

- electrostatic bonds ⎫
- covalent bonds   ⎬ — primary bonds
- metallic bonds   ⎭
- Van der Waal's forces — secondary bonds

Electrostatic Theory

According to this theory electrostatic forces in the form of an electrical
double layer are formed at the adhesive-adherend interface. These forces account
for resistance to separation. This theory gains support from the fact that electrical discharges have been noticed when an adhesive is peeled from a substrate.[6]

Diffusion Theory

This theory suggests that adhesion is developed through the interdiffusion
of molecules in the adhesive and adherend. The diffusion theory is primarily applicable when both the adhesive and adherend are polymers with long-chain
molecules capable of movement. Solvent cementing or heat welding of thermoplastics is considered to be due to diffusion of molecules.[6]

Weak-Boundary Layer Theory

According to this theory, first described by Bikerman, when bond failure
seems to be at the interface, usually a cohesive break or a weak boundary layer
is the real event.[8] Weak-boundary layers can originate from the adhesive, the adherend, the environment, or a combination of any of these three factors. Weak-
boundary layers can occur in the adhesive or adherend if an impurity concentrates near the bonding surface and forms a weak attachment to the substrate.
When failure takes place it is the weak-boundary layer that fails, although failure
appears to take place at the adhesive-adherend interface.[6]

Polyethylene and metal oxides are examples of the weak-boundary-layer effect. Polyethylene has a weak, low-molecular-weight constituent that is evenly
distributed throughout the polymer. This weak-boundary layer is present at the
interface and contributes to low failing stress when polyethylene is used as an
adhesive or adherend. Some metal oxides are weakly attached to their base
metals. Failure of adhesive joints made with these materials occurs cohesively
within the oxide. Certain oxides, such as aluminum oxide, are very strong and
do not significantly impair joint strength. Weak-boundary layers, such as are
found in polyethylene and metal oxides, can be removed or strengthened by various surface treatments. Weak-boundary layers formed from the bonding en-

vironment, generally air, are very common. When the adhesive does not wet the substrate, as shown in Figure 1-1, a weak-boundary layer is trapped at the interface, causing a reduction in joint strength.[6]

REQUIREMENTS FOR A GOOD BOND

The basic requirements for a good adhesive bond are the following:[6]

- cleanliness of surfaces
- wetting
- proper choice of adhesive
- good joint design

Cleanliness

To obtain a good adhesive bond it is important to start with a clean adherend surface. Foreign materials such as dirt, oil, moisture, and weak-oxide layers, must be removed, or else the adhesive will bond to these weak-boundary layers rather than to the substrate. There are various surface treatments that may remove or strengthen the weak-boundary layers. A number of such treatments will be discussed in Chapter 4. These treatments generally involve physical or chemical processes, or a combination of both.[6]

Wetting

This is discussed briefly above (see Figure 1-1). The result of good wetting is greater contact area between the adherends and the adhesive over which the forces of adhesion may act.[6]

Solidification

The liquid adhesive, once applied, must be capable of being converted into a solid in any one of three ways. The method by which solidification occurs depends on the choice of adhesive. The ways in which liquid adhesives are converted to solids are:[6]

- chemical reaction by any combination of heat, pressure, and curing agents
- cooling from a molten liquid to a solvent
- drying as a result of solvent evaporation

Adhesive Choice

A number of considerations must be taken into account when selecting the adhesive to be used for a particular application. Some of these factors are listed in Table 1-1. The general areas of concern to the design engineer when selecting adhesives to be used are (1) the material to be bonded, (2) service requirements, (3) production requirements, and (4) cost.[6]

Table 1-1: Factors Influencing Adhesive Selection
(Modified after Reference 9)

- Capability of bonding specific adherends (see "internal chemical factors" below)

- Service requirements
  Stress (tension, shear, impact, peel, cleavage, fatigue)
  Chemical factors
    External (effect of chemical agents, including solvents, acids, alkalies, etc.)
    Internal (effect of adherend on adhesive, e.g., exuded plasticizers in certain plastics and rubbers; effect of adhesive on the adherend, e.g., crazing, staining, etc.

- Environmental factors
  Weathering
  Light (important only with translucent adherends)
  Oxidation
  Moisture
  Salt spray
  Temperature extremes (including thermal cycling)
  Biological factors (bacteria, fungi, vermin, rodents)

- Specialized functional requirements (thermal or electrical conductance, etc.)

- Production requirements
  Application method
  Bonding range
  Blocking
  Curing condition (time versus temperature)
  Storage stability
  Working life
  Coverage

- Cost

- Health and safety hazards

Joint Design

The adhesive joint should be designed to take advantage of the desirable properties of adhesives and to minimize their shortcomings. Such design considerations will be discussed in detail in Chapter 3. Although satisfactory adhesive-bonded assemblies have been made from joints designed for mechanical fastening, optimum results are obtained only in assemblies specially designed for adhesive bonding.[6]

MECHANISM OF BOND FAILURE

Adhesive joints may fail adhesively or cohesively. *Adhesive failure* is interfacial bond failure between the adhesive and the adherend. *Cohesive failure* occurs when fracture allows a layer of adhesive to remain on both surfaces. When the adherend fails before the adhesive, it is known as a cohesive failure of the substrate. The various modes of failure are illustrated in Figure 1-2. Cohesive failure within the adhesive or one of the adherends is the ideal type of failure because, with this type the maximum strength of the materials in the joint has been reached. In analyzing an adhesive joint that has been tested to destruction

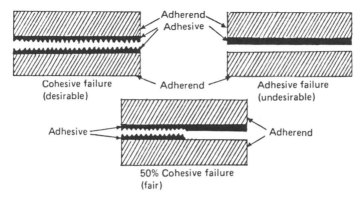

Figure 1-2: Cohesive and adhesive bond failure.[6,10]

the mode of failure is often expressed as a percentage cohesive or adhesive failure, as shown in the figure. The ideal failure is a 100% cohesive failure.[6,10]

Failure mode should not be used as the only criterion for a useful joint, however. Some adhesive-adherend combinations may fail in adhesion, but exhibit greater strength than a similar joint bonded with a weaker adhesive that fails cohesively. The ultimate strength of a joint is a more important criterion than the mode of joint failure. An analysis of failure mode, nevertheless, can be an extremely useful tool in determining whether the failure was due to a weak-boundary layer or to improper surface preparation.[6]

The exact cause of premature adhesive failure is very difficult to determine. If the adhesive does not wet the surface of the substrate completely, the bond is certain to be less than optimum. Internal stresses occur in adhesive joints because of a natural tendency of the adhesive to shrink during setting, and because of differences in physical properties between the adhesive and the substrate. The coefficient of thermal expansion of adhesive and adherend should be as close as possible to minimize the stresses that may develop during thermal cycling or after cooling from an elevated-temperature cure. Fillers are often used to modify the thermal-expansion characteristics of adhesives and limit internal stresses. Another way to accommodate these stresses is to use relatively elastic adhesives.[6]

The types of stress acting on completed bonds, their orientation to the adhesive, and the rate at which they are applied, are important factors in determining the durability of the bond. Sustained loads can cause premature failure in service, even though similar unloaded joints may exhibit adequate strength when tested after aging. Some adhesives break down rapidly under dead load, especially after exposure to heat or moisture. Most adhesives have poor resistance to peel or cleavage loads. A number of adhesives are sensitive to the rate at which the joint is stressed. Rigid, brittle adhesives sometimes have excellent tensile or shear strength but have very poor impact strength. Operating environmental factors, such as those shown in Table 1-1, are capable of degrading an adhesive joint in various ways. If more than one environmental factor (e.g., heat and moisture) is acting, their combined effect can be expected to produce a synergistic result in reducing adhesive strength. Whenever possible, candidate adhesive joints should be evaluated under simulated operating loads in the actual environment the joint is supposed to encounter.[6]

ADHESIVES AND SEALANTS CONSUMPTION

Tables 1-2 and 1-3 present data extracted from a recent *Adhesives Age* article based on the study, "U.S. Adhesives and Sealants Market", published by Predicasts, Cleveland, Ohio. This study predicts that the 1982 consumption of 9.1 billion pounds of adhesives and sealants will increase to about 15 billion pounds by 1995, and that the dollar value will rise from $6.5 billion to $25 billion in 1995.[11]

Table 1-2: Adhesives and Sealants Markets—1982
(Modified after Reference 11)

Adhesives and Sealants	MM Pounds	MM Dollars
Paper and packaging	4,630 (51%)	2,535 (39%)
Wood and construction	1,985 (22%)	1,300 (20%)
Manufacturing	2,033 (22%)	1,885 (29%)
Home and auto (domestic)	440 (05%)	780 (12%)
Total	9,088 (100%)	6,500 (100%)

Table 1-3: Usage of Various Types of Adhesives and Sealants—1982
(Modified after Reference 11)

Type	Amount, MM Pounds
Natural	
Starch	2,337
Sodium silicate	615
Dextrin	310
Rubber	234
Animal	160
Flour	109
Other	255
Sealants	530
Total	4,550 (50.1%)
General synthetic	
Phenolic	1,055
Hot melt	533
Vinyl	773
Rubber	794
Urea	592
Sealants	310
Total	4,057 (55.6%)
Synthetic engineering	
Acrylic adhesives and sealants	182
Polyurethane adhesives and sealants	119
Epoxy adhesives	33
Polyester adhesives	16
Cyanoacrylate adhesives	1
Anaerobic adhesives and sealants	3
Silicone sealants	37
Other adhesives	50
Other sealants (primarily polysulfides)	40
Total	481 (5.3%)

REFERENCES

1. Modified from ASTM D 907-82, Standard Definitions of Terms Relating to Adhesives, published in Volume 15.06-Adhesives, *1984 Annual Book of ASTM Standards*.
2. Modified from Chapter 1, Introduction, scope, exclusions, methodology, *Structural Adhesives, with Emphasis on Aerospace Applications, Treatise on Adhesion and Adhesives*, Volume 4. A Report of the ad hoc Committee on Structural Adhesives for Aerospace Use, National Materials Advisory Board, Marcel Dekker (1976).
3. Staff written, Joining techniques, Section 4, *Machine Design*, Fastening and Joining Reference Issue, 48 (26):155–162 (November 18, 1976).
4. Sharpe, L.H., The materials, processes and design methods for assembly with adhesives, *Machine Design*, 38 (19):179–200 (August 19, 1966).
5. Society of Manufacturing Engineers, *Adhesives in Modern Manufacturing*, (E.J. Bruno, ed.) (1970).
6. Petrie, E.M., Plastics and adhesives as adhesives, *Handbook of Plastics and Elastomers*, (C.A. Harper, ed.), McGraw-Hill, NY (1975).
7. De Lollis, N.J., *Adhesives for Metals–Theory and Technology*, Industrial Press, NY (1970).
8. Bikerman, J.J., Causes of poor adhesion, *Industrial and Engineering Chemistry*, 59 (9):40–44 (Sept., 1967).
9. Koehn, G.W., Design manual on adhesives, *Machine Design*, (April, 1954).
10. Cagle, C.V., *Adhesive Bonding, Techniques and Applications*, McGraw-Hill, NY (1968).
11. Barker, A., Adhesive consumption may rise 60% by volume by 1995, *Adhesives Age*, 27 (1):32–34 (January 1984).

2

Definitions of Terms

COMMENTS

The definitions given in this chapter are taken from a number of sources, which have been cited. They include the latest ASTM Committee D-14 and ISO TC 61 definitions. In most cases several definitions are given. A few definitions found in the literature have been used to supplement or replace more complicated definitions.

This chapter follows the introduction because of the importance of the definition of technical terms frequently encountered in the text.

DEFINITIONS

A-stage. An early stage in the reaction of certain thermosetting resins in which the material is fusible and still soluble in certain liquids. Sometimes referred to as Resol. (See also *B-stage* and *C-stage*.)[1]

Adhesive. A material which is adhesive-resistant and applicable as a non-sticking surface coating; release agent.[2]

Adhere, v. To cause two surfaces to be held together by adhesion (q.v.).[1] To be in a state of adherence.[3]

Adherence. The state in which two surfaces are held together by interfacial forces.[3]

Adherend. A body which is held to another body by an adhesive (a narrower term than *substrate*, q.v.).[1]

Adhesion. The state in which two surfaces are held together by interfacial forces which may consist of valence forces, or interlocking action, or both. (See also *adhesion, mechanical* and *adhesion, specific*.)[1] The state in which two surfaces are held together by chemical, or physical forces, or both, with the aid of an adhesive.[3] The attraction between substances which, when they are brought into contact, makes it necessary to do work to separate them.[4]

Adhesion, mechanical. Adhesion between surfaces in which the adhesive holds the parts together by interlocking action. (See also *adhesion, specific.*)[1]

Adhesion, specific (sometimes called *adhesion, molecular*). Adhesion between surfaces which are held together by valence forces of the same type as those which give rise to cohesion. (see also *adhesion, mechanical.*)[1]

Adhesive. A substance capable of holding materials together by surface attachment. *Adhesive* is the general term and includes, among others, *cement, glue, mucilage,* and *paste* (q.v.). All of these terms are used interchangeably. Various descriptive adjectives are applied to the term to indicate characteristics such as physical form, chemical type, materials bonded, and conditions of use.[1] A substance capable of bonding materials together by adhesion.[3] Any material that causes one body to stick or adhere to another.[5]

Adhesive, anaerobic. An adhesive that cures spontaneously in the absence of oxygen and catalyzed by metallic ions.[3] An adhesive that cures only in the absence of air after being confined between assembled parts. An example is dimethacrylate adhesive used for bonding assembly parts, locking screws and bolts, retaining gears and other shaft-mounted parts, and sealing threads and flanges.[6]

Adhesive, assembly. An adhesive that can be used for bonding parts together, such as in the manufacture of a boat, airplane, furniture, etc. This term is commonly used in the wood industry to distinguish such adhesives, formerly called "joint glues," from those used in making plywood, sometimes called "veneer glues." It is applied to adhesives used in fabricating finished structures or goods, or subassemblies thereof, as differentiated from adhesives used in the production of sheet materials, such as plywood or laminates. (Reference 1, modified).

Adhesive, bonded. (Note the comma, meaning *bonded adhesive*, not *adhesive bonded*.) A solvent solution of resins, sometimes containing plasticizers, which dries at room temperature.

Adhesive, cold-setting. An adhesive that sets at temperatures below $68°F$ $(20°C)$.[1]

Adhesive, contact. An adhesive that is apparently dry to the touch and which will adhere to itself instantaneously upon contact; also called *contact bond adhesive*.[1] An adhesive applied to both adherends and allowed to become dry, which develops a bond when the adherends are brought together without sustained pressure.[3]

Adhesive, dispersion (or emulsion). A two-phase system with one phase (the adhesive material) in a liquid suspension.[3]

Adhesive, encapsulated. An adhesive in which the particles or droplets of one of the reactive components are enclosed in a protective film (microcapsules) to prevent cure until the film is destroyed by suitable means.[3]

Adhesive, film. An adhesive in film form, with or without a carrier, usually set by means of heat and pressure.[3] The main advantage is uniformity of glueline thickness.

Adhesive, film-supported. An adhesive material incorporating a carrier that remains in the bond when the adhesive is employed. The carrier support material is usually composed of organic and/or inorganic fibers which may be in woven form.[2]

Adhesive, film-unsupported. An adhesive material in film form without a carrier support.[2]

Adhesive, foamed. An adhesive, the apparent density of which has been decreased substantially by the presence of numerous gaseous cells dispersed throughout its mass.[1]

Adhesive, foaming. An adhesive designed to foam *in situ*, after application, in order to provide extensive gap-filling properties.[3]

Adhesive, gap-filling. An adhesive subject to low shrinkage in setting, employed as sealant.[2]

Adhesive, heat-activated. A dry adhesive that is rendered tacky or fluid by application of heat, or heat and pressure, to the assembly.[1,3]

Adhesive, heat-sealing. A thermoplastic film adhesive which is melted between the adherend surfaces by heat application to one or both of the adjacent adherend surfaces.[2]

Adhesive, hot-melt. An adhesive that is applied in a molten state and forms a bond on cooling to a solid state.[1] A bonding agent which achieves a solid state and resultant strength by cooling, as contrasted with other adhesives which achieve the solid state through evaporation of solvents or chemical cure. A thermoplastic resin which functions as an adhesive when melted between substrate and cooled.[7]

Adhesive, hot setting. An adhesive that requires a temperature at or above 212°F (100°C) to set it.[1]

Adhesive, intermediate-temperature setting. An adhesive that sets in the temperature range from 87° to 211°F (31° to 99°C).[1]

Adhesive, latex. An emulsion of rubber or thermoplastic rubber in water.[2]

Adhesive, multiple-layer. A film adhesive, usually supported with a different adhesive composition on each side, designed to bond dissimilar materials such as the core-to-face bond of a sandwich composite structure. (Reference 1, modified).

Adhesive, one-component. An adhesive material incorporating a latent hardener or catalyst activated by heat. Usually refers to thermosetting materials, but also describes anaerobic, hot-melt adhesive, or those dependent on solvent loss for adherence.[2] Thermosetting one-component adhesives require heat to cure.

Adhesive, pressure-sensitive. A viscoelastic material which, in solvent-free form, remains permanently tacky at room temperature. Such material will adhere instantaneously and tenaciously to most solid surfaces with the application of very slight manual pressure.[1] These adhesives are frequently used on tapes. An adhesive which, in the dry state, is aggressively and permanently tacky at room temperature and firmly adheres to a variety of dissimilar surfaces upon contact without the need for more than finger or hand pressure.[8]

Adhesive, room-temperature setting. An adhesive that sets in the temperature range from 68° to 86°F (20° to 30°C), in accordance with the limits for Standard Room Temperature specified in ASTM Methods D 618.[1,9]

Adhesive, separate-application. An adhesive consisting of two parts, one part being applied to one adherend and the other part to the other, the two then being brought together to form a joint. (Reference 1, modified). *Acrylics* are examples of this type.

Adhesive, solvent. An adhesive having a volatile organic liquid as a vehicle. This term excludes water-based adhesives.[1]

Adhesive, solvent-activated. A dry adhesive or adherend that is rendered tacky just prior to use by application of a solvent.[3]

Adhesive spread. See *spread.*

Adhesive, structural. A bonding agent used for transferring required loads between adherends exposed to service environments typical for the structure involved.[1] An adhesive of proven reliability in engineering structural applications in which the bond can be stressed to a high proportion of its maximum failing load for long periods without failure.[3] A material employed to form high-strength bonds in structural assemblies which perform load-bearing functions, and which may be used in extreme service conditions, e.g., high- and low-temperature exposure.[2]

Adhesive, two-component. An adhesive supplied in two parts which are mixed before application.[2] Such adhesives usually cure at room temperature.

Adhesive, warm-setting. A term that is sometimes used as a synonym for *intermediate-temperature-setting adhesive* (q.v.).[1]

Amylaceous, adj.—Pertaining to, or of, the nature of starch; starchy.[1]

Anodize. To coat a metal with a protective film by subjecting it to electrolytic action as the anode of a cell.

Assembly (for adhesive). A group of materials or parts, including adhesive, which has been placed together for bonding, or which has been bonded together.[1,3]

Autoclave. A closed container which provides controlled heat and pressure conditions.[2]

B-stage. An intermediate stage in the reaction of certain thermosetting resins in which the material softens when heated to a rubbery state and swells when in contact with certain liquids, but may not entirely fuse or dissolve in some of the solvents which will dissolve resins in the A-stage. The resin in an uncured thermosetting adhesive is usually in this stage. Sometimes referred to as Resitol.[1,2]

Backing. The flexible supporting material for an adhesive. Pressure-sensitive adhesives are commonly backed with paper, plastic films, fabric, or metal foil; heat-curing thermosetting adhesives are often supported on glass cloth backing.[2]

Bag molding (blanket molding). A method of molding or bonding involving the application of fluid pressure, usually by means of air, steam, water, or vacuum, to a flexible cover which, sometimes in conjunction with the rigid die, completely encloses the material to be bonded.[1]

Bag, vacuum. A flexible bag in which pressure may be applied to an assembly inside the bag by means of evacuation of the bag.[3]

Binder. A component of an adhesive composition that is primarily responsible for the adhesive forces which hold two bodies together.[1]

Bite, n. The penetration or dissolution of adherend surfaces by an adhesive.[2]

Blister. An elevation of the surface of an adherend, somewhat resembling in shape a blister on the human skin; its boundaries may be indefinitely outlined and it may have burst and become flattened. A blister may be caused by insufficient adhesive, inadequate curing time, temperature or pressure, or trapped air, water, or solvent vapor.[1]

Blocked curing agent. A curing agent or hardener rendered unreactive, which can be reactivated as desired by physical or chemical means.[1]

Blocking. An undesired adhesion between touching layers of a material, such as occurs under moderate pressure during storage or use.[1] An unintentional adherence between materials.[3]

Blushing. The condensation of atmospheric moisture at the bond-line interface.[10]

Body. The consistency of an adhesive which is a function of viscosity, plasticity, and rheological factors.[2]

Bond, n. The union or joining of materials by adhesives. (Reference 1, modified). The attachment at the interface between an adhesive and an adherend.[3]

Bond, v. To unite or join materials by means of an adhesive. (Reference 1, modified).

Bond line. See *glue line.*

Bond strength. The unit load (force) supplied in tension, compression, flexure, peel, impact, cleavage, or shear, required to break an adhesive assembly, with failure occurring in or near the plane of the bond (the interface). The term *adherence* (q.v.) is frequently used in place of bond strength. (Reference 1, modified).

Bond, structural. See *structural bond.*

C-stage. The final stage in the reaction of certain thermosetting resins in which the material is relatively insoluble and infusible. Certain thermosetting resins in a fully cured adhesive layer are in this stage. Sometimes referred to as *resite.*[1]

Catalyst. A substance that markedly speeds up the cure of an adhesive when added in minor quantity compared to the amounts of the primary reactants.[1] Material which promotes cross linking in a polymer or accelerates drying.[2]

Caul, n. A sheet of material employed singly or in pairs in hot or cold pressing of assemblies being bonded. Cauls are used to protect either the faces of the assembly, or the press platens, or both against marring and staining to prevent sticking, to facilitate loading, to impart a desired surface tension or finish, or to provide uniform surface distribution. Cauls may be made of any suitable material, such as aluminum, stainless steel, hardboard, fiberboard, or plastic. The length and width dimensions are generally the same as those of the platen of the press where it is used.[1]

Cement, n. (See *adhesive* and *solvent cement*.) A synonym for *adhesive*; a mixture of water with finely powdered lime and clay which hardens and adheres to suitable aggregates to form concrete or mortar; an inorganic paste with adhesive properties.[2]

Cement, v. To bond with a cement.

Cohesion. The state in which the particles of a single substance are held together by primary or secondary valence forces. In adhesives, cohesion is the state in which the particles of the adhesive or adherend are held together.[1]

Cold pressing. A bonding operation in which an assembly is subjected to pressure without the application of heat.[1]

Collagen. The protein derived from bone and skin used to prepare animal glue and gelatin.[2]

Colophony. The resin obtained from various species of pine trees.[2]

Condensation. A chemical reaction in which two or more molecules combine with the separation of water or some other simple compound. If a polymer is formed, the process is called *polycondensation*.[1]

Consistency. That property of a liquid adhesive by virtue of which it tends to resist deformation. Consistency is not a fundamental property, but is comprised of viscosity, plasticity, and other phenomena.[1] The term is usually applied to materials whose deformations are not proportional to applied stresses.[2]

Contact angle: The angle (θ) between a substrate plane and the free surface of a liquid droplet at the line of contact with the substrate.[11]

Contact bonding. The deposition of cohesive materials on both adherend surfaces and their assembly under pressure.[2]

Copolymer. See *polymer*.

Copolymerization. See *polymerization*.

Core. The honeycomb structure used in sandwich panel construction; innermost portion of a multilayer adherend assembly.[2]

Corrosion. The chemical reaction between the adhesive or contaminant and the adherend surfaces, due to reactive components in the adhesive film, leading to deterioration of the bond strength. (Reference 2, modified).

Cottoning. The formation of web-like filaments of adhesive between the applicator and substrate surface.[2]

Coverage. The spreading power of an adhesive over the surface area of the adherend. (Reference 2, modified).

Crazing. Fine cracks that may extend in a network on, under the surface of, or through a layer of adhesive.[i]

Creep. The dimensional change with time of a material under load, following initial instantaneous elastic or rapid deformation. Creep at room temperature is sometimes called *cold flow.*[1]

Cross-linking (Crosslinking). The union of adjacent molecules of uncured adhesive (often existing as long polymer chains) by catalytic or curing agents.[2]

Cure, v. To change the physical properties of an adhesive, usually thermosetting, by chemical reaction, which may be condensation, polymerization, vulcanization, or crosslinking. It is usually accomplished by the action of heat and catalyst, alone or in combination, with or without pressure. (Reference 1, modified).

Cure (curing) temperature. The temperature to which an adhesive or an assembly is subjected to cure the adhesive.[2]

Cure (curing) time. The period of time necessary for an adhesive or an assembly to cure under specified conditions of temperature, pressure, or both.[2]

Curing agent (hardener). A substance or mixture of substances added to an adhesive to promote or control the curing reaction. An agent which does not enter into the reaction is known as a *catalytic hardener* or *catalyst.* A *reactive curing agent* or *hardener* is generally used in much greater amounts than a catalyst, and actually enters into the reaction.[6]

Degrease. To remove oil and grease from adherend surfaces.[2]

Delamination. The separation of layers in a laminate because of failure of the adhesive, either in the adhesive itself, or at the interface between the adhesive and the adherend.[1]

Dextrin. A water-based product derived from the acidification and/or roasting of starch.[2]

Dielectric curing. The use of a high-frequency electric field through a joint to cure a synthetic thermosetting adhesive. A curing process for wood and other nonconductive joint materials. Curing results from the heat generated by the resonance of the molecules within the adhesive due to the imposed field.[2]

Diluent. An ingredient usually added to an adhesive to reduce the concentration of bonding materials.[1] A liquid additive whose sole function is to reduce the concentration of solids and the viscosity of an adhesive composition.[2] Also called *thinner,* which is deprecated by some workers.

Diluent, reactive. A low-viscosity liquid added to a high viscosity solvent-free thermosetting adhesive which reacts chemically with the adhesive during curing. The advantage of lowered viscosity is gained with minimum loss of other properties.[3]

Doctor bar or blade. A scraper mechanism that regulates the amount of adhesive on the spreader rolls, or on the surface being coated.[1] A mechanism (bar or blade) on application equipment for spreading a material evenly on the application rolls or on the surface being coated, thereby controlling its thickness. (Reference 3, modified).

Doctor roll. A roller mechanism that revolves at a different surface speed, in a direction opposite to that of the spreader roll, resulting in a wiping action to control the amount of adhesive supplied to the spreader roll.[1,2]

Double spread. See *spread.*

Dry, v. To change the physical state of an adhesive or an adherend by the loss of solvent constituents by evaporation, absorption, or both.[1]

Elasticity, modulus of. The ratio of stress to strain in elastically deformed material.[2]

Elastomer. A macromolecular material which, at room temperature, is capable of recovering substantially in size and shape after removal of deforming force.[1]

Emulsion. A stable dispersion of two or more immiscible liquids held in suspension by small percentages of substances called emulsifiers.[12]

Extender. A substance, generally having some adhesive action, added to an adhesive to reduce the amount of the primary binder required per unit area.[1] Another function is to reduce costs.[3] Such materials also improve void-filling properties and reduce crazing.[2]

Failure, adherend. Joint failure by cohesive failure of the adherend.[2]

Failure, adhesive. Rupture of an adhesive bond at the interface between the adhesive and adherend. (Reference 1, modified). Rupture of an adhesive bond in which the separation appears visually to be at the adhesive/adherend interface.[3]

Failure, cohesive. Rupture of an adhesive bond in such a way that the separation appears to be within the adhesive. (Reference 1, modified). Rupture of an adhesive bond in which the separation appears visually to be in the adhesive or the adherend. (Reference 2, modified).

Failure, contact. The failure of an adhesive joint, as a result of incomplete contact during assembly, between adherend and adhesive surfaces or between adhesive surfaces.[2]

Failure, wood. The rupturing of wood fibers in strength tests on bonded specimens, usually expressed as the percentage of the total area involved which shows such failure.[1] This is a form of adherend failure.

Fatigue. A condition of stress from repeated flexing or impact force upon the adhesive-adherend interface; weakening of material caused by repetitive loading and unloading.[2]

Faying surface. The surface of an adherend which makes contact with another adherend.[2]

Feathering. The tapering of an adherend on one side to form a wedge section, as used in a scarf joint.[2]

Filler. A relatively nonadhesive substance added to an adhesive to improve its working properties, permanence, strength, or other qualities.[1]

Filler sheet. A sheet of deformable or resilient material that, when placed between the assembly to be bonded and the pressure applicator, or when distributed within a stack of assemblies, aids in providing uniform application of pressure over the area to be bonded.[1]

Fillet. That portion of an adhesive which fills the corner or angle formed where two adherends are joined.[1] The term for junction of the outer skin and inner core in honeycomb assemblies.[2]

Flow. Movement of an adhesive during the bonding process before the adhesive is set.[1] (See also *cold flow* under *creep*.)

Gel, n. A semisolid system consisting of a network of solid aggregates in which liquid is held.[1]

Gel, v. To reach a gel condition or state.

Gelation. Formation of a gel.[1]

Glue, n. Originally a hard gelatin obtained from hides, tendons, cartilage, bones, etc. of animals. Also, an adhesive prepared from this substance by heating with water. Through general use the term is now synonymous with the term "adhesive."[1] The term is most commonly used for wood adhesives, however.

Glue line (bond line). The layer of adhesive which attaches two adherends.[1] The interface between an adhesive and an adherend.[3]

Green strength (grab). The ability of an adhesive to hold two surfaces together when first brought into contact and before the adhesive develops its ultimate bonding properties when fully cured.[13]

Gum. Any of a class of colloidal substances exuded by, or prepared from plants. Sticky when moist, they are composed of complex carbohydrates and organic acids which are soluble or swell in water.[1]

Hardener. A substance or mixture of substances added to an adhesive to promote or control the curing reaction by taking part in it by catalysis or cross-linking. The term is also used to designate a substance added to control the degree of hardness of the cured film. (Reference 1, modified).

Heat reactivation. The use of heat to effect adhesive activity, e.g., hot-melt adhesive; completion of the curing process of a B-staged resin.[2]

Heat seal. The use of heat reactivation to prepare a joint with a thermoplastic material present, as a thin layer, on the adherends; bringing adherend surfaces to their melting point and bonding under pressure.[2]

Heteropolymerization. See *polymerization*.

Honeycomb core. A sheet material, which may be metal, foamed into cells (usually hexagonal) and used for sandwich construction in structural assemblies, especially in aircraft construction. (Reference 2, modified).

Impact shock. See strength, impact.

Inhibitor. A substance that slows down a chemical reaction. Inhibitors are sometimes used in certain adhesives to prolong storage or working life.[1]

Interface. The contact area between adherend and adhesive surfaces.[2]

Jig. A form used to hold a bonded assembly until the adhesive has cured. A supporting frame for the production of laminate shapes under pressure. (Reference 2, modified).

Joint. The location at which two adherends are held together with a layer of adhesive.[1]

Joint, butt. A joint made by bonding two surfaces that are perpendicular to the main surface of the adherends.[3]

Joint, lap. A joint made by placing one adherend partly over another and bonding together the portions of the adherends. Double lap joints involve the overlapping by opposing faces of one adherend.[2]

Joint, scarf. A joint made by cutting away similar segments of two adherends at an angle less than 45 degrees to the major axis of two adherends and bonding the adherends with the cut areas fitted together to be coplanar.[1,3]

Joint, starved. A joint that has an insufficient amount of adhesive to produce a satisfactory bond. This condition may result from too thin a spread to fill the gap between the adherends, excessive penetration of the adhesive into the adherend (when porous), too short an assembly time, or the use of excessive pressure. (Reference 1, modified).

Laminate, n. A product made by bonding together two or more layers of material or materials.[1]

Laminate, v. To unite layers of material with adhesive.[1]

Laminate, cross, n. A laminate in which some of the layers of material are oriented at right angles to the remaining layers with respect to the grain or strongest direction in tension. (Reference 1, modified).

Laminate, parallel, n. A laminate in which the grain of all layers of material are oriented approximately parallel to each other.[2] A laminate in which all of the layers of material are oriented approximately parallel with respect to the grain or strongest direction in tension. (Reference 1, modified).

Latex. A stable dispersion of a polymeric material in an essentially aqueous medium.[14]

Legging. The drawing of filaments or strings when adhesive-bonded substrates are separated.[1]

Mastic. A high-viscosity, low-cost adhesive, either latex or solvent-based, used in industrial applications, such as in applying wall boards and floor tiles. Mastics are applied by knife, trowel, or pressure guns and become immobile on loss of solvent or water.

Matrix. The part of an adhesive which surrounds or engulfs embedded filler or reinforcing particles and filaments.[1]

Modifier. Any chemically inert ingredient added to an adhesive formulation that changes its properties.[1]

Modulus. See *elasticity, modulus of* and *rigidity, modulus of.*

Monomer. A relatively simple compound which can react to form a polymer.[1]

Mucilage. An adhesive prepared from a gum and water. Also, in a more general sense, a liquid adhesive which has a low order of bonding strength.[1] Mucilages are used in schools and offices for applications usually involving paper products.

Newtonian fluid. A fluid in which the shearing rate is directly proportional to the applied torque.[2]

Novalak. A phenolic-aldehydic resin that, unless a source of methylene groups is added, remains permanently thermoplastic.[1]

Open time. See *open assembly time* under *time, assembly.*

Paste, n. An adhesive composition having a characteristic plastic-type consistency, that is, a high order of yield value, such as that of an adhesive prepared by heating a mixture of starch and water and subsequently cooling the hydrolyzed product. (Reference 1, modified).

Penetration. The passage of an adhesive into an adherend.[2]

Permanence. The resistance of an adhesive bond to deteriorating influences.[1]

Photographing. See *telegraphing.*

Pick-up roll. A spreading device where the roll for picking up the adhesive runs in a reservoir of adhesive.[1]

Plasticity. A property of adhesives that allows the material to be deformed continuously and permanently without rupture upon the application of a force that exceeds the yield value of the material.[1]

Plasticizer. A material, such as a high-boiling point organic solvent, incorporated in an adhesive to increase its flexibility, workability, or distensibility. The addition of the plasticizer may cause a reduction in melt viscosity, lower the temperature of the second-order transition, or lower the elastic modulus of the solidified adhesive.[1,2]

Polycondensation. See *condensation* and *polymer.*

Polymer. A compound formed by the reaction of simple molecules having functional groups which permit their combination to proceed to higher molecular weights under suitable conditions. Polymers may be formed by polymerization (addition polymers) or polycondensation (condensation polymers). When two or more different monomers are involved, the product is a *copolymer.* (Reference 1, modified).

Polymerization. A chemical reaction in which the molecules of a monomer are linked together to form large molecules whose molecular weight is a multiple of that in the original substance. When two or more different monomers are involved, the process is called *copolymerization* or *heteropolymerization.* (Reference 1, modified).

Porosity. The ability of an adherend to absorb an adhesive.[2]

Postcure, v. To expose an adhesive assembly to an additional cure, following the initial cure, for the purpose of modifying specific properties.[1]

Post-vulcanization bonding. Conventional adhesive bonding of previously vulcanized elastomeric adherends.

Pot life (working life). The period of time during which an adhesive or resin prepared for application after mixing with catalyst, solvent, or other compounding ingredients, remains usable.[1,3] The effective working time for an adhesive after preparation; interval before the adhesive system becomes unusable through an increase in viscosity or curing.[2]

Prebond treatment. See *surface preparation.*

Pressure-sensitive adhesives (PSA's). Adhesive materials which bond to adherend surfaces at room temperature immediately as low pressure is applied.[3] Adhesives which require only pressure application to effect permanent adhesion to an adherend.[2]

Primer. A coating applied to a surface of an adherend prior to the application of an adhesive to improve adhesion and/or durability of the bond.[3]

Qualification test. A series of tests conducted by the government procuring activity, or an agent thereof, to determine conformance of materials, or materials systems, to the requirements of a specification which normally results in a Qualified Products List (QPL) under the specification. Qualification under a specification generally requires conformance to all tests in the specification. It may, however, be limited to conformance to a specific type, or class, or both.[1]

QPL. See *qualification test.*

Release agent. An adhesive material which prevents bond formation.[2]

Release paper. A sheet, serving as a protectant, or carrier, or both, for an adhesive film or mass which is easily removed from the film or mass prior to use.[1]

Resin. A solid, semisolid, or pseudosolid organic material that has an indefinite and often high molecular weight, exhibits a tendency to flow when subjected to stress, usually has a softening or melting range, and usually fractures conchoidally. A *liquid resin* is an organic polymeric liquid which, when converted to its final state for use, becomes a resin.[1]

Resinoid. Any of the class of thermosetting synthetic resins, either in their initial temporarily infusible state, or in their final infusible state.[1]

Resite. An alternative term for *C-stage* (q.v.).[1]

Resitol. An alternative term for *B-stage* (q.v.).[1]

Resol. An alternative term for *A-stage* (q.v.).[1]

Retrogradation. A change of starch pastes from low to high consistency on aging.[1]

Rosin. A resin obtained as a residue in the distillation of crude turpentine from the sap of the pine tree (gum rosin), or from an extract of the stumps and other parts of the tree (wood rosin).[1]

Sagging. Run or flow-off of adhesive from an adherend surface due to application of excess or low-viscosity material.[1]

Sandwich panel. An assembly composed of metal skins (facings) bonded to both sides of a lightweight core.[2]

Sealant. A gap-filling material to prevent excessive absorption of adhesive, or penetration of liquid or gaseous substances.[2]

Self-vulcanizing, adj. Pertaining to an adhesive that undergoes vulcanization without the application of heat.[1]

Service conditions. The environmental conditions to which a bonded structure is exposed, e.g., heat, cold, humidity, radiation, vibration, etc.[2]

Set, v. To convert an adhesive into a fixed or hardened state by chemical or physical action, such as condensation, polymerization, oxidation, vulcanization, gelation, hydration, or evaporation of volatile constituents.[1] The term is most commonly used with thermoplastic adhesives, unless a chemical process, such as polymerization, is involved.

Shear, tensile. The apparent stress applied to an adhesive in a lap joint.[2]

Shelf life. See *storage life.*

Shortness. A qualitative term that describes an adhesive that does not string cotton, or otherwise form filaments or threads during application.[1]

Shrinkage. The volume reduction occurring during adhesive curing, sometimes expressed as a percentage volume or linear shrinkage; size reduction of adhesive layer due to solvent loss or catalytic reaction.[2]

Single-spread. See *spread.*

Sizing. The process of applying a material on a surface in order to fill pores and thus reduce the absorption of the subsequently applied adhesive or coating, or to otherwise modify the surface properties of the substrate to improve the adhesion. Also, the material used for this purpose. The latter is sometimes called *size.*[1]

Skinning. The formation of a dry surface layer (skin) on an adhesive coating following too rapid evaporation of the solvent vehicle.[2]

Slip. The ability of an adhesive to accommodate adherend movement or repositioning after application to adherend surfaces.[2]

Slippage. The movement of adherends with respect to each other during the bonding process.[1]

Slip sheet interliner. A sheet or film used to cover an adhesive during handling; protective film for a film adhesive.[2]

Softener. A plasticizing additive to reduce adhesive embrittlement; component of elastomeric films to increase their flexibility.[2]

Solids content. The percentage by weight of the nonvolatile matter in an adhesive.[1]

Solvent bonding. See *solvent welding.*

Solvent cement. An adhesive utilizing an organic solvent as the means of depositing the adhesive constituent.[2] An adhesive which dissolves the plastics being joined, forming strong intermolecular bonds, then evaporating.[6] An adhesive made by dissolving a plastic resin or compound in a suitable solvent or mixture of solvents. The solvent cement dissolves the surfaces (of the pipe and fittings) to form a bond between the mating surfaces, provided the proper cement is used for the particular materials and proper techniques are followed.[15]

Solvent cement joint. A joint made by using a solvent cement to unite the components.[15]

Solvent cementing. See *solvent welding.*

Solvent joint. A joint made by using a solvent to unite the components.[15]

Solvent reactivation. The application of solvent to a dry adhesive layer to regenerate its wetting properties.[2]

Solvent welding (solvent bonding) (solvent cementing). The process of joining articles made of thermoplastic resins by applying a solvent capable of softening the surfaces to be joined, and pressing the softened surfaces together. Adhesion is attained by means of evaporation of the solvent, absorption of the solvent into adjacent material and/or polymerization of the solvent cement. ABS, acrylics, cellulosics, polycarbonates, polystyrenes, and vinyls are plastics which may be joined in this way.[6] The formation of a joint in which a self bond between the polymeric components is promoted by the temporary presence of a solvent, in the absence of an extraneous adhesive.[16]

Spread (adhesive spread). The quantity of adhesive per unit joint area applied to an adherend, usually expressed in points of adhesive per thousand square feet of joint area.[1]

(1) *Single spread* refers to the application of adhesive to only one adherend of a joint.

(2) *Double spread* refers to application of adhesive to both adherends of a joint.

Squeeze-out. Excess adhesive pressed out at the bond line due to pressure applied in the adherends. (Reference 1, modified).

Stabilizer. An adhesive additive which prevents or minimizes change in properties, e.g., by adherend absorption, demulsification, or rapid chemical reaction.

Storage life. The period of time during which a packaged adhesive can be stored under specified temperature conditions and remain suitable for use. Sometimes called *shelf life.*[1] Refrigerated storage often extends storage life considerably.

Strength, cleavage. The tensile load expressed as force per unit of width of bond required to cause cleavage separation of a test specimen of unit length.[2]

Strength, dry. The strength of an adhesive joint determined immediately after drying under specified conditions, or after a period of conditioning in the standard laboratory atmosphere.[1]

Strength, fatigue. The maximum load that a joint will sustain when subjected to repeated stress application after drying, or after a conditioning period under specified conditions.[2]

Strength, impact. Ability of an adhesive material to resist shock by a sudden physical blow directed against it. Impact shock is the transmission of stress to an adhesive interface by sudden vibration or jarring blow of the assembly, measured in work units per unit area.[2]

Strength, longitudinal shear (lap-joint strength). The force necessary to rupture an adhesive joint by means of stress applied parallel to the plane of the bond.[3]

Strength, peel. The force per unit width necessary to bring an adhesive to the point of failure and/or to maintain a specified rate of failure by means of a stress applied in a peeling mode.[2]

Strength, shear. The resistance of an adhesive joint to shearing stresses; the force per unit area sheared, at failure.[2]

Strength, tensile. The resistance of an adhesive joint to tensile stress; the force per unit area under tension, at failure.[2]

Strength, wet. The strength of an adhesive joint determined immediately after removal from a liquid in which it has been immersed under specified conditions of time, temperature, and pressure. The term is commonly used to designate strength after immersion in water. In latex adhesives the term is also used to describe the joint strength when the adherends are brought together with the adhesive still in the wet state.[1]

Stringiness. The property of an adhesive that results in the formation of filaments or threads when adhesive transfer surfaces are separated. Transfer surfaces include rolls, picker plates, stencils, etc.[1] The complete breakoff of adhesive film when it is divided between transfer rollers, stencils, picker plates, etc; uneven transfer of an adhesive to an adherend surface.[2]

Structural adhesive. See *adhesive, structural.*

Structural bond. A bond which stresses the adherend to the yield point, thereby taking full advantage of the strength of the adherend. On the basis of this definition, a dextrin adhesive used with paper (e.g., postage stamps, envelopes, etc.) which causes failure of the paper, forms a structural bond. The stronger the adherend, the greater the demands placed on the adhesive. Thus, few adhesives qualify as "structural" for metals. A further requirement for a structural adhesive is that it be able to stress the adherend to its yield point after exposure to its intended environment.[17]

Substrate. A material upon the surface of which an adhesive-containing substance is spread for any purpose, such as bonding or coating. A broader term than adherend, q.v.[1]

Surface preparation (surface treatment). A physical or chemical preparation, or both, of an adherend to render it suitable for adhesive joining.[1] The term *prebond treatment* is sometimes used, but is deprecated.[2]

Syneresis. The exudation by gels of small amounts of liquid on standing.[1]

Tack. The property of an adhesive that enables it to form a bond of measurable strength immediately after adhesive and adherend are brought into contact under low pressure.[1]

Tack, aggressive. See *tack, dry.*

Tack, dry, n. The property of certain adhesives, particularly nonvulcanizing rubber adhesives, to adhere on contact to themselves at some stage in the evaporation of volatile constituents, even though they seem dry to the touch. Sometimes called *aggressive tack*.[1,3] The self-adhesion property of certain adhesives which are touch-dry (a stage in the evaporation of volatile constituents).[2]

Tack-dry (tacky-dry), adj. The state of an adhesive which has lost sufficient volatiles (by evaporation or absorption into the adherend) to leave it in the required sticky (tacky) condition.[3]

Tackifier. An additive intended to improve the stickiness of a cast adhesive film; usually a constituent of rubber-based and synthetic resin adhesives.[2]

Tack range (tack stage). The period of time in which an adhesive will remain in the tacky-dry condition after application to an adherend under specified conditions of temperature and humidity.[1,2]

Tacky-dry. See *tack-dry*.

Tape. A film form of adhesive which may be supported on carrier material.[2]

Teeth. The resultant surface irregularities or projections formed by the breaking of filaments or strings which may form when adhesive-bonded substrates are separated.[1]

Telegraphing. A condition in a laminate or other type of composite construction in which irregularities, imperfections, or patterns of an inner layer are visibly transmitted to the surface. Telegraphing is occasionally referred to as *photographing*.[1] The visible transmission of faults, imperfections, and patterned striations occurring in an inner layer of a laminate structure.[2]

Temperature, curing. The temperature to which an adhesive or an assembly is subjected to cure the adhesive. The temperature attained by the adhesive in the process of curing it (adhesive curing temperature) may differ from the temperature of the atmosphere surrounding the assembly (assembly curing temperature).[1]

Temperature, drying. The temperature to which an adhesive on an adherend, or in an assembly, or the assembly itself, is subjected to dry the adhesive. The temperature attained by the adhesive in the process of drying it (adhesive drying temperature) may differ from the temperature of the atmosphere surrounding the assembly (assembly drying temperature).[1]

Temperature, maturing. The temperature, for a given time and bonding procedure, which produces required characteristics in components bonded with ceramic adhesives.[2]

Temperature, setting. The temperature to which an adhesive or an assembly is subjected to set the adhesive. The temperature attained by the adhesive in the process of setting (adhesive setting temperature) may differ from the temperature of the atmosphere surrounding the assembly (assembly setting temperature).[1]

Tests, destructive. Tests involving the destruction of assemblies in order to evaluate the maximum performance of the adhesive bond.[2]

Tests, nondestructive. Inspection tests for the evaluation of bond quality without damaging the assembly, e.g., ultrasonics, visual inspection, etc.[2]

Thermoplastic, adj. Capable of being repeatedly softened by heat and hardened by cooling.[1]

Thermoplastic, n. A material that will repeatedly soften when heated and harden when cooled.[1]

Thermoset (thermosetting), adj. Having the property of undergoing a chemical reaction by the action of heat, catalysis, ultraviolet light, etc. leading to a relatively infusible state. (Reference 1, modified).

Thermoset, n. A material that has the property of undergoing, or has undergone a chemical reaction by the action of heat, catalysis, ultraviolet light, etc. leading to a relatively infusible state. (Reference 1, modified).

Thinner, n. A volatile liquid added to an adhesive to modify the consistency or other properties.[1] (See *diluent*.)

Thixotropic, adj. A term applied to materials having the property of thixotropy (q.v.).

Thixotropy. A property of adhesive systems of thinning upon isothermal agitation and thickening upon subsequent rest.[1] A property of materials which display a reduction in viscosity when a shearing action is applied. Some adhesive systems become thinner in consistency on agitation and thicker again when left undisturbed.[2]

Time, assembly. The time interval between the spreading of the adhesive on the adherend and the application of pressure, or heat, or both, to the assembly. For assemblies involving multiple layers or parts, the assembly time begins with the spreading of the adhesive on the first adherend. Assembly time is the sum of the open and closed assembly times. *Open assembly time* is the time interval between the spreading of the adhesive on the adherend and the completion of assembly of the parts for bonding. During this period the adhesive-coated surfaces are exposed to the air before being brought into contact. *Closed assembly time* is the time interval between completion of assembly of the parts for bonding and the application of pressure, or heat, or both, to cure or set the adhesive. (Reference 1, modified and Reference 3).

Time, curing. The period of time during which an assembly is subjected to heat or pressure, or both, to cure the adhesive. Further cure may take place after removal of the assembly from the conditions of heat, or pressure, or both.[1]

Time, drying. The period of time during which an adhesive on an adherend or an assembly is allowed to dry with or without the application of heat, or pressure, or both.[1]

Time, joint conditioning. The time interval between the removal of the joint from the conditions of heat, pressure, or both, used to accomplish bonding and the attainment of approximately maximum bond strength. Sometimes called *joint aging time*.[1]

Time, setting. The period of time during which an assembly is subjected to heat, pressure, or both, to set the adhesive.[1]

Vehicle. The carrier medium (liquid) for an adhesive material which improves its ease of application to adherends; solvent component of an adhesive.[2]

Viscosity. The ratio of the shear stress existing between laminae of moving fluid and the rate of shear between these laminae. A fluid exhibits *Newtonian behavior* when the rate of shear is proportional to the shear stress. A fluid exhibits *non-Newtonian behavior* when an increase or decrease in the rate of shear is not accompanied by a proportional increase or decrease in the shear stress.[1] A measure of the resistance to flow of a liquid. For Newtonian liquids, the shear rate is proportional to the shear stress between laminae of moving fluid; for non-Newtonian liquids it is not proportional.[2]

Viscosity coefficient (coefficient of viscosity). The shearing stress tangentially applied that will induce a velocity gradient. A material has a viscosity coefficient of one poise when a shearing stress of one dyne per square centimeter produces a velocity gradient of (1 cm/s)/cm.[1]

Vulcanization. A chemical reaction in which the physical properties of a rubber are changed in the direction of decreased plastic flow, less surface tackiness, and increased tensile strength by reacting it with sulfur or other suitable agents.[1] The crosslinking of an adhesive material by means of heat or catalysis; the chemical reaction of rubber with sulfur or other agents to alter its physical properties, e.g., to cause less tackiness, reduced plastic flow, and increased tensile strength.[2]

Vulcanize, v. To subject to *vulcanization*, q.v.[1]

Warp, n. A significant variation from the original, true, or plane surface of a material. (Reference 1, modified). A distortion of an adherend surface.[2]

Webbing. Filaments or threads that may form when adhesive transfer surfaces are separated. Transfer surfaces may be rolls, picker plates, stencils, etc.[1]

Weldbonding. A process in which a joint is formed by spot welding through an uncured adhesive bond line, or by flowing an adhesive into a spot-welded joint.[18]

Wetting. The process in which a liquid spontaneously adheres to and spreads on a solid surface. A surface is said to be completely wet by a liquid if the contact angle (θ) is zero, and incompletely wet if it is a finite angle. Surfaces are commonly regarded as unwettable if the angle exceeds 90 degrees.[11]

Wood veneer. A thin sheet of wood, generally within the thickness range 0.01 to 0.25 inch (0.3 to 6.3 mm) to be used in a laminate.[1]

Working life. See *pot life.*

Yield value. The stress, either normal or shear, at which a marked increase in deformation occurs without an increase in load.[1]

REFERENCES

1. American Society for Testing and Materials (ASTM), ASTM D 907-82, Standard Definitions of Terms Relating to Adhesives, 1982, published in volume 15.06 (Adhesives), *1984 Annual Book of ASTM Standards.*

2. Shields, J., *Adhesives Handbook*, 2nd Edition, Newnes-Butterworths, London (1976) (Available from Butterworths, Woburn, MA 01801). The 3rd Edition was published by Butterworths, London, in 1984.)

3. International Organization for Standardization (ISO), International Stan-American National Standards Institue (ANSI), 1430 Broadway, NY, NY 10018.

4. Wake, W.C., *Adhesion and the Formulation of Adhesives*, 2nd Edition, Applied Science Publishers Ltd. Available from Elsevier Science Publishing Co., NY, NY (1982).

5. Parker, R.S.R. and Taylor, P., *Adhesion and Adhesives*, Pergamon Press, London (1966).

6. Whittington, L.R., *Whittington's Dictionary of Plastics*, 2nd Edition, Technomic Publishing Co., Westport, CT (1978). (Sponsored by the Society of Plastics Engineers).

7. Bandaruk, W., Adhesive Bonding with Graphite Fibers as an Internal Electrical-Resistance Heat Source, pp 267–278 in *Applied Polymer Symposia No. 3, Structural Adhesive Bonding*, 1966, (M.J. Bodnar, ed.), Wiley-Interscience, NY. (Paper presented at Symposium sponsored by Picatinny Arsenal and held at Stevens Institute of Technology, Hoboken, NJ, September 14–16, 1965).

8. Hodgson, M.E., Pressure Sensitive Adhesives and Their Applications, *Adhesion 3*, (K.W. Allen, ed.), Applied Science Publishers, Ltd. Available from International Ideas, Inc., Philadelphia, PA.

9. American Society for Testing and Materials (ASTM), ASTM D 618-61 (1981)[€1], Standard Methods of Conditioning Plastics and Electrical Insulating Materials for Testing (1981), published in volume 8.01 (Plastics), *1983 Annual Book of ASTM Standards*.

10. *Adhesives in Modern Manufacturing*, (E.J. Bruno, ed.), Selection of Adhesives. Society of Manufacturing Engineers (SME), Dearborn, MI (1970).

11. Gent, A.N. and Hamed, G.R. (University of Akron), *Science of Adhesion*. Rept. NSF/OIR-83009, prepared for National Science Foundation, Washington, D.C. 108 pp. (1983). Available from NTIS as PB 84-176734.

12. *The Condensed Chemical Dictionary*, 8th Edition, Reinhold, NY (1971).

13. Synthetic Surfaces, Inc., Scotch Plains, NJ, *The Importance of Green Strength in Adhesive Selection*, Tech Sheet 472A (1982).

14. Blackley, D.C., *High Polymer Latices*, Vol. 1, Macleren & Sons, London (1966).

15. American Society for Testing and Materials (ASTM), ASTM F 412-82a, Standard Definitions of Terms Relating to Plastic Piping Systems, 1982, published in volume 8.04 (Plastic Pipe and Building Products), *1983 Annual Book of ASTM Standards*.

16. Titow, W.V., Solvent Welding of Plastics, *Adhesion 3*, (K.W. Allen, ed.), Applied Science Publishers, Ltd. Available from International Ideas, Inc., Philadelphia, PA.

17. DeLollis, N.J., *Adhesives, Adherends, Adhesion*, Robert E. Krieger Publishing Co., Huntington, NY, 1980. (Based on original first edition, *Adhesives for Metals*, 1970).

18. Bolger, J.C., Single-Component Epoxy Adhesives for Sheet Steel and Aluminum Bonding, pp 369–387 in *Applied Polymer Symposia No. 32, Durability of Adhesive Bonded Structures*, 1977, (M.J. Bodnar, ed.), Wiley-Interscience, NY. (Paper presented at symposium at Picatinny Arsenal, Dover, NJ, October 27–29, 1976).

3

Joint Design and Design Criteria

BASIC PRINCIPLES

Joints for adhesive bonding must be designed particularly for the use of adhesives. The habit of beginning with a design used for another method of fastening and modifying it slightly for adhesive bonding is a poor one, often leading to disastrous results. The aim of joint design is to obtain maximum strength for a given area of bond. In designing joints specifically for adhesive bonding the basic characteristics of adhesives must dictate design of joints. Adhesive bonds act over areas and not a single point. For this reason the joint should be designed with the objective of minimizing concentration of stress.

The selection of joint design is influenced by limitations in production facilities, production costs, and the desired final appearance of the part. The strength of an adhesive joint is determined primarily by (1) the mechanical properties of adherend and adhesive, (2) residual internal stresses, (3) degree of true interfacial contact, and (4) the joint geometry. Each of these factors has a strong influence on joint performance. The design engineer must be concerned with the elimination of stress concentrations which reduce the strength and useful life of the joint. Localized stresses are not always apparent and may occur as a result of differential thermal expansion of the adhesive and adherends. Another cause is shrinkage of adhesive during cure, when volatiles are given off. These volatiles may become entrapped. Internal stresses decrease as adhesive thickness decreases, reducing the tendency to trap volatiles. Air can also become entrapped at the interface if the adhesive has too high a viscosity, does not flow easily as it undergoes curing, or if it does not wet the substrate.[1]

TYPES OF STRESS

Figure 3-1 shows five types of stress found in adhesive joints. Any combination of these stresses may be encountered in an adhesive application. These stresses are as follows:

31

(1) *Compression.* When loaded in pure compression, a joint is less likely to fail than when loaded in any other manner, but compression-loaded joints are limited in application.

(2) *Shear.* This type of loading imposes an even stress across the whole bonded area, utilizing the joint area to the best advantage and providing an economical joint that is most resistant to joint failure. Whenever possible, most of the load should be transmitted through the joint as a shear load.[2]

(3) *Tension.* The strengths of joints loaded in tension or shear are comparable. As in shear, the stress is evenly distributed over the joint area, but it is not always possible to be sure that other stresses are not present. If the applied load is offset to any degree, the advantage of an evenly distributed stress is lost and the joint is more likely to undergo failure. The adherends should be thick with this type of joint and not likely to deflect to any appreciable degree under the applied load. Such a situation will result in nonuniform stress.[2] Tensile stress develops when forces acting perpendicular to the plane of the joint are distributed uniformly over the entire area of the bond. The types of stress likely to result when other than completely axial loads are applied are *cleavage* and *peel*. Since adhesives generally have poor resistance to cleavage and peel, joints designed to load the adhesive in tension should have physical restraints to ensure axial loading.[3]

(4) *Peel.* One or both of the adherends must be flexible in this type of loading. A very high stress is applied to the boundary line of the joint, and unless the joint is wide or the load small, failure of the bond will occur. This type of loading is to be avoided if possible.[2]

(5) *Cleavage.* Cleavage is somewhat similar to peel and occurs when forces at one end of a rigid bonded assembly act to split the adherends apart.[2] Cleavage may be considered as a situation in which an offset tensile force or a moment has been applied. The stress is not evenly distributed (as is the case with tension) but is concentrated on one side of the joint. A sufficiently large area is needed to accommodate this stress, resulting in a more costly joint.[2]

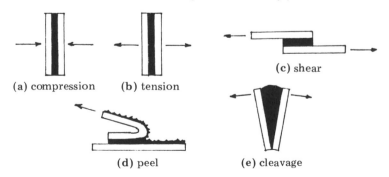

(a) compression   (b) tension   (c) shear
(d) peel   (e) cleavage

Figure 3-1: Types of stresses in adhesive joints. Adapted from Shields.[2]

METHODS OF IMPROVING JOINT EFFICIENCY

As mentioned above, joints should be specifically designed for adhesive bonding. Figure 3-2 illustrates various types of adhesive joints used for flat adherends. Adhesive bonds designed to follow the following general principles will result in maximum effectiveness:[4]

- The bonded area should be as large as possible within allowable geometry and weight constraints

- A maximum percentage of the bonded area should contribute to the strength of the joint

- The adhesive should be stressed in the direction of its maximum strength

- Stress should be minimized in the direction in which the adhesive is weakest

Thermosetting adhesives, such as epoxies, are relatively rigid and exhibit high tensile and shear strength under both dynamic loading and static loading. Such adhesives also have good fatigue resistance. Rigid brittle adhesives are not recommended for bonds stressed in peel or cleavage, however. Elastomeric adhesives, on the other hand, have low tensile or shear strength, but these adhesives develop high peel or cleavage strength. Adhesives that possess high tensile and shear strength over short periods of static stress give poor results over longer periods or under vibrating stresses.[6]

Types of loads and joints which concentrate stresses in small areas or on edges should be avoided. Joints which stress the adhesive in shear are preferable since adhesives generally show considerable strength under this type of stress. Sudden applications of load, such as during impact, require the use of elastic or resilient adhesives to absorb the shock. Brittle adhesives will ordinarily fail under such conditions.[6]

Joint Design Criteria

The bonded area should be large enough to resist the greatest force that the joint will meet in service. The calculation of stress in the adhesive joint is not a reliable way of determining the exact dimensions required. It is relatively difficult to decide on an allowable stress. The strength of the bond is affected by environmental conditions, age, temperature of cure, composition and size of adherends, and the thickness of the adhesive layer.[2]

The stress in the adhesive is ordinarily a combination of various stresses. The relative flexibility of the adhesive to that of the adherends has a pronounced effect on the stress distribution. Figure 3-3 is a typical example of a simple lap joint under tensile loading. Study of Figure 3-3(c) shows that most of the stress is concentrated at the ends of the lap. The greater part of the lap (adjacent to the center) carries a comparatively low stress. Therefore, if the overlap length is doubled, the load-carrying capability of the joint is increased by a relatively low percentage. The greatest gain in strength is obtained by increasing the joint width.[2]

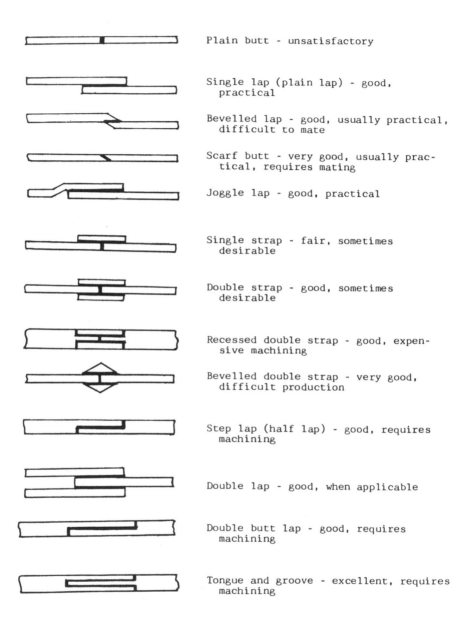

Plain butt - unsatisfactory

Single lap (plain lap) - good, practical

Bevelled lap - good, usually practical, difficult to mate

Scarf butt - very good, usually practical, requires mating

Joggle lap - good, practical

Single strap - fair, sometimes desirable

Double strap - good, sometimes desirable

Recessed double strap - good, expensive machining

Bevelled double strap - very good, difficult production

Step lap (half lap) - good, requires machining

Double lap - good, when applicable

Double butt lap - good, requires machining

Tongue and groove - excellent, requires machining

Figure 3-2: Types of joints used in adhesive-bonding flat adherends. Adapted from References 3, 5, and 6.

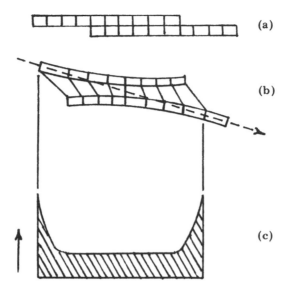

Figure 3-3: Tensile force on lap joint showing (a) unloaded joint, (b) joint under stress, and (c) stress distribution in adhesive. Modified after Reference 2.

The single lap joint shown in Figure 3-4 is typical of most adhesive joints. Increasing the width of the joint results in a proportionate increase in strength, while increasing the overlap length (L) beyond a certain limit has very little effect, as seen in Figure 3-5.[2]

In addition to overlap length and width, the strength of the lap joint is dependent on the yield strength of the adherend. The modulus and thickness of the adherend determine its yield strength, which should not exceed the joint strength. The yield strength of thin metal adherends can be exceeded where an adhesive with a high tensile strength is employed with a relatively small joint overlap. Figures 3-5 and 3-6 show the relationship between shear strength, adherend thickness, and overlap length.[2]

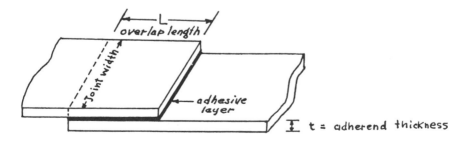

Figure 3-4: Single lap joint.

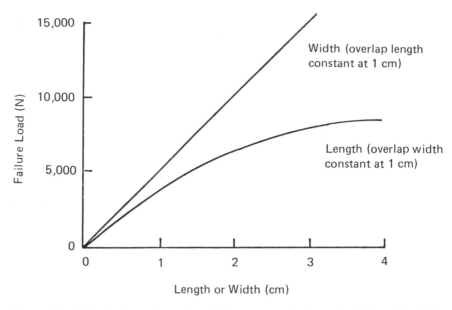

Figure 3-5: Effect of overlap and width on strength of a typical joint. Modified after Reference 2.

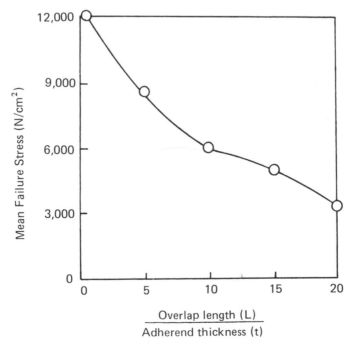

Figure 3-6: Correlation between shear strength and L/t ratio. Modified after Reference 2.

The fall-off in the effective load-bearing capacity of the overlap joint is usually expressed as a correlation between shear strength and the L/t ratio, as seen in Figure 3-6, but sometimes the ratios t/L or $t^{1/2}/L$ are used. The t/L ratio is often called the "joint factor."[2] Many variables that have significant effects on the strength of an adhesive are related by the L/t curve. Some of these are: adherend modulus, adhesive, test temperature, bond-line thickness, and joint configuration. The L/t curve is generally used for each variable that may enter the design, and data are presented to the designer as families of L/t curves.[7] One variable commonly plotted is load or stress against L/t.

Typical Joint Designs

The ideal adhesive-bonded joint is one in which, under all practical loading conditions, the adhesive is stressed in the direction in which it most resists failure. Figure 3-2 shows a number of types of joints used in bonding *flat adherends*. These will be discussed briefly.[3]

Butt joints—These joints are not able to withstand bending forces because under such forces, the adhesive would undergo cleavage stress. If the adherends are too thick to design simple overlap joints, modified butt joints can be designed. Such joints reduce the cleavage effect caused by side loading. Tongue-and-groove joints are self-aligning and provide a reservoir for the adhesive. Scarf butt joints keep the axis of loading in line with the joint and require no extensive machining.[3]

Lap joints—These are the most commonly used adhesive joints. They are simple to make, can be used with thin adherends, and stress the adhesive in its strongest direction. The simple lap joint, however, is offset and the shear forces are not in line, as seen in Figure 3-3.[3] It can be seen in this stress distribution curve that most of the stress (cleavage stress) is concentrated at the ends of the lap. The greater part of the overlap (adjacent to the center) carries a comparatively low stress. If the overlap length is increased by 100%, the load-carrying capability is increased by a much lower percentage. The most effective way to increase the bond strength is to increase the joint width.[2] Modifications of lap-joint designs which improve efficiency include:[3]

- Redesigning the joint to bring the load on the adherends in line
- Making the adherends more rigid (thicker) near the bond area (Figure 3-7)
- Making the edges of the bonded area more flexible for better conformance, thereby minimizing peel

Modifications of lap joints are shown in Figure 3-2.

Joggle lap joints—This is the easiest design for aligning loads. This joint can be made by simply bending the adherends. It also provides a surface to which it is easy to apply pressure.[3]

Double lap joints—These joints have a balanced construction which is subjected to bending only if loads in the double side are not balanced.[3]

Beveled lap joints—These joints are also more efficient than plain lap joints.

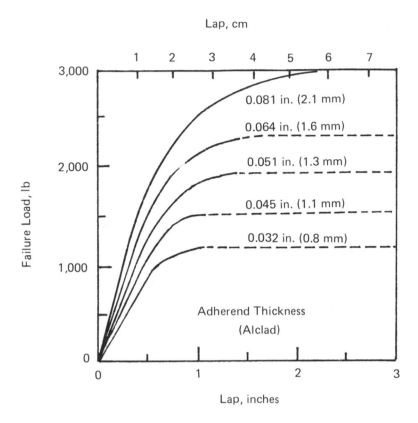

Lap, cm

Figure 3-7: Interrelation of failure loads, depth of lap, and adherend thickness for lap joints with a specific adhesive and adherend. Modified after References 3 and 8.

The beveled edges allow conformance of the adherends during loading, with a resultant reduction of cleavage stress on the ends of the joint.[3]

Strap joints—These joints keep the operating loads aligned and are generally used where overlap joints are impractical because of adherend thickness. As in the case of the lap joint, the *single strap* is subject to cleavage stress under bending forces. The *double strap joint* is superior when bending stresses are met. The *beveled double strap* and *recessed double strap* are the best joint designs to resist bending forces. These joints both require expensive machining, however.[3]

Peeling of Adhesive Joints

When thin members are bonded to thicker sheets, operating loads generally tend to peel the thin member from its base, as shown in Figure 3-8.[3] *Riveting* may provide extra strength at the ends of the bond, but the use of rivets may result in stress concentrations. *Beading* the end of the joint is helpful, but not always feasible. An increase in peel strength will result from *increasing the width* of the end of the joint. Finally, increasing the stiffness of the adherends is often quite effective. The stiffer the adherends, the smaller the deflection of the joint for a given force, and the smaller the peel stresses.[9]

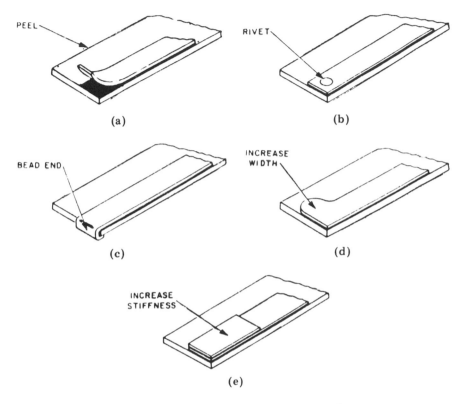

Figure 3-8: Designs that minimize peel.[9]

Stiffening Joints

In many cases, thin sheets of adherend are rigidized by bonding stiffening members to the sheet. When such sheets are flexed, the bonded joints are subjected to cleavage stress. Figure 3-9 illustrates design methods used for reducing cleavage stress on stiffening joints.

Cylindrical Joints

Several recommended designs for rod and tube joints are shown in Figures 3-10 and 3-11. These designs are preferable to simple butt joints because of (1) their resistance to bending forces and subsequent cleavage, and (2) their increase in bonding area. In the case of tubular forms (Figure 3-11) the bonding area is small unless the tube walls are very heavy. Most of these joints require a machining operation.[3]

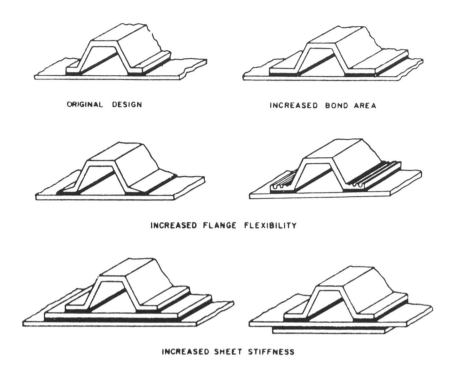

ORIGINAL DESIGN   INCREASED BOND AREA

INCREASED FLANGE FLEXIBILITY

INCREASED SHEET STIFFNESS

Figure 3-9: Methods of minimizing peel for stiffening sections (flange joints).[9]

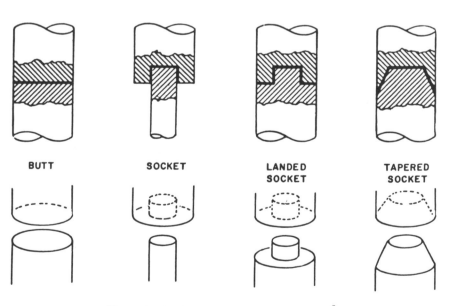

BUTT   SOCKET   LANDED   TAPERED
                              SOCKET   SOCKET

Figure 3-10: Straight joints for solid bars.[9]

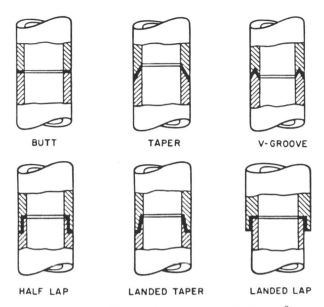

Figure 3-11: Straight joints for tubular forms.[9]

Angle and Corner Joints

Angle and corner joints for flat adherends are illustrated in Figures 3-12 and 3-13. In both cases, the butt joints are susceptible to cleavage under bending stress. Among the angle joints (Figure 3-12) the *dado joint* is probably the best, provided that the reduction in section required for the recess is acceptable. This design is less subject to cleavage stress than the right-angle butt joints (also called "L" angle joints), and is easier to form. The double right-angle butt joint is also called the "T" angle joint. Corner joints (Figure 3-13) for flat adherends are best designed to use fixtures. For solid rods and tubular forms, fixtures are always required.[9]

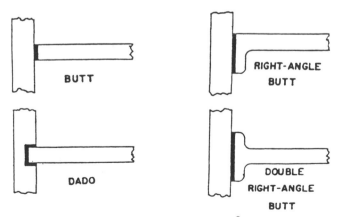

Figure 3-12: Angle joints.[9]

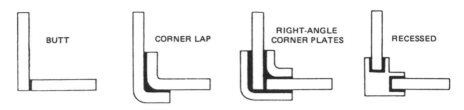

Figure 3-13: Corner joints.[9]

JOINTS FOR PLASTICS AND ELASTOMERS

Flexible Materials

Thin or flexible polymeric substrates may be joined using a *simple* or *modified lap joint*. The *double strap joint* shown in Figure 3-2 is best, but time-consuming to form. The strap material should be fabricated from the same material as the parts to be joined. If this is not possible, it should have approximately equivalent strength, flexibility, and thickness. The adhesive should have the same degree of flexibility as the adherends. If the sections to be bonded are relatively thick, a *scarf joint*, also shown in Figure 3-2, is acceptable. The length of the scarf should be at least four times the thickness, as shown in Figure 3-14.[3]

Figure 3-14: Recommended scarf joint configuration for flexible plastics and elastomers.

Figure 3-15 shows several types of joints for rubber under tension. The horizontal white lines are equidistant when the joints are unstressed. It is obvious that the *scarf joint* is least subject to stress concentration with materials of equal modulus, and the *double scarf joint* is the best for materials of unequal modulus. These designs offer the best resistance to peel and, all other factors being equal, represent the best choices.[9]

When bonding elastic material, forces on the elastomer during cure should be carefully controlled, since too much pressure will cause residual stresses at the bond interface. Stress concentrations may also be minimized in rubber-to-metal joints by elimination of sharp corners and by the use of metal adherends sufficiently thick to prevent peel stresses that may arise with thinner-gage metals. As with all joint designs, polymeric joints should avoid peel stresses. Figure 3-16 illustrates methods of bonding flexible substrates so that the adhesive will be stressed in its strongest direction.[9]

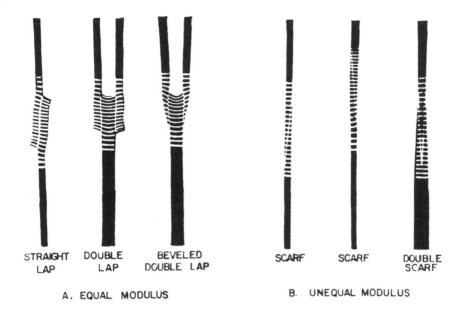

STRAIGHT   DOUBLE   BEVELED   SCARF   SCARF   DOUBLE
LAP   LAP   DOUBLE LAP   SCARF

A. EQUAL MODULUS   B. UNEQUAL MODULUS

Figure 3-15: Joints for rubber under stress.[9]

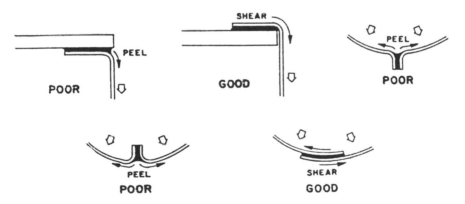

Figure 3-16: Joints for flexible materials.[9]

Rigid Plastics

In the case of rigid plastics, the greatest problems are in *reinforced plastics* which are often anisotropic, having directional strength properties. Joints made from such substrates should be designed to stress both the adhesive and the adherend in the direction of greatest strength. Laminates should be stressed parallel to the laminations to prevent delamination of the substrate.[3]

Single and joggle lap joints (Figure 3-2) are more likely to cause delamination than scarf or bevelled lap joints. Strap joints may be used to support bending loads.[3]

WOOD JOINTS

> This material was drawn, with permission, entirely from an excellent book authored by R.S. Miller and copyrighted by Franklin Chemical Industries, a major manufacturer of wood glues.[10]

Destructive forces on wood joints may be (1) applied from the exterior of a joint, or (2) induced by internal forces. *External forces* may be those applied by sitting on a chair, or by racking caused by pushing a case across a rough floor. *Internal forces* may be from shrinking of wood parts, or the steady pull of poorly mated parts forced together by clamping pressure. Proper joint selection will alleviate the problem with the external forces, while correction for problems with internal forces may be concerned with good manufacturing practice, such as selection of wood species, uniformity of moisture content, and joint accuracy. As a general rule, as far as joint design is concerned, it is best to select the simplest joints that will do the job satisfactorily.

Success in gluing wood depends in part on a knowledge of the structure of wood. The tendency of wood to shrink or swell across the grain is a major problem in construction work. The manner in which the lumber is sawed from the log will determine its behavior. The outer boards cut from a log have the greatest tendency to shrink due to the fact that they contain the most sap and have the flattest grain. In addition, the sap side (the side lying toward the outside of the log) shrinks and cups the most when drying, as a result of having the highest moisture content. This tendency to shrink can be minimized by employing *kiln drying*, a technique by which wood is artificially seasoned in sealed chambers under controlled heat and moisture.

An assembly joint should be designed so that most of the glued area is either tangential or radial grain. With most wood species, joints between side-grain (Figure 3-17A) and flat-grain surfaces can be made as strong as the wood itself in shear parallel to the grain, in tension across the grain, and in cleavage. Side grain can be held to prevent any movement by using dowels (Figure 3-17C), splines (Figure 3-17D), or milled surfaces (Figure 3-17E). The tongue-and-grooved joint (Figure 3-17F) and other shaped joints have the advantage of larger gluing surfaces than the straight joints. However, in practice they do not provide higher strength with most woods. The theoretical advantage is often lost, wholly or partly, because the shaped joints are more difficult to machine than straight plain joints to provide a perfect part fit. Tongue-and-groove joints are, however, frequently used on side panels, or in raised panel doors. If the panel insert is allowed to float, that is, if it is smaller than the encompassing frame, there will be less tendency for the expansion and contraction (from moisture changes) of the panel to stress the joints in the enclosing frame.

It is almost impossible to make end-butt joints (Figure 3-18A) sufficiently strong or permanent to meet the normal requirements of furniture joining. With the most careful gluing, not more than about 25% of the tensile strength of the wood parallel to the grain can be obtained in butt joints. In order to approximate the tensile strength of certain wood species, a scarf, serrated or other form of joint that approaches a side-grain surface must be used. The plain scarf (Figure 3-18B) is the easiest to glue and entails fewer machining difficulties than the many-angle forms. It is also difficult to obtain the proper joint strength in end-

to-side joints (Figure 3-18C) which are further subjected in use to unusually severe stresses as a result of unequal dimensional changes in the two members of
the joint as their moisture content changes. It is necessary to use irregular shapes
of joints, dowels (Figure 3-18D), tenons, glue blocks (Figure 3-18E), or other
devices to reinforce such a joint in order to bring side grain into contact with
side grain, or to secure larger gluing surfaces.

Detailed instructions on how to make these and many other types of wood
joints are given in R.S. Miller's 1980 book.[10]

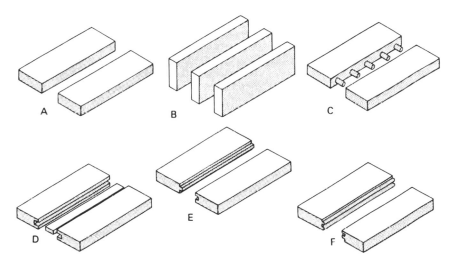

Figure 3-17: Common edge-glued and face-glued wood joints.[10] Copyright
Franklin Chemical Industries, Inc. Used with permission.

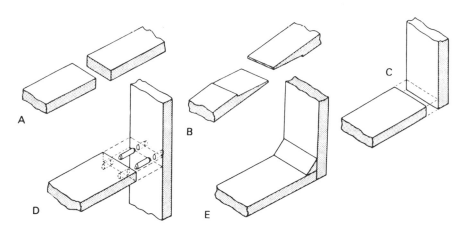

Figure 3-18: End-to-end and end-to-side butt joints and methods of reinforcement.[10] Copyright Franklin Chemical Industries, Inc. Used with permission.

Glue or Corner Blocks

Glue or corner blocks are small square or triangular pieces of wood used to strengthen and support the two adjoining surfaces of a butt joint. While this joint reinforcement features a strong face-to-face gluing surface, varying moisture contents will cause a differential wood movement, since the grain direction of the block and the substrate are at right angles to each other. This will highly stress the joint. Because of this fact, a number of short blocks is preferable to one long one.

Mortise and Tenon Joints

The mortise-and-tenon is a very good joint, stronger and more widely used than the butt joint. It is one of the best techniques used in fine furniture making. Figure 3-19 shows a number of common mortise-and tenon joints.

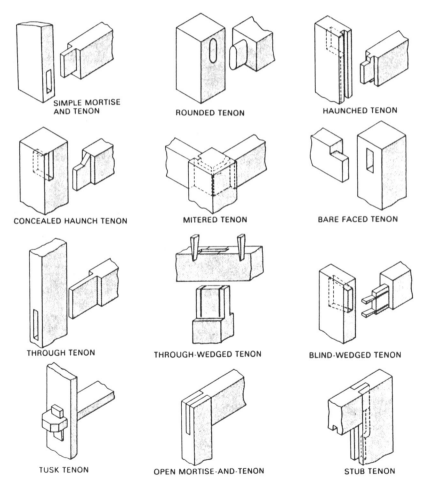

SIMPLE MORTISE AND TENON

ROUNDED TENON

HAUNCHED TENON

CONCEALED HAUNCH TENON

MITERED TENON

BARE FACED TENON

THROUGH TENON

THROUGH-WEDGED TENON

BLIND-WEDGED TENON

TUSK TENON

OPEN MORTISE-AND-TENON

STUB TENON

Figure 3-19: Common mortise-and-tenon joints.[10] Copyright Franklin Chemical Industries, Inc. Used with permission.

Dovetail Joints

The interlocking of two pieces of wood by a special fan-shaped cutting is called a dovetail joint. It is used extensively in making fine furniture, drawers, and in projects where good appearance and strength are desired. A dovetail joint has considerable strength because of the flare of the projections, technically known as pins, on the ends of the boards which fit exactly into similarly shaped dovetails. Figure 3-20 shows four basic dovetail joints.

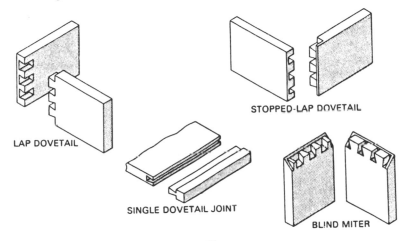

STOPPED-LAP DOVETAIL

LAP DOVETAIL

SINGLE DOVETAIL JOINT

BLIND MITER

Figure 3-20: Common dovetail joints.[10] Copyright Franklin Chemical Industries, Inc. Used with permission.

Dado Joints

A dado joint is formed when one piece of wood is set into a groove or dado cut into another. There are many variations of a dado joint (Figure 3-21) used in cabinet work and furniture making. Actually a dado is a poor joint for gluing, as it has the same deficiency as an end-to-side grain joint—low glue joint strength. The dovetail dado does have more strength than a plain dado, as the shear strength of the wood prevents the rupture of the joint.

Lap Joints

Typical lap joints are shown in Figure 3-22. The desired visual effect of such joints is that the two thicknesses overlap within a single thickness. A well-made lap joint is a strong joint in that the glued surface is a face-to-face surface.

Miter Joints

The miter joint (Figure 3-23) is primarily for show. For example, it may be used for an uninterrupted wood grain around edges (side to top to side of a cabinet), or at corners (picture frame). The joining ends or edges are usually cut at angles of 45 degrees, then glued and clamped. The 45-degree glued miter joint is stronger than the end-to-end joint, but not nearly as strong as a face-to-face glued joint. For a high-quality mitered joint, modifications such as splines, keys, feathers, or dowels may be used, as shown in Figure 3-24.

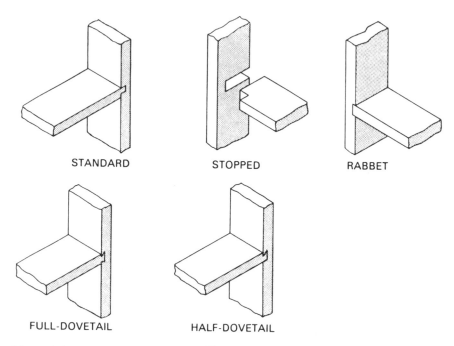

Figure 3-21: Common dado joints.[10] Copyright Franklin Chemical Industries, Inc. Used with permission.

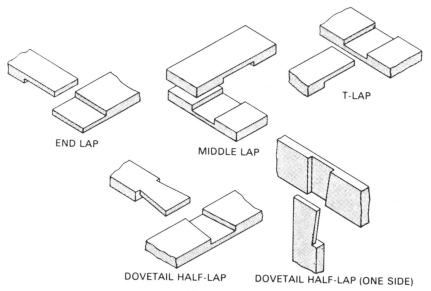

Figure 3-22: Common lap joints.[10] Copyright Franklin Chemical Industries, Inc. Used with permission.

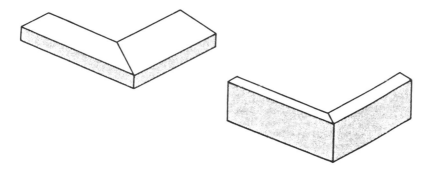

Figure 3-23: Flat and edge miter joints.[10] Copyright Franklin Chemical Industries, Inc. Used with permission.

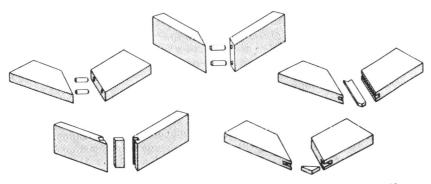

Figure 3-24: Flat and edge miters with dowels, splines, and feathers.[10] Copyright Franklin Chemical Industries, Inc. Used with permission.

Lapped Miter Joints

These joints, shown in Figure 3-25 are generally used only if it is desirable to have the miter show on one side. This joint will give the appearance of a conventional miter from one side, yet will be considerably stronger because of the increase in the area of contacting surfaces.

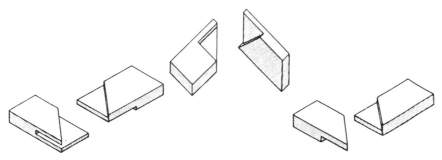

Figure 3.25: Lapped miter joints.[10] Copyright Franklin Chemical Industries, Inc. Used with permission.

Corner Joints

Figure 3-26 shows various types of corner joints that can be used in wood-working, ranging from a simple rabbet to the complex lock miter joint. The rabbet joint conceals open end grain and also reduces the twisting tendency of a joint. The backs of most cases, cabinets, bookcases, and chests are joined with the end grain facing the back.

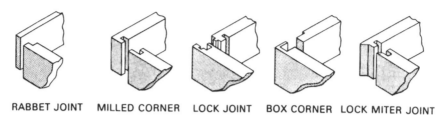

RABBET JOINT   MILLED CORNER   LOCK JOINT   BOX CORNER   LOCK MITER JOINT

Figure 3-26: Common corner joints.[10] Copyright Franklin Chemical Industries, Inc. Used with permission.

METAL JOINTS

Information on metal joints is found in the general discussion at the beginning of this chapter. The emphasis in this book is on metal-to-metal bonds.

DIFFERENT ADHEREND TYPES IN JOINTS

In cases where joints differ in rigidity, their thicknesses should be adjusted so that they will be equally rigid. In other words, the more flexible adherends should be made thicker. The *double scarf lap joint* (shown in Figure 3-15B), with the adherends tapered in proportion to their moduli of elasticity, is recommended. With very large differences in rigidity, as in rubber-to-steel bonds, this adjustment cannot be carried out.[9]

Where differences in the *coefficients of thermal expansions* of adherends may cause problems, including warpage of the final assembly, elastic adhesives are often useful. It is also important to use careful joint design and balance the construction. In the case of common straight joints, a *double lap* or *double strap joint* is recommended (see Figures 3-2 and 3-15). In laminating sheets, balance can be achieved by fabricating a "sandwich" so that one of the adherends is bonded to both sides of the other. An alternative solution is to break the continuity of the sheet with the lower coefficient of thermal expansion. For example, when bonding wood to metal, the wood can be made into several smaller panels before fabrication of the joint.

STRESS ANALYSIS OF ADHESIVE JOINTS

This is an extremely complicated subject and will only be touched upon here. The ultimate objective is to develop a design method for bonded construction based on the principles of mechanics and rational engineering design so that joint behavior can be predicted.

Theoretical Analysis of Stresses and Strains

Most theoretical analyses have been carried out on single or double lap joints, which are the primary types of joints used for determining the strength of adhesives. Properly designed joints stress the adhesive in shear. Adhesives are especially weak in peel, and are also weak under tensile loads applied normal to the plane of the joint. The earliest theoretical lap-joint work involved simplifying assumptions: that the joint was a simple overlap type; that both adherends were made of the same metal and had the same geometry; that the adherends and adhesive behaved elastically; that bending or peeling stresses were not involved; that thermal expansion or residual stresses were ignored; and that deflections were small.[11]

Recent theoretical studies have become much more complex. New computer-assisted techniques permit the use of finite-element-matrix-theory type approaches. The effects of important variables are being determined by parametric studies. More complex joints are also being studied. New adherend materials, including advanced filamentary composites, are also being evaluated. The elastic, low-deflection, constant temperature behavior of scarf and stepped-lap joints has been replaced by elastic-plastic, large-deflection behavior, combined with thermal expansion differences, or curing shrinkage-induced residual stresses.

Experimental Analyses

Typically, the yardstick for qualitatively measuring the internal resistance of an adhesive bond to an external load has been the determination of the strain distribution in the adhesive and adherends. This is a difficult task. Even in simple lap joints the actual stress-strain distributions under load are extremely complex combinations of shear and tensile stresses, and are very prone to disturbance by nonuniform material characteristics, stress concentrations or localized partial failures, creep and plastic yielding, etc. It is extremely difficult to accurately measure the strains in adhesive joints with such small glueline thicknesses and such relatively inaccessible adhesive. Extensometers, strain gages, and photoelasticity are being used with limited success.[11]

Failure Analyses

The function of a structural adhesive joint is to transmit an external load to the structural member. If the joint fails to function as it is intended, it will undergo damage or failure. The damage could be actual fracture of the structure, excessive elastic deformation, or excessive inelastic flow. The criteria for what constitutes structural failure depend on the performance requirements of the joint. The fundamental problem in the mechanics of adhesives and joints is to obtain some relationship between the loads applied to the joint and a parameter that will adequately describe the criteria for structural failure. The most common criterion for such failure of lap-type joints is actual fracture of the joint.

For a given combination of adherend and adhesive, the stress analyst must decide what the mode or theory of failure would be if the applied loads become large enough to cause failure. The decision as to which theory would realistically determine the mode of failure is usually based on past experience, or upon some form of experimental evidence.[11]

The next step is to determine a relationship between the applied load and a parameter that will describe the failure of the joint. Such parameters might be stress, strain, strain energy, etc. Finally, when maximum tolerable stresses have been obtained, the allowable stress values or factors of safety are decided upon to allow for factors such as long- and short-term loading, fatigue loading, special environmental conditions, and other special considerations. This step is ordinarily based on experience, engineering judgment, and legal, government, or commercial specifications.[11]

Methods of Stress Analysis

Theory of Volkersen—In 1938, Volkersen analyzed the distribution of shearing stresses in the adhesive layers of a lap joint. Volkersen's model is useful only with very stiff adhesives which do not bend on loading the joint. A dimensionless stress concentration factor is found to depend on the geometry and physical parameters of the joint. By introducing further simplifications, certain reasonable geometric conditions, and identical adherends, a simple formula is obtained:[12]

$$\Delta = \frac{GL^2}{Etd}$$

where G = shear modulus of the adhesive

E = Young's modulus of the adherend

d = thickness of the adherend

L = length of the overlap

t = thickness of the adherend

DeBruyne has suggested that, when all other variables are kept constant, the quantity $\sqrt{t}/L$ with dimension (length)$^{-1/2}$, the "joint factor" derived from the above equation, is useful in correlating joint strengths.[12]

Volkersen's theory predicts that the shear stresses in the adhesive layer reach a maximum at each end of the overlap, when the bonded plates are in pure tension. Photoelastic analysis of these composite structures show that stresses are uniform in the central part of the model adhesive, but high near the edges of the steel plate used in the analysis (Figure 3-2). Stress distributions at the end were found to be independent of the length of the overlap when its length was at least three times the thickness of the adhesive layer.[12]

Theory of Goland and Reissner—In Volkersen's theory, the so-called "tearing" or "peeling" stresses were ignored. Goland and Reissner[13] took the bending deformation of the adherends into account, as well as the transverse strains in the adhesive and the associated tearing stresses. These workers showed that the maximum tearing and shear stresses reach asymptotic values for large overlap lengths. Provided the system remains linearly elastic, the joint strength reaches a limiting value with increasing overlap length. In actual practice, however, a limiting strength is obtained because the adherends are loaded to their ultimate strength.[13]

REFERENCES

1. Design Considerations, *Adhesives in Modern Manufacturing*, published by the Society of Manufacturing Engineers (SME), edited by E.J. Bruno (1970).

2. Shields, J., *Adhesives Handbook*, 2nd Edition, Newnes-Butterworth, London (1976). (The 3rd Edition was published by Butterworths, London, 1984.)

3. Petrie, E.M., Plastics and Elastomers as Adhesives, *Handbook of Plastics and Elastomers*, (C.A. Harper, ed.), McGraw-Hill, NY, NY (1975).

4. Kubo, J.T., Joints, *Advanced Composites Design Guide, Vol. 1—Design*, 3rd Edition, 2nd Revision, September 1976. Prepared for Air Force Flight Dynamics Laboratory, WPAFB, Ohio, by Rockwell International Corp.

5. Sharpe, L.M., The Materials, Processes, and Design Methods for Assembling with Adhesives, Design Guide, *Machine Design*, 38 (19):179–200 (August 18, 1966).

6. Kubo, J.T., Joints, *Advanced Composites Design Guide*, 3rd Edition, 3rd Revision, January 1977. Prepared for Air Force Flight Dynamics Laboratory, WPAFB, Ohio, by Rockwell International Corp.

7. Lunsford, L.R., Stress Analysis of Bonded Joints, *Applied Polymer Symposia No. 3, Structural Adhesive Bonding*, presented at a symposium sponsored by Picatinny Arsenal and held at Stevens Institute of Technology, September 14-16, 1965, pp 57-73, Wiley-Interscience (1966).

8. Perry, H.A., Room Temperature Setting Adhesives for Metals and Plastics, *Adhesion and Adhesion Fundamentals and Practices*, edited by J.E. Rutzler and R.L. Savage, Society of Chemical Industry, London (1954).

9. Military Handbook MIL-HDBK-691A, *Adhesives* (May 17, 1965).

10. Miller, R.S., *Home Construction Projects with Adhesives and Glues*, Franklin Chemical Industries, Columbus, Ohio, 1983. (This book was published in 1980 in almost identical form under the title *Adhesives and Glues—How to Choose and Use Them*).

11. Design, Analysis and Test Methods, Vol. 4, *Treatise on Adhesion and Adhesives*, Structural Adhesives with Emphasis on Aerospace Applications, a Report of the ad hoc committee on Structural Adhesives for Aerospace Use, National Materials Advisory Board, National Research Council, Marcel Dekker, NY (1976).

12. Salomon, G., Introduction, *Adhesion and Adhesives*, 2nd Edition, Vol. 2, *Applications*, Edited by R. Houwink and G. Salomon, Elsevier, Amsterdam (1967).

13. Goland, M. and Reissner, E., The Stresses in Cemented Joints, *Journal of Applied Mechanics*, 2 (1):A-17—A-27 (March 1944).

4

Surface Preparation of Adherends

INTRODUCTION

Since adhesive bonding is a surface phenomenon, surface preparation prior to adhesive bonding is the keystone of successful bonding. Surface preparation, or surface pretreatment as it is sometimes called, is carried out to provide adherend surfaces receptive to the development of strong, durable adhesive joints. It is desirable, although not always practical, to have the basic adherend material exposed directly to the adhesive, with no intervening layer of oxide film, paint, chromate coating, phosphate coating, nor silicone release agent. Such layers are called "weak boundary layers," and in their presence the adhesive never directly contacts the adherend surface.[1]

The selection of exactly what surface preparation method to use for a particular adherend requires careful evaluation. A number of factors, some obvious and some not, influence the choice. The size of component parts and the availability of equipment and facilities are in the obvious category. Less obvious factors include the rapid depletion of active chemicals in an immersion bath, or the accumulation of foreign materials in the bath which give rise to weak boundary layers.[1]

In order to minimize the production of scrap, it is important to have rigid process control. In the case of metals, the optimum surface preparation to provide durability and uniform quality is ordinarily a chemical immersion or spray process. In the case of very large parts application is often carried out by the use of reagents in paste form. The choice of the process to be used is desirable only if the production scheme or the entire system will accommodate it. Cost must also be considered and balanced against the requirement for reliability, maintainability, and criticality of the joint. Compromises must be made in the entire system. Good process control can be obtained only by preparation of a complete process specification, including detailed surface preparation requirements.[1]

To place surface preparation in its proper perspective, the adherend-to-organic material (i.e., adhesive) must be considered from design through fabrica-

54

tion. Joint design, adhesive selection, and processing all must be considered. Each of these factors are interdependent. The use of an optimum surface preparation is of little value if an unsuitable adhesive is used, if the bond is not properly processed, or if the joint design involves peel or cleavage stress.[2]

Proper surface preparation should help ensure that the weakest link in an adhesive joint is within the adhesive or organic material layer, not at its interface with the adherend. In adhesive bonds, this type of fracture is known as *cohesive*, and a layer of adhesive remains on both adherends, as shown in Figure 1-2. When fracture or failure occurs at the adhesive-to-adherend interfact, it is called *adhesive*. As will be seen in Chapter 12, many ASTM test methods prescribe reporting the mode of failure or fracture as a certain percentage cohesive and a certain percentage adhesive. From a surface preparation viewpoint, the ideal type of failure of a bonded joint or test specimen is one with 100 percent cohesive failure.[2]

Table 4-1 illustrates the importance of proper surface preparation for five different metal adherends, using two different adhesives and a variety of surface treatments.[3] Similarly, Figure 4-1 shows how various surface treatments improve the durability of aluminum under adverse conditions. Surface treatment is particularly important on durability of aluminum. The exact ranking order of results may change in studies of this type, depending on the adhesive primer and type of alloy being used.[4]

Table 4-1: Effect of Metal Substrate Surface Preparation in
Adhesive-Bonded Joints[3]

Adherend	Treatment	Adhesive	Shear strength psi	MPa
Aluminum	As received	Epoxy	444	3.06
Aluminum	Vapor degreased	Epoxy	837	5.77
Aluminum	Grit blast	Epoxy	1751	12.1
Aluminum	Acid etch	Epoxy	2756	19.0
Aluminum	As received	Vinyl-phenolic	2442	16.8
Aluminum	Degreased	Vinyl-phenolic	2741	18.9
Aluminum	Acid etch	Vinyl-phenolic	5173	35.7
Stainless steel	As received	Vinyl-phenolic	5215	36.0
Stainless steel	Degreased	Vinyl-phenolic	6306	43.5
Stainless steel	Acid etch	Vinyl-phenolic	7056	48.7
Cold-rolled steel	As received	Epoxy	2900	20.0
Cold-rolled steel	Vapor degreased	Epoxy	2910	19.9
Cold-rolled steel	Grit blast	Epoxy	4260	29.6
Cold-rolled steel	Acid etch	Epoxy	4470	30.8
Copper	Vapor degreased	Epoxy	1790	12.3
Copper	Acid etch	Epoxy	2330	16.1
Titanium	As received	Vinyl-phenolic	1356	9.35
Titanium	Degreased	Vinyl-phenolic	3180	21.4
Titanium	Acid etch	Vinyl-phenolic	6743	46.5
Titanium	Acid etch	Epoxy	3183	21.8
Titanium	Liquid pickle	Epoxy	3317	22.9
Titanium	Liquid hone	Epoxy	3900	26.9
Titanium	Hydrofluorosi- licic acid etch	Epoxy	4005	27.6

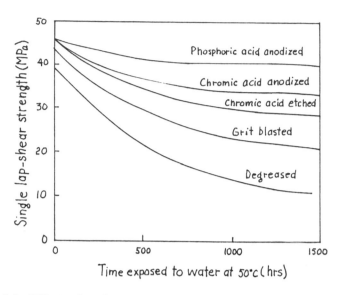

Figure 4-1: Effect of surface pretreatment on the performance of aluminum-alloy epoxy joints subjected to accelerated aging in water at 50°C (122°F). Modified after Reference 4.

In the case of aluminum, optimum bond strength results when a treatment consisting of vapor degreasing, abrading, alkaline cleaning and acid etching is used.[3]

CLEANING

Before adhesive bonding, it is essential to thoroughly clean adherends, since unclean adherends will not be receptive to adhesion, regardless of the quality of materials used, or the exacting control of the application process. Proper surface preparation is extremely important in assuring strong and lasting bonds. For many adherends, surface preparation requirements go far beyond simple cleanliness.

After cleaning to remove obvious surface contamination, including soil, grease, oil, fingermarks, etc. specific chemical or physical treatments are usually required to produce a surface receptive to adhesion. Polytetrafluorethylene (PTFE) and other fluorinated polymers are good examples of surfaces having this requirement. Adhesion to clean pure PTFE is close to zero and it is only after severe chemical treatment which alters the surface chemically and physically that the surface is receptive to adhesion. Magnesium is somewhat similar, requiring considerable treatment beyond simple cleaning. In the case of magnesium, however, very good *initial* adhesion is possible with clean magnesium. However, because of the chemically active nature of the metal, the permanency of this adhesion is poor. With many organic materials, such as adhesives, reactions occur at the metal-to-organic material interface, producing by-products having low cohesive properties. These by-products subsequently degrade the initially good adhesion to an unacceptable level. For this reason it is necessary to use chemical or anodic treatments after cleaning magnesium. Such treatments result in the formation of thin inorganic chemical films which adhere tenaciously to the magnesium and, at the same time, are receptive to adhesion to organic materials such as adhesives. The film laid down in this manner also functions as a barrier between the metal and organic material and prevents any reaction between them. The term "cleaning" has come to mean removal of dirt, contaminants, and oils, and also specific chemical surface treatments for adhesive bonding.[2]

General Sequence of Cleaning

Any surface preparation requires completion of one, two, or all three of the following operations:[2]

- Solvent cleaning
- Intermediate cleaning
- Chemical treatment

Priming, discussed below, may also be carried out in some cases to ensure superior durable bonds under particularly adverse environments.

Solvent cleaning—Solvent cleaning is the process of removing soil from a surface with an organic solvent without physically or chemically altering the material being cleaned. This includes various methods, such as vapor degreasing, spraying, immersion, and mechanical or ultrasonic scrubbing. Solvent cleaning can be an end in itself, as in the case of vapor degreasing aluminum honeycomb core before bonding. It may also be a preliminary step in a series of cleaning and chemical treatment operations.[2]

The four basic solvent cleaning procedures are:[2]

(1) Vapor degreasing

(2) Ultrasonic vapor degreasing

(3) Ultrasonic cleaning with liquid rinse

(4) Solvent wipe, immersion or spray

Vapor degreasing—This is a solvent cleaning procedure for removing soluble soils, particularly oils, greases, and waxes, as well as chips and particulate matter adhering to the soils, from a variety of metallic and nonmetallic parts. The principle of vapor degreasing is to contact the part with hot solvent vapors. These vapors condense on the part at a sufficient rate to form liquid flow which dissolves and washes the soil from the parts as the condensed solvent drains by gravity. Vapor degreasing requires both the proper type of solvent and degreasing equipment. The solvents used must have certain properties, including the following:[2]

(1) High solvency of oils, greases, and other soils

(2) Nonflammable, nonexplosive, and nonreactive under conditions of use

(3) High vapor density, compared to air, and low rate of diffusion into air, to reduce loss

(4) Low heat of vaporization and specific heat to maximize condensation and minimize heat consumption

(5) Chemical stability and noncorrosiveness

(6) Safety in operation

(7) Boiling point low enough for easy distillation and high enough for easy condensation

(8) Conformance to air pollution control legislation

The eight common vapor degreasing solvents are:

(1) Methyl chloroform (1,1,1-trichloroethane)

(2) Methylene chloride

(3) Perchloroethylene

(4) Trichloroethylene

(5) Trichlorotrifluoroethane

(6) Trichlorotrifluoroethane-acetone azeotrope

(7) Trichlorotrifluoroethane-ethyl alcohol azeotrope

(8) Trichlorotrifluoroethane-methylene chloride azeotrope

These materials have a wide range of boiling points, from a low of $102°F$ ($39°C$) for methylene chloride to a high of $250°F$ ($121°C$) for perchloroethylene. Trichloroethylene and perchloroethylene are the solvents most commonly used for vapor degreasing, particularly the former. Considerable detail on the equipment required and the vapor degreasing process is given in Reference 2.

Ultrasonic vapor degreasing[2]—Vapor degreasers are available with ultrasonic transducers built into the clean solvent rinse tank. The parts are initially cleaned by either the vapor rinse or by immersion in boiling solvent. They are then immersed for ultrasonic scrubbing, followed by rinsing with vapor or spray plus vapor. During ultrasonic scrubbing high-frequency inaudible sound waves (over 18,000 cycles per second) are transmitted through the solvent to the part, producing rapid agitation and cavitation. The cavitation (repeated formation and implosion or collapsing of tiny bubbles in the solvent) transmits considerable energy to the parts and to the soil on them. Particulate materials, insolubles, and strongly adherent soils are quickly removed from a part, even on remote surfaces and blind holes. The ultrasonic frequency and intensity for optimum cleaning must be selected by test; it depends on the type of part being cleaned, soil removed, and particular solvent in the system. Some ultrasonic degreasers have variable frequency and power controls. The most common frequency range for ultrasonic cleaning is from 20,000 to 50,000 cycles per second. Power density may vary widely, but 2.5 to 10 watts per square inch is common.

Ultrasonic cleaning with liquid rinse[2]—This is a common procedure for relatively high-quality cleaning utilizing ultrasonic energy to scrub the parts and a liquid solvent to rinse away the residue and loosened particulate matter. This procedure, rather than using the vapor degreasing technique for precleaning and final rinsing, utilizes manual application of liquid solvents. The process is not limited to any particular solvents and, indeed, organic solvents need not be used. It is widely used with aqueous solutions: surfactants, detergents, alkaline, and acid cleaners. The only real limitations are that the cleaning fluid must not attack the cleaning equipment, fluids must not foam excessively and the fluids must cavitate adequately for efficient cleaning. This process is not as efficient as vapor rinse.

Solvent wipe, immersion or spray[2]–This process is suitable for many surface preparation applications and pretreatments. One or a combination of these techniques may be used. A large number of solvents are recommended. *Solvent wiping* is the most portable and versatile of these methods, but is also the least controllable. There is always a danger of incomplete removal of soil, spreading soil around in a uniform manner so its presence is not readily visible, and contaminating a surface with unclean wiping materials. For general cleaning, wiping materials should be clean, freshly laundered cotton rags, new cheese cloth, or cellulose tissues. For special super-clean applications where cleaning must be accomplished in a controlled "clean" room, lint-free specially processed polyurethane foam wiping materials are available (from Sills and Associates, Glendale, CA). The solvent should be used only once, and it should be poured onto the wiping material. The wiping material should never be immersed in the solvent. Solvent containers with small openings should be used that only permit pouring of the solvent. The surface to be cleaned should be wiped systematically with the solvent-soaked cloth or tissue. The wiping material should be discarded and the surface cleaned again with new solvent and cloth or tissue. This cycle should be repeated until there is no evidence of soil on either the cloth or the cleaned surface.

Solvent immersion is useful for batch or piece cleaning of relatively small parts. The immersion and soaking in solvent is often sufficient to remove light soil. Scrubbing may be required for heavier soils. The most efficient scrubbing method is ultrasonic, discussed above. Other scrubbing techniques include tumbling, solvent agitation, brushing, and wiping. After the parts are soaked and scrubbed, the parts must be rinsed. The quality of cleaning produced by the immersion process depends primarily on the final rinse. The *solvent spray* cleaning method is efficient because of the scrubbing effect produced by the impingement of high-speed solvent particles on the surface being cleaned. The solvent impinges on the surface in sufficient quantity to cause flow and drainage, which washes away the loosened soil. Also, since only clean solvent is being added to the surface, and scrubbing and rinsing occur, there is no danger of contamination, as there is with the immersion process.

Safety[2]–Four safety factors must be considered in all solvent cleaning operations: toxicity, flammability, hazardous incompatibility, and equipment. The solvents must be handled in a manner preventing toxic exposure of the operator. Where flammable solvents are used, they must be stored, handled and used in a manner preventing any possibility of ignition. In a few cases, knowledge of the hazardous incompatibility of the solvents, cleaning equipment and materials to be cleaned is essential. Safe equipment and proper operation is also critical. Snogren lists Maximum Acceptable Concentrations (MACs), which term is synonymous with Threshold Limit Values (TLVs) for a number of solvents used in cleaning. These values are given in parts per million (ppm). Examples are: acetone—1,000 ppm; methyl alcohol—200 ppm; methyl chloroform—500 ppm. Obviously methyl alcohol, which requires less concentration to incapacitate than methyl chloroform and acetone, is the solvent with the greatest exposure danger. It should be pointed out that TLVs are merely guides in the control of health hazards and represent conditions under which it is believed that nearly all workers may be repeatedly exposed, day after day, without adverse effect.

All flammable solvents should be stored in metal containers, such as safety cans, and must be used in metal containers. Snogren gives flammability limits in terms of percent by volume in air (lower and upper) for the solvents commonly used. Some of the solvents he lists are nonflammable. The flammable solvents must be used only in areas ventilated to prevent accumulation of vapors and fumes. Other obvious precautions must also be taken. Unstabilized methyl chloroform, trichloroethylene and perchloroethylene are subject to chemical reaction on contact with oxygen or moisture to form acid by-products. These acids are highly corrosive to metals. Only stabilized grades of these solvents should be used. Strong alkalies such as caustic soda may react with trichloroethylene to form explosive mixtures (dichloroacetylene). Fluorocarbon solvents may react violently with highly reactive alkaline earth metals.[2]

Intermediate cleaning[2]—Intermediate cleaning is the process of removing soil from a surface with physical, mechanical, or chemical means without altering the material chemically. Small amounts of parent material may be removed in this process. Examples include *grit blasting, wire brushing, sanding, abrasive scrubbing*, and *alkaline* or *detergent cleaning*. Solvent cleaning should always be carried out before this step. Intermediate cleaning operations may be an end in themselves. Examples of such a situation include cleaning stainless steel with uninhibited alkaline cleaner, or detergent scrubbing epoxy laminates. Detailed intermediate cleaning procedures are given below for a number of adherends.[2]

Chemical treatment[2]—Chemical treatment is the process of treating a clean surface by chemical means. The chemical nature of the surface is changed to improve its adhesion qualities. Solvent cleaning should always precede chemical treatment, and frequently intermediate cleaning should be used in between. Chemical treatments, such as acid etch procedures, are given below for a number of adherends.[2]

PRIMING

An adhesive primer is usually a dilute solution of an adhesive in an organic solvent applied to a dried film thickness of 0.00006 to 0.002 inch (0.0015 to 0.05 mm). Some of its functions are as follows:

- Improve wetting[5]
- Protect the adherend surface from oxidation after cleaning, extending the time that may elapse between surface preparation and ad-

hesive application. (Such an extension may increase the usable time for aluminum adherends from 12 hours to up to 6 months)[2]

- Help inhibit corrosion (corrosion inhibiting primers—CIPs)[5]

- Modify the properties of the adhesive to improve certain characteristics, such as peel[6]

- Serve as a barrier coat to prevent unfavorable reactions between adhesive and adherend[5]

- Hold adhesive films or adherend in place during assembly. This type of primer retains tack, or develops tack at room or elevated temperatures[6]

The use of primers provides (1) more flexible manufacturing scheduling, (2) high reliability of joints, (3) less rigorous cure conditions, (4) wider latitude in choice of adhesive system, and (5) more durable joints.[5] Primers are usually not fully cured during their initial application. They are dried at room temperature and some are force-dried for 30 to 60 minutes at 150°F (65.5°C). These steps, frequently called "flashing," provide a dry nontacky surface which can be protected from contamination and physical damage by wrapping with clean paper, sealing in polyethylene bags, or covering with a nontransferring adhesive-backed paper.[2]

When primers are desirable the manufacturer's literature will ordinarily specify the best primer to use. The primers, like the adhesives, are usually proprietary in nature and are made to match the adhesives.[6]

ACTIVATED-GAS SURFACE TREATMENT OF POLYMERS (PLASMA TREATMENT)

In 1966 Schonhorn and Hansen[7] reported on a highly effective treatment for the surface preparation of low-surface-energy polymers for adhesive bonding. The techniques consisted largely of exposing the polymer surface to an inert gas plasma at reduced pressure generated by electrodeless glow discharge (i.e., radio-frequency field). For polyethylene only very short treatment times were necessary (ca. 9 seconds), while larger contact times were required for other polymers such as PTFE.[8] Plasma gases used (O_2, He, N_2) can be selected to include a wide assortment of chemical reactions. In the process atoms are expelled from the surface of the polymer to produce a strong, wettable crosslinked skin.[3] Commercial equipment for this process is made by Branson International Plasma Corporation, 31172 Huntwood, P.O. Box 4136, Haywood, CA 94544. Plasma-treated polymers usually form adhesive bonds anywhere from two to four times stronger than bonds formed by traditional chemical or mechanical preparation. Table 4-2 illustrates the effectiveness of this process for a large number of polymers. Picatinny Arsenal workers have reported studies on a large number of polymers using activated gas plasma treatment.[9] One of the more recent studies covered newer plastics (VALOX aromatic polyester, polyethersulfone, polyarylsulfone, polyphenylene sulfide, E-CTFE fluoropolymer, nylon 11, nylon 6/12, and nylon 12).[12]

Table 4-2: Typical Adhesive-Strength Improvement for Aluminum-Plastic Lap Shear Sandwich Specimens Bonded with Epon-Versamid 140 (70/30) Epoxy Adhesive[9]

Plastic Material	Control (psi)	Control (MPa)	After Various Plasma Treatments * (psi)	(MPa)	(psi)	(MPa)
High-density polyethylene	315 ± 38	2.17 ± 0.26	1160 ± 136	8.0 ± 0.93 to	3500 ± 68	24.3 ± 0.47
Low-density polyethylene	372 ± 52	2.57 ± 0.36	1250 ± 33	8.62 ± 0.22 to	1466 ± 106	10.11 ± 0.73
Nylon 6	846 ± 166	5.83 ± 1.15	1220 ± 120	8.41 ± 0.83 to	3956 ± 195	27.3 ± 1.34
Polystyrene	566 ± 17	3.90 ± 0.12	3118 ± 278	21.50 ± 1.92 to	4015 ± 85	27.7 ± 0.59
MYLAR A (PET film)	530 ± 51	3.66 ± 0.35	1215 ± 29	8.38 ± 0.20 to	1660 ± 40	11.45 ± 0.28
MYLAR D (PET film)	618 ± 25	4.26 ± 0.17	1185 ± 83	8.17 ± 0.57 to	1216 ± 522	8.39 ± 3.60
TEDLAR PVF**	278 ± 2	1.92 ± 0.014	1200 ± 80	8.28 ± 0.55 to	1370 ± 80	9.45 ± 0.55
LEXAN polycarbonate	410 ± 10	2.83 ± 0.69	660 ± 27	4.58 ± 0.19 to	928 ± 66	6.40 ± 0.46
Polypropylene	370	2.55	200	1.38	3080 ± 180	21.2 ± 1.25
Cellulose acetate butyrate	1090 ± 27	7.52 ± 0.19	455 ± 95	3.14 ± 1.34 to	2516 ± 22	17.4 ± 0.15
Acetal copolymer	118 ± 4	0.81 ± 0.007	186 ± 58	1.28 ± 0.40 to	258 ± 16	1.78 ± 0.11

*   O₂, He, and N₂ for various periods ranging from 30 seconds to 60 minutes

** Polyvinyl fluoride

METHODS FOR EVALUATING EFFECTIVENESS OF SURFACE PREPARATION

Before actual bonding, the subjective "water-break" test, or the quantitative and objective contact-angle test may be carried out. *After* bonding, the effectiveness of surface preparation may be determined by observing the mode of failure of the adhesive joint (Figure 4-1).

Water-Break Test

This test depends on the observation that a clean surface (one that is chemically active or polar) will hold a *continuous film* of water, rather than a series of isolated droplets. This is known as a water-break-free condition. A break in the water film indicates a soiled or contaminated area. Distilled water should be used in the test, and a drainage time of about 30 seconds should be allowed. Any trace of residual cleaning solution should be removed or a false conclusion may be made. If a water-break-free condition is not observed on the treated surface, it should not be used for bonding. The surface should be recleaned until the test is passed. If continued failures occur, the treating process itself should be analyzed to determine the cause of the problem.[2]

Contact-Angle Test

Wettability may also be determined by measuring the contact angle between the polymer surface and a drop of a reference liquid, such as distilled water. A small contact angle indicates that the liquid is wetting the polymer effectively, while large contact angles show that the wetting is poor. Every surface has a *critical surface tension*, γ_c, of wetting. Liquids with surface-free energies below γ_c will have zero contact angles, and will wet the surface completely, while liquids with surface-free energies above γ_c will have finite contact angles. The critical surface tension is in units of dynes/cm at 20°C. Contact angles for untreated materials vary from 37° to 48° for relatively polar materials such as nylon to highs of 100° and 97° for the nonpolar, unbondable silicone and polyethylene resins. After exposure to activated argon plasma, contact angles are reduced to 40° for PMMA and to 19° or less for nylon, polystyrene, polyethylene, and RTV silicone.[6] Zisman has written a very comprehensive chapter on surface-tension phenomena relating to adhesion.[11]

SURFACE EXPOSURE TIME (SET)

SET is the time elapsed between the surface preparation and actual bonding. After parts have been subjected to surface preparation, they must be protected from contamination during transportation and storage. The clean surface should never be touched with bare hands or soiled gloves. If more than a few hours are required between cleaning and priming, the parts should be covered, or, for still longer periods, wrapped in clean kraft paper until the priming can be carried out. After priming, the dried primer surfaces, if not to be bonded immediately, should *again* be protected by wrapping in kraft paper. Whether or not a primer is used, these steps should be carried out. The period of time for which the parts can be safely stored in this way will vary, depending on the nature of the adherends, the adhesive, the surface-preparation method, and the ultimate bond strength required.[12,13]

Picatinny Arsenal (now the U.S. Army Armament Research and Development Center) has published a number of reports and papers on SET, using both shear tests and peel tests. Peel tests, in particular the roller peel test, have been found to be more sensitive to variation in surface preparation than shear tests. Peel tests show that, in general, increasing SET tends to reduce the critical joint strength. Up to 30 days may elapse between surface preparation and actual bonding if the faying surfaces are protected and the relative humidity is kept at about 50%, without serious loss in joining strength. Temperatures and relative humidities above normal will cause deterioration in shorter periods of time.[12,13]

SURFACE PREPARATION OF METALS

The general methods described above are all applicable, but the processes listed below have been found to provide strong reproducible bonds and fit easily into the bonding operation.

Aluminum

Chemical treatments have usually been found to be the most effective with aluminum, especially where long-term environmental exposure is required. The sulfuric acid-dichromate etch (FPL etch, named after Forest Products Laboratory, USDA) has been used successfully for many years. Other important methods include chromate conversion coating and anodizing. Corrosion protection is particularly important with aluminum. Corrosion-resistant adhesive primers (CRAP), as well as anodic and chromate conversion coatings help prevent corrosive failure of adhesion.[2]

FPL immersion etch (improved method)[14]—

(a) Degrease, solvent wipe, or both. (Acetone can be used to remove gross contaminants. Perchloroethylene vapor is good for degreasing)

(b) Immerse for 10 minutes at 155°F (68°C) in the following solution: 30 parts by weight (pbw) water, 10 pbw sulfuric acid (sp. gr. 1.84) and 1 pbw sodium dichromate ($Na_2Cr_2O_7 \cdot 2H_2O$)

(c) Rinse with water not over 150°F (65.5°C)

(d) Air-dry, dry in an oven, or use infrared lamps at not over 150°F (65.5°C)

CAUTION: In making up the solution, add the acid to 60% of the water, stir in the dichromate, and then add the rest of the water. NEVER add the water to the concentrated acid!

Parts should ordinarily be bonded within 48 hours. This method is designated Method A in ASTM D 2651-79.[14] Modifications (Method C—ASTM D 2651-79) include increasing the concentration of the sodium dichromate-sulfuric acid solution to as high as 17 parts water, 7 parts sulfuric acid, and 2 parts dichromate. *Alkaline degreasing solutions* may also be used instead of, or in addition to, vapor degreasing, and prior to the sodium dichromate-sulfuric acid solution treatment (known as acid-etch). Although commercial proprietary solutions are available, the following solution may be used: 3.0 pbw sodium metasilicate, 1.5 pbw trisodium pyrophosphate, 1.5 pbw sodium hydroxide, 0.5 pbw Nacconal NR (available from Stepan Chemical Co., 22 West Frontage, Northfield, IL 60093) or equivalent, and 133.0 pbw water.

FPL Paste Etch—This method is used for secondary bonding of parts that contain previously bonded areas, for repair of bonded assemblies, or when the size of parts make immersion impractical. The parts should be bonded in the range of 70°-90°F (21°-32°C). A paste is prepared using the sulfuric acid-sodium dichromate solution described above, in finely divided silica (available from Stepan Chemical Co., 22 West Frontage, Northfield, IL 60093) or Fuller's earth and applied to the surface. The paste should not be allowed to dry after application by brush. Polypropylene brushes should be used because of their chemical resistance. The paste should be allowed to remain in place for 20-25 minutes. Extra coats may be applied to prevent the paste from drying out or turning green. A clean cheesecloth moistened with water should be used to remove all traces of the paste at the end of the exposure period. Finally, a dry cheesecloth should be used. Water may be sprayed on if desired. Drying should be carried out at a maximum of 150°F (66°C). As might be expected, bond strengths obtained by this technique are somewhat lower than those obtained by immersion. This method is designated Method F in ASTM D 2641-79.[14]

Chromate-Free Etch Process—One of the less desirable chemicals used for the processing of aluminum prior to adhesive bonding is the acid chromate etching solution. This material is not only toxic and hazardous during use, but is highly injurious if released into surrounding water supplies. Although costly, it is both possible and practical to remove essentially all traces of chromate from industrial waste effluents. Russell and Garnis at Picatinny Arsenal found that an etching solution recommended for the precleaning of aluminum prior to resistance welding gave excellent results. This solution consisted of nitric acid and sodium sulfate (N-S). In later modifications, a "P" etch was developed containing sulfuric acid, sodium sulfate, nitric acid, and ferric sulfate. However, it was found that the presence of nitric acid resulted in the evolution of oxides of nitrogen when aluminum is treated. These oxides are toxic and must be vented. In an effort to eliminate the necessity for venting the toxic etching fumes, a new etchant composition called "P2" was developed which does not give off any appreciable fumes and which results in bonds of good strength and improved durability.[15,16,94]

Degreasing or solvent cleaning may be carried out prior to using the P2 etch. Another option is alkaline cleaning. The composition of the etch solution is as follows: 370 g concentrated sulfuric acid and 150 g 75% ferric sulfate diluted with deionized water to make 1 liter.

The acid is added to approximately half the required water with constant stirring. The ferric sulfate is then added and mixing is continued. Water is then added to bring the volume to 1 liter. The solution is heated to 145°-155°F (63°-68°C) and the parts are immersed in this solution for 12 minutes. Rinsing follows in agitated tap water for 2 minutes. A second rinse, also at room temperature, of deionized water is then sprayed on the part to rinse off the tap water.[15,17] This *sulfuric acid-ferric sulfate etch* yields bonds at least equal to those made using the sulfuric acid-dichromate (FPL) etch. When used as a deoxidizer prior to phosphoric acid anodizing (PAA) (see below), the results are essentially equal to those using the sulfuric acid-dichromate etch.[17] In a variation of this process, the final rinse is for 1 to 3 minutes in demineralized water at ambient to 160°F (71°C), followed by drying in ambient air up to 160°F (71°C).[17]

Anodization—This process is sometimes used for bare (nonclad) aluminum machined or chem-milled parts that must be protected against corrosion. Anodic coatings used include chromic, sulfuric and phosphoric acid types. *Anodizing* is a process involving electrolytic treatment of metals in which stable films or coatings are formed on the surface of the metals. Anodic coatings can be formed on aluminum alloys in a wide variety of electrolytes, using either alternating or direct current, or combinations of both. *Phosphoric acid anodizing (PAA)*, currently the most widely used method, is described in ASTM D 3933-80.[18] In short, the process involves immersing the parts in a 9–12% solution of phosphoric acid at 67°–82°F (19°–28°C) at anywhere from 9–16 volts under direct current (DC) for 20–25 minutes, rinsing, drying, inspecting, and then priming. Durability data obtained by this method is slightly better than that obtained by using an etch.[17] Phosphoric acid anodizing (PAA) was studied by Boeing Aircraft in the early 1960's and introduced commercially in 1974. This process is less critically dependent than etching on processing variables such as the time between treatment and rinsing. It is also possible to use polarized light as a quality control test for the anodizing pretreatment. The oxide layer formed by this process is much thicker and the "whiskers" longer with PAA than for chromic acid etching, although the thickness of the anodic oxide is dependent on the nature of the aluminum alloy being protected. Phosphoric acid anodizing produces surfaces that are more resistant to hydration than those produced with other anodizing methods, including chromic acid.[19]

Phosphoric acid anodizing is also known to give much more consistent results in durability studies than chromic acid etching. Figure 4-2 shows results using the Boeing wedge test.[19]

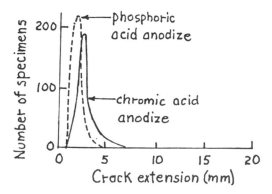

Figure 4-2: Consistency of the durability provided by phosphoric acid anodizing versus chromic acid anodizing, as measured by the Boeing Wedge Test. Modified after Reference 19.

Beryllium

This metal and its alloys must be heated with care when handling or processing produces dust, chips, scale, slivers, mists, or fumes, since air-borne particles of beryllium and beryllium oxide are extremely toxic, with serious latent effects. Abrasives and chemicals used with beryllium must be properly disposed of.[2] One procedure is to degrease with trichloroethylene, followed by immersion in the following solution for 5–10 minutes at 20°C:[20]

Sodium hydroxide	20–30 pbw
Distilled water	170–180 pbw

Rinse in distilled water after washing in tap water and oven-dry for 10 minutes at 121°-177°C.

CAUTION: Beryllium reacts quickly with methyl alcohol, Freon, perchloroethylene, methyl ethyl ketone/Freon, and can be pitted by long exposure to tap water containing chlorides or sulfates.[20]

A proprietary coating used to provide a corrosion-resistant barrier is Berylcoat "D", currently available from Brush Wellman, Inc., 17876 St. Clair Avenue, Cleveland, OH 44110.

Brass

Brass is an alloy of copper and zinc. Sandblasting or other mechanical means of surface preparation may be used. The following procedure combines mechanical and chemical treatment:[21]

(1) Abrasive blast, using either dry or wet methods. Particle size is not especially critical

(2) Rinse with deionized water

(3) Treat with a 5% solution of sodium dichromate in deionized water

(4) Rinse in deionized water

(5) Dry

Another method is the following:[20,22]

(1) Degrease in trichloroethylene

(2) Immerse for 5 minutes at 20°C in:

Zinc oxide	20 pbw
Sulfuric acid, concentrated	460 pbw
Nitric acid, 67%, sp. gr. 1.41	360 pbw

(3) Rinse in water below 65°C

(4) Re-etch in the acid solution for 5 minutes at 49°C

(5) Rinse in distilled water after washing

(6) Dry in air

(Temperatures of washing and drying must not exceed 65°C)

Bronze

Bronze is an alloy of copper and tin. The surface treatment involving *zinc oxide*, *sulfuric acid*, and *nitric acid* given above for brass is satisfactory for bronze.

Cadmium

Cadmium is ordinarily used as a coating on steel. It can best be made bondable by electroplating with silver or nickel. Another procedure is the following:[23]

(1) Degrease or solvent clean with trichloroethylene

(2) Scour with a commercial, nonchlorinated abrasive cleaner such as Bab-O or Ajax

(3) Rinse with distilled water

(4) Dry with clean, filtered air at room temperature

It may be desirable to use a primer or sealant. Adhesive choice is particularly important with cadmium coatings.[23]

Copper and Copper Alloys

Copper is used in three basic forms: pure, alloyed with zinc (brass), and alloyed with tin (bronze). Copper has a tendency to form brittle amine compounds with curing agents from some adhesive systems. The most successful surface treatments are *black oxide* or Ebonol C, given below, and chromate *conversion coatings*, which are especially recommended when the adhesive is slightly corrosive to copper.[2]

Black oxide coating—This method is intended for relatively pure copper alloys containing over 95% copper. It tends to leave a stable surface. It is also recommended for use when the adhesive contains chlorides. This is Method E in ASTM D 2651-79.[14]

(a) Degrease

(b) Immerse for 30 seconds at room temperature in the following solution:

Nitric acid (70%, technical)	10 pbv (parts by volume)
Water	90 pbv

(c) Rinse in running water and transfer immediately to the following solution without allowing the parts to dry.

(d) Immerse for 1-2 minutes at 98°C (208.4°F) in the following solution:

Ebonol C (available from Enthane, Inc., 218 W. Frontage Road, West Haven, CT 06510)—24 oz to make 1 gallon of solution. This solution should *not* be boiled

(e) Rinse in running cold tap water

(f) Air-dry

(g) Bond as soon as possible within the working day

Sodium dichromate-sulfuric acid process[14]–This method has been found by some to be superior to the ferric chloride method.

(a) Wash with acetone

(b) Immerse for 10 minutes at 150°F (66°C) in a solution containing:

Ferric sulfate (commercial)	1 pbw (454 g)
Sulfuric acid, 95%	0.75 pbw (340 g)
Deionized water	8 pbw (3.784 liters)

(c) Rinse in deionized water at room temperature

(d) Immerse in a solution containing:

Sodium dichromate	1 pbw
Sulfuric acid, 95%	2 pbw
Deionized water	17 pbw

until a bright clean surface is obtained

(e) Rinse with deionized water

(f) Dip in concentrated ammonium hydroxide (sp. gr. 0.88)

(g) Wash in tap water

(h) Rinse in deionized water

(i) Air-dry at room temperature (max. 104°F or 40°C)

Steps (f) through (h) are optional

Gold

Use methods given for platinum.

Magnesium and Magnesium Alloys

The surface preparation methods for magnesium alloys are closely associated with corrosion prevention. Because magnesium is highly reactive, corrosion-preventive coatings must be applied for most service applications. The major problem is to apply sufficient thickness of coating to prevent corrosion, but not so thick that the bond fails cohesively in the coating.

Alkaline-detergent solution (ASTM D 2651, Method A)[14]–

(a) Degrease (use caution, since contaminants, such as metal particles, oils, etc. may result in a fire or explosion hazard)

(b) Immerse for 10 minutes at 60°-71°C (140°-160°F) in an alkaline-detergent solution, as follows:

Sodium metasilicate	2.5 pbw
Trisodium pyrophosphate	1.1 pbw

Sodium hydroxide	1.1 pbw
Nacconal NR or equivalent	0.3 pbw
Water	95 pbw

(c) Rinse thoroughly

(d) Dry at not over 60°C (140°F)

Hot chromic acid (ASTM D 2651, Method B)[14] –

(a) Degrease (see caution for alkaline detergent solution)

(b) Immerse for 10 minutes at 71° to 87.8°C (140°-190°F) in the following solution:

Chromic acid (CrO_3)	1 pbw
Water	4 pbw

(c) Rinse thoroughly

(d) Dry at not over 60°C (140°F)

This method can be combined with the alkaline detergent solution method above (ASTM D 2651, Method A) to give improved bond strengths. The hot chromic acid solution should follow the alkaline detergent method, with a water rinse in between.

Anodic and other corrosion-preventive treatments (ASTM D 2651, Method D)[14]–Light anodic treatment and various corrosion-preventive treatments produce good surfaces for adhesive bonding. These treatments were developed by magnesium alloy producers, such as Dow Chemical Company, Midland, MI (Dow 17 and Dow 7) and others. Details are available from ASM Handbook, Vol. II[24] and MIL-M-45202, Type I, Classes 1, 2, and 3.[25]

Conversion coatings (ASTM D 2651, Method E)[14]–Some surface dichromate conversion coatings and wash primers designed for corrosion protection can be used for adhesive bonding. Preliminary tests should be carried out to determine the suitability of the process before acceptance. Details are found in ASM Metals Handbook, Vol. II [24] and MIL-M-3171.[26]

Nickel and Nickel Alloys

Abrasive cleaning[2] –

(a) Solvent clean, preferably with vapor degreasing

(b) Abrade with 180 to 240-grit paper, or grit-blast with aluminum oxide 40-mesh abrasive

(c) Solvent clean again as in (a)

Nitric acid etch[20] –

(a) Vapor degrease in trichloroethylene

(b) Etch for 4-6 seconds at RT (~68°F) in concentrated nitric acid (sp. gr. 1.41)

(c) Wash in cold and hot water, followed by a distilled-water rinse

(d) Air-dry at 40°C (104°F)

Sulfuric-nitric acid pickle[2,27]—

(a) Immerse parts for 5–20 seconds in the following solution at room
     temperature:

Sodium chloride	30 g
Sulfuric acid (60°Be')	1.50 liters
Nitric acid (40°Be')	2.25 liters
Water	1.00 liter

(b) Rinse in cold water

(c) Immerse in a 1–2% ammonia solution for a few seconds

(d) Rinse thoroughly in distilled water

(e) Dry at temperatures up to 150°F

Manual cleaning for plated parts[20]—Thin nickel-plated parts should not be
etched or sanded. A recommended practice is light scouring with a nonchlorinated commercial cleaner, rinsing with distilled water, drying under 120°F
(49°C), and then priming as soon as possible.

Nickel-base alloy treatments—Monel (nickel-copper), Inconel (nickel-iron-
chromium) and Duranickel (primarily nickel) are the major types of nickel-base
alloys. According to Keith, et al[28] procedures recommended for stainless steel
will give at least satisfactory results with these alloys.

Platinum

Abrasive cleaning[2]—

(a) Solvent clean, preferably by vapor degreasing (avoid immersion and
     wiping, unless these steps can be followed by vapor rinsing or
     spraying)

(b) Abrade lightly with 180–240 grit paper

(c) Solvent clean again as in (a)

Abrasive scouring[2]—

(a) Scrub surfaces with distilled water and nonchlorinated scouring
     powder

(b) Rinse thoroughly with distilled water (surfaces should be water-
     break-free)

Silver

Use the methods given above for platinum, or use the following methods:
Chromate conversion coating[2]—

(a) Solvent clean

(b) Immerse for 5-10 seconds in the following solution at RT:

Kenvert No. 14 powder	³⁄₈ oz
Water	1.0 gal

(c) Rinse in distilled water

(d) Dry at temperatures to 150°F (66°C)

Kenvert No. 14 is described in Bulletin No. 14, *A Powder for Producing Iridescent Chromate Films on Zinc, Cadmium, Zinc-Base, Die Castings, Hot Dipped Galvanizing, Copper, Brass and Silver*, Conversion Chemical Corp., Rockville, CT, March 1969.

Degrease-abrade-prime[3]–

(a) Vapor degrease

(b) Abrade lightly with emery cloth

(c) Prime coat with Hughson's Chemlok 507 (silicone-type polymer) diluted with 5-10 volumes of methanol or ethanol

Steel

Compared to aluminum, relatively little research has been carried out on optimum surface treatments for steel preparatory to adhesive bonding. Steel alloys are usually used where high temperatures and other rigid requirements are expected. As a result problem areas such as corrosion, thermal degradation of adhesives and primers, and moisture effects become important. Surface preparation is even more critical in bonding steel than in many other metals. Initial adhesion may be good, but bonds degrade rapidly under rigid environmental exposure. For this reason *primers* are desirable. Many steel alloys will form surface oxides rapidly, and drying cycles after cleaning are therefore critical. Alcohol rinses after water rinses tend to accelerate drying and reduce undesirable surface layer formation. *Mild steel* (carbon steel) may require no more extensive treatment than degreasing and abrasion to give excellent adhesive bonds. Tests should be carried out with the actual adhesive to be used to determine whether a chemical etch or other treatment is essential.[1,23,29]

Brockmann[30] suggests that, in contrast to aluminum and titanium alloys, where the surfaces are usually treated by chemical methods, etching procedures for the different types of steel, except for *stainless steel*, are not recommended. According to Brockmann, the best results are usually obtained by using mechanical roughening techniques, such as grinding with corundum (Al_2O_3) as the grit material. He feels the only chemical treatment required is for *stainless steel* and involves etching in an aqueous solution containing oxalic acid and sulfuric acid (see below).

DeLollis[6] suggests that the most common method for preparing steel surfaces for bonding is probably sandblasting, preceded by vapor degreasing to remove any oily film present. Since a clean steel surface is easily oxidized, it is particularly important that the compressed air source used in sandblasting be dry and free of oil. While some methods call for rinsing in a clean volatile solvent such as methyl ethyl ketone or isopropyl alcohol, it should be remembered that these solvents frequently contain small amounts of water that may cause subsequent rusting. (Xylene or toluene are preferable solvents). Furthermore, when volatile solvents evaporate from a metal surface the surface is chilled. In a humid environment such an effect may result in condensation of moisture, which nullifies the cleaning effort. After sandblasting, the best treatment is to blow the particulate matter off the surface with clean, dry air. Solvent wiping should be used only as a last resort.[6]

DeLollis[6] suggests that vapor honing or vapor blast can be used for small steel parts without danger of warping or of any significant change in tolerance. A wetting agent and rust inhibitor are usually present in the abrasive-water suspension used in this method. These agents must be rinsed off thoroughly in clean water. After this step, a clean, dry, water-compatible solvent must be used to remove the water. Finally, clean, dry compressed air or nitrogen should be used to blow off the solvent.

Abrasive treatments may result in warping of thin sheet-metal stock. In this type of application any of the acid-etch solutions used with stainless steel can be used with carbon steels. They must be used carefully, however, since they would react more rapidly with carbon steel than with stainless steel.[6]

Chemical treatments for carbon steel to be used only when abrasive equipment is not available, are described below:

Acid Treatment[1] –

(a) Prepare a solution of:

Orthophosphoric acid, conc. (sp. gr. 1.73)	1 pbw
Ethyl alcohol, denatured	1 or 2 pbw

or

Hydrochloric acid, conc.	1 pbw
Distilled water	1 pbw

(CAUTION: Add the acid to the water)

(b) Degrease in a vapor bath of trichloroethylene

(c) Immerse for 10 minutes in the phosphoric acid solution at 140°F (60°C) or for 5-10 minutes at 68°F (20°C) in the hydrochloric acid solution

(d) Remove the black residue formed on the metal with a clean, stiff brush while holding the metal under cold, running distilled or deionized water

(e) Dry the metal by heating at 250°F (120°C) for 1 hour. If the steel part cannot be stored immediately in an area of low humidity (<30% RH) apply the adhesive immediately after cleaning

Potassium iodide–phosphoric acid method[31]–This is a relatively recently developed surface treatment based on the oxidation of the steel to produce a continuous epsilon-oxide (Fe_2O_3). It is reported to give strong durable bonds and to improve the salt-spray and humidity resistance of polymer-coated steel. The steps are:

(a) Vapor degrease

(b) Alkaline-clean in a solution containing 3% trisodium phosphate and 3% sodium carbonate for 5 minutes at 180°F (82°C).

(c) Rinse in deionized water

(d) Immerse in a solution of 50 g potassium iodide (KI) per liter of 1:1 conc. phosphoric acid:water v/v at 200±20°F (93±11°C) for 2-10 minutes

(e) Rinse in deionized water

(f) Dry at 60°-160°F (11°-71°C)

Stainless Steel

Stainless steel, or corrosion-resistant (CRES) steel, has a high chromium content (11% or higher) as the primary alloying element. There have been a large number of surface preparation methods reported in the literature. An excellent report prepared in 1968 by Battelle Memorial Institute workers[32] gives details of twelve surface preparation methods and discusses the relative advantages and disadvantages of each. In addition to mechanical methods, strong acids and strong alkalies are used. A wet-abrasive blast with a 200-grit abrasive, followed by thorough rinsing to remove the residue, is an acceptable procedure for some uses, but does not produce high bond strengths. Strong *acid* treatments are usually used for general bonding to produce strong bonds with most adhesives. *Passivation* in nitric acid solution and concentrated sulfuric acid-saturated sodium dichromate solution both produce high-strength bonds, but with low or marginal peel strengths. Such joints may fail under vibration stress, particularly when a thin stainless steel sheet is bonded with low-peel-strength adhesives. The *acid-etch process* outlined below can be used to treat Types 301 and 302 stainless steel. This process results in a heavy black smut formation on the surface. This material must be removed if maximum adhesion is to be obtained. The acid-etch process produces bonds with high peel and shear strength. The 400 series of straight-chromium stainless steels should be handled in the same manner as the plain carbon steels. The precipitation hardening (PH) stainless steels each present an individual problem. Processes must be adopted or developed for each one.[1,2,21]

Acid etch (for Types **301** and **302** stainless steel)[33]—This method has been used successfully with 3M Company's AF-126-2 nitrile-epoxy film adhesive and Scotch-Weld EC 2414R modified epoxy paste adhesive. The film adhesive was cured at 50 psi (345 kPa) and 250°F (121°C) for 1 hour. The paste adhesive was cured at 8 psi (55 kPa) and 250°F (121°C) for 40 minutes. Stressed durability tests gave excellent results at 140°F (60°C) and 95% RH. The procedure is as follows:

(a) *Degrease*—Wash with acetone to remove grease, oils, markings and solvent-removable soils. Follow with vapor degreasing with a suitable safety solvent.

(b) *Sulfuric acid etch*—Immerse the parts for 4 minutes at 140°F (60°C) in a solution of 25–35% by volume of sulfuric acid (sp. gr. 1.84) in deionized water. *Do not start timing until gassing is evident.* A piece of 1020 steel may be rubbed across the parts to start the etching process.

(c) *Rinse* in running tap water at service-water temperature for 2 minutes.

(d) *Smut removal*—Immerse parts for 5 minutes at 150°F (66°C) in a solution of 22–28 pbw of sulfuric acid (sp. gr. 1.84) and 2–3 pbw of sodium dichromate in deionized water.

(e) *Rinse* in running tap water at service-water temperature for 2 minutes.

(f) *Dry* the parts at 140°F (60°C) for 30 minutes in a preheated air-circulating oven.

(g) *Packaging*—Wrap the parts in clean kraft paper until ready to bond.

Oxalic-sulfuric acid process for maximum heat resistance[34]—

(a) Immerse for 10 minutes at 90°C (194°F) in the following solution:

Oxalic acid	18.5 g
Sulfuric acid, conc.	10 ml (18.4 g)
Water	150 ml

(b) Remove the solution and brush the black smut deposit with a good stiff brush. Nylon is suggested

(c) Rinse with clean running water

(d) Dry in a stream of hot air

(e) Return to desiccator

Hughson Chemicals[35] recommends soaking parts in a commercial steel cleaner of high alkalinity (as Prebond 700, available from Bloomingdale Dept., American Cyanamid Co., Havre de Grace, MD), before immersing in the oxalic-sulfuric acid solution.

Bromophosphate treatment—This treatment is a modification of the iodo-phosphate treatment described above for carbon steel. No chloride ion is used in the method. KBr is used instead of KI, and 1–5% by volume of concentrated H_2SO_4 should be added to the treating solution in the case of difficult-to-etch CRES steels to ensure the removal of the existing surface layers.[31]

(a) Vapor degrease

(b) Alkaline-clean in a solution of 3% trisodium phosphate and 3% sodium carbonate for 5 minutes at 180°F (82°C)

(c) Rinse in deionized water

(d) Immerse in a solution of 50 g potassium bromide (KBr) per liter of 1:1 concentrated phosphoric acid:water for 2–10 minutes at 200±20°F. In the case of difficult-to-etch CRES steels, to ensure removal of the existing surface layer, use 50 g KBr per liter of 1:9:10 concentrated sulfuric acid:concentrated phosphoric acid:water

(e) Rinse in deionized water and dry at 60°-160°F (15.6°-71°C)

This method can be adapted to spraying techniques.

ASTM suggested methods—ASTM D2651[14] lists seven methods of surface preparation for stainless steel. Method A is *abrasion.*

Tin

Solvent cleaning with methylene chloroform or trichloroethylene is recommended prior to abrading. Scraping, fine sanding or scouring are suitable methods of abrading, which should be followed by solvent cleaning.[2] This is one of the few metals for which abrasion may be used without being followed by an acid etch.[23]

Titanium

Titanium is a space-age metal being used widely in aerospace applications requiring high strength-to-weight ratios at elevated temperatures up to 600°F (316°C).[6] Adhesive-bonded helicopter rotor blades consisting of titanium skins in the form of sandwich panels have been in use since the 1950's. The earliest surface treatments for these titanium surfaces were based on cleaning and etching in alkaline mixtures. These processes provided good joint strength. However, they were sensitive to the chemical composition of the titanium alloy and, within the same alloy, were sometimes affected by batch-to-batch variations. At a later date an etch for pickling stainless steel was used. This etch was based on mixtures of nitric and hydrofluoric acids. However, when intermediate-curing-temperature adhesives were developed, these adhesives presented some difficulties. They did not give as high strengths with titanium as they did with aluminum alloys. As a result, the phosphate-fluoride process was developed (see below). This process had been the most widely used surface preparation for titanium for aerospace applications. It was described in ASTM D 2651[14] and MIL-A-9067.[36]

In the late 1960's Army helicopters in Southeast Asia began to develop severe debonding problems in sandwich panels of titanium and glass-reinforced epoxy composite skins bonded to aluminum honeycomb core. These failures were thought to be caused by the ingress of moisture to the interface, and the failure of the joints was considered to have been accelerated due to the combined effects of moisture and stress. As a result of this problem a number of research programs were undertaken to explain the mechanism of failure in hot/humid environments and to improve the environmental resistance of titanium adhesive-bonded joints. One such program resulted in the development of a modified phosphate-fluoride process by Picatinny Arsenal workers, Wegman, et al.[37] At a later date two commercial surface preparations, PASA JELL and VAST became available as alternatives. These procedures are described below. With the need for superior adhesive bonds of titanium joints in aerospace and other applications, many surface preparation methods have been developed in the last ten or twelve years.[38]

One of the problems requiring close control with titanium processing is *hydrogen embrittlement*. The formation of hydrogen gas is inherent in the acid etching and anodizing processes, and hydrogen pick-up on surfaces of titanium can cause embrittlement. Extreme caution must be used when treating titanium with acid etchants that evolve hydrogen. Immersion times must be closely controlled and minimized.[2]

Stabilized phosphate-fluoride treatment[1]–This method, developed by Wegman at Picatinny Arsenal, is an improvement over the basic phosphate-fluoride method described in MIL-A-9067. The improvement is obtained by the addition of *sodium sulfate* in the pickle. The method is reported to give good initial bond strength and excellent durability under adverse conditions, including high temperature/high humidity (140°F or 60°C and 95% RH under load). In this method the proper crystalline structure is established by the phosphate-fluoride process which is then stabilized by the incorporation of sodium within the crystalline structure.

(a) *Vapor-degrease* or *clean* with acetone

(b) *Alkaline-clean*: Immerse parts in alkaline cleaner (non-silicated), 5-10% by volume, for 5 minutes at 150°F (66°C). A suitable formulation is Oakite HD 126 1.5 oz per gallon (11.2 g/liter) in deionized water

(c) *Rinse* in running tap water at 105°F (40°C) for 2 minutes

(d) *Modified HF Acid-Nitric Acid Pickle*: Immerse for 2 minutes at room temperature in the following solution:

Hydrofluoric acid, 70%	2-3 fl oz/gal
Sodium sulfate, anhydrous	3.0 oz/gal
Nitric acid, 70%	40-50 oz/gal
Water, deionized	to make 1 gal

(e) Rinse in running tap water at service temperature;

(f) *Phosphate-fluoride etch*: Soak parts for 2 minutes at room temperature in a solution containing:

Trisodium phosphate	6.5-7.0 oz/gal
Potassium fluoride	2.5-3.0 oz/gal
Hydrofluoric acid, 70%	2.2-2.5 oz/gal
Water, deionized	to make 1 gal

(g) Rinse in running tap water at service temperature for 2 minutes

(h) *Hot-water soak*: Immerse in deionized water at 150°F (66°C) for 15 minutes

(i) *Final rinse*: Rinse in water at room temperature to 160°F (71°C) for 15 minutes

(j) Dry at 140°F (60°C) for 30 minutes in a preheated air-circulating oven

(k) Wrap the parts in clean kraft paper until ready to bond

This method gives excellent durability for both 6,4 titanium and chemically pure (CP) titanium. The former, however, shows a loss of lap-shear strength after 5 years outdoor weathering. The CP titanium does not show this effect.

Alkaline cleaning[33]–Use steps (a)(b)(c)(j) and (k) only, under the procedure for Stabilized Phosphate-Fluoride Method above. The results give poorer durability than that method, however.

Alkaline etch[33]–

(a) *Vapor degrease* or *clean* with acetone

(b) *Alkaline clean*: Immerse parts in non-silicated alkaline cleaner for 5 minutes at 150°F (66°C). A suitable formulation is 1.5 oz/gal (11.2 g/liter) of Oakite HD 246 in deionized water. Oakite products are available from Oakite Products, Inc., 50 Valley Road, Berkeley Heights, NJ 07922

(c) Rinse in running tap water at 105°F (40°C) for 2 minutes

(d) Rinse in running deionized water for 1 minute at service temperature

(e) *Alkaline etch*: Immerse for 5-10 minutes at 180°-200°F (82°-93°C) in a solution of 4 lb/gal (479 g/liter) of TURCO 5578. TURCO products are available from Turco Products Division, Purex Corp., Box 6200, Carson, CA 90749

(f) Rinse under running tap water at service temperature for 2 minutes

(g) Rinse under running deionized water at service temperature for 1 minute

(h) Dry at 140°F (60°C) for 30 minutes in a preheated air-circulating oven

(i) Wrap the parts in clean kraft paper until ready for bonding

Pasa Jell treatment[38]–Pasa Jell is a proprietary chemical marketed by Semco Division, Products Research and Chemical Corp., 5454 San Fernando Road, Glendale, CA 91203. This formulation is available either as a thixotropic paste suitable for brush application, or as an immersion solution for tank treatment. Pasa Jell 107 is a blend of mineral acids, activators and inhibitors inorgically thickened to permit application in localized areas. The approximate chemical constituents are 40% nitric acid, 10% combined fluorides, 10% chromic acid, 1% couplers, and the balance water. The immersion process requires nonmetallic tanks made of PVC, polyethylene, or polypropylene. A recommended mixture uses 1:1 dilution for 12 minutes. With the thixotropic Pasa Jell paste, a reaction time of 10-15 minutes assures durable bonds. Pasa Jell is sometimes used in combination with the alkaline etch material Turco 5578 mentioned above under *Alkaline Etch*. The process produces an amorphous-looking oxide containing O, Ti, N, Si, C, F, and Cr. The oxide has an anatose structure stable up to 175°C (347°F) and converts to rutile at 350°C (662°F).[38]

VAST Process[38,39]–VAST is an acronym for Vought Abrasive Surface Treatment and is a development of Vought Systems of LTV Aerospace Corp., Dallas,

TX. In the VAST process the titanium is blasted in a specially designed chamber with a slurry of fine abrasive containing fluorosilicic acid under high pressure. The aluminum oxide particles are about 280 mesh in size and the acid concentration is maintained at 2%. The process produces a grey smut on the surface of 6A1-4V-Ti alloy, which must be removed by a rinse in 5% nitric acid. The joint strength resulting is superior to that provided by the unmodified phosphate fluoride process, but is slightly lower than that after the Turco 5578 alkaline etch. The film produced is crystalline, having an anatose structure containing Ti, O, Si, F, Pb, and C. The oxide is stable up to 175°C (347°F), but starts converting to rutile at higher temperatures. The VAST process, because of its need for special equipment, has found limited use.

The process details are as follows:[1]

(a) Wipe surface with methyl ethyl ketone

(b) Alkaline clean with Turco 5578, 5 oz/gal (37.3 g/liter)

(c) Rinse with deionized water at room temperature

(d) Use VAST process for 5-10 minutes in a suitable chamber. The slurry consists of 2,000 ml of 2% hydrofluorosilicic acid (H_2SiF_6) plus 500 ml of 240-grit purified aluminum oxide (Al_2O_3). The white aluminum oxide available from the Carborundum Co., Electro Minerals Division, Niagara Falls, NY 14302, is acceptable

(e) Rinse with tap water spray at room temperature

(f) Immerse for 1 minute in 5% nitric acid solution (optional, depending on titanium alloy)

(g) Rinse in deionized water at room temperature

(h) Air-dry

(i) Bond within 4 days after treatment to be safe, although experiments have shown no changes up to 9 days

Alkaline-peroxide etch (RAE etch)[38]—When titanium is immersed in alkaline hydrogen peroxide solutions, depending on the concentration of sodium hydroxide and hydrogen peroxide, the metal is either etched or oxidized. The concentrations which produce grey oxides produce adhesive-wettable surfaces. A recommended mixture is 2% caustic soda and 2.2% hydrogen peroxide. Exposure to the oxidizing solution may be at room temperature, but 10-36 hours are required under these conditions to produce high-strength durable joints. Good bonding surfaces are produced within 20 minutes at 50°-70°C (122°-158°F). The oxidized adherends are washed with plain and acidified water before rinsing with acetone and drying at 100°C (212°F).

High bond strengths (50-55 MPa) (7,250-7,975 psi) are produced if the titanium surface is subjected to alumina blasting prior to treatment. The alkaline-peroxide etch has many advantages over the acid-based treatments. The chemical constituents are less toxic; the treatment does not require acid-resistant containers; the process is free from hydrogen pick-up; waste is easily disposed of. The process, however, is limited to batch production because of the high instability of the hydrogen peroxide at the high exposure temperature. The process was developed jointly by British Aerospace and the Royal Aircraft Establishment (RAE). Durability is reported to be excellent with 91–92% of initial joint strength (control) being retained after 4 years in a warm/wet environment (Innisfail).

Recent studies—In 1978 a tri-service (Army, Navy, Air Force) study program was begun on the effect of surface pretreatments of titanium on the durability of adhesive-bonded joints. Martin Marietta Laboratory performed the surface characterization studies. Ten surface-preparation methods were studied. The complete results were summarized in two Army restricted distribution reports in 1983.[40,41]

Tungsten and Alloys

Hydrofluoric-nitric-sulfuric acid method[22]—

(a) Degrease in a vapor bath of trichloroethylene

(b) Abrade the surface with medium-grit emery paper

(c) Degrease again as in (a)

(d) Using equipment constructed of fluoropolymer resins, polyethylene or polypropylene, prepare the following solution:

Hydrofluoric acid, 60%, sp. gr. 1.18	5 pbw
Nitric acid, conc., sp. gr. 1.41	30 pbw
Sulfuric acid, conc., sp. gr. 1.84	50 pbw
Distilled water	15 pbw

Blend the hydrofluoric acid and the nitric acid with water. Then *slowly* add the sulfuric acid, stirring constantly with a Teflon[R] or polyethylene rod. Add a few drops of 20% *hydrogen peroxide*.

(e) Immerse for 1–5 minutes in the above solution at room temperature

(f) Rinse under tap water

(g) Finish rinsing in distilled water

(h) Dry in an oven at 160°–180°F (71°–82°C) for 10–15 minutes

Uranium

For any adhesive to bond well with this metal, the uranium must be freshly pickled and dried. The dark-colored, loosely-adherent surface oxide layer must be removed. Cleaning is satisfactory when the metal surface becomes bright and shiny.

Abrasive method[1]—This method, developed by Picatinny Arsenal workers, involves degreasing the uranium in a vapor bath of trichloroethylene and then sanding the bonding surfaces in a pool of the adhesive to be used for bonding. This is done to prevent further oxidation of the uranium block by exposure to air after sanding, and also to prevent contamination of the surrounding area with radioactive particles. Polyamide-epoxy and polyurethane adhesives are recommended. The problem of oxidation remains, however, when only the two bonding surfaces of the uranium block are cleaned and adhesive coated. Under these conditions, the oxide layer spreads from the uncleaned side of the uranium until the bonding surfaces themselves are completely oxidized. By sanding all six sides of each uranium block and then coating with a protective coating of adhesive, the problem is solved.

Acetic acid-hydrochloric acid method—This method is used with aluminum-filled adhesives when no primer or surface coating is to be applied.[1]

(a) Pickle uranium bars clean in 1:1 nitric acid-water

(b) Rinse briefly in distilled water

(c) Immerse in 9:1 acetic acid-hydrochloric acid bath for 3 minutes. Note that 200 ml of this bath can accommodate no more than 4 bars of $4\frac{1}{2}''$ x $1''$ x $\frac{1}{8}''$ dimensions without a rigorous reaction taking place and a black film forming on the uranium surface

(d) Rinse briefly in distilled water

(c) Rinse in acetone

(f) Air-dry

Nitric acid bath—This is to be used when a primer or surface coating is to be applied.[1]

(a) Pickle-clean in 1:1 nitric acid-water

(b) Rinse in acetone

(c) Immerse for 10 minutes in coating bath of 1.0 g purified stearic acid dissolved in 95-99 ml of acetone and 1-5 ml of nitric acid

(d) Air-dry

(e) Rinse with carbon tetrachloride spray

(f) Air-dry

(g) Store in distilled water in a polyethylene bottle at $60°C$ $(140°F)$

Zinc and Alloys

The most common use of zinc is in *galvanized metals*. Zinc surfaces are usually prepared mechanically. One mechanical and two chemical methods follow.

Abrasion (for general-purpose bonding)[20]—

(a) Grit- or vapor-blast with 100-grit emery cloth

(b) Vapor-degrease in trichloroethylene (TCE)

(c) Dry at least 2 hours at RT, or 15 minutes at $200°F$ $(93°C)$ to remove all traces of TCE

Acid etch[22]—

(a) Vapor degrease in trichloroethylene

(b) Abrade with medium-grit emery paper

(c) Repeat the degreasing in (a)

(d) Etch for 2–4 minutes at RT in the following:

Hydrochloric acid, conc., or acetic acid	15 pbv
Distilled water	85 pbv

(e) Rinse with warm tap water

(f) Finish with distilled or deionized water

(g) Dry in an oven at $150°$–$160°F$ ($65.6°$–$71°C$)

Sulfuric acid/dichromate etch[20]–

(a) Vapor degrease in trichloroethylene

(b) Etch for 3–6 minutes at $38°C$ ($100°F$) in:

Sulfuric acid, conc., sp. gr. 1.84	2 pbw
Sodium dichromate, crystalline	1 pbw
Distilled water	8 pbw

(c) Rinse in running tap water

(d) Rinse in distilled water

(e) Dry in air at $40°C$ ($104°F$)

Conversion coatings–Phosphate and chromate conversion coatings are available for zinc from commercial sources as proprietary materials.

Weldbonding Metals

Chemical etching is essential to assure high-strength weld-bond joints. There are differences in the surface requirement after cleaning for welding and bonding. In order to achieve Class A resistance welds it is necessary to have a chemically active surface which may have a high surface resistance. Final selection of the surface preparation method should be based on the end use of the hardware and consideration of the relative importance of weld quality and adhesive joint strength.[42]

Lockheed workers Kizer and Grosko[43] have studied weldbonding using four different surface preparation techniques on aluminum. A technique developed for *aluminum* by the Northrop Corporation for the Air Force follows. It was found to be superior to the previously recommended FPL etch plus 60-minute dichromate seal.[44,45]

(a) *Vapor degrease* in trichloroethane vapor for 60 seconds, followed by spray rinse of condensed trichloroethane fluid for an additional 60 seconds. All parts must be free of water prior to vapor degreasing.

(b) *Alkaline clean* 10–15 minutes at $155°±10°F$ ($69°±5.5°C$) in Turco 4215-S solution, 6–8 oz/gal (44.8–46.9 g/liter). (Turco 4215-S is available from Turco Products Div., Purex Corporation, Carson, CA.)

(c) Immediately *spray-rinse* in cold deionized water for at least 5 minutes and inspect for water-break-free condition. Should water breaks occur, repeat the above steps.

(d) Immerse the alkaline-cleaned parts in a *deoxidizer* solution consisting of 8-16% by volume of nitric acid and 2.7-3.3 oz/gal (20.1-24.6 g/liter) Amchem 7 for 7-10 minutes at room temperature. When chemical addition is required to maintain the strength of the solution, use Amchem 17 instead of Amchem 7. (These products are available from Amchem Products, Inc., Ambler, PA.)

(e) Immediately *spray-rinse* in cold deionized water for at least 5 minutes.

(f) *Anodize* at 1.5±0.2 volts for 20-25 minutes in a solution of 1.35±0.15 oz/gal (10.1±1.1 g/liter) phosphoric acid and 1.35±0.15 oz/gal (10.1±1.1 oz/gal) sodium dichromate. Anodizing should be conducted in a room-temperature solution using a "Ripple-free" DC power supply.

(g) Immediately *spray-rinse* in cold deionized water for at least 5 minutes.

(h) *Oven-dry* 30-60 minutes at 140°-150°F (60°-66°C).

(i) Cleaned parts must be handled only with clean white cotton gloves and may be stored for periods up to 21 days prior to weld bonding by wrapping in chemically neutral paper.

For *titanium*, Grumman Aerospace Corporation has developed two surface treatments for the Air Force. They are given below:[46]

Vapor Honing/Pasa Jell 107M Procedure—

(a) Remove organic contaminants by methyl ethyl ketone (MEK) *solvent rinse*.

(b) Remove inorganic contaminants by immersion in *non-etch alkali* (hot Oakite 164) at 140°F (60°C). (This material is available from Oakite Products, Inc., 50 Valley Road, Berkeley Heights, NJ.)

(c) Immerse in *Pasa Jell 107M solution* at ambient temperature for 10 minutes. (This is a proprietary solution containing HNO_3, $Na_2Cr_2O_7$, H_2SiF_6 and proprietary surfactants supplied by Semco Division, Products Research and Chemical Corp., Glendale, CA.)

(d) Use *VAST process*[47] of surface impingement in suitable chamber. The slurry consists of 2% hydrofluorosilicic acid (H_2SiF_6) and 500 ml of 240-grit purified aluminum oxide (Al_2O_3), such as white aluminum oxide available from the Carborundum Co., Electro-Minerals Div., Niagara Falls, NY.

(e) *Rinse* with tap-water spray at room temperature

(f) Immerse 1 minute in 5% *nitric acid solution* (optional, depending on titanium alloy)

(g) *Rinse* in deionized water at room temperature

(h) *Air-dry*

(i) *Bond* within 4 days after treatment. (Experiments have shown, however, that no changes have taken place up to 9 days.)

SURFACE PREPARATION OF PLASTICS

The following procedures list commonly recommended methods for specific plastic adherends. These procedures provide strong reproducible bonds which fit readily into the bonding operation. The general methods described above should also be noted.

Thermoplastics

Thermoplastic surfaces, unlike thermosetting, ordinarily require physical or chemical modification to achieve acceptable bonding. This is especially true in the case of crystalline thermoplastics such as the polyolefins (primarily poly-ethylene and polypropylene), linear polyesters, and fluoropolymers. Methods used to improve the bonding characteristics of these surfaces include:

- Oxidation by means of chemical or flame treatment (The former is the most commonly used method.)

- Electrical discharge to leave a more reactive surface

- Ionized inert gas, which strengthens the surface by crosslinking and leaves it more reactive (See discussion above under Activated-Gas Surface Treatment of Polymers.)

- Metal ion treatment

The surface-preparation methods suggested below are recommended for conventional adhesive bonding. Solvent cements and thermal welding do not require chemical treatment of the plastics surface. As with metallic substrates, the effects of plastic surface treatments decrease with time, so it is important to carry out the priming or bonding steps as soon as possible after surface preparation.[3]

Acetal copolymer (Celcon®)—This material, made by Celanese Plastics Co., is a highly crystalline copolymer with excellent solvent and chemical resistance. Two suggested surface-preparation methods are as follows:

Chromic-acid etch[48]—

(a) Wipe parts with acetone

(b) Air-dry

(c) Etch the faying surfaces for 10–15 seconds in a solution made up as follows:

Sulfuric acid	400 pbw
Potassium dichromate	11 pbw
Water, deionized	44 pbw

(d) Flush with tap water

(e) Rinse with deionized water

(f) Oven-dry at 140°F (60°C)

Hydrochloric-acid etch[49]–Immerse parts for 5 minutes in conc. HCl at room temperature. Use a glass rod to move fresh acid into contact with the Celcon. One-third to one-half mil of Celcon is removed each minute of the etching process. Rinse and then dry at room temperature for 4 hours. Heat-formed or machined Celcon should be stress-relieved before etching. A specially developed primer used in very thin coats gives excellent results.

Acetal homopolymer (Delrin®)–This is a highly crystalline polymer made by the Du Pont Company. It has excellent solvent resistance. Two methods are recommended as follows:

Chromic-acid etch[50]–

(a) Wipe with acetone or methyl ethyl ketone

(b) Immerse in the following chromic acid solution for 10–20 seconds at room temperature (68°–86°F):

Potassium dichromate $K_2Cr_2O_7$	75 g (15 pbw)
Tap water	120 g (24 pbw)
Sulfuric acid, conc.	1,500 g (300 pbw)

Dissolve the potassium dichromate in the clear tap water, then add the sulfuric acid in increments of about 200 g, stirring after each addition.

(c) Rinse in running tap water for at least 3 minutes

(d) Rinse in distilled or deionized water

(e) Dry in an air-circulating oven at 100°F (38°C) for about 1 hour

"Satinizing"[51-53]–This is a patented process (U.S. Patent 3,235,426) described by the Du Pont Company for preparing their Delrin acetal homopolymer for painting, metallizing and adhesive bonding. In this process a mildly acidic solution produces uniformly distributed anchor points on the adherend surface. Adhesives bond mechanically to these anchor points, resulting in strong adhesion to the surface of the homopolymer. The following formulation is first prepared:

Kieselguhr (pyrogenic silica)	0.5% by wt
1,4-Dioxane	3.0% by wt
Para-toluene sulfonic acid (p-TSA)	0.3% by wt
Perchloroethylene	96.2% by wt

No preferential treatment is required for mixing these components. The first three ingredients may be worked into a *premix*, to which the perchloroethylene may be added whenever a batch is needed for "satinizing." Since the dioxane is flammable, it should be used only in areas suitable for work with combustible products. The ingredients contained in the "satinizing" solution are hazardous because of the flammability of the dioxane and the toxicity of the vapors of the dioxane, perchloroethylene and formaldehyde. Ventilation equipment for use in this process must be carefully designed. Both perchloroethylene and 1,4-dioxane vapors will be evolved from the dip bath and formaldehyde will be liberated from the part surface as a result of the etching process. Ventilation should be incorporated at the dip bath to remove those vapors in such a way that they will not be drawn across the breathing zone of an operator. Steps 1–3 below should be carried out in explosion-proof equipment because of the flammability of the 1,4-dioxane. Excessive air velocity should be avoided to prevent unnecessarily high losses of etching solution.

The liberation of formaldehyde fumes from the part surface as well as the continued evaporation of liquids (perchloroethylene and 1,4-dioxane) will continue through steps (2) and (3). These steps, therefore, will require ventilation to remove hazardous vapors as well. Exhaust equipment should be designed in such a way that condensation of water, which could subsequently find its way back to the equipment used in steps (1) and (2), is prevented.

The parts are conveyed through the following four-step operation:

(1) *Dip*—Immerse the Delrin article in the bath described above for 10–30 seconds at 175°–250°F (79.4°–121°C). The part is cleaned simultaneously by the degreasing action of the perchloroethylene. The parts should not touch each other or other objects. The bath area should be ventilated to remove formaldehyde fumes. An efficient stirrer should be used to keep the kieselguhr solid assistant in suspension and to insure homogenity.

(2) *Chemical action*—Transfer the Delrin parts to an air oven adjusted to 100°–250°F (38°–121°C). In this step the solvents from step (1) are evaporated and rapid etching is induced. The depth of the etch is controlled concurrently by the temperature and time of exposure in the oven. At 250°F (121°C) adequate etching should take no longer than 1 minute. Lower oven temperatures require longer exposure times. The oven area should be ventilated to remove formaldehyde fumes.

(3) *Water rinse*—Remove the Delrin parts from the oven and spray with hot water at 160°–175°F (71°–79.4°C). This stops the etching action.

(4) *Air-dry*—Dry in an oven with air, with the drying rate adjusted to accommodate production and floor space requirements. A short, high-temperature (250°F) (121°C) drying step for several minutes will reduce the tendency of the part to continue to release formaldehyde vapors. The part, after drying, is ready for bonding.

Acrylonitrile-butadiene-styrene (ABS)—This terpolymer may be prepared by the following techniques:

Abrasion with primer[54]—

(a) Sand with a medium-grit sandpaper

(b) Wipe free of dust

(c) Dry in an oven at 160°F (71°C) for 2 hours

(d) Prime with Dow Corning A-4094 or General Electric SS-4101

Warm chromic-acid etch[2]—

(a) Degrease in acetone

(b) Etch in the following solution for 20 minutes:

Sulfuric acid, conc.	26 pbw
Potassium dichromate	3 pbw
Water	11 pbw

(c) Rinse in tap water

(d) Rinse in distilled water

(e) Dry in warm air

Allyl diglycol carbonate (CR-39) (PPG Industries)—This is a castable resin developed in 1940. It is sold in monomer form and is used to produce highly transparent plastics for optical applications. The following surface treatment is recommended.[20]

(a) Degrease in methanol, isopropanol, or detergent

(b) Abrade by grit or vapor blast, or use 100-grit emery cloth. Sand or steel shot are suitable abrasives

(c) Solvent degrease in methanol or isopropanol

Cellulosics [cellulose acetate, cellulose acetate butyrate (CAB), cellulose nitrate, cellulose propionate, ethyl cellulose]—Cellulosics are ordinarily solvent cemented, unless they are to be bonded to dissimilar adherends, using conventional adhesives. In such cases the following surface preparation method may be used:

Abrasion and cleaning[1,55]—

(a) Solvent-degrease in methyl or isopropyl alcohol

(b) Grit or vapor-blast, or use 220-grit emery cloth

(c) Repeat step (a)

(d) Heat 1 hour at 200°F (93°C) and apply adhesive while still hot

Ethylene-chlorotrifluorethylene copolymer (E-CTFE) (Halar®-Allied Corporation)[1]—This fluoropolymer has excellent chemical resistance. Its surface is difficult to prepare because of its high surface tension and non-polar condition.

(a) Wipe with acetone

(b) Treat with sodium naphthalene complex for 15 minutes at room temperature. Sodium naphthalene solutions are commercially available. (When using the complex solutions, they should be kept in a tightly stoppered glass container to exclude air and moisture. Extreme care should be taken in handling the solutions, since they are hazardous due to the explosive nature of the sodium. Directions are given in the manufacturers' literature for making up these solutions, which contain sodium metal, naphthalene and tetrahydrofuran in varying proportions.)

(c) Remove from solution with metal tongs

(d) Wash with acetone to remove excess organic material

(e) Wash with distilled or deionized water to remove the last traces of metallic salts from the treated surface

(f) Dry in an air-circulating oven at $99°\pm5°F$ $(37°\pm3°C)$ for about 1 hour

Other methods used to prepare the surface of E-CTFE include flame oxidation, corona discharge (activated gases), and exposure to electric discharge from a spark coil.

Ethylene-tetrafluoroethylene copolymer (ETFE)[1]–This high-temperature-resistant fluoropolymer made by the Du Pont Company as Tefzel® can be easily processed by conventional methods, including extrusion and injection molding. It can be joined by screw assemblies, snap-fit, press-fit, cold or hot heading, spin-welding, and heat-bonding. No information has been published, to the author's knowledge, on adhesive bonding. One type of high-performance film (Type CE) is made cementable with a transparent treatment for bonding. This material is claimed to be receptive to many types of adhesives.

Ethylene-vinyl acetate (EVA)[56]–This copolymer is really an elastoplastic, with properties intermediate between that of a plastic and an elastomer. A suggested surface preparation method is as follows:

SIRA method[20]–

(a) Degrease in methanol

(b) Prime with epoxy adhesive

(c) Fuse into the surface by heating for 30 minutes at $100°C$ $(212°F)$

Fluorinated ethylene-propylene copolymer (FEP)–This fluoropolymer is supplied by the Du Pont Company as Teflon FEP. This is a true thermoplastic and is a copolymer of tetrafluoroethylene and hexafluoropropylene. FEP can be heat sealed, heat bonded, laminated, and combined with many materials. It can be used by itself as a hot-melt adhesive. FEP can be surface prepared using the same techniques as given above under E-CTFE.

Ionomer[1]–Ionomers are polymers in which ionized carboxyl groups create ionic crosslinks in the intermolecular structure. The properties are similar to crosslinked thermosets, even though these resins are processed at conventional temperatures like other thermoplastic resins. Ionomers are supplied by the Du Pont Company as Surlyn resins. These resins, like polyethylene, do not develop high-strength adhesive bonds. The following surface preparation technique is suggested:

Abrasion[20]–

(a) Solvent-degrease in acetone or methyl ethyl ketone

(b) Grit- or vapor-blast, or use 100-grit emery cloth. 180-grit alumina is a suitable abrasive.

(c) Repeat step (a)

Nylon (polyamide)[1]–Both solvent cementing and adhesive bonding are used with nylon (nylon 6,6 is the type most commonly used). The solvents, however, are quite different from those used for cellulosics and polystyrene. Nylon molding resins are not ordinarily cemented by conventional solvent-type cements because of their lack of solubility in ordinary non-toxic solvents. Nylon is frequently bonded to metals, introducing problems not found in nylon-to-nylon bonds. Conventional adhesives must be used in nylon-metal bonds. Nylons are relatively crystalline thermoplastics that are not considered solvent-sensitive. For this reason relatively strong solvents must be used in surface preparation. Solvents recommended for cleaning include:

Acetone (B.P. 56.5°C)

Methyl ethyl ketone (B.P. 79.6°C)

Perchloroethylene (Perclene®) (B.P. 121°C)

1,1,1-Trichloroethylene (B.P. 74.1°C)

M-17

Freon TMC (B.P. 36.2°C)

Detergents have also been used instead of organic solvents in cleaning nylon. Nylon, like polycarbonate, will readily absorb moisture from the air. For this reason, it is desirable to keep the humidity low before bonding.

Abrasive method[57]–

(a) Wash with acetone

(b) Dry

(c) Hand-sand with 120-grit abrasive cloth until the gloss is removed

(d) Remove sanding dust with a short-haired stiff brush

Bonding or priming should be carried out as soon as possible. Parts should be warmed slightly before bonding. Primers recommended include nitrile-phenolic, resorcinol-formaldehyde, vinyl phenolic, silane, etc. Epoxies have been used on metallic surfaces, with the nylon being bonded to the epoxy.[58]

Perfluoroalkoxy resins (PFA)[1]–This is a relatively new class of melt-processable fluoroplastics introduced by Du Pont in 1972. They combine high-temperature performance with a capability for thermoplastic processing. Surface preparation is as given for E-CTFE above.

Phenylene-oxide-based resins (Noryl®) (Polyaryl ethers)–Noryl is the General Electric Co. designation for polystyrene-modified polyphenylene oxide.

These resins have outstanding dimensional stability at high temperatures, broad temperature use range, and other desirable properties. Solvent cementing is the usual method of joining. General Electric's recommendation for surface preparation is sanding or the following acid etch procedure.[59,60]

(a) Solvent-clean with isopropanol or aqueous solutions of most commercial detergents

(b) Sand *or* chromic-acid etch. If chromic-acid etch is used, etch 1 minute at 176°F (80°C) in the following solution:

Sulfuric acid, conc.	375 g
Potassium dichromate	18.5 g
Water	30 g

(c) Rinse in distilled water

(d) Dry

Note: Studies made with the original *unmodified* polyphenylene oxide etched in the above solution for 3 minutes at 150°F showed that the chromic acid solution became ineffective after its first use. The evaluation was made with AF-30 nitrile-phenolic adhesive and modified T-peel tests. If it is desired to avoid this possible problem, vacuum blasting, followed by an acetone wipe, can be used.[1]

Polyarylate (Ardel®—Union Carbide Corp.)—This new engineering thermoplastic announced in 1980 is an aromatic polyester based on bisphenol A and phthalic acid. According to the manufacturer (personal communication) its surface preparation is similar to that of polysulfone (see Polysulfone below).

Polyaryl Sulfone (Astrel 360 plastic)[1]—This relatively new engineering thermoplastic supplied by the 3M Company is structurally useful at temperatures up to 500°F (260°C). The manufacturer's literature mentions welding by thermal, ultrasonic, or solvent techniques. Air Force studies evaluated sandblast, acid etch, and solvent-wash methods, with the latter recommended for steel-plastic-steel bonds.

Solvent wash—Triple-wash (3 successive washes) in a 65/35 mixture by volume of Freon PCA and reagent-grade isopropyl alcohol. This mixture is also available from Du Pont as Freon TP-35, except that Freon TF solvent is used instead of Freon PCA. The latter is an ultrapure version of TF and essentially becomes TF as soon as it is opened, according to a personal communication with the Du Pont Company.

Sandblast—

(a) Ultrasonic-clean with an alkaline-etching solution

(b) Cold-water rinse

(c) Alcohol-wash

(d) Sandblast with 150-mesh silica sand

(e) Alcohol-wash

(f) Dry with dry nitrogen

Acid etch—

(a) Ultrasonic-clean in an alkaline cleaning solution

(b) Cold-water wash

(c) Bathe 15 minutes at 150°-160°F (66°-71°C) in an acid-etch solution containing:

Sodium dichromate	3.4% by wt
Sulfuric acid, conc.	96.6% by wt

(d) Cold-water wash

(e) Dry at 150°F (66°C) in air-circulating oven

Polycarbonate (supplied as Lexan®–General Electric Co., Merlon–Mobay Chemical Co., and Tuffak–Rohm and Haas Co.)[1]–Solvent cementing is the recommended method for bonding this resin to itself, or to plastics soluble in the same solvents. The latter include *cellulose acetate butyrate, acrylics*, or *polyurethane*. However, adhesive bonding must be used in bonding to metals, wood, rubber, and most other plastics. Polycarbonate is hygroscopic, like nylon. For this reason it is important to keep the humidity low before bonding.

In surface preparation of polycarbonate for adhesive bonding the solvents recommended for cleaning are methyl alcohol, isopropyl alcohol, petroleum ether, heptane, VM&P naphtha, and white kerosene. Ketones, toluol, trichloroethylene, and benzol should not be used, since polycarbonate is incompatible with these solvents which cause crazing or cracking. A number of other cleaning solvents, including paint thinners, are similar to ketones in that they cause crazing or cracking, so care should be used in the selection of solvents used for cleaning. Light solutions (1%) of detergents, such as Alkanox or Joy are also suggested.[1,61]

Flame treatment[62]—

(a) Wipe with ethyl alcohol to remove dirt and grease

(b) Pass the part through the oxidizing portions of the flame of a Bernz-O-Matic propane torch

(c) Treatment is complete when both sides are polished to a high gloss, free of scratches and visible flaws. The process usually requires 5–6 passes on both sides

(d) Allow the part to cool for 5–10 minutes before bonding

Abrasion[57]—

(a) Wipe with ethyl alcohol, methyl alcohol, or other acceptable solvent or cleaner mentioned above

(b) Air-dry

(c) Sand with fine-grit (120-grit, 400-grit maximum) abrasive cloth or sandpaper

(d) Remove sanding dust with a clean dry cloth or a short-haired stiff brush

(e) Repeat solvent wipe

General Electric procedure for use with RTV silicone adhesives[13]–RTV (room-temperature-curing) silicone adhesives are recommended for applications requiring excellent bond strength, a high service temperature, and good thermal expansion. Silicone adhesives may be used with all standard polycarbonate resins, including glass-reinforced grades. For optimum performance, General Electric recommends the following procedure:

(a) Lightly abrade mating surfaces with fine emery

(b) Clean surface of grease or foreign matter with methyl alcohol or other compatible cleaning solvent (see list above)

(c) Prime abraded surface with GE Silicone Primer SS-4004 and allow a minimum of one hour dry time. Unprimed surfaces reduce the bond strength significantly

(d) Apply silicone adhesive in desired thickness. The final bond thickness may range from 0.005 in. to 0.030 in.

(e) Assemble

Polychlorotrifluoroethylene (PCTFE) (KEL-F–3M Co.) (Plaskon CTFE–Allied Chemical Corp.)[1]–This resin, also called trifluoromonochloroethylene, is a fluoropolymer with excellent chemical resistance. It offers problems in surface preparation because of its high surface tension and non-polar surface. Surface preparation is as given above for E-CTFE.

Polyester (saturated)[1]–Thermoplastic polyesters are available as polyethylene terephthalate (PET), used primarily in films or fibers as Mylar, Dacron and Celanar, and polybutylene terephthalate (PBT), used in molding as Valox (General Electric Co.), Celanex (Celanese Plastics) and Tenite (Eastman Chemicals). Glass-filled grades of these resins are also available. Thermoplastic polyesters are generally resistant to solvents and chemicals, including alcohol, ethers, and aliphatic hydrocarbons, most chlorinated hydrocarbons, aqueous salt solutions and aqueous acid and basic solutions under specified conditions. Specific surface preparations for the above thermoplastic polyesters are:

Treatments for Valox® thermoplastic polyester[65,66]–

(1) Abrade-degrease

(a) Sand lightly with 240-grit sandpaper

(b) Degrease with toluol or trichloroethylene. Toluol is recommended for the unreinforced Valox

(2) Plasma treatment[64]–Activated gas plasma, using oxygen, argon, or

water vapor plasma, has given good bonds 3 to 4 times stronger than those obtained with the above method. [See section on Activated-Gas Surface Treatment of Polymers (Plasma Treatment) above.]

Treatments for Celanex® thermoplastic polyester[66]—

(1) Abrade-degrease

   (a) Sandpaper

   (b) Degrease by wiping with a solvent such as acetone

*Treatments for Tenite polyterephthalate[67]—*Dip parts in methylene chloride for 10-15 seconds before bonding, preferably with polyurethane adhesives.

Polyetheretherketone (PEEK)[68]—This is a new material available from ICI Americas, Inc. PEEK is an aromatic polyether high-performance thermoplastic polymer developed primarily for coating and insulation of high-performance wiring, although it can also be molded. It is available in unreinforced and glass-reinforced grades. Acetone has a slight crazing effect on PEEK, but isopropyl alcohol, toluene and trichloroethylene are known to have no effect, so they may be used in cleaning.

Surfaces to be bonded should be clean, dry, and free from grease. This may be achieved by wiping the joint surfaces with "Genklene" (available from ICI) or a similar degreasing solvent. In addition, roughening of the surfaces with an abrasive will aid in giving a more reliable bond. Degreasing should be repeated after abrading. Bond strengths may be further increased by more vigorous surface activation procedures, in particular, flame treatment with a blue (oxidizing) flame, or treatment with a chromic acid etching solution. Table 4-3 shows results obtained with these various treatments when bonding PEEK to aluminum with Araldite 100, a general-purpose epoxy adhesive.

Table 4-3: Effect of Various Surface Treatments on PEEK

Araldite epoxy adhesive, cured 17 hours at 40°C; lap joint 13 mm (0.5 in) wide with 13 mm overlap. Total bond area 170 mm² (0.25 in²), Substrate Aluminum. Testing method: tested in shear at 20 mm/min (0.8 in/min).

Surface Treatment	Bond Strength, kg
Abrading + degreasing	40
Abrading + degreasing + flame oxidation	48
Abrading + degreasing + chromic acid etch	37

Polyethersulfone (PES) (Victrex—ICI Americas, Inc.)[69]—This is another new high-performance engineering thermoplastic. Methyl alcohol, ethyl alcohol, isopropanol and low-boiling petroleum ether are solvents that may be safely used in cleaning. Acetone, methyl ethyl ketone, perchloroethylene, tetrahydrofuran, toluene and methylene chloride should *not* be used, since they have an adverse effect on the plastics.

Polyethersulfone can be bonded to itself by solvent cementing adhesives, or ultrasonic welding. Mating surfaces should be degreased with carbon tetrachloride before adhesive bonding.[70]

Polyetherimide (Ultem® –General Electric Co.)[71]–Polyetherimide is a new amorphous engineering thermoplastic developed by GE. It has a number of desirable properties, including a high heat-distortion temperature. Unlike most amorphous resins polyetherimide is resistant to a wide range of chemical media, including most hydrocarbons, alcohols, and fully halogenated solvents. Only partially halogenated hydrocarbons such as methylene chloride and trichloroethane will dissolve polyetherimide, and these compounds should not be used for cleaning.

Polyethylene (PE)[1]–Polyethylene has a non-polar, non-porous, inert surface. For this reason, adhesives cannot link chemically or mechanically to untreated polyethylene surfaces. Since polyethylene is relatively inert to most solvents, solvent cementing cannot be used. Therefore, for bonds to itself or to other materials, adhesive bonding with suitable surface preparation must be used. There are a number of surface treatments in use, including chemical, electronic, flame, and primer treatments. Oxidation treatments are the most successful. These include immersion in a chromic-acid solution, exposure to corona discharge, flame oxidation, immersion in an aqueous solution of chlorine, or exposure to chlorine gas in the presence of ultraviolet light. The chromic acid oxidation method is probably the most convenient for use with molded plastic parts of diverse contour.

Chromic-acid etch[1]–

(a) Wipe with acetone, MEK or xylene

(b) Immerse in the following chromic-acid solution for 60–90 minutes at room temperature, or 30–60 seconds at 160°F (71°C):

Potassium dichromate ($K_2Cr_2O_7$)	75 g 15 pbw
Tap water	120 g 24 pbw
Sulfuric acid, conc.	1,500 g 300 pbw

Dissolve the potassium dichromate in the clean tap water, then add the sulfuric acid in increments of about 200 g, stirring after each addition

(c) Rinse in running tap water for at least 3 minutes

(d) Dry in an air-circulating oven at 100°F (38°C) for about 1 hour

Oxidizing flame method[61]–This method utilizes an oxyacetylene burner which is passed over the faying surface until it appears glossy. To make certain that too much oxide is not on the surface, a light scouring with soap and water should be carried out. The surface should then be washed with distilled water and dried at room temperature. Flame treatment is fast and provides bond strength greater than the chromic acid treatment described above. However, it requires very careful control to prevent heat warpage. The procedure is safest with pieces of thick cross section.

Gas plasma treatment[61]–This treatment is very effective and is recommended for use with very complex geometric surfaces which cannot be sanded or flame-treated adequately, and where very strong bonds are required. (See section on Activated-Gas Surface Treatment of Polymers above.)

Polymethylmethacrylate (PMMA) (Plexiglas–Rohm and Haas Co.) (Lucite–Du Pont Co.)–In bonding PMMA to itself, solvent cementing is ordinarily used. However, when bonding PMMA to a dissimilar substrate, conventional adhesive bonding must be used. Surface treatment is as follows:[1]

(a) Wipe with methanol, acetone, MEK, trichloroethylene, isopropanol, or detergent

(b) Abrade with fine-grit (180–400 grit) sandpaper, or use abrasive scouring with small amounts of water, dry-grit blasting, or wet abrasive blasting

(c) Wipe with a clean, dry cloth to remove particles

(d) Repeat solvent wipe

Polymethylpentene (TPX)[72]–This material is now supplied by Mitsui Petrochemical Industries, Ltd. It is a copolymer of 4-methylpentene and is useful up to 400°F (204°C). Chemical behavior is similar to other polyolefins. It is attacked by strong oxidizing agents. Some light hydrocarbons and chlorinated solvents can cause swelling and subsequent strength loss. Surface preparation methods recommended are:

Degreasing–abrasion (for general-purpose bonding)[20]–

(a) Solvent-clean in acetone

(b) Grit- or vapor-blast, or use 100-grit emery cloth

(c) Solvent-clean again in acetone

Chromic-acid etch[20]–

(a) Solvent-clean in acetone

(b) Immerse for 1 hour at 60°C (140°F) in:

Sulfuric acid, conc.	26 pbw
Potassium dichromate	3 pbw
Water	11 pbw

(c) Rinse in tap water

(d) Rinse in distilled water

(e) Dry in warm air

Acid-permanganate etch[20]–

(a) Solvent clean in acetone

(b) Immerse for 5-10 minutes at 194°F (90°C) in a saturated solution of potassium permanganate acidified with sulfuric acid

(c) Rinse in tap water

(d) Rinse in distilled water

(e) Dry in warm air

Polyphenylene sulfide (PPS) (Ryton–Phillips Chemical Co.)—This crystalline polymer has a high melting point (550°F) (288°C), outstanding chemical resistance, thermal stability and nonflammability. It has no known solvents below 375°-400°F (191°-204°C). The surface may be prepared as follows:

Solvent-sandblast-solvent [73]—

(a) Solvent degrease in acetone

(b) Sandblast

(c) Solvent-degrease again in acetone

Picatinny Arsenal method [74]—

(a) Wipe the faying surfaces with ethanol-soaked lintless paper

(b) Sand with 120-grit sandpaper

(c) Clean off the dust with a stiff, bristled brush

Polypropylene (PP)—This polyolefin is very similar to polyethylene, and the surface treatment is generally similar to that described for polyethylene. The thermal treatment, however, is somewhat more rigorous than prescribed for polyethylene. While the treatment for polyethylene is 60-90 minutes at room temperature, or 30-60 seconds at 160°F (71°C), polypropylene should be treated for 1-2 minutes at 160°F (71°C).

Chromic-acid etch [50,73]—

(a) Wipe with acetone, MEK, or xylene

(b) Immerse in the following chromic-acid solution for 1-2 minutes at 160°F (71°C)

Potassium dichromate ($K_2Cr_2O_7$)	75 g	15 pbw
Tap water	120 g	24 pbw
Sulfuric acid, conc.	1,500 g	300 pbw

Dissolve the potassium dichromate in the clean tap water, then add the sulfuric acid in increments of about 200 g, stirring after each addition

(c) Rinse in running tap water for at least 3 minutes

(d) Rinse in distilled or deionized water

(e) Dry in an air-circulating oven at 100°F (38°C) for about 1 hour

Oxidizing-flame method[61]–As in Polyethylene.

Gas-plasma method[61]–As in Polyethylene.

Polystyrene (PS)[61]–This material is ordinarily bonded to itself by solvent cementing, although conventional adhesive bonding, thermal welding, and electromagnetic bonding have been used. Conventional adhesive bonding may be necessary for bonding polystyrene to a dissimilar material. Surface preparation methods use as follows:

Abrading or sanding[58]–

(a) Degrease with methyl or isopropyl alcohol

(b) Abrade with 200-grit sandpaper and remove dust particles

Sodium dichromate-sulfuric acid process[58]–

(a) Degrease with isopropyl or methyl alcohol

(b) Immerse for 3–4 minutes in the following solution maintained at 210°-220°F (99°-104°C):

Sulfuric acid, conc.	90 pbw
Sodium dichromate	10 pbw

(c) Rinse thoroughly with distilled water

(d) Dry below 120°F (49°C)

Note: This is a somewhat unusual modification of the chromic-acid etching process. Since no water is used to dilute the acid, a much higher immersion temperature may be used without boiling off the water. The usual high-temperature immersion in the conventional chromic acid solution is about 160°F (71°C), while 210°-220°F (99°-104°C) is used here.

Nonimmersion etch process[58]–If polystyrene parts are to be used in high-frequency electrical applications, it may be desirable that only the faying surfaces be treated as follows:

(a) Degrease with methyl or isopropyl alcohol

(b) Apply to the faying surfaces the following thixotropic paste:

Sulfuric acid, conc.	3 pbw
Powdered potassium	1 pbw

Add Cab-O-Sil (Cabot) fused silica earth as required to obtain a thixotropic paste

(c) Heat parts to 180°F (82°C) and hold for 3–4 minutes with the paste on the surface

(d) Rinse thoroughly with distilled water

(e) Dry below 150°F (66°C)

Polysulfone (Udel—Union Carbide Corp.)[61]—Polysulfone is a thermoplastic with very high strength and one of the highest service temperatures (340°F or 171°C) of any melt-processible thermoplastic. It is highly resistant to mineral acids, alkalies, salts, detergent solutions, oils and alcohol. On the other hand, it is attacked by polar organic solvents such as ketones, chlorinated hydrocarbons, and aromatic hydrocarbons. Polysulfone stress cracks easily and is considered notch-sensitive. It can be solvent cemented, conventionally adhesive bonded or ultrasonically welded. Air Force-sponsored studies evaluated three surface-preparation techniques: (1) sandblast, (2) acid-etch, and (3) solvent-wash. The *acid-etch* procedure was found to give the best results. (See procedures for Polyaryl Sulfone above for details of these processes.)

Another method, using a grit-blast and ultrasonic cleaning, was used by General Electric in an Air Force-sponsored study. In this method polysulfone parts were prepared for bonding to niobium with AF-42 epoxy nylon film adhesive. The procedure is as follows:[75]

(a) Degrease in alcohol

(b) Grit-blast with 27–50 micron aluminum oxide (Al_2O_3). Use S.S. White Airbrasive Equipment, or equivalent, available from S.S. White Industrial Products, Pennwalt Corp., Piscataway, NJ 08851.

(c) Clean in an ultrasonic cleaner as follows:

(1) Immerse parts in Neutra-Clean 7 (available from Shipley Co., Inc., Newton, MA 02162) (12 oz/gal) (90 g/liter)

(2) Rinse with tap water

(3) Rinse with distilled water

(4) Rinse with isopropyl alcohol for at least 30 seconds

(5) Flash with dry nitrogen

(6) Dry with warm air at 150°F (66°C) maximum

Other methods used include vapor degreasing in methanol, sanding or vapor blasting, and alcohol-wipe and/or light sanding, plus vinyl, urethane, or silicone primer.

Polytetrafluoroethylene (PTFE)[1]—This well-known polymer is available from Du Pont as Teflon TFE, from Allied Corporation as Halon TFE, from ICI Americas as Fluon, and from Pennwalt Corp., as Tetran TFE. PTFE does not melt like a true thermoplastic, but "sinters." As with other fluoroplastics, PTFE is relatively inert chemically. It is prepared for bonding by the technique given above for E-CTFE.

Polyvinyl chloride (PVC)—The procedure given here is intended for use with the straight homopolymer rigid compounds. These compounds contain up to 5% plasticizer, making them difficult to bond with epoxy and other non-rubber types of adhesives. The homopolymer is not readily soluble, making it difficult to bond by solvent-cementing techniques. Thermal welding, especially hot-gas welding, is commonly used in joining rigid PVC. In general, a solvent is always used to remove plasticizer, grease and dirt. Abrasion may or may not be used. If it is, sandpaper (various grits), vapor-blasting, steel wool, or scouring powder are used. The following procedure is suggested.[1]

(a) Wipe with solvent, such as methanol, low-boiling petroleum ether, MEK, toluene, or trichloroethylene

(b) Abrade with medium-grit (200-grit) sandpaper

(c) Blow off the dust

(d) Repeat solvent-wipe

(e) Dry

(f) For maximum strength, prime with nitrile-phenolic adhesive, or bond immediately

Polyvinyl fluoride (PVF) (Tedlar®—Du Pont Co.) (film only)—Du Pont has two types of Tedlar film, Type B, with both sides already treated for bonding, and Type S, which is untreated and is used as a release film. Type B is used in laminating to metals, wood and other materials and requires no further surface preparation for adhesive bonding, according to Du Pont. However, there are published methods for preparing the untreated film. They are as follows:

Alkaline etch[58]—

(a) Degrease with isopropyl alcohol

(b) Immerse for 2-5 minutes in Prebond 700 alkaline compound, 8 oz/gal (59.7 g/liter) maintained at 160°-170°F (71°-77°C)

(c) Rinse

(d) Dry below 120°F (49°C)

Sodium-naphthalene etch[76,77]—This treatment, described above under E-CTFE, may also be used.

Other methods[1]—PVF can be adhesively bonded with adhesive systems recommended for *PVC*, but may require added surface preparation for good adhesion. Activated-gas plasma treatment has also been used successfully (see discussion under Activated-Gas Surface Treatment of Polymers above).

Polyvinylidene fluoride (PVDF) (Kynar—Pennwalt Corp.)[1]—Polyvinylidene fluoride is a high-performance, high-molecular-weight homopolymer. It is resistant to most acids and bases, aliphatics, aromatics, alcohols, and chlorinated solvents. Strongly polar solvents such as ketones and esters cause partial solvation, especially at high temperatures. Kynar is serviceable over the temperature range of -80° to 300°F (-62° to 149°C). Recommended procedures are as follows:

Solvent clean—abrasion[2]—

(a) Solvent clean with isopropyl or methyl alcohol

(b) Abrade with any of the following methods:

(1) *Sand*, hand or machine, using 180–320 grit aluminum oxide paper

(2) *Abrasive scour*, using tap water and a fine, non-chlorinated scouring powder. Rinse with tap water, then distilled water

(3) *Grit-blast, dry*, using nonmetallic grit, such as flintstone, silica, silicon carbide or aluminum oxide

(4) *Abrasive blast*, using 3:1 slurry of water and 220–325 grit aluminum oxide. Rinse in distilled water

(c) Solvent clean again as in (a)

Activated gas surface treatment[1]–This technique, discussed above under Activated-Gas Surface Treatment of Polymers, has also been used successfully.

Styrene-acrylonitrile (SAN) (Lustran—Monsanto Chemical Co.) (Tyril—Dow Chemical Co.)[1]–This copolymer is usually solvent cemented with solvents similar to those used for polystyrene, although the solvents usable are more restricted. In applications where solvent cements cannot be used, as in bonding to metals, the procedures suggested for Polystyrene above may be used.[1] Trichloroethylene and gasoline have also been used with no further treatment.[78]

Thermosets

Most thermoset plastics are relatively easy to bond. Solvent cements obviously cannot be used since these materials are not soluble. In some cases, solvent solutions can be used to join thermoplastics to thermosets. In most cases, however, conventional adhesive bonding is the only practical non-mechanical way to join a thermoset to a thermoset, or to non-plastic materials. Ultrasonic bonding can be used only with ultrasonic adhesives. Frequently a mold-release agent is present in thermoset materials and must be removed before adhesive bonding. Mold-release agents are usually removed after a detergent wash, solvent-wash, or solvent-wipe, followed by light sanding to break the smooth surface glaze. A final solvent-wipe with clean solvent and clean lint-free cloth or paper tissue is usually used. Solvents used include acetone, methyl ethyl ketone (MEK), toluene, low-boiling petroleum ether, trichloroethylene, and isopropyl alcohol. Fine abrasive paper (sand, carborundum, or aluminum oxide-abrasive grit/metal, sand or oxide) or metal wool (steel, aluminum or brass) or steel shot are frequently used for abrasion of the surface of thermosets.[61]

Diallyl phthalate (DAP)[58]–This resin has exceptional electrical insulating properties, high-temperature stability, and good resistance to most chemicals and moisture. Surfaces are hard and tough and pick up very little moisture. DAP parts are ordinarily molded or glass laminates and are not easy to bond adhesively. At present, *sanding* or *buffing* appear to be the best surface treatments. Only filled resins are available. Besides glass, fillers include minerals, Orlon®, and asbestos. Surface preparation is as follows:

(a) Wipe with acetone, MEK, toluene, low-boiling petroleum ether, trichloroethylene, or isopropyl alcohol

(b) Sand-, grit-, or vapor-blast or use steel wool

(c) Wipe with a clean dry cloth to remove grit and particles

(d) Repeat solvent-wipe in (a)

Epoxies[58]—Epoxy resins are among the workhorses of thermoset resins. They have a wide range of properties. They show good dimensional stability and good electrical properties and mechanical strength. They have good creep resistance and are useful over a wide temperature range (-100° to 800°F) (-73° to 427°C). These resins are easy to adhesive-bond, requiring only a clean dry surface, usually solvent-cleaned and sanded. Both filled and unfilled grades are available. Fillers include minerals, glass, silica glass, and glass microballoons. Epoxy laminates should be treated as covered below under Reinforced Thermosets (Laminates). A recommended surface treatment is as follows:

(a) Wipe with acetone or other suitable solvent (see list under diallyl phthalate) or vapor degrease

(b) Abrade by scouring or other treatment suggested above under thermosets

(c) Wipe with a clean dry cloth to remove grit

(d) Repeat solvent treatment in (a)

(e) Apply a silane-type primer, such as Hughson Chemical Company's Chemlok 607, diluted with 5–10 volumes of methanol or ethanol. This step is *optional*

Melamine-formaldehyde (melamines)[58]—The amino resins, like urea-formaldehyde, are hard, rigid, and abrasion-resistant. They are self-extinguishing and have superior electrical properties. They have excellent dimensional stability and good creep resistance. They are noted for their high impact strength and resistance to water and solvents. Only filled resins are available. The fillers include cellulose, rag, and glass. A suggested surface preparation is as follows:

(a) Scrub with an abrasive household detergent

(b) Rinse with tap water, then deionized water

(c) Dry

(d) Sand

(e) Wipe with isopropyl alcohol

(f) Dry

(g) Prime or bond

The procedure given above for Epoxies may also be used.

Phenol-formaldehyde (phenolics)[58]—Phenolics have an excellent combination of high physical strength and high-temperature resistance, good electrical properties, and good dimensional stability. Phenolics are widely bonded, not only on molded parts, but also as laminates and castings. Both filled and unfilled types are available. Fillers are asbestos, cellulose, wood, flour, and glass. The method of surface preparation given above under Epoxies is suggested.

Polyester[58]–These unsaturated, non-linear resins are similar to epoxies and phenolics in their surface-preparation requirements, basically requiring only sanding for good results. Polyesters have good resistance to oils and solvents. The method of surface preparation given for epoxies is suggested.

Polyimide[1]–This is a class of polymeric materials distinguished by exceptional thermal stability. They are available commercially in several forms, including fabricated parts and shapes, molding resins, films, and coatings for wire and fabrics. Some polyimides, *condensation polymers*, are essentially linear, with structures similar to those of thermoplastics. Others, *addition polymers*, used as engineering resins in parts (moldings and laminates), are thermosetting. Du Pont's Vespel® precision parts are available in graphite-PTFE-and MoS_2-filled bearing compositions, as well as unfilled. (Du Pont markets the parts only. The resin is not available commercially.) These parts will withstand temperatures up to 900°F (482°C) for short periods. Du Pont's recommendations for Vespel parts are as follows:

Du Pont Co. recommendations for Vespel parts[79]–

(a) Remove surface contamination, such as dirt and oil with solvents. Clean by refluxing in perchloroethylene or trichloroethylene, or clean ultrasonically in Freon TF (trichlorotrifluoroethane).

(b) Mechanically abrade with a wet or dry abrasive blast. Use a light abrading (approx. 50-100 microinches) to maintain a uniform adhesive thickness. Such abrading can be obtained with an air- or air-liquid abrasive blast.

(c) Remove residual particles with the solvent in (a)

(d) Dry

Sodium-hydroxide etch[20]–

(a) Degrease in acetone

(b) Etch for 1 minute at 60°-90°C (140°-194°F) in:

Sodium hydroxide	5 pbw
Water	95 pbw

(c) Rinse in cold water

(d) Dry in hot air

Polyurethane[58]–This resin, like polyimide, is sometimes thermoplastic. It is formed from the reaction of a polyisocyanate with compounds containing a reactive hydrogen. The plastic may be flexible or rigid, or somewhere in between, and can be obtained in different densities and forms (i.e., sheets, molding and casting resins, etc.). Polyurethanes are unexcelled as cryogenic materials and have excellent adhesion and skid resistance, good chemical resistance, and superior impact resistance, but are limited at elevated temperatures (250°F or 121°C). Polyurethanes, or urethanes, as they are frequently called, also have good electrical properties. Cleaning usually involves light sanding and a dry bonding surface. A primer (urethane or silane) will usually improve adhesion. Polyurethane is usually considered to have better cohesive than adhesive properties. For this reason, the interface may be the "weak link." A suggested surface preparation is as follows:

(a) Wipe with acetone or MEK

(b) Abrade with 100-grit emery cloth, sandpaper, or steel wool

(c) Wipe with a clean dry cloth to remove grit and particles

(d) Wipe again with solvent in (a)

(e) Dry

(f) Use urethane or silicone primer to improve adhesion

Silicone Resins[2,80]—Silicone resins are provided in several forms. They are used as thermally stable electrical insulation resins, paint vehicles, molding compounds, laminates, impregnating varnishes, encapsulating materials, and as baked-on release agents. Recommended surface preparation is as follows:

(a) Wipe with acetone or other solvent listed in the general discussion under thermosets

(b) Sand-, grit-, or vapor-blast, or use steel wool

(c) Wipe with a clean dry cloth to remove grit and particles

(d) Repeat solvent wipe in (a)

Silicone laminates should be treated as covered under Reinforced Plastics/thermosets below.

Urea-formaldehyde (U-F)[2,58]—Like melamine-formaldehyde resins, these amino resins are hard, rigid, abrasion-resistant materials. They have excellent dimensional stability and good solvent and creep resistance. They are self-extinguishing and have superior electrical properties. They are noted for high impact strength and resistance to water and solvents. Only cellulose-filled resins are available. The following surface preparation method is recommended:

(a) Scrub with an abrasive household detergent

(b) Rinse with tap water, then deionized water

(c) Dry

(d) Sand

(e) Wipe with isopropyl alcohol

(f) Dry

(g) Prime or bond

The procedure given above for epoxies may be used.

Reinforced Plastics/Composites

This category may be broken down into two subdivisions: thermosets and thermoplastics. Until recently, the thermosets were by far the most important; thermoplastics are catching up fast.

Reinforced plastics/thermosets[1]–The reinforced thermosets of most importance are the glass-reinforced plastics (GRP). The methods most often used for preparing GRP laminates for bonding all involve removal of the original resin surface area yielding a roughened faying surface for bonding. Removal of surface gloss is usually sufficient to dispose of surface soils, exuded resin impurities, absorbed gases or vapors, and release agents used during the manufacture of the GRP. The methods of surface preparation most commonly used for laminated plastics are as follows:

Tear-ply (peel-ply) method[1]–During the manufacture of the GRP laminate, either one or both outer surfaces of the laminate are made of a layer of an adhesive material, the outer layer of which can later be stripped or peeled off easily. Nylon and Dacron® are frequently used for this purpose. This ply is cured as an integral part of the laminate. The resin used in the construction of the GRP adheres poorly to the nylon or Dacron cloth incorporated in the outer layer or layers, permitting the nylon or Dacron cloth to be peeled uniformly in one piece. The texture of the nylon or Dacron fabric layer is reproduced in the outer layer of the laminate. When the adhesive-bonding operation is to be performed, a thin knife is slid under a ply, and the desired distance peeled and trimmed. Loose particles are removed by brushing or blowing the surface with clean filtered air. There is very little chance of surface contamination with tear ply. The tear ply is applied only to the areas that are to be adhesive bonded. This technique is preferred over machine- or hand-sanding because of the reduced danger of surface contamination. Figure 4-3 shows how the tear ply is used.

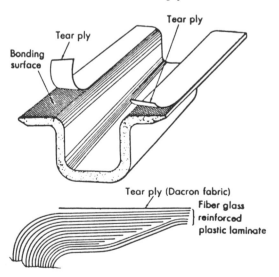

Figure 4-3: Structural reinforced plastic laminate with a tear ply.[2] (Used with permission of Palmerton Publishing Co., Inc.)

Sanding[1]—The surface of laminated plastics may be sanded lightly with medium-grit (80 to 120 grit) emery paper or other abrasive paper capable of roughening the surface without substantially damaging the reinforcing fibers. It is desirable to have the direction of bonding parallel to the surface fiber orientation in order to minimize damage to these fibers. This technique works best on large faying surfaces where the edges, rounded as a result of sanding, can be trimmed, or where the edge effects are negligible. Sanding dust may be removed with a short-haired stiff brush. If water-break-free surfaces are not obtained, the procedure should be repeated.

Grit-blasting or sandblasting may also be used if available. A specially-trained operator is needed to produce uniformity within pieces by sandblasting. A clean, uniform-size grit is essential for proper surface preparation with sandblasting. This may be a problem when treating GRP surfaces, and for this reason this technique is rarely used.

After sanding or sandblasting, the surface is sometimes wiped with solvent, such as MEK, acetone, toluene, trichloroethylene, Freon TF, Freon TMC, or M-17, depending on the known mold lubricants. In some cases, solvent is used before and after abrading the surface. If water-break-free surfaces are not obtained, the procedure should be repeated.

Picatinny Arsenal workers found that glass-reinforced plastic laminates prepared for bonding by hand- and machine-sanding can be stored up to 30 days at 73°F and 50% RH with no adverse effect on bond strength. Machine sanding gave slightly better results than hand sanding. Tear ply and sanding gave about the same results, but the tear-ply method provides less danger of surface contamination. Variations in bond strength are more likely to occur as a result of changes in sanding techniques than because of the difference in methods.

The Picatinny workers found that surface exposure time (SET) for GRP laminates varied with the adhesive used. In general, bond quality diminished with increased SET. The best overall adhesive evaluated was epoxy film adhesive, which was found to be the least sensitive to the method of surface preparation used. In general, the best results were obtained when GRP laminates were bonded within four hours after sanding. If absolutely necessary, bonding can be carried out after periods of time up to 14 days SET with only moderate strength loss.

Manual scouring[81]—In this method, the faying surface is scrubbed with tap water and an abrasive household-type cleaner to remove contaminants and release agents. Clean cloths or nonmetallic bristle brushes will facilitate the scouring. The surface is rinsed with clean running tap water followed by a rinse in distilled water, and then dried at 130°-150°F (54°-66°C). The parts should exhibit a water-break-free surface. If they do not, the above procedure should be repeated.

Solvent-soak and abrading[81]—If water-break-free surfaces cannot be obtained by tear-ply, sandblasting, or manual scouring, the plastic laminates should be treated as follows:

(a) Soak for 48 hours in reagent-grade acetone

(b) Dry for 3-4 hours at 190°-200°F (88°-104°C)

(c) Sand lightly with 200-grit sandpaper

(d) Remove particles by air-blast or vacuum

(e) Check for water-break-free condition. If achieved, the parts may be bonded. If not, resoak the laminates in acetone for an additional 24 hours minimum and repeat the rest of the procedure. Parts that do not pass the water-break-free test should be rejected.

Reinforced thermoplastics (glass-reinforced)[1]–Surface preparation of glass-reinforced thermoplastics is not necessarily similar to the technique prescribed for the base unreinforced resin. In general, there is a drop of up to 50% in the expected bond strength with glass reinforcement. The reason for this drop is not fully known, although it may be the result of the finishes used on the glass. In many cases solvent cementing will be used. Examples are aqueous phenol for glass-reinforced nylon 6 and methylene chloride for polystyrene.

Plastic Foams[2]

Plastic foams are generally treated by the same surface-preparation methods used for the base polymer. This is particularly true for the thermoplastic "structural" foams, such as polycarbonate, modified polyphenylene oxide, etc. In some cases these thermoplastic foams may be solvent cemented. The thermosetting foam most commonly encountered is polyurethane, which is available in rigid, semi-rigid, and flexible form. The rigid foams are obtainable in a wide range of densities. Obviously the denser grades, with their increased surface area of actual material exposed, will be easier to prepare by the usual technique of sanding lightly and then removing the resultant dust by vacuum or blowing-off. The cellular structure of plastic foams offers an advantage in surface preparation in that a physical interlocking of the two faying surfaces is possible in a foam-to-foam bond. With rigid polyurethane foams a conventional adhesive bond may not be necessary, since foam "poured-in-place" or "foamed-in-place" will adhere tenaciously to most substrates during curing.

SURFACE PREPARATION OF RUBBERS

Until recently the superiority of vulcanized bonded rubber over adhesively bonded rubber has been generally recognized. Almost all types of elastomers can be bonded to a large variety of substrates by the process known as vulcanization. The strength of similar bonds by adhesive bonding, or *post-vulcanization (PV) bonding,* as it is also called, was originally believed to be much weaker and less resistant to environmental exposure. Recent published work has demonstrated that this is not true, with some exceptions. The advantages of PV bonding are:[82]

- A variety of vulcanized elastomers can be bonded

- Common elastomer surface treatments can be used

- Bonded systems possess good environmental resistance

- Bonds are comparable to vulcanization bonding

- The process is applicable to rubber-to-metal parts as well as to rubber-to-rubber assemblies

- Large sizes or complicated shapes may be PV bonded, which would not be economically feasible by the "bond-in-the-press" technique

- Vulcanization bonding is impractical for field repair where the required equipment is usually not available

Table 4-4 shows bond values obtained with seven elastomers, comparing the bond values obtained with both vulcanization bonding and post-vulcanization (adhesive) bonding, using a 300°F (149°C) cure. In most cases the results are comparable. EPDM was the only elastomer where a significant decrease in bond strength was obtained by post-vulcanization bonding.

Table 4-4: Comparison of Vulcanization and Post-Vulcanization Bonding*[82]

	. . . . Tear Strength—Peel. . , , , , , .			
	Vulcanization .		Post-Vulcanization	
	(pli)	(N/m)	(pli)	(N/m)
Natural rubber	45	7,880	43	7,529
Natural rubber (B)	75	11,132	56	9,806
SBR	189	33,094	100	17,513
Neoprene	93	16,284	116	20,312
Butyl rubber	101	17,685	66	11,623
Nitrile rubber	95	16,635	136	23,814
EPDM	130	22,763	28**	4,903

*Adhesive used was Hughson Chemicals Chemlok 234B.
**Failure occurred at the rubber-cement interface. All other values were for rubber-tearing bonds.

Solvent washing and *abrading* are common surface treatments for most elastomers, but *chemical treatment* is often recommended for maximum strength and other properties. Many synthetic and natural rubbers require "cyclizing" with concentrated sulfuric acid until hairline cracks appear on the surface. Some rubbers require primers for optimum bonding.

Mechanical abrasion is usually accompanied by sanding or buffing the surface with 80-140 grit sandpaper or a buffing wheel. The dust from the buffing process, or other contamination, is usually removed by means of a clean cloth dipped in a suitable solvent. The solvent must be reasonably compatible with the chemical type of rubber being cleaned. If the solvent is very strongly incompatible with the rubber involved, or if too much is used, the rubber will swell excessively, may curl unacceptably, or may be degraded. Particular care must be taken not to trap solvent in a system that is totally closed, or else the rubber may be damaged. On the other hand, a mild wipe of a somewhat aggressive solvent may help tackify the rubber surface. Methyl ethyl ketone (MEK) and toluene are solvents commonly used for cleaning elastomers. MEK is a strong solvent for the fluoroelastomers, and excessive curling has been experienced in using MEK in any quantity for this type of elastomer. If such curling occurs, a more compatible solvent, such as toluene, may be substituted. Chlorinated solvents, such as trichloroethylene (TCE), perchloroethane, and 1,1,1-trichloroethane are also used. Cyclization, mentioned above, and chlorination are other commonly used surface-preparation methods and will be described below under Neoprene.

Recent work has shown that the bondability of elastomers is a function of polarity. The less polar EPDM and butyl elastomers are more difficult to bond than the more polar nitrile, neoprene, SBR, and natural rubber.[82]

The surface-preparation methods for the various rubbers are given below.

Neoprene (Polychloroprene) (CR)[1]

Neoprene is the generic name for polymers of chloroprene (2-chloro-1,3-butadiene) manufactured since 1931 by Du Pont. There are 22 types of solid neoprene and 13 types of neoprene latex.

Abrasive Treatment[1,2] —

(a) Scrape the surface with a sharp blade to remove gross layers of wax, sulfur, and other compounding ingredients which may have floated to the surface

(b) Solvent-wipe with ethyl, isopropyl, or methyl alcohol, MEK, or toluene

(c) Uniformly abrade surfaces with 80-120 grit abrasive paper. Machine sanding with a "jitterbug" oscillating sander is preferred over hand sanding because the machine sanding produces a more uniform surface with less effort. A buffing wheel may also be used.

(d) Solvent-wipe again as in (b) to remove particles

Cyclization[2,83]—This process is often applied preparatory to bonding with flexiblized epoxy adhesives. It has been used successfully in many rubbers, in addition to neoprene. A commonly used cyclization procedure is as follows:

(a) Scrape surface with a sharp blade to remove gross layers of wax, sulfur, and other compounding ingredients which may have floated to the surface

(b) Solvent-wipe with ethyl, isopropyl, or methyl alcohol, MEK, or toluene

(c) Immerse rubber surface in conc. sulfuric acid (sp. gr. 1.84) for 5-45 minutes. The optimum time must be determined experimentally

(d) Rinse thoroughly with tap water, preferably hot

(e) Rinse thoroughly with distilled water

(f) Neutralize by immersing for 5-10 minutes in a 10-20% solution of ammonium hydroxide (sodium hydroxide is also used)

(g) Rinse thoroughly with tap water

(h) Rinse thoroughly with distilled water

(i)   Dry at temperatures up to 150°C)

(j)   Flex the resultant brittle surface of the rubber with clean rubber or plastic gloves so that a finely-cracked appearance is produced

Cyclizing hardens and slightly oxidizes the surface, thereby permitting the necessary wetting of the adhesive bond. The finely cracked surface indicates that the rubber is ready for bonding. Light lacy lines on the surface indicate insufficient immersion time. Deep, coarse cracks and a thick crusty surface indicate excessive immersion. If immersion is not feasible, the acid may be made up into a thick paste by the addition of barium sulfate (barytes) or Cab-O-Sil (G.L. Cabot Corp., Boston, MA) or use 100 pbw conc. sulfuric acid: 5 pbw Santocel C (Monsanto Co.), or equivalent. A stainless steel or other acid-resistant spatula should be used to apply the paste. After the paste is applied, the rest of the treatment [steps (d) through (j)] should be carried out.

Cyclization (Boeing-Vertol Modification)[84]–Boeing-Vertol, in a Picatinny Arsenal-sponsored study, developed a much milder method, as follows:

(a)   Clean in toluene

(b)   Force-dry 1 hour at 140°F (60°C)

(c)   Immerse 2 minutes in conc. sulfuric acid

(d)   Rinse in tap water

(e)   Force-dry 1 hour at 140°F (60°C)

Chlorination[2]–

(a)   Scrub neoprene in 120°F (49°-60°C) nonionic detergent solution (2-3% by weight)

(b)   Rinse thoroughly in tap water

(c)   Rinse thoroughly in distilled water

(d)   Air-dry

(e)   Immerse 1½ to 3 minutes in the following solution at room temperature (prepare solution just prior to use, adding ingredients in the order listed):

Distilled water	97.0 pbw
Sodium hypochlorite (as Clorox or Purex)	0.3 pbw
Hydrochloric acid (sp. gr. 1.20)	0.3 pbw

(f)   Rinse thoroughly with distilled water and dry at temperatures up to 150°F (66°C)

Activated gas plasma–This method, described generally above under Activated-Gas Surface Treatment of Polymers, has been used successfully with neoprene. The best results were obtained by Picatinny Arsenal studies with ammonia, air and nitrogen plasmas at 30 minutes exposure.[85]

Ethylene-Propylene-Diene Terpolymer (EPDM)

This rubber, along with the copolymer of ethylene and propylene (EPM), forms a classification called *ethylene-propylene rubbers*. However, since EPDM, the terpolymer, is in much greater use than EPM, the former is frequently called ethylene-propylene rubber. Its outstanding property is very high resistance to ozone and weathering. Commercial production was begun in 1963, and this is now considered one of the fastest growing rubbers. Being less polar, EPDM is relatively difficult to bond.[61]

Abrasive treatment[2]–As in Neoprene; acetone or MEK are recommended as solvents.

Cyclization[1]–As in Neoprene.

Silicone Rubber (Polydimethylsiloxane)

These rubbers are completely synthetic materials with wide applications for many uses. They comprise two types, heat-vulcanizing and room-temperature vulcanizing (RTV). Both types have unique properties unobtainable with organic rubbers, particularly where superior endurance and extended life expectancy are required. Silicone elastomers maintain their usefulness from -150° to 500°F (-101° to 316°C). Resistance to oxidation, oils and chemicals is high and stability against weathering is good.[61]

Solvent cleaning[2]–

(a) Sand with a medium-grit sandpaper

(b) Solvent-wipe surfaces with acetone, MEK, ethyl, methyl, or isopropyl alcohol, or toluene

Soap-and-water wash[62]–A simple washing with a mild (Ivory) soap and water, followed by thorough rinsing, was found to give satisfactory results in a Picatinny Arsenal study.

Primers[9,86]–Priming the silicone rubber with Hughson Chemical Company's Chemlok 607 adhesive in methanol solvent has been found to give good results after cleaning with acetone or methanol. The primer should be dried at 395°F (200°C) for 10-15 minutes. The Chemlok 607 can also be used as the adhesive following the priming. Chemlok silanol-type primers have also been developed. Other satisfactory primers include Union Carbide's Type AP-133, supplied in denatured alcohol at 5% solids. This primer should be air-dried at room temperature for 15-40 minutes before bonding.

Activated-gas plasma[87]–This method, described generally above under Activated Gas Surface Treatment of Polymers, has been used successfully with RTV vinyl-addition-type silicone rubber, using excited oxygen plasma. This treatment permitted the use of an epoxy adhesive for bonding. Usually cured RTV silicones are considered unbondable by conventional adhesives such as epoxies.

Butyl Rubber (IIR)

Butyl rubber has been commercially produced since 1942 and is a well-established specialty elastomer with a wide range of applications. It is a copolymer of isobutylene and isoprene. Its special properties are: low gas permeability, thermal stability, ozone and weathering resistance, vibration damping, and higher coefficient of friction and chemical and moisture resistance.[61] Being less polar, butyl rubber is relatively difficult to bond.

Abrasive treatment[2]–As in Neoprene; toluene is recommended as a solvent.

Cyclization[2]–As in Neoprene. Some authorities do not recommend cyclization for butyl rubber.[83,88]

Chlorination[2]–As in Neoprene. Sharpe[83] does not recommend chlorination for butyl rubber.

Primers–Prime with butyl rubber adhesive in an aliphatic solvent,[22] or use silicone primers such as Union Carbide's AP-133, Y-4310, or Y-5042.

Activated-gas plasma–This method, described generally above under Activated-Gas Surface Treatment of Polymers, has been used successfully with butyl rubber. Very good results were obtained with air and nitrogen plasmas at 30 minutes in Picatinny Arsenal studies.[85]

Chlorobutyl Rubber (CIIR)

This modified butyl rubber was introduced in 1960. It contains about 1.2 wt % of chlorine and has greater vulcanization flexibility and enhanced cure compatibility than other general-purpose elastomers.[1]

Abrasive treatment[2]–As in Neoprene.

Cyclization[2]–As in Neoprene.

Chlorination[2]–As in Neoprene.

Other methods[2]–See Table 4-5 for a comparison of results obtained by a number of different surface-preparation methods.

Table 4-5: Peel Strength of Epoxy-Polyamide Resin Adhesive to Chlorobutyl Rubber Pretreated by Various Methods[2]

Surface Treatment	lb/in width	N/m
Cyclization	18.0	3152
Solvent-clean and chlorinate	14.0	2452
Solvent-clean, abrade and chlorinate	16.0	2802
Detergent-clean and chlorinate	30.0*	5254
Ethanol-wipe	7.0	1226
Abrade and ethanol-wipe	7.0	1226
MEK-wipe	6.0	1051
Abrade and MEK-wipe	6.5	1138

*Broke rubber.

Chlorosulfonated Polyethylene (CSM) (Hypalon® –Du Pont Co.)

This synthetic rubber, also known as chlorosulfonyl polyethylene, introduced in 1952, is characterized by ozone resistance, light stability, heat resistance, weathering, resistance to deterioration by corrosive chemicals, and good oil resistance. Presently available types contain from 25–43% chlorine and 1.0–1.4% sulfur.[1]

Abrasive treatment[2]–As in Neoprene; acetone or MEK are recommended as solvents.

Primers[84]–

(a) Wipe with toluene

(b) Force-dry 1 hour at 140°F (60°C)

(c) Prime with an adhesive primer such as 3M Co. Scotch-Cast XR-5001 or Bloomingdale BR 1009-08 (polyamide epoxy in solvent blend)

Nitrile Rubber (Butadiene-Acrylonitrile) (NBR)

Nitrile rubbers are copolymers of butadiene and acrylonitrile and are frequently referred to as Buna N. Their properties vary with the acrylonitrile content. Nitrile rubbers exhibit a high degree of resistance to attack by oils at both normal and elevated temperatures.[1]

Abrasive treatment[2]–As in Neoprene; methanol or toluene are recommended as solvents.

Cyclization[2]–As in Neoprene, except that the exposure should be for 10–15 minutes.

Chlorination[2]–As in Neoprene.

Primers[1,84,89]–Primers have been used successfully with Buna N rubber. Those that have been particularly recommended include:

Hughson's Chemlok AP-131 Primer for bonding to steel

Hughson's Chemlok 205 (adhesive) for bonding to aluminum

Hughson's Chemlok 220 (adhesive) for bonding to aluminum

3M Company's Scotchcast XR-5001 for bonding to aluminum

Depending on the adhesive to be used, better results may be obtained by using a toluene wipe, with or without sanding, before priming. One recommended procedure is as follows:[1]

(a) Wipe with toluene

(b) Force-dry at 140°F (60°C)

(c) Prime with an adhesive primer such as those listed above

Activated-gas plasma–This method, described generally above under Activated Gas Surface Treatment of Polymers, has been used successfully with Buna N rubber. The best results were obtained with ammonia, air, and nitrogen plasmas at 30 minutes in Picatinny Arsenal studies.[85]

Polyurethane Elastomers[61]

This category covers thermosetting materials derived from the reaction of an isocyanate with a hydroxy compound. Their unique elastomeric properties are: exceptionally high abrasion resistance at moderate temperatures, excellent oil and solvent resistance, very high tear and tensile strength, and high hardness with good mechanical strength. Both polyester (AU) and polyether (EU) types are available, as are thermoplastic polyester urethane. They are usually completely soluble in strong polar solvents such as tetrahydrofuran and dimethyl formamide, and at high temperatures they can be molded, extruded or calendered.[61] Treatments are given below.

Abrasive treatment—As in Neoprene; methanol is recommended as a solvent.

Primers[89]—A number of primers can be used, including Hughson's Chemlok primers AP-134, AP-131, 4-5224, and Union Carbide's A 1100 (a chlorosilane). Methanol can be used as a degreasing agent prior to priming.

Synthetic Natural Rubber (Polyisoprene) (IR)

This rubber approximates the chemical composition of natural rubber (NR). Its predominant structure is cis-1,4-polyisoprene, the same as that of natural rubber. In general, the synthetic polyisoprenes are lower in modulus (more flexible) and higher in elongation than the natural rubber. Synthetic polyisoprene was first introduced commercially in 1960.[61] Suggested surface treatments are as follows:

Abrasive treatment[2]—As in Neoprene; methanol or isopropanol are suggested as solvents.

Cyclization[2]—As in Neoprene, except that exposure should be for only 5-10 minutes, since this is not an acid-resistant rubber. Some workers suggest much shorter periods (10–45 seconds).

Chlorination[2]—As in Neoprene.

Styrene-Butadiene Rubber (SBR) (Buna S)

This type of rubber, once called GR-S, is one of the most important synthetic rubbers. It is a copolymer of styrene and butadiene.[61] Treatments are as follows:

Abrasive treatment[2]—As in Neoprene; toluene is recommended as a solvent. Excessive toluene, however, causes the rubber to swell. A 20-minute drying will restore the part to its original dimensions.

Cyclization[2]—As in Neoprene, except that the exposure time should be 10-15 minutes.

Chlorination[2]—As in Neoprene.

Primers[20,89]—Recommended primers include butadiene-styrene adhesive in aliphatic solvent and Hughson Chemical Company's Chemlok 205 adhesive in methyl isobutyl ketone (MIBK) diluent.

Activated-gas plasma[85]—This method, described generally above under Activated-Gas Surface Treatment of Polymers, has been used with some success with Buna S rubber. The best results were obtained using helium, air, and nitrogen plasmas in Picatinny Arsenal studies.

Polybutadiene (Butadiene Rubber) (BR)

Although this type of rubber was first produced in Europe in the early 1930's, it was almost unknown in the U.S. until 1960. Presently it is second to SBR in usage.[61] Treatments are as follows:

Abrasive treatment[2]—As in Neoprene; methanol is recommended as a solvent.

Cyclization[2]—As in Neoprene, except that the exposure time should be 10-15 minutes, since this is a non-acid resistant rubber.

Chlorination[2]—As in Neoprene.

Solvent wipe[20]—A simple solvent wipe with methanol may be used for general-purpose bonding.

Fluorosilicone Elastomers

These rubbers have become the second largest in volume of the fluoroelastomer types. The non-silicone-containing fluoroelastomers are discussed below. Fluorosilicone rubbers retain most of the useful qualities of the regular silicone rubbers and have improved resistance to many fluids, except for ketones and phosphate esters. They are most useful when low-temperature resistance is required, in addition to fluid resistance.[61] A recommended treatment is a *solvent wipe* with methyl, ethyl, or isopropyl alcohol, or with toluene.

Fluorocarbon Elastomers

These materials, also called fluoroelastomers, include Kel-F elastomers and Fluorel (both 3M Company). Viton (Du Pont Company) and Fluorel are copolymers of vinylidene fluoride and hexafluoropropylene (FPM). Kel-F is a copolymer of chlorotrifluoroethylene and vinylidene fluoride and is called polychlorotrifluoroethylene (CFM). It was first developed in 1954. Viton and Fluorel are the most important members of this group today. These elastomers have excellent resistance to ozone, oxidation, weathering, heat, aliphatic and aromatic hydrocarbons, and alcohols. They are also highly impermeable to gases, have good strength, electrical resistivity, and resistance to abrasion, water, acids, and halogenated hydrocarbons. This type usually includes the fluorosilicone elastomers which are discussed above.[61] Treatments suggested are as follows:

Sodium etch[90]—For optimum bond strength use this procedure, described above under Ethylene-Chlorotrifluoroethylene Copolymer (E-CTFE).

Dry abrasion[90]—For relatively low adhesion with low environmental resistance:

    (a) Wipe or spray with, or immerse in 1,1-trichloroethane, Freon TMC or M-17, acetone, MEK, toluene, ethyl or isopropyl alcohol

    (b) Abrade lightly and uniformly with 180-320 grit abrasive paper

    (c) Repeat (a)

Epichlorohydrin Elastomer (Hydrin—B.F. Goodrich Chemical Co.) (Herclor—Hercules, Inc.)

This elastomer homopolymer (CO) and the copolymer (ECO) combine exceptional resistance to aromatic and halogenated hydrocarbons with high resistance to ozone, weathering, gas permeability, compression set, compact and tear. These compounds have good tensile strength, resilience, and resistance to abrasion, water, acids, and alkalies. They also have good low-temperature and heat resistance. Epichlorohydrin elastomers of both types require a simple cleaning with acetone (*not* MEK), alcohols, or aromatics, such as toluene. No abrasion is necessary.[61]

Polysulfide Rubber (PTR)

The first introduction of a polysulfide rubber for commercial application was in 1930 when Thiokol Chemical Company developed Thiokol Type A. Since then a number of other types of Thiokol have been developed. The types commonly used now are Types FA and ST. Thiokols are highly impermeable to gases, and have excellent resistance to ozone, oxidation, weathering and aliphatic and aromatic hydrocarbons.[61] Treatments are as follows:

Abrasive treatment[2]—As in Neoprene; methanol is recommended as a solvent.

Chlorination[20]—

(a) Degrease in methanol

(b) Immerse overnight in strong chlorine water

(c) Wash

(d) Dry

Snogren,[2] however, does not recommend chlorination for treating polysulfide.

Primers—Silicone primers recommended by Hughson Chemicals for use with polysulfide sealants include Hughson's Chemlok AP-131, Y-4310, and Y-5254.[89]

Polypropylene Oxide (Propylene Oxide Rubber) (PO) (PAREL 58—Hercules, Inc.)

This elastomer is a sulfur-vulcanizable copolymer of propylene oxide and allyl glycidyl ether. Its vulcanizates are particularly attractive for dynamic uses where high resilience, excellent flex life, and flexibility at extremely low temperatures are required. It performs similarly to natural rubber in these applications, but, in addition, has the added advantages of: (1) good resistance to aging at high temperatures, (2) good ozone resistance, and (3) moderate resistance to loss of properties in contact with fuels, and/or some solvents. The surfaces of this elastomer should be prepared by solvent wiping with TCE, xylene, toluene, or other appropriate solvent.[91]

Polyacrylate (Polyacrylic Rubber) (AMC) (ANM)

Polyacrylic elastomers are noted primarily for their high resistance to heat, ozone, oxidation, weathering, aliphatic hydrocarbons, and sulfur-bearing oils. They also have good resilience and gas impermeability and moderate strength. Abrasion resistance is not as good as with nitrile rubber or SBR.[61] Surface preparation is as follows:

Dry abrasion[90]—As in Fluorocarbon Elastomers above; methanol has also been used as a solvent.

Thermoplastic Rubber (Thermoplastic Elastomer)[1,92]

This is a new class of polymers combining the end-use properties of vulcanized elastomers with the processing advantages of thermoplastics. Because of their unique molecular configurations they may be processed with the same techniques utilized with other thermoplastics, but the mechanical properties of the final articles are essentially indistinguishable from those of similar articles fabricated from conventional vulcanized elastomers.

There are several types, including polyester (Hytrel–Du Pont Co.), polystyrene-butadiene-polystyrene block copolymers (Kraton–Shell Development Co.), polystyrene-isoprene-polystyrene block copolymers (Solprene–Phillips Chemical Co.), polyolefin (TPR thermoplastic rubber–Uniroyal, Inc.), (Somel thermoplastic elastomer–Du Pont Co.), and urethane, mentioned above under Polyurethane Elastomers.

Since these materials are thermoplastic and may, therefore be soluble in a wide range of organic solvents, it is quite possible that solvent cementing would be the most appropriate method of bonding the thermoplastic rubbers to themselves or to mutually-compatible plastics. Solvent cementing might work with Kraton, but not with TPR and Hytrel, which do not have reactive bonding sites. In cases where the thermoplastic rubber must be bonded to metals or other non-plastic or non-rubber materials, neither solvent cementing nor fusion techniques can be considered. In such cases, conventional adhesives must be used. In the case of Hytrel polyester elastomer, Du Pont recommends cleaning with MEK and then using commercial primers. Abrasion is not recommended. Polyurethane rubber, on the other hand, has reactive sites and is cementable with solvents.

SURFACE PREPARATION OF WOOD AND WOOD PRODUCTS[93]

Careful machining is essential in preparing wood for gluing. To obtain the strongest joints the wood surfaces should be machined smooth and true with sharp tools, and be essentially free from machine marks, chipped or loosened wood grain, and other surface irregularities. To obtain uniform distribution of gluing pressure, each lamination or ply of plywood should be of uniform thickness. When possible, machining should be carried out immediately preceding gluing, so that the surfaces will be kept clean and are not distorted by moisture changes. Where the four sides of a piece are to be glued, it is preferable to glue in two operations and machine just before each operation.

Surfaces made by saws are usually rougher than those made by planers, jointers, and other machines operated with cutter heads. Modern saws, freshly sharpened, well aligned, and skillfully operated are capable of producing surfaces adequate for gluing many products without further preparation, thereby providing a saving in time and labor. However, where the saws are not well-maintained, the resultant sawed surfaces provide glue joints that are weaker and more conspicuous than those between well-planed or jointed surfaces. Consequently, if inconspicuous glue joints of maximum strength are required, planed or jointed surfaces are generally more reliable. Machine marks caused by feeding the stock through a planer too fast for the speed of the knife prevent complete contact of the joint faces when glued. Machine marks in cores of thinly-veneered panels are likely to show through the finished surface. Unequal thickness or width, which causes uneven distribution of gluing pressure and usually results in weak joints, may be due to the grinding, setting or wearing of machine knives. Knives that are dull or improperly set or ground may produce a burnished surface that interferes with gluing or formation of the strongest glue bonds.

Wood surfaces are sometimes intentionally roughened by tooth planing, scratching, or *sanding* with coarse sandpaper, in the belief that rough surfaces provide better glued joints. However, comparative strength tests at the USDA Forest Products Laboratory failed to show such an advantage. Also, studies of the penetration of glue into the wood have shown the theoretical benefit of the roughened surfaces to be improbable. *Light sanding*, on the other hand, has been helpful in preparing for gluing such surfaces as resin-impregnated wood laminated paper/plastic, plywood that has been glazed from dull tools, or by being pressed excessively against smooth, hard surfaces.

Significant developments in sanding equipment have been reported in recent years. Advantages of so-called abrasive planing in preparing wood for gluing produce deeper cuts in a single pass, close tolerances and improved surface quality for gluing.[93]

SURFACE PREPARATION OF MISCELLANEOUS MATERIALS

Asbestos (Rigid)[20]

(a) Abrade with 100-grit emery cloth

(b) Remove dust

(c) Degrease in acetone

(d) Dry in air to allow solvent to evaporate

An alternative procedure is to prime with diluted adhesive or low-viscosity resin ester.

Brick and Fired Non-Glazed Building Materials[3]

(a) Degrease with methyl ethyl ketone

(b) Abrade surface with a wire brush

(c) Remove all dust and contaminants

Carbon and Graphite (For General-Purpose Bonding)[2]

(a) Abrade with 220-grit emery cloth

(b) Remove dust

(c) Solvent degrease in acetone

Glass (Non-Optical)[3,20]

Abrasive treatment (for general-purpose bonding)—

(a) Grit-blast with aluminum oxide or carborundum and water slurry (1 part by volume of 220 to 325-grit slurry of aluminum oxide or carborundum to 3 parts by volume of distilled water)

(b) Degrease in acetone or detergent

(c) Dry 30 minutes at 100°C (212°F). (The drying process improves the bond strength.)

(d) Apply the adhesive before the glass cools to room temperature

Acid etch (for maximum strength)—

(a) Clean in acetone or detergent

(b) Immerse for 10-15 minutes at 20°C (68°F) in:

Sodium dichromate	7 pbw
Sulfuric acid, conc.	400 pbw
Water	7 pbw

(c) Rinse in tap water

(d) Rinse in distilled water

(e) Dry thoroughly

Primers[2]—Adhesion to clean glass may be promoted by the use of silicone primers. The selection of primers depends on the particular adhesive system used. The addition of silane additives to the adhesive system also improves adhesion to glass.

Glass (Optical)[6]

Optical glass should never be subject to any acid or alkaline etching or leaching action.

(a) Clean in ultrasonic equipment with a detergent solution, water, alcohol sequence

(b) Air- or oven-dry at less than 40°C

If the optical glass is to be stored for any length of time, glass containers such as petri dishes should be cleaned and dried, using the above sequence. The optical glass can then be safely stored in the cleaned glass container.

Ceramics (Unglazed)[2,20]

For unglazed ceramics such as alumina, silica, etc.:

(a) Grit-blast with aluminum oxide or carborundum and water slurry (1 part by volume of 220-325 grit slurry of aluminum oxide or carborundum to 3 parts by volume of distilled water

(b) Degrease in acetone or detergent

Ceramics (Glazed)[20]

For glazed ceramics such as porcelain:

(a) Solvent degrease in acetone or wash in warm aqueous detergent

(b) Rinse

(c) Dry

Concrete

Portland cement type[2]–

(a) If the concrete is contaminated with oil or grease, scrub with a detergent solution (approx. 2% in water) and thoroughly rinse, or clean with any appropriate solvent.

(b) If the concrete is contaminated with old paint, plant growth, soot or other soil, it should be *dry-grit-blasted* with a non-metallic grit, such as flintstone, silica, silicone carbide, or aluminum oxide. For old concrete, the surface should be blasted or cut down until good solid material is exposed (one-sixteenth inch or more). If there is a heavy layer of loose material it is advisable to first remove the bulk of this material by vacuuming or sweeping. Dust and loose particles should be cleaned off by blowing with clean filtered air.

(c) If grit-blasting is not used, the cement should be *acid etched* with a solution of 15% by weight of concentrated hydrochloric acid in water. One gallon of this solution will treat 5 sq. yds (4.2 sq. meters).

Bituminous type[2]–

(a) Scrub with a 2% by weight solution of detergent

(b) Rinse with a high-pressure hose until the surface no longer feels slippery

(c) In cases where limestone, dolomite, or other carbonate aggregate is present in the bituminous concrete, improved adhesion will be obtained by using the acid-etch solution outlined above for portland cement concrete

(d) In cases where an excessively heavy cake of oil and grime has been deposited, use a combination of a detergent wash and mechanical cleaning

Painted Surfaces[2,3]

To obtain maximum-strength bonds the paint surface must be removed. Even if the bond between the adhesive and the two surfaces, paint and the other adherend, is good, the strength of the joint will be no better than the strength of the bond between the surface that has been painted and the paint. For temporary or other bonds that do not require maximum strength, clean the painted surface with a detergent solution, abrade with a medium-grit emery cloth, and wash again with detergent. Alternate procedures are as follows after loose paint has been removed:

(a) Solvent cleaning[2]–Immerse, spray, or wipe in either of the following solutions, which degrease the surface and leave a dull finish to the paint:

Formulation No. 1

Methylene chloride	15% by volume
n-Butanol	35% by volume
Mineral spirits	48% by volume
Methyl ethyl ketone	2% by volume

Formulation No. 2

Equal parts of methylene chloride, n-butanol, isopropyl alcohol, toluene, and methyl acetate

(b) Scrub surfaces with an alkaline cleaner, such as a solution of 2-4% by weight of trisodium phosphate (TSP) in hot water. Rinse with clean water.

REFERENCES

1. Landrock, A.H., *Processing Handbook on Surface Preparation for Adhesive Bonding*, Picatinny Arsenal Technical Report 4883, Picatinny Arsenal, Dover, NJ (December 1975).
2. Snogren, R.C., *Handbook of Surface Preparation*, Palmerton Publishing Co., New York, NY (1974).
3. Petrie, E.M., Plastics and Elastomers as Adhesives, *Handbook of Plastics and Elastomers*, (C.A. Harper, ed.), McGraw-Hill, NY (1975).
4. Kinloch, A.J., Introduction, *Durability of Structural Adhesives*, (A.J. Kinloch, ed.), Applied Science Publishers, London and NY (1983).
5. Akers, S.C., The Function of Adhesive Primers in Adhesive Bonding of Aircraft Structures, pp 23-28 in Applied Polymer Symposia No. 19, *Processing for Adhesives Bonded Structures*, (M.J. Bodnar, ed.), Interscience Publishers, NY (1972).
6. DeLollis, N.J., *Adhesives, Adherends, Adhesion*, Robert E. Krieger Publishing Co., Huntington, NY, (1980). (This book is the 2 nd edition of a 1970 publication under the name *Adhesives for Metals—Theory and Technology*.)
7. Schonhorn, M. and Hansen, R.H., A New Technique for Preparing Low Surface Energy Polymers for Adhesive Bonding, Polymer Letters, *Journal of Polymer Science*, B, vol. 4, pp 203-209 (1966).
8. Dukes, W.A. and Kinloch, A.J., Preparing Low-Energy Surfaces for Bonding, *Developments in Adhesives*—1 (W.C. Wake, ed.), Applied Science Publishers, London (1977).
9. Hall, J.R., et al., Activated Gas Plasma Surface Treatment of Polymers for Adhesive Bonding, *Journal of Applied Polymer Science*, vol. 13, pp 2085-2096 (1969).
10. Ross, M.C., *Activated Gas Plasma Surface Treatment of Newer Structural Plastics for Adhesive Bonding*, U.S. Army Armament Research and Development Command, Large Caliber Weapon System Laboratory, Technical Report ARLCD-TR-77088, ARRADCOM, Dover, NJ (August 1978).
11. Zisman, W.A., Influence of Constitution of Adhesion, *Handbook of Adhesives*, 2nd Edition (I. Skeist, ed.), Van Nostrand Reinhold, NY (1977).
12. Wegman, R.F., et al., *Effects of Varying Processing Parameters in the Fabrication of Adhesive-Bonded Structures—Part 2, Important Considera-*

tions for the Bonding Process, Picatinny Arsenal Technical Report 3999, Picatinny Arsenal, Dover, NJ (July 1970).

13. Tanner, W.C., Manufacturing Processes with Adhesive Bonding, pp 1-21 in Applied Polymer Symposia No. 19, *Processing for Adhesives Bonded Structures*, (M.J. Bodnar, ed.), Interscience Publishers, NY (1972).

14. American Society for Testing and Materials (ASTM), ASTM D 2651-79, Standard Recommended Practice for Preparation of Metal Surfaces for Adhesive Bonding, vol. 15.06, *1983 Annual Book of ASTM Standards.*

15. Russell, W.J. and Garnis, E.A., *Chromate-Free Method of Preparing Aluminum Surfaces for Adhesive Bonding: An Etchant Composition of Low Toxicity*, U.S. Army Armament Research and Development Command, Large Caliber Weapon Systems Laboratory, Technical Report ARLCD-TR-78001, ARRADCOM, Dover, NJ (May 1978).

16. Russell, W.J., Chromate-Free Process for Preparing Aluminum for Adhesive Bonding, pp 105-117 in *Journal of Applied Polymer Science*, Applied Polymer Symposia No. 32 (M.J. Bodnar, ed.), Interscience Publishers (1977).

17. Rogers, N.L. and Russell, W., *Evaluation of Nonchromated Etch for Aluminum Alloys (P2-Etch)*, U.S. Army Armament Research and Development Command, Large Caliber Weapon Systems Laboratory, Contractor Report (Bell Helicopter Textron), ARLCD-CR-80008, ARRADCOM, Dover, NJ (April 1980).

18. American Society for Testing and Materials (ASTM), ASTM D 3933-80, Standard Practice for Preparation of Aluminum Surfaces for Structural Adhesives Bonding (Phosphoric Acid Anodizing), vol. 15.06, *1983 Annual Book of ASTM Standards.*

19. Brewis, D.M., Aluminum Adherends, *Durability of Structural Adhesives*, J. Kinloch, ed.), Applied Science Publishers, London and New York (1983).

20. Shields, J., *Adhesives Handbook*, 2nd Edition, Newnes-Butterworth, London (1976). (The 3rd Edition was published by Butterworths, London, in 1984.)

21. Rogers, N.L., Surface Preparation of Metals for Adhesive Bonding, pp 327-340 in Applied Polymer Symposia No. 3, *Structural Adhesives Bonding*, (M.J. Bodnar, ed.) Interscience Publishers (1966).

22. Guttman, W.H., *Concise Guide to Structural Adhesives*, Reinhold, NY (1961).

23. Cagle, C.V., Surface Preparation for Bonding Beryllium and Other Adherends, *Handbook of Adhesive Bonding*, (C.V. Cagle, ed.), McGraw-Hill, NY (1973).

24. American Society for Metals (ASM), Metals Handbook, 9th edition, vol. 2, *Properties and Selections: Nonferrous Alloys and Pure Metals* (1979).

25. Military Specification, MIL-M-45202C, Magnesium Alloy, Anodic Treatment of (April 1, 1981).

26. Military Specification, MIL-M-3171C, Magnesium Alloy, Processes for Pretreatment and Prevention of Corrosion on (March 14, 1974).

27. *Heating and Pickling Huntington Alloys*, Huntington Alloy Products Division, The International Nickel Co., Inc., Huntington, WV (1968).

28. Keith, R.E., et al., *Adhesive Bonding of Nickel and Nickel-Base Alloys*, NASA TMX-53428 (October 1965).

29. Devine, A.T., *Adhesive Bonded Steel: Bond Durability as Related to Selected Surface Treatments*, U.S. Army Armament Research and Development Command, Large Caliber Weapon Systems Laboratory, Technical Report ARLCD-TR-77027 (December 1977).

30. Brockman, W., Steel Adherends, *Durability of Structural Adhesives* (A.J. Kinloch, ed.), Applied Science Publishers (1983).
31. Vazirani, H.N., Surface Preparation of Steel for Adhesive Bonding, *Journal of Adhesion* 1:222–232 (July 1969).
32. Keith, R.E., et al., *Adhesive Bonding of Stainless Steels Including Precipitation Hardening Stainless Steels*, NASA TMX-53574 (April 1968). (Available from NTIS as AD 653 526.)
33. Slota, S.A. and Wegman, R.F., *Durability of Adhesive Bonds to Various Adherends*, Picatinny Arsenal Technical Report 4917 (June 1976).
34. Atkins, R.W., et al., Explosives Research and Development Establishment, Ministry of Defence (UK), Waltham Abbey, Essex, England, *An Investigation into the Influence of Surface Pre-Treatment by Particle Impact on the Strength of Adhesive Joints Between Like Steel Surfaces*, ERDE-TR-120 (January 1973). (Available from NTIS as AD 771 003).
35. Hughson Chemical Co., Division of Lord Corporation, Erie, PA, *Preparation of Substrates for Bonding*, Technical Bulletin 7101A (May 1970).
36. Military Specification, MIL-A-9067C, Adhesive Bonding, Process and Inspection Requirements for (16 March 1961). (This specification was cancelled for Air Force use by Notice 1, dated 9 December 1974.)
37. Wegman, R.F., et al., *Evaluation of the Adhesive Bonding Processes Used in Helicopter Manufacture, Part I. Durability of Adhesive Bonds Obtained as a Result of Processes Used in the UH-1 Helicopter*, Picatinny Arsenal Technical Report 4186, Picatinny Arsenal, Dover, NJ, September 1971.
38. Mahoon, A., Titanium Adherends, *Durability of Structural Adhesives*, (A.J. Kinloch, ed.), Applied Science Publishers, London and NY (1983).
39. Hohman, A.E., and Lively, G.W., *Surface Treatment of Titanium and Titanium Alloys*, U.S. Patent 3,891,456, filed October 17, 1973.
40. Wegman, R.F., et al., *Effect of Titanium Surface Pretreatments on the Durability of Adhesive Bonded Joints—Part I*, U.S. Army Armament Research and Development Command, Fire Control and Small Caliber Weapons Laboratory, Technical ARSCD-TR-83005, ARRADCOM, Dover, NJ (May 1983).
41. Wegman, R.F., et al., *Effect of Titanium Surface Pretreatments on the Durability of Adhesive Bonded Joints—Part II*, U.S. Army Armament Research and Development Command, Fire Control and Small Caliber Weapons Laboratory, Technical ARSCD-TR-83005, ARRADCOM, Dover, NJ (April 1983).
42. Beemer, R.D., Introduction to Weld Bonding, *SAMPE Quarterly*, 5:37–41 (October 1973).
43. Kizer, J.A., and Grosko, J.J., Development of the Weldbond Joining Process for Aircraft Structures, pp 353–370 in Applied Polymer Symposia No. 19, *Processing for Adhesives Bonded Structures*, (M.J. Bodnar, ed.), Interscience Publishers, NY (1972).
44. Bowen, B.B., et al., *Improved Surface Treatments for Weldbonding Aluminum*, AFML-TR-159 (October 1976).
45. Wu, K.C. and Bowen, B.B., Advanced Aluminum Weldbonding Manufacturing Methods, Preprint Book, pp 536–554, and 22nd National SAMPE Symposium, vol. 22, San Diego, CA (April 16–28, 1977).
46. Mahon, J., et al., *Manufacturing Methods for Resistance Spotweld—Adhesive Bond Joining of Titanium*, AFML-TR-76-21 (March 1976).
47. Lively, G.W. and Hohman, A.E., Development of a Mechanical-Chemical Surface Treatment for Titanium Alloys for Adhesive Bonding, Proceed-

ings, pp 145-155, 5th National SAMPE Technical Conference, Kiamesha Lake, NY (October 9-11, 1973).

48. Ross, M.C., et al., *Effects of Varying Processing Parameters on the Fabrication of Adhesives–Bonded Structures, Part X, Adhesive Bonding Structural Plastics*, Picatinny Arsenal Technical Report 4318 (July 1972).

49. Smith, D.R., How to Prepare the Surface of Metals and Non-Metals for Adhesive Bonding, *Adhesives Age*, 10 (3):25-31 (March 1967).

50. Bodnar, M.J., and Powers, W.J., Adhesive Bonding of the Newer Plastics, *Plastics Technology*, 4 (8):721-725 (August 1958).

51. King, A.F., Du Pont Company, Technical Services Laboratory, Plastics Department, *Adhesives for Bonding Du Pont Plastics*, TR-152 (August 1966).

52. Du Pont Company, Polychemicals Department, Wilmington, DE, *Delrin Acetal Resin Post Molding Operations–"Satinizing" Surface Treatment*, Brochure A-22794, undated.

53. Du Pont Company, Wilmington, DE, *Delrin Acetal Resins Design Handbook*, booklet, A-67041 (1967).

54. Wegman, R.F., *Adhesive Bonding of Silicone Rubber to Kralastic for the Law Launcher*, Picatinny Arsenal Technical Memorandum 1865, Picatinny Arsenal, Dover, NJ (March 1969).

55. Salomon, G., Introduction, *Adhesion and Adhesives–Vol 2–Applications*, 2nd Edition (R. Houwink and G. Salomon, eds), Elsevier Publishing Co. (1967).

56. Various Du Pont bulletins which will not be cited here because of the lack of positive information on surface preparation for adhesive bonding.

57. Ross, M.C., et al., *Effects of Varying Processing Parameters in the Fabrication of Adhesive-Bonded Structures, Part X. Adhesive Bonding of Structural Plastics*, Picatinny Arsenal Technical Report 4318, Picatinny Arsenal, Dover, NJ (July 1972).

58. Cagle, C.V., Bonding Plastic Materials, *Handbook of Adhesive Bonding* (C. V. Cagle, ed.), McGraw-Hill, NY (1973).

59. Abolins, V. and Eickert, J., Adhesive Bonding and Solvent Cementing of Polyphenylene Oxide, *Adhesive Age*, 10 (7):22-27 (July 1967).

60. Reinhard, D., *Adhesive Bonding of Noryl Resins*, Noryl Processing Guide, NPG 12, undated, General Electric Co., Plastics Department, Selkirk, NY.

61. Landrock, A.H., *Effects of Varying Processing Parameters in the Fabrication of Adhesive-Bonded Structures, Part XVIII. Adhesive Bonding and Related Joining Methods for Structural Plastics–Literature Survey*, Picatinny Arsenal Technical Report 4424, Picatinny Arsenal, Dover, NJ (November 1972).

62. Devine, A.T., et al., *Bonding Lexan Polycarbonate to Silicone Rubber for Gas Masks*, Picatinny Arsenal Technical Report 3930, Picatinny Arsenal, Dover, NJ (April 1970).

63. General Electric Co., Plastics Department, Pittsfield, MA, *Lexan Polycarbonate Resin Design*, Brochure CDC 536, undated.

64. Personal Communication, Mr. Donald Croft, General Electric Co., Products Section, Pittsfield, MA (3 January 1975).

65. General Electric Co., Plastics Department, Pittsfield, MA, *Valox Thermoplastic Polyester*, Brochure, VAL-5A, undated.

66. Celanese Plastics Co., Newark, NJ, *Celanex Thermoplastic Polyester–Properties and Processing*, Bulletin JIA (May 1974).

67. Personal Communication, R. Paul Rich, Eastman Chemical Products, Inc. (March 17, 1972).

68. ICI Petrochemicals and Plastics Division, PEEK Aromatic Polymer, Provisional Data Sheet PK 11, Adhesive Bonding (1979).

69. Victrex PES Aromatic Polymer—Data for Design—Unfilled Grades, Technical Service Note VX101, 3rd edition (1978).

70. Victrex PES Aromatic Polymer—Adhesive for Use with Victrex Polyethersulfone, No. VX TS 1.78 (August 1978).

71. Rimsa, S.B. and Serafty, I.W., *Polyetherimide—A New High Performance Thermoplastic*, Preprint Book, pp 646-648, 40th Annual Technical Conference and Exhibition, Society of Plastics Engineers, San Francisco, CA (May 10-13, 1982).

72. Ohi, H., Mitsui Petrochemical Industries, Ltd., Tokyo, Japan, Polymethylpentene, *Modern Plastics Encyclopedia 1983-84*, McGraw-Hill, NY, NY.

73. Anderson, M.D. and Bodnar, M.J., Surface Preparation of Plastics for Adhesive Bonding, *Adhesives Age*, 7 (11):26-32 (November 1964).

74. Ross, M.C., *A Preliminary Study of Adhesive Bonding of Newer Structural Plastics*, Picatinny Arsenal Technical Memorandum 2204 (April 1976).

75. Austin, J.F., et al., *Oxygen Concentrator Bonding Program*, Summary Report, General Electric Co., Lynn, MA, April–August 1969. ASD-TR-70-42, December 1970. Contract F 33657-68-C-1076 with ASD, Wright-Patterson AFB, OH.

76. Hysol Division, The Dexter Corporation, Olean, NY, *Preparing the Surface for Adhesive Bonding*, Bulletin G1-600, undated.

77. Katz, I. (revised and edited by C.V. Cagle), *Adhesive Materials—Their Properties and Usage*, Fisher Publishing Co., Long Beach, CA (1971).

78. Personal Communications, Mr. Edward Olstein, Dow Chemical Co., Midland, MI (24 January 1975).

79. Du Pont Company, Plastics Department, Experiment Station, Wilmington, DE, *Adhesive Procedure for Vespel Precision Parts*, unnumbered release, 6 pp, revised December 1973.

80. Trimineur, R.J., Silicone, *Modern Plastics Encyclopedia* 1974-1975, pp 113-114, McGraw-Hill, NY.

81. Cagle, C.V., *Adhesive Bonding: Techniques and Applications*, McGraw-Hill, NY (1968).

82. Spearman, B.P. and Hutchinson, J.D., Post Vulcanization Bonding Techniques, *Adhesives Age*, 7 (4):30-33 (April 1974).

83. Sharpe, L.H., Adhesive Bonding, Fastening and Joining Reference Issue, *Machine Design*, 14 (21):119-120, (September 11, 1969).

84. McIntyre, R.T., et al., *Effects of Varying Processing Parameters in the Fabrication of Adhesive Bonded Structures, Part VI, Production Methods*, Picatinny Arsenal Technical Report 4162, Picatinny Arsenal, Dover, NJ (February 1971).

85. Westerdahl, C.A., et al., *Activated Gas Plasma Surface Treatment of Polymers for Adhesive Bonding—Part III*, Picatinny Arsenal Technical Report 4279, Picatinny Arsenal, Dover, NJ (February 1972).

86. International Plasma Corporation (IPC), Product Bulletin 2402.

87. DeLollis, N.J., and Montoya, O., Bondability of RTV Silicone Rubber, *Journal of Adhesion*, 3 (1):57-67 (September 1967).

88. Gaughan, J.E., Bonding Elastomeric Compounds, *Handbook of Adhesive Bonding*, (C.V. Cagle, ed.), McGraw-Hill, NY (1973).

89. Various Hughson Chemicals (Division of Lord Corporation, Erie, PA) publications, 1973 and 1974 on Chemlok primers and adhesives.

90. Snogren, R.C., Selecting Surface Preparation Processes for Adhesive Bonding, Sealing and Coating, paper presented at Design Engineering Confer-

ence and Show, American Society of Mechanical Engineers, held at Chicago, IL, April 22-25, 1968. Paper 68-DE-45. Also published as Selection of Surface Preparation Processes, Parts 1 and 2, *Adhesives Age*, 7 and 8 (July and August 1969).

91. Hercules, Inc., *Parel Elastomer*, Bulletin DRP-101A.
92. Personal communication, Mr. Dominick Bianca, Du Pont Company, Wilmington, DE, February 21, 1975.
93. Selbo, M.L., *Adhesive Bonding of Wood*, U.S. Dept. of Agriculture, Technical Bulletin 1512 (1975).
94. Wegman, R.F., et al., *The Function of the P2 Etch in Treating Aluminum Alloys for Adhesive Bonding*, preprint book, pp 273-281, 29th National SAMPE Symposium, Vol. 29, Reno, Nevada (April 3-5, 1984).

5

Adhesive Types and Their Properties
and Applications

INTRODUCTION

There are a number of different classifications of adhesives, each with its own advantages. No one classification is universally recognized, however. Classifications include function, chemical composition, physical form and application. It is hoped that the following material will be useful to the reader in understanding the interrelationships between the multiplicity of adhesive types currently available. Following the classification outline describing these classification categories is a discussion of the properties and applications of the major adhesive and sealant types.

CLASSIFICATION

Classification by Function

Structural adhesives—These are materials of high strength and performance (see Glossary, Chapter 2, for more comprehensive definitions). Their primary function is to hold structures together and to be capable of resisting high loads.[1]

Nonstructural adhesives—These adhesives are not required to support substantial loads, but merely hold lightweight materials in place. This type is sometimes called "holding adhesives." Examples include adhesive/sealants (sealing adhesives) which are primarily intended to fill gaps, and mucilages, which are used to adhere paper to paper in office applications.[1]

Classification by Physical Form[1]

The physical state of the adhesive generally determines how it is to be applied.

Liquid adhesives—This type is easily handled by means of mechanical spreaders or by spray and brush.

126

Paste adhesives—Paste adhesives have high viscosities to allow application on vertical surfaces with little danger of sag or drip. These bodied adhesives also serve as gap fillers between two poorly mated surfaces.

Tape and film adhesives—These are poor gap fillers, but provide a uniformly thick bond line, and offer the advantage of no need for metering and easy dispensing. Adhesive films are available as pure sheets of adhesive, or with cloth or paper reinforcement.

Powder or granule adhesives—These must be heated or solvent-activated to be made liquid and applicable.

Classification by Mode of Application and Setting[1]

Adhesives are often classified by their mode of application. Depending on viscosity, adhesives are *sprayable*, *brushable*, or *trowelable*. Heavy-bodied adhesive pastes and mastics are considered extrudable, and may be applied by syringe, caulking gun, or pneumatic pumping equipment.

Another way to distinguish between adhesive types is in the manner by which they flow or solidify. As shown in Table 5.1, some adhesives solidify simply by losing solvent, while others harden as a result of heat activation or chemical reaction. Pressure-sensitive systems flow under pressure and are stable when pressure is absent.

Table 5-1: Adhesives Classified by Activation and Cure Requirements[1,2]

Requirement	Types Available	Forms Used	Remarks
Heat	Room-temperature to 450°F (232°C)-types available; 250°-350°F (121°-177°C)-types most common for structural adhesives.	Formulated in all forms; liquid most common.	Application of heat will usually increase bond strength of any adhesive, including room-temperature-curing types.
Pressure	Contact to 500 psi types available; 25-200 psi (0.17-1.38 MPa) types most common for structural adhesives.	Formulated in all forms; liquid and powder most common.	Pressure types usually have greater strength, except for modified epoxies.
Time (RT-Curing)	Types requiring a few seconds to a week are available, ½ to 24 hr types are most common for structural adhesives.	Formulated in all forms.	Time required varies with pressure and temperature applied and immediate strength.
Catalyst (RT-Curing)	Extremely varied in terms of chemical catalyst required; may also contain thinners, etc.	Two components: paste + liquid or liquid + liquid.	This type may sometimes require elevated temperature (<212°F) (100°C) and/or pressure instead of, or in addition to, a chemical agent.
Vulcanizing	Varied types requiring addition of a chemical agent (usually sulfur); may also contain a catalyst.	Two liquid components.	Premixed types requiring 250°-350°F (121°-177°C) for vulcanization are available.
Reactivation	Types requiring heat or solvent or second coating of adhesive.	Dry film or previously applied liquid.	Heat type is best for non-porous surfaces and/or maximum strength.

Classification by Specific Adherends or Applications[1]

Adhesives may be classified according to their end use, as follows:

Adherends (substrates) bonded

Metal adhesives
Wood adhesives
Vinyl adhesives

Environments for which intended

Acid-resistant adhesives
Heat-resistant adhesives
Weatherable adhesives
Cryogenic adhesives
Non-critical (general purpose) adhesives

Classification by Chemical Composition

The classification describes synthetic adhesives as thermosetting, thermoplastic, elastomeric, or combinations of these types (alloys).[1,2]

Thermosetting adhesives—These are materials that cannot be heated and softened repeatedly after the initial cure. Curing takes place by chemical reactions at room or elevated temperature, depending on the type. Some thermosetting adhesives require considerable pressure, while others require only contact pressure. Solvents are sometimes used to facilitate application. This type is only commonly available as solventless liquids, pastes and solids.[1]

Thermosetting adhesives are provided as one-part and two-part systems. The *one-part systems* usually require elevated-temperature cure and have a limited shelf life. The *two-part* systems have longer shelf lives and can usually be cured slowly at room temperature, or somewhat faster at moderately higher temperatures. A disadvantage is their need for careful metering and mixing to make sure the prescribed proportions are used and the resultant mix is homogeneous. Once the adhesive is mixed, the working life is limited.[1]

Because thermosetting-resin adhesives, when cured, are densely cross-linked their resistance to heat and solvents is good, and they show little elastic deformation under load at elevated temperatures. Bonds are ordinarily good to 200-500°F (93-260°C). Creep strength is good and peel strength fair. The major application is for stressed joints at somewhat elevated temperatures. Most materials can be bonded with thermosetting adhesives, but the emphasis is on structural applications.[1]

Thermosetting adhesives include:[1]

Cyanoacrylates	Epoxy
Polyester	Polyimide
Urea-formaldehyde	Polybenzimidazole
Melamine-formaldehyde	Acrylic inc ;
Resorcinol- and phenol-formaldehyde	Acrylic acid diester

Thermoplastic adhesives—These materials do not cross-link during cure, so they can be resoftened repeatedly with heat. They are single-component systems that harden upon cooling from a melt *(hot melts)*, or by evaporation of a solvent or water vehicle. *Wood glues* are thermoplastic emulsions that are common household items. They harden by evaporation of water from an emulsion. Most thermoplastic adhesives are applied in a vehicle.[1]

Thermoplastic adhesives are not ordinarily recommended for use over 150°F (66°C), although they can be used up to 200°F (90°C) in some applications. These materials have poor creep resistance and fair peel strength. They are used mostly in unstressed joints and designs with caps, overlaps and stiffeners. The materials most commonly bonded are nonmetallics, especially wood, leather, cork, and paper.[1,2] With the exception of some of the newer hot-melt adhesives, thermoplastic adhesives are not generally used for structural applications.

Thermoplastic adhesives include:[1]

Cellulose acetate	Polyvinyl acetals
Cellulose acetate butyrate	Polyvinyl alcohol
Cellulose nitrate	Polyamide
Polyvinyl acetate	Acrylic
Vinyl vinylidene	Phenoxy

Elastomeric adhesives—These materials are based on synthetic or naturally occurring polymers having superior toughness and elongation. Elastomeric adhesives may be supplied as solvent solutions, latex cements, dispersions, pressure-sensitive tapes, and single- or multiple-part solventless liquids or pastes. Curing varies, depending on the type and form of adhesive. These adhesives can be formulated for a wide variety of applications, but they are generally used for their high degree of flexibility and superior peel strength.[1,2]

Elastomeric adhesives are usually supplied in liquid form, although some elastomeric films are available. Most of these adhesives are solvent dispersions or water emulsions. Temperature environments up to 150-400°F (66-204°C) are practical. Elastomeric adhesives never melt completely. Bond strengths are relatively low, but flexibility is excellent. These adhesives are used in unstressed joints on lightweight materials, so they cannot be considered structural adhesives. They are particularly advantageous for joints in flexure. Most of these adhesives are modified with synthetic resins for bonding rubber, fabric, foil, paper, leather, and plastic films. They are also used as tapes.[1,2]

Elastomeric adhesives include:[1,2]

Natural rubber	Styrene-butadiene rubber
Reclaimed rubber	Polyurethane
Butyl rubber	Polysulfide
Polyisobutylene	Silicone
Nitrile rubber	Neoprene

Adhesive alloys—These are made by combining resins of two different chemical groups chosen from the thermosetting, thermoplastic, or elastomeric groups. The thermosetting resin, chosen for its high strength, is plasticized by the second resin, making the alloy tougher, more flexible, and more resistant to impact.[2] The adhesives utilize the most important properties of each component material. They are commonly available as solvent solutions and as supported and unsupported films.[2]

Except for some epoxy types, heat and pressure are usually required for curing. Most alloy adhesives are solvent dispersions or 100% solids. These adhesives have a balanced combination of properties and generally are stronger over wider temperature ranges than other adhesives. They are definitely structural adhesives and are used where the highest and strictest end-use conditions must be met, sometimes regardless of cost, as in military applications.[1,2]

Materials bonded include metals, ceramics, glass, and thermosetting plastics. Applications are primarily for high strengths and high temperatures.[1,2]

Adhesive alloys used include:[1]

Epoxy-phenolic	Neoprene-phenolic
Epoxy-polysulfide	Vinyl-phenolic
Epoxy-nylon	Polyvinyl acetal-phenolic
Nitrile-phenolic	

Natural vs. Synthetic Adhesives

This is another type of broad classification.

Natural adhesives—This term is used to include vegetables- and animal-base adhesives and natural gums. These include casein, blood, albumin, hide, bone, fish, starch, resin, shellac, asphalt, and inorganic adhesives (sodium silicate, etc.). Their use, except for the inorganic adhesives, is mostly limited to paper, paperboard, foil and light wood. They are inexpensive, easy to apply and have a long shelf life. These adhesives develop tack quickly, but have low strength properties. Most are water-soluble and use water as a solvent. They are supplied as liquids or dry powders to be mixed with water. Some are organic solvent dispersions.[2]

Synthetic Adhesives—This term is usually used to apply to all adhesives other than natural adhesives, i.e., elastomeric, thermoplastic, thermosetting, and alloys. All structural adhesives are synthetic.

SME Classification

A Society of Manufacturing Engineers publication has provided a useful classification, in depth, of adhesives, as follows:[3]

Chemically Reactive Types

Catalytic plural components — chemical cure

Epoxy	Polysulfide
Phenolic	Polyurethane

Resorcinol-formaldehyde   Silicones
Polyester

Catalytic plural components — moisture cure

Silicones   Cyanoacrylate
Polyurethane   Epoxy (one-container type)
Polysulfides

Heat-activated systems (one-part), (may be solid films)

Polybenzimidazole (PBI)   Polyvinyl acetals
Polyimide (PI)   Vinyl-phenolics
Epoxy   Vinyl-epoxies
Nylon   Urethanes
Phenolic

Evaporative or Diffusion Adhesives

Solvent-based systems

Natural rubber   Acrylics
Reclaimed rubber   Miscellaneous
Synthetic rubbers   Cellulose esters
   Nitrile rubber   Asphalt
   Neoprene (polychloroprene)   Polyamides
   Butyl rubber   Phenoxy resins
   Styrene-butadiene rubber (SBR)   Bisphenol-A polycarbonate
Phenolics   Polysulfone
Urethanes
Vinyl resins
   Polyvinyl acetate
   Vinyl-phenolics
   Polyvinyl alkyd ethers
   Polystyrene

Water-based systems

Natural rubber   Miscellaneous adhesives
Reclaimed rubber   Natural products (animal
Synthetic rubber   glue, starch, soya, blood glue,
Vinyl resins   casein, cellulose derivatives)
Acrylics   Carboxylic-containing copolymers

Hot-Melt Adhesives

Ethylene-vinyl acetate (EVA) and polyolefin resins
Polyamide (nylon) and polyester resins
Other hot melts
   Polyester-amides
   Thermoplastic elastomers

Delayed-Tack Adhesives

Styrene-butadiene copolymers   Polystyrene
Polyvinyl acetate   Polyamides

Tape and Film Adhesives

Vinyl-phenolics   Elastomer-epoxies
Epoxy-phenolics   (as nitrile-epoxies)
Nitrile-phenolics   Aromatic polymers
Nylon-epoxies   (PI and PBI)

Pressure-Sensitive Adhesives

Classification by Rayner[4]

This is another useful classification. It somewhat resembles the classification
by "chemical composition" described above.

Thermosetting Resin Adhesives

Urea-formaldehyde   Polyesters
Melamine-formaldehyde   Silicones
Phenol-formaldehyde   Furanes
Resorcinol-formaldehyde   Acrylics
Epoxy   Soluble nylons
Polyisocyanate (polyurethane) carbon   Polyaromatics (PI and PBI)*

*These are really thermoplastics, but are often grouped with
thermosets because of their high melting points.

Thermoplastic Resin Adhesives

Cellulose adhesives (cellulosics)
Polyvinyl adhesives
 Polyvinyl ester adhesives (esp. polyvinyl acetate)
 Polyvinyl acetal adhesives
 Polyvinyl alcohol adhesives
 Polyvinyl alkyl ether adhesives (some are elastomers)
 Polystyrene adhesives
Acrylic resin adhesives
 Acrylic esters
  Acrylic acid diesters (incl. anaerobic sealants)
  Cyanoacrylates
 Acrylic copolymers
Polyamide resin and nylon adhesives (including nylon adhesives with
 traces of phenolic)
Miscellaneous thermoplastic adhesives
 Polycarbonates
 Polyacetals
 Polyethylene
 Polysulfide (sometimes considered thermoplastics, but these are
  really rubbers)

Other hot-melt thermoplastic adhesives

Ethyl cellulose	Polyisobutylene
Polyvinyl acetate	Hydrocarbon resins
Ethylene vinyl acetate (EVA)	Polypropylene
Ethylene-ethyl acrylate (EEA)	Polyamides
Butyl methacrylates	Polyesters
Polystyrene and copolymers	Phenoxies

Two-Polymer Adhesives (Alloys)

Vinyl-phenolics	Nylon-epoxies
Epoxy-phenolics	Elastomer-epoxies
Nitrile-phenolics	Neoprene-phenolics
Epoxy-polysulfide	

Additional Classification

The following grouping has been found by the author to be convenient. Some of the adhesive types have already been listed.

Adhesives/Sealants (discussed in following chapter)

Hardening types (including flexible materials)

Non-hardening types

Primers

Microencapsulated Adhesives

Conductive Adhesives

Electrically conductive adhesives

Thermally conductive adhesives

Pre-Mixed Frozen Adhesives

Anaerobic adhesives

Fast-Setting Adhesives (Cyanoacrylates)

Elastomeric Adhesives (including pressure-sensitive adhesives)

Natural Glues

Vegetable glues

Starch	Soybean glue
Dextrins	Rosin

Animal glues

Casein	Animal glues
Blood adhesives (blood glues)	(bone and hide glues)
Shellac	Fish glues

Inorganic glues (adhesives, cements)

Soluble silicates	Litharge cements
Phosphate cements	Sulfur cements

Basic salt (Sorel cements)   Hydraulic cements

ADHESIVE COMPOSITION

Adhesives in use today generally have the following composition.[1]

Adhesive Base or Binder

This is the primary component and has the function of holding the substrates together. The binder is generally the component from which the name of the adhesive is derived. An epoxy adhesive may have many components, but the main component is the binder, *epoxy resin.*

Hardener (for Thermosetting Adhesives)

This is a substance added to an adhesive to promote the curing reaction by taking part in it by catalysis or crosslinking. Two-part adhesive systems generally have one part which is the base (described above) and a second part, which is the hardener. Upon mixing, a chemical reaction takes place which causes the adhesive to solidify. A *catalyst* is sometimes incorporated into a adhesive formulation to speed up the reaction between base and hardener. Very small amounts of catalyst are used, compared to the principal components (base and hardener).

Solvents

Solvents are sometimes needed to disperse the adhesive to a spreadable consistency. Solvents used with synthetic resins and elastomers are generally organic in nature. Often a mixture of solvents is needed to achieve the desired properties.

Diluents

These are liquid ingredients added to an adhesive to reduce the concentration of solid bonding materials. Diluents are added principally to lower the viscosity and modify the processing conditions of some adhesives. *Reactive diluents* react with the binder during cure, become part of the cured product, and do not evaporate as does a solvent.

Fillers

Fillers are relatively nonadhesive substances added to the adhesive to improve its working properties, strength, permanence, or other qualities. Fillers are also used to reduce materials costs. Considerable changes can be made in the properties of an adhesive by selective use of fillers. Fillers are used to modify adhesives to govern such properties as thermal expansion, electrical and thermal conductivity, shrinkage, and heat resistance.

Carriers or Reinforcements

These are usually thin fabrics or paper used to support the semicured thermosetting adhesive composition to provide a tape or film. The carrier also serves as a bond-line spacer and reinforcement for the adhesive.

INDIVIDUAL ADHESIVE TYPES AND THEIR CHARACTERISTICS

The above discussions have covered the general relationships between the various adhesive classifications. The reader will be aware that there is a considerable amount of overlap between the various general groupings. The following discussion will cover individual adhesive types in some detail, ignoring the classification. As a matter of convenience, the arrangement is in alphabetical order. Some general categories are also listed, such as alloys, aromatic polymer adhesives, conductive adhesive, delayed-tack adhesives, elastomeric adhesives, anaerobic adhesives, film and tape adhesives, hot-melt adhesives, inorganic glues, microencapsulated adhesives, rubber-base adhesives, solvent-based system, thermoplastic resin adhesives, thermosetting-resin adhesives, and water-based adhesives.

Acrylics

A number of acrylic resins are used for bonding cloth, plastics, leather, and, in some cases, metal foils. The acrylic monomers most commonly used in adhesives are ethyl acrylate, methyl acrylate, methacrylic acid, acrylic acid, acrylamide, and acrylonitrile. The polymers or copolymers are soluble in common organic solvents and can be supplied in much the same manner as other solvent-based systems. In addition, the polymers are soluble in the monomers, and when a catalyst is added, polymerization results, providing good bonding to glass and to plastic surfaces of similar composition, such as polymethyl methacrylate.[3]

A variety of acrylic copolymers are prepared by emulsion polymerization.[3]

Within the past decade various acrylic adhesives called "reactive adhesives," "modified acrylics," "second-generation acrylics," or "reactive-fluid adhesives" have become available. These are formulations that polymerize *in the glueline* to become an integral part of an adhesive assembly. The term "first-generation acrylic adhesives" covered adhesives developed a number of years ago. These materials used solutions of polymers, usually rubber, in methacrylate monomers, and involved polymerization of these monomers in the presence of reinforcing resin. The much newer "second-generation acrylics" are based on a combination of different modifying polymers for acrylics and a surface activator. The modifying polymer reinforces and toughens the bond and provides a reactive chemical site which acts as a catalyst in the presence of special activators. Adhesion takes place when the monomers and activator graft-polymerize in the modifying polymer in the glueline.

In commercial form, the "second-generation" acrylic adhesive system consists of two components, each being a 100%-solids composition in fluid form, and each reacting to form the adhesive film. Curing is by a free-radical reaction. As a result, these materials do not require careful metering and accurate mixing for full performance. Other advantages are:[5,6]

- Tolerance for oily and otherwise poorly prepared surfaces.

- Rapid bonding at room temperature, which can be speeded up further with mild heat or the use of accelerators.

- Low shrinkage during cure.

- High peel and impact strength, combined with excellent shear strength.

- Good environmental resistance and elevated-temperature properties (up to 350°F or 177°C).

Excellent bonds can be obtained with a wide variety of substrates. Aluminum, brass, copper, stainless steel, and carbon steel are easily bonded to similar or dissimilar metals. Most plastics, including glass-reinforced grades, can also be bonded, along with wood, glass, cement-asbestos board, and hardboard. Some types will bond cured elastomers. Typically a thin layer (0.001 inch or 0.0025 mm) of the activator is applied to one of the adherends and a layer of the adhesive (0.001 to 0.01 inch) (0.026 to 0.26 mm) is placed on the other adherend. The two substrates are then pressed together and secured until adequate handling strength develops. Most second-generation acrylic adhesives cure to this point in 2 to 20 minutes, but some cure in as little as 10 seconds. In all cases, cure is complete in 24 hours or less. These adhesives provide excellent shear, peel and impact strengths at temperatures from −160 to +250°F (−107 to +121°C). Short exposures up to 350°F (177°C) are possible.[5,6]

Bonds made with second-generation acrylics will resist immersion in isooctane, motor oil, aircraft hydraulic fluid, 10% sodium chloride solution, distilled water, ethyl alcohol, and dilute mineral acids and alkalies. They are *not* resistant to concentrated acids, alkalies, and acetone. Resistance to weathering, including salt spray environments, is also excellent.[5,6]

Second-generation adhesives can be used to replace spot welding where immediate handling of joined metal parts is required. Another broad area of application is for bonding dissimilar substrates, including metals and other materials with different coefficients of expansion.[5,6]

For additional acrylic adhesive applications, see *Anaerobic adhesives* and *Cyanocrylate adhesives* below.

Allyl Diglycol Carbonate (CR-39) (See *Polyester Adhesives*)

Alloyed (Two-Polymer) Adhesives

These adhesives are assuming increasing importance as structural adhesives in recent years, especially in metal bonding. They comprise a thermosetting and a thermoplastic polymer, including certain elastomers. Although each component has adhesive properties by itself, considered overall, the conjoint system forms a stronger and more versatile adhesive. The two-polymer systems have been particularly successful as film and tape adhesives (which see). The physical properties of each component polymer are modified by the addition of the other and for this reason the heat resistance of one may be increased, while that of the other is reduced. Similarly, the toughness of one may be increased by sacrificing the flexibility of the other. It is therefore possible to formulate a variety of adhesives with a wide range of characteristics by simply varying the ratio of one polymer to the other. In most widely used two-polymer adhesives, the thermosetting component is a phenolic. Phenolic resins are generally compatible, although not easily miscible, with a number of thermoplastic polymers. Particularly good compatibility is shown between conventional alcohol-soluble phenolic resins and polyvinyl esters and acetals. Epoxies are of growing importance in two polymer adhesive systems. The most important thermoplastic components are the *polyvinyl acetals* (polyvinyl formal and butyral) and *synthetic rubber*, particularly nitrile rubber. Soluble nylons are also becoming important.[1]

Five of the most important two-polymer adhesives used in films and tapes are discussed under that heading below. These are vinyl-phenolics, epoxy-phenolics, nitrile-phenolics, nylon-epoxies, and elastomer-epoxies. Neoprene phenolics are available in solvent solutions and in supported and unsupported films. These adhesives are used to bond a variety of substrates. Curing is under heat and pressure 302-500°F (150-260°C) and 40-260 psi for 15-30 minutes for film and 194°F (90°C) and contact to 102 psi (0.7 MPa) pressure for 15-30 minutes for the liquid, after drying at 176°F (90°C). Because of their high resistance to creep and most service environments, neoprene-phenolic joints can withstand prolonged stress. Fatigue and impact strengths are excellent, but shear strength is lower than that of other modified phenolic adhesives. Epoxy polysulfides are available as two-part liquids or pastes that cure at room temperature or higher to rubbery solids that provide bonds with excellent flexibility and chemical resistance. These adhesives bond well to many substrates. Shear strengths and elevated-temperature properties are low, but resistance to peel and low temperatures is good. Of the five alloy tape adhesives mentioned above, *vinyl-phenolic* is also available in solvent solutions and emulsions, liquid and coreacting powder. *Epoxy-phenolic* is also available as a two-part-paste. Solvent blends of this material are usually force-dried at 176-194°F (80-90°C) for 20 minutes before assembly of adherends. Curing is generally for 30 minutes at 203°F (95°C) with contact pressure, followed by 30 minutes to 2 hours at about 329°F (165°C) and 10-58 psi (0.07-0.4 MPa) pressure. The post curing provides optimum strength at elevated temperatures. *Nitrile-phenolic* and *nylon-epoxy* adhesives are also available as solvent solutions as well as in film form. The *nitrile-phenolic* film is cured at 302-500°F (150-260°C) for 15-30 minutes with bonding pressures from 67-261 psi (0.12-1.8 MPa). The liquid alloy is dried at 176°F (80°C) and cured for 15-30 minutes at 194°F (90°C) and contact to 101 psi (0.7 MPa) pressure. The *nylon-epoxy* paste is cured for 3 days at 68°F (20°C) to 1 day at 302°F (150°C) under bonding pressure from 23-46 psi (0.11-0.32 MPa). Cure temperatures for some formulations can be increased to 392°F (200°C) with corresponding reduction in cure time (4 hours). No volatiles are released during cure, so that large areas can be bonded without venting.[1,7]

Anaerobic Adhesive/Sealants

This adhesive type has become available in recent years and has been promoted for use as a sealant. The adhesives used are acrylate acid diesters (polyester-acrylic). They are essentially monomeric, thin liquids which polymerize to form a tough plastic bond wben confined between closely fitting metal joints. Contact with air before use keeps the monomeric adhesive liquid. Metal surfaces accelerate the polymerization in the absence of air (anaerobic conditions). These materials will bond all common metals, glass, ceramics, and thermoset plastics to each other. Phenolic plastics and some plated metals, such as cadmium and zinc, require a primer such as ferric chloride. Polymerization is essentially a free-radical-type addition polymerization.[8-10]

The most important use of anaerobic adhesive/sealants is as liquid lock washers for screws and bolts. Because of their strong penetrating ability they can be applied either before or after assembly. The prevailing torque for the strongest grades is many times greater than that of locknuts and lockscrews. Cure speed is largely dependent on the parts being joined. There are three basic cure-speed types, fast (5 minutes to 2 hours), medium (2 to 6 hours) and slow 6 to 24 hours), all at room temperature without primer. The application of heat will speed up the cure. As a rule, these adhesives will cure *outside the connection* if the temperature exceeds 200°F (93°C), despite the presence of inhibiting air. Heat cures up to 300°F (149°C) are practical. Anaerobic adhesives can be cured faster with accelerators or primers, especially on inactive surfaces (nonmetals). The recommended primers are degreasing solvents which, on evaporation, leave a light deposit of a catalyst to speed up curing.[8-10]

These anaerobic adhesives fill all surface irregularities and tolerance gaps and effectively seal clearances up to 0.030 inches (0.76 mm). They can be applied by high-speed applications in moving production lines. The cured film has excellent chemical resistance to most liquids and gases within an operating temperature range of −65 to 450°F (−54 to 232°C).[8-10]

Anaerobic *structural* adhesives combining urethane-modified acrylic technology have been developed for more exacting applications. These adhesives can be formulated to meet the requirements of Federal Specification MMM-A-132 (which see in Chapter 12). While anaerobic sealants and thread-locking products are designed to take normal tensile and shear loading, these products find primary application in shear loading. Anaerobic adhesives can now withstand continuous aging at 450°F (232°C). Salt-spray resistance is also excellent.[8-10]

An excellent, although perhaps outdated source of information is a chapter on anaerobic adhesives by Burnham et al. of Loctite.[11]

Aromatic Polymer Adhesives (Polyaromatics)

In recent years considerable progress has been made in improving the thermal and oxidative stabilities or organic resins at high temperatures. Heat-resistant resins and polymers have been developed as adhesives to meet the needs of the aircraft industry (supersonic aircraft) and space vehicles (missiles, satellites, rockets) where resistance to temperatures approaching 600°F (316°C) is required throughout the life of bonded assemblies based on metals and reinforced plastic composites. The oxidative stability of organic polymers is improved by the incorporation into the molecule of aromatic units such as benzene rings and heterocyclic rings, such as substituted imide, imidazole and thiazole. The most important resins currently available for use as adhesives in high-temperature structural applications are the polyimides (PI) and polybenzimidazoles (PBI), both described below. These resins are supplied as pre-polymers containing open heterocyclic rings, which are soluble and fusible. At elevated temperatures the prepolymers undergo condensation reactions leading to ring closure and the formation of insoluble and infusible cured resins.[7]

These adhesives are available in film and tape form. They show better bond strengths above 500°F (260°C) in air than epoxy-phenolic, although the latter gives better strength retention after exposure to water or other polar liquids at lower temperatures. The major disadvantages are their high cost, generally 10 or more times that of epoxy-based adhesives, the difficulty in handling or curing, and the problems involved in the elimination of volatiles during cure in order to obtain a void-free bond. A long and careful series of cure and post-cure steps at progressively increasing temperatures up to the 600-700°F (316-371°C) level, coupled with intermittent application and release of high-clamping pressure, is required to obtain optimum results. Currently only the polyimides can be used in the 500-800°F (260-427°C) service temperature range.[12]

Asphalt

Asphalt is a low-cost thermoplastic material which is highly temperature-dependent. The addition of a thermoplastic rubber at 1-5% by weight greatly reduces the dependence of viscosity on temperature. Useful operating tempera-ture ranges can often be doubled in this manner. The addition of a thermoplastic rubber, such as butyl rubber or polyisobutylene, at 10-30% by weight produces a truly thermoplastic product with elasticity, resilience and high cohesive strength. Such mixtures are useful in sealants. Asphalt emulsions are used to raise solids contents, improve water resistance, and lower the cost of laminating adhesives. Such adhesives are used in laminating paper and other packing mate-rials where a water-barrier layer is required. Another important application is in roofing and flooring adhesives.[3,13,14]

Butyl Rubber Adhesives

Butyl rubber is an elastomeric polymer used widely in the adhesives and sealants area, both as primary binders and as tackifiers and modifiers. Butyl rubber is a copolymer of isobutylene with a small amount of isoprene.[15] These materials have relatively low strength and tend to creep under load. They are useful in packaging applications where advantage can be taken of their low permeability to gases, vapors and moisture. Butyl rubber is also used as an adhesive sealant. It is generally used in solvent solution.[1,3] Table 5-2 (p. 145) summarizes some of the important properties of butyl rubber adhesives.

Cellulose Ester Adhesives

These include *cellulose acetate, cellulose acetate butyrate, cellulose cap-rate*, and *cellulose nitrate* (nitrocellulose or pyroxylin). Cellulose esters are used for bonding leather, paper, and wood. While not generally used with metals, specific nonporous substrates such as cellophane (regenerated cellulose) and glass are sometimes bonded with cellulose nitrate or other cellulose esters applied from solution.[3,7]

Cellulose acetate and cellulose acetate butyrate are water-clear and more heat-resistant, but less water resistant, than cellulose nitrate. Cellulose acetate butyrate has better heat and water resistance than cellulose acetate and is compatible with a wide range of plasticizers. Cellulose nitrate is tough, develops strength rapidly, is water resistant, bonds to many surfaces, and discolors in sunlight. The dried adhesive (nitrocellulose) is highly flammable.[1,4]

Cellulose Ether Adhesives

These include *ethyl cellulose, hydroxy ethyl cellulose, methyl cellulose, sodium carboxy methyl cellulose* and *benzyl cellulose.* Ethyl and benzyl cellulose can be used as hot-melt adhesives. Methyl cellulose is a tough material, completely non-toxic, tasteless and odorless, which makes it a valuable adhesive for food packages. It is capable of forming high-viscosity solutions at very low concentrations, so it is very useful as a thickening agent in water-soluble adhesives. Hydroxy ethyl cellulose and sodium carboxy methyl cellulose can also be used as thickeners. The cellulose ethers have fair to good resistance to dry heat. Water resistance varies from excellent for benzyl cellulose to poor for methyl cellulose.[1,4,7]

Conductive Adhesives

Appropriate fillers have been used to produce adhesives with high thermal or electrical conductivity for specialized applications. The basic resins used include epoxies, urethanes, silicones, and polysulfones. Epoxies, however, are the most widely used resins.[16]

Electrically conductive adhesives (chip-bonding adhesives)—Synthetic resins are made electrically conductive by the addition of either metallic fillers or conductive carbons. The carbon can be either an amorphous carbon, such as acetylene black, or a finely divided graphite. Usually finely divided silver flake is used in conductive epoxies and in conductive coatings. Silver has the advantage of having moderately conductive salts and oxides, so that a slight amount of oxidation or tarnishing can be tolerated. The resistivity techniques give much lower values than methods involving thin gluelines, such as ASTM D 2739 (which see in Chapter 12), where interfacial resistance plays an important role.[16]

Silver is preferable to *gold* as a filler because it is less costly and has lower resistivity. Under conditions of high humidity and DC potential, however, silver is reported to undergo electrolytic migration to the surface of the adhesive. Silver-coated copper microspheres do not migrate, nor does gold. The highest silver loading possible is about 85% by weight. Silver loadings lower than the optimum (about 65%) cause conductivities to drop sharply, but offer higher adhesive strengths. *Carbon (graphite)* gives fairly low conductivities. Aside from silver and gold, other metallic fillers used are *nickel, aluminum* and *copper.* Each of these metals presents particular compounding problems. Silver is often used in flake form and it is, therefore, difficult to achieve particle-to-particle contact, as with spherical metal particles. Unfortunately the stearate coating on silver flake tends to outgas at elevated temperatures and may contaminate critical parts, such as in microelectronic applications. Some silver products are uncoated and do not evolve outgassing products. Copper and aluminum form oxide films, which reduce electrical conductivity by hampering particle-to-particle contact.[16]

Electrically conductive adhesives are used in microelectronic assemblies. These applications include attachment to fine lead wires to printed circuits, electroplating bases, metallization on ceramic substrates, grounding metal chassis, bonding wire leads to header pins, bonding components to plated-through holes on printed circuits, wave-guide tuning, and hole patching. Conductive adhesives are used as substitutes for spot welding when welding temperatures build up excess resistance at the weld because of oxide formation. Another application is in ferroelectric devices used to bond electrode terminals to the crystals in stacks. These adhesives replace solders and welds where crystals tend to be deposited by soldering and welding temperatures. Bonding of battery terminals is another application useful when soldering temperatures may be harmful. Conductive adhesives form joints with sufficient strength so that they can be used as structural adhesives where electrical continuity, in addition to bond strength, is required, as in shielded assemblies.[17]

Sharpe et al. have recently written an excellent comprehensive review of electrically conductive adhesives.[18]

Thermally conductive adhesives—De Lollis[17] has described the use of these adhesives in electrical/electronic assemblies where the temperature rise due to evolution of heat from tubes, resistors, transformers, etc. in high-density circuits is often critical and a cause for concern. Design considerations for these applications must include thermally conductive parts for removing heat from the circuitry involved. This circuitry may or may not be encapsulated. In confined circuitry, as on a printed-circuit board, non-encapsulated heat sinks bonded in place, is one solution. In this case, aluminum is usually the preferred heat-sink material because of its light weight and high thermal conductivity. If good dielectric properties are required, a high concentration of inorganic or mineral fillers can be used.

A typical thermally conductive epoxy system used as an adhesive, as well as for other purposes, has a thermal conductivity of 7.55 Btu in/ft^{2}hr°F (0.0026 cal/cm sec °C and a volume resistivity of 1.5×10^{15} ohm-cm (1.5×10^{17} ohm-m). Fillers used include alumina (aluminum oxide), beryllia (beryllium oxide), other unspecified inorganic oxides, boron nitride, and silica. Theoretically, boron nitride would be an excellent thermally conductive filler, but weight loadings with this filler cannot go much over 40% in epoxy resins, and the resultant products are always thixotropic pastes. Beryllia powder has an excellent thermal conductivity by itself, but when mixed with a resin binder its conductivity drops drastically. This material is highly toxic and high in cost. Alumina is quite commonly used.[16]

Bolger and Morano,[88] in a very recent discussion, have described both electrically and thermally conductive adhesives.

Cyanoacrylate Adhesives

These so-called "wonder" adhesives are marginally thermosetting materials which form strong thermosetting bonds between many materials without the need for heat or an added catalyst. They are particularly useful in bonding metal to nonmetal. Lap-shear strengths of 2000 psi (13.7 MPa) have been reported. The resistance of these adhesives to moisture, however, is still somewhat low.[3]

These materials were first introduced commercially in 1958 by Eastman Chemicals. They set very quickly when squeezed out to thin films between many types of adherends. As with other acrylics, the monomers are liquids of low viscosity which polymerize very easily in the presence of a slightly basic surface having on it adsorbed water. Polymerization is ionic. The resulting polymers have different properties, depending on the alkyl group. The methyl ester (methyl 2 - cyanoacrylate) is the one most commonly used. This material is formulated with a thickener (to prevent starved joints from being formed) and a plasticizer to make it more resistant to shock loading. The thickener can be a polymer of the same monomer. An essential feature is a stabilizer to prevent polymerization in the container, which is usually of polyethylene.[19]

The polymerization of cyanoacrylates, being inhibited by a low pH (high acidity), does not proceed satisfactorily on acid surfaces such as wood, and the suggested incorporation of poly-N-vinyl pyridine or polyethyleneamine, or even simple amines, presumably serves the dual purpose of thickening the liquid and increasing the pH.[19]

Adhesives based on higher homologs than the methyl form have been in use for a number of years. These include the ethyl, propyl and butyl esters of cyanoacrylic acid. Moisture resistance of the methyl-2 cyanoacrylate is only fair. Ethyl cyanoacrylate has been shown to form stronger bonds than the methyl form between several different types of plastic surfaces. The higher homologs, however, generally do not form bonds as strong as the methyl form.[20]

The most important step in the successful application of a cyanoacrylate adhesive is the application of a thin adhesive film between two well-mated surfaces. The thinner the film the faster the rate of bond formation and the higher the bond strength. Bond strengths, of course, are dependent upon proper surface preparation.[20]

In general, aging properties of the cyanoacrylates are good. Rubber-to-rubber and rubber-to-metal bonds have weathered outdoors for over seven years. Test bonds have also passed stringent water-immersion and salt-spray tests. Plastic-to-plastic and plastic-to-rubber bonds have aged satisfactorily for three to five years. Metal-to-metal bonds generally age rather poorly, except under optimum conditions where little glue line is exposed to moisture. Solvent resistance is also generally good. Dilute alkaline solutions weaken the bond considerably, while dilute acid solutions weaken it to a lesser degree. Impact resistance is generally poor, primarily because of the thin, inflexible bond. This is especially true with two rigid substrates, such as metals. The methyl cyanoacrylate bond melts at approximately 330°F (165°C). Prolonged exposure to temperatures in this range results in a gradual but permanent breakdown of the bond. Generally, the upper temperature limit for continuous exposure is about 170°F (77°C). At low temperatures bonds remain intact at least down to −65°F (-54°C). Grades of cyanoacrylates with specialized improved properties are now available. One such grade has improved heat resistance to 475°F (246°C), high viscosity, and very fast setting ability.[20]

Among the advantages of the cyanoacrylates are the following:[20]

- Very fast bond formation
- High bond strength with thin glue line

- No added catalyst or mixing needed
- No solvent to evaporate during bond formation
- Contact pressure usually sufficient
- Very low shrinkage
- Economical because of minute quantities needed, although volume price is high.

The military specification covering cyanoacrylate adhesives is MIL-A-46050, described in Chapter 12. This specification covers five types and four classes of adhesives and four types of surface activators. The latest revision of this specification was issued in February 1983.

A useful book collection of information on cyanoacrylate adhesives was edited by Lee in 1981.[21]

The Loctite Corporation has very recently developed an elastomer-modified cyanoacrylate adhesive called "Black Max" which is reported to achieve improved strength, resiliency and fast fixturing at the expense of a rather limited shelf life of four months from the date of shipment. This adhesive cures to fixturing strength in two minutes, to 80% strength in 24 hours, and to full strength in 72 hours. On aluminum its average strength is 2400 psi (16.6 MPa) after full RT-cure, vs. 900 psi (6.2 MPa) for a typical epoxy adhesive and 550 psi (3.8 MPa) for "instant" adhesives. After 240 hours of ASTM tensile-shear thermal-cycling tests this adhesive improved its strength to 3,100 psi (21.3 MPa) for "instant" adhesive. Loctite claims that this adhesive is consistently 20 times stronger than epoxies on aluminum, 10 times stronger on neoprene, four times stronger on steel, and twice as strong on epoxy/glass after the ASTM tests. This adhesive is designed for assembly-line cure.[22]

Delayed-Tack Adhesives

In the packaging field acrylics are often used for *delayed-tack* adhesive coatings for labels. Copolymer dispersions of acrylic ester with vinyl acetate, vinyl chloride, or styrene are usually employed for this application. The backing material, usually paper, is coated after the dispersion has been modified accordingly. The coated papers are tack-free under normal conditions so that the sheets and cuttings can be rolled up or stacked. These adhesives consist primarily of one or a mixture of polymer film formers in dispersion form and one or several crystalline plasticizers. The plasticizer is usually employed in the form of a dispersion with very small particle size. Resin solutions or dispersions are used as additives for obtaining certain adhesive effects. The adhesive coating must be dried at a temperature below the melting point of the plasticizer in order to obtain a tack-free product.[23]

Labels produced in this way are applied in the following manner: The adhesive coat is heated directly by infrared radiation or hot air, or by hot plates from the reverse side, to a temperature above the melting point of the plasticizer. The polymer is plasticized, i.e. the coating is tackified by the liquified plasticizer, which is present in excess. In this condition the label can be bonded to the substrate by applying slight pressure. The adhesion to glass, metals, PVC, wood, etc. is durable even after the plasticizer has recrystallized.[23]

Other polymers used for delayed-tack adhesives include styrene-butadiene copolymers, polyvinyl acetate, polystyrene, and polyamides. The solid (crystalline) plasticizers include dicyclohexyl phthalate, diphenyl phthalate, N-cyclohexyl-p-toluenesulfonamide, and o/p-toluenesulfonamide. Because of the range of melting points available, adhesives with different heat-activation temperatures are possible. Delayed-tack adhesives are commonly used for coating on paper for labels on bread packages and cans.[23]

Elastomeric Adhesives

Most of these adhesives are natural or synthetic rubber-based materials, usually with excellent peel strength, but low shear strength. Their resiliency provides good fatigue and impact properties. Except for silicone, which has high temperature resistance, their use is generally restricted to temperatures of 150–200°F (66–93°C), and creep under load occurs at room temperature. The basic types of elastomeric adhesives used for nonstructural applications are shown in Table 5.2. These systems are generally supplied as solvent solutions, latex, cements, and pressure-sensitive tapes. Solvent solutions and latex cements require removal of the liquid vehicle from the adhesive before bonding can be carried out. This is accomplished by simple evaporation or forced heating. Some of the stronger or more environmentally resistant rubber-based adhesives require an elevated-temperature cure. Usually only slight pressure is required with pressure-sensitive adhesives (PSA's) to obtain a satisfactory bond. These adhesives (which see) are permanently tacky and flow under pressure to provide intimate contact with the adherend surface. In addition to pressure-sensitive adhesives, elastomers are used in the construction industry for mastic compounds. Neoprene and reclaimed rubber mastics are used to bond gypsum board and plywood flooring to wood framing members. The mastic systems cure by evaporation of solvent through the porous substrates. Elastomer-adhesive formulation is particularly complex because of the need for antioxidants and tackifiers.[1]

Table 5.2 summarizes the properties and characteristics of elastomeric adhesives for non-structural applications. Individual elastomeric adhesive types are discussed below in this listing under their own headings.

Epoxy Adhesives

In the unhardened state the chemical structure for an epoxy resin is characterized by the epoxide group, shown as follows:

$$-C \overset{\displaystyle O}{\underset{\textstyle }{\diagup \diagdown}} C-$$

All epoxy compounds contain two or more of these groups. Epoxy resins may vary from low-viscosity liquids to high melting-point solids. At least 25 types are known. At least 68 curing agents, including the more commonly used amines, primary and secondary amines, and anhydrides are used. Only a few of these are used widely in adhesive formulations.[17]

Of all the thermosetting plastics, the epoxies are probably more widely used than any other plastic in different application areas. There are resin/hardener systems (two-part) which cure at room temperature, as well as one-part systems which require extreme heat cures to develop optimum properties. The proper

Table 5-2: Properties of Elastomeric Adhesives in Nonstructural Applications (Modified after Petrie[1])

	Natural Rubber (polyisoprene)	Adhesives Reclaimed Rubber	Butyl Rubber	Polyisobutylene
Description	Solvent solutions, latexes and vulcanizing type	Solvent solutions, some water dispersions. Most are black; some gray and red	Solvent system, latex	Solvent solution
Curing Method	Solvent evaporation, vulcanizing type by heat or RT (two-part)	Evaporation of solvent	Solvent evaporation, chemical cross-linking with curing agents and heat	Evaporation of solvent
Usual Adherends	Natural rubber, masonite, wood, felt, fabric, paper, metal	Rubber, sponge rubber, fabric, leather, wood, metal, painted metal, building materials	Rubber, metals	Plastic film, rubber, metal foil, paper
Advantages	Excellent resilience, moisture and water resistance	Low cost, applied very easily with roller coating, spraying, dipping or brushing; gains strength rapidly after joining; excellent moisture and water resistance.	Excellent aging characteristics; chemically cross-linked materials have good thermal properties	Good aging characteristics; used as tackifiers in other adhesives. Also provide softness and flexibility and improve adhesion by "wetting out" substrates.
Limitations	Becomes quite brittle with age; poor resistance to organic solvents; does not bond well to metals	Becomes quite brittle with age; poor resistance to organic solvents.	Metals should be treated with an appropriate primer before bonding. Attacked by hydrocarbons.	Attacked by hydrocarbons; poor thermal resistance
Special Characteristics	Excellent tack, good strength, shear strength 30–180 psi (0.21–1.23 MPa); peel strength 0.56 pli (98.1 N/m). Surface can be tack-free to touch and yet bond to similarly coated surface.	Low cost widely used. Peel strength higher than natural rubber; failure occurs under relatively low constant loads.	Low permeability to gases, good resistance to water and chemicals; poor resistance to oils; low strength.	Sticky, low-strength bonds; copolymers can be cured to improve adhesion, environmental resistance and elasticity; good aging; poor thermal resistance; attacked by solvents.

(continued)

Table 5-2: (continued)

	Adhesives		
	Nitrile Rubber	Styrene-Butadiene Rubber (SBR)	Polyurethane
Description	Latexes and solvent solutions compounded with resins, metallic oxides, fillers, etc.	Solvent solutions and latexes. Because tack is low, rubber resin is compounded with tackifiers and plasticizing oils	Two-part liquid or paste
Curing Method	Evaporation of solvent and/or heat pressure	Evaporation of solvent	RT or higher
Usual Adherends	Rubber (particularly nitrile), metal, vinyl plastics	Fabrics, foils, plastic film laminate, rubber and sponge rubber, wood	Plastics, metals, rubber
Advantages	Most stable synthetic rubber adhesive, excellent oil resistance, easily modified by addition of thermosetting resins	Good heat aging and water resistance; uniform appearance, nonstaining light color; disperses easily in hydrocarbon solvents; low cost	Excellent adhesion at cryogenic temperatures and excellent retention of elasticity and shock resistance at these temperatures
Limitations	Does not bond well to natural rubber or butyl rubber	Strength characteristics poor; tendency to creep, lack of tack requires a tackifier for use in adhesives	Poor resistance to hydrolytic degradation (reversion), even in the polyether type
Special Characteristics	Most versatile rubber adhesive. Superior resistance to oil and hydrocarbon solvents. Inferior in tack range, but most dry tack-free, an advantage in pre-coated assemblies. Shear strength of 150 to 2,000 psi (1.03 to 13.8 MPa), higher than neoprene, if cured	Usually better aging properties than natural or reclaimed rubber. Low dead-load strength; bond strength similar to reclaimed rubber. Useful temperature range from −40°F (−40°C) to 60°F (71°C)	Excellent tensile shear strength from −400° to 200°F (−240° to 93°C); poor resistance to moisture before and after cure; good adhesion to plastics

(continued)

Table 5-2: (continued)

	Polysulfide (Thiokols)	Silicone	Neoprene (Chloroprene)
Description	Two-part liquid or paste	Solvent solution: heat or RT-curing and pressure sensitive; and RT-vulcanizing solventless pastes	Latexes and solvent solutions, often compounded with resins, metallic oxides, fillers, etc.
Curing Method	RT or higher	Solvent evaporation, RT or elevated temperature	Evaporation of solvent
Usual Adherends	Metals, wood, plastics	Metals, glass, paper, plastics and rubber, including silicone and butyl rubber and fluorocarbons	Metals, leather, fabric, plastics, rubber (particularly neoprene), wood, building materials
Advantages	Resistance to water, organic solvents, greases, oils, salt water; aging and weathering excellent. Low-temperature flexibility superior (flexible down to –80°F (–62°C)	Retain flexibility (peel) over a wide temperature range. Resistant to moisture, hot water, oxidation, corona discharge, and weathering	Good resistance to water, salt spray, biodeterioration, aliphatic hydrocarbons, acetone and ethyl alcohol, lubricants, weak acids and alkalies. Shows good shear and peel strengths
Limitations	High-temperature resistance poor. Usually softens at 158° to 201°F (70° to 94°C), with little strength retention above 248°F (120°C)	Some forms (acid-curing) may corrode electrical equipment	Unsuitable for contact with aromatic and chlorinated hydrocarbons, certain ketones, and strong oxidizing agents. Cold flow at shear strengths over 2 psi (2.9 MPa)
Special characteristics	Resistant to wide range of solvents, oils and greases, good gas impermeability; resistant to water, sunlight, ozone; retains flexibility over a wide temperature range; not suitable for permanent load-bearing applications	Of primary interest in pressure-sensitive type used for tape. High strengths for other forms are reported from –130°F to 500°F (–73°C to 260°C); limited service to 700°F (371°C). Excellent dielectric properties	Superior to other rubber adhesives in most respects—strength, quick-setting; maximum temperature to 200°F (93°C), sometimes 350°F (177°C); aging good; resistant to light, weather, mild acids and oils

selection of various hardeners, resins, modifiers and fillers can provide the desired properties for any particular application. Because of this wide versatility and their basic adhesive qualities, epoxies provide excellent structural adhesives which can be engineered to widely different specifications. Since epoxies are 100% reactive and no volatiles are produced during cure, essentially no shrinkage occurs during polymerization. Epoxy adhesives can be formulated to meet a wide variety of bonding requirements. Systems can be designed to operate at temperatures as low as $-250°F$ $(-157°C)$ and as high as $400°F$ $(204°C)$.[24]

Epoxy adhesives have good adhesion to most materials, in addition to excellent cohesive strength (good attraction to itself). Epoxy adhesives also have excellent chemical resistance and good elevated-temperature capabilities. As with many other structural adhesives, to obtain optimum strength, particularly under adverse conditions, proper surface preparation must be used on the substrates. Epoxies give good to excellent bonds to steel, aluminum, brass, copper and most other metals. Similar results are obtained with thermosetting and thermoplastic plastics and with glass, wood, concrete, paper, cloth, and ceramics. The material to be bonded will usually dictate the formulation to be used. Unfortunately, epoxy adhesives have relatively low peel strengths.[24]

One-part epoxy adhesives include solvent-free liquid resins, solutions in solvent, liquid resin pastes, fusible powders, sticks, pellets and paste, supported and unsupported films, and preformed shapes to fit a particular joint. *Two-part epoxy adhesives* usually are comprised of the resin and curing agent, which are mixed just prior to use. The components may be liquids, putties, or liquid and hardener powder. They may also contain plasticizers, reactive diluents, fillers and resinous modifiers. The processing conditions are determined by the curing agent employed. In general, two-part systems are mixed, applied within the recommended pot life (a few minutes to several hours), and are cured at room temperature for up to 24 hours, or at elevated temperatures to reduce the cure time. Typical heat cures range from 3 hours at $140°F$ $(60°C)$ to 20 minutes at $212°F$ $(100°C)$.[7]

With an aliphatic amine (e.g. diethylenetriamine) curing agent at room temperature, the resin is sufficiently cured in 4-12 hours to permit handling of the bonded assembly. Full strength develops over several days. A compromise between cure rate and pot life must be made where too rapid a cure at room temperature results in the formulation of an unspreadable mixture in the mixing pot. Heat build-up (exothermic reaction) can be restricted by lowering the temperature of the mixture, limiting the size of the batch, or by using shallow mixing containers. Actions such as these will extend the pot life. Contact bonding pressures usually suffice, but small pressures from 2.2-2.9 psi (0.016-0.02 MPa) provide more uniform joints with maximum strength. One-part systems incorporate a hardening agent which requires heat to activate curing. A period of 30 minutes at $212°F$ $(100°C)$ is typical.[7]

Hardening agents used with epoxy adhesives—Hardeners used in curing bisphenol-A epoxy resins, the type most commonly used in adhesives, include the following:[4]

- *Aliphatic polyamine hardeners.* These are used in adhesive systems capable of curing at normal or slightly elevated temperatures. The most important examples are diethylenetriamine (DETA), triethylenetetramine (TETA), and diethylenepropylamine (DEAPA).

- *Fatty Polyamides (Versamids®, General Mills, Inc.).* These are condensation products of polyamines and unsaturated fatty acids. They are high-melting linear polyamides of the nylon type, containing carboxyl end groups and amide groups along the chain. The amount of hardener required for curing is large and the proportion not critical. These materials are used to impart *flexibility*, as well as for curing. Fatty polyamides are probably the most widely used epoxy curing agents.

- *Aromatic polyamine hardeners.* These mostly solid hardeners include metaphenylenediamine (MPD), diaminodiphenylmethane (MDA), and diaminodiphenyl sulfone (DDS). In general, these hardeners provide poorer bond strengths and are more sensitive to temperature cycling than the aliphatic amines. Shrinkage is also high.

- *Anhydride hardeners.* These materials are organic polycarboxylic anhydrides. Most require severe curing cycles. They provide thermal stability superior to that of the amines. Anhydride-cured epoxies are often brittle and require a flexibilizer, which will unfortunately result in reduced heat and chemical resistance.

- *Boron trifluoride hardeners.* Boron trifluoride monoethylamine (BF-MFA) melts at 203°F (95°C) and is used in one-part adhesives.

- *Miscellaneous curing agents.* The most important is dicyandiamide (DICY), used frequently in metal bonding. This material melts at about 392°F (200°C) and is nonreactive at room temperature, so it is convenient for use in a one-package adhesive in the form of a powder or rod.

Epoxy-Phenolic Adhesives

These relatively expensive adhesives account for only a small fraction of the current usage of structural adhesives. They are used primarily for military applications designed for service between 300 and 500°F (149 and 260°C). Epoxy-phenolics are blends of thermosetting phenolic and epoxy resins. They are supplied as viscous liquids, which may contain solvents, or as glass-cloth or fabric-supported films or tapes. They are often modified with fillers and thermal stabilizers.[12]

Solvent blends are usually force-dried at 176-194°F (80-90°C) for 20 minutes before assembly of adherends. Curing is generally for 30 minutes at 203°F (95°C) with contact pressure, followed by 30 minutes to 2 hours at about 329°F (165°C) and 10-58 psi (0.07-0.4 MPa) pressure. Postcuring is used to obtain optimum curing at elevated temperatures.[7]

Applications are for high-temperature structural bonding of metals (including copper and alloys, titanium, galvanized iron and magnesium, glass and ceramics, and phenolic composites. Epoxy-phenolics are also used in bonding honeycomb sandwich composites. Liquid types are often used as primers for tapes.

These materials display excellent shear and tensile strength over a wide temperature range. Films give better strengths than liquid systems. Peel and impact strengths are usually poor.[7]

Strength data for epoxy-phenolic film and tape adhesives are given below under *Film and Tape Adhesives*. With these materials, resistance is good to weathering, aging, water, weak acids, aromatic fuels, glycols and hydrocarbon solvents. The service-temperature range is usually −76 to 500°F (−60 to 200°C), but special formulations are suitable for cryogenic temperatures down to −436°F (−260°C).[7]

Epoxy-Polysulfide Adhesives

These materials are the products of a reaction between an epoxy resin and liquid polysulfide polymer, usually catalyzed by an added tertiary amine.[7] They are available as two-part liquids or pastes that are usually cured at room temperature or higher to rubbery solids that provide bonds with excellent flexibility and chemical resistance. Epoxy-polysulfides bond well to many different substrates. Shear strength and elevated-temperature properties are low, but resistance to peel and low temperature is good.[1,7]

Curing is usually for 24 hours at 68°F (20°C), or up to 20 minutes at 212°F (100°C). Bonding pressures are low, in the range of 10–23 psi (0.07–0.16 MPa). A disagreeable sulfur odor forms during processing, making ventilation important.[7] Resistance to water, salt spray, hydrocarbon fuels, alcohols and ketones is good. Weathering properties are excellent. Epoxy-polysulfide adhesives are suitable for use down to −148°F (−100°C) and below. Some blends have been used down to liquid-nitrogen temperatures of −325°F (−198.5°C). The maximum service temperature is about 122°F (50°C) to 160–180°F (71–82°C).[25] The moisture resistance of bonds is very good, but may be poorer for stressed bonds. Some formulations will corrode copper adherends.[7]

Applications of epoxy-polysulfide adhesives are primarily for structural assemblies requiring some degree of resilience. Epoxy-polysulfides are used in bonding concrete for floors, airport runways, bridges and other concrete structures, metals, glass and ceramics, wood, rubber, and some plastics. They are particularly good for outdoor applications where temperature extremes (freeze-thaw cycles) will be met.[7]

Epoxy-polysulfides can be heavily filled without adversely affecting their properties.[25]

Film and Tape Adhesives (See also *Alloyed Adhesives* above)

A number of high-strength structural adhesives are currently supplied in film and tape form. Although the bond strengths provided by both film and tape and one-component pastes are generally similar, there are several advantages of the film and tape:[3]

- Provides uniform, controlled glue line thickness.

- Speed and ease of application (a clean, solvent-free operation is facilitated).

- Two-sided films can be prepared for use in lightweight sandwich construction. The honeycomb side will provide good filleting, while the skin side will provide high peel strength. If one side of the film is tacky, it is easier to align the assembly to be bonded.

In some film adhesives a cover or knitted fabric that may be used to support the polymer film will also carry part of the load and will provide improved bond strength by more efficient distribution of the applied forces. Film adhesives are produced in two forms, *unsupported*, or alternatively, *supported* on a flexible carrier such as glass, cloth, nylon or paper. The carrier will usually have little effect on adhesive properties.[3] The adhesive polymer is usually elastomeric, blended with curing agents, fillers and other ingredients and is usually extruded, calendered, or cast into 5–15 mil-thick *unsupported* films. This type is called *film adhesive* When the mixture is cast or calendered onto a mesh support, such as woven or nonwoven mesh of glass or other fibers, the resulting product is called *tape adhesive*. Films and tapes may be either soft and tacky, or stiff and dry. They may be room temperature storable, or may require refrigeration between manufacture and time of use. Most film and tape adhesives are cured at elevated temperatures and pressures. Film and tape adhesives differ from paste and liquid adhesives in that the former contain a high proportion of high-molecular-weight polymer. The 100%-solids paste and liquid adhesives contain only low-molecular-weight resins to permit them to remain fluid and usable. The film and tape adhesives contain components that permit them to be much tougher and more resilient than paste adhesives. Figures 5-1 and 5-2 compare typical tensile shear data for a number of adhesive types. It should be noted that the best film and tape types have higher peak values and broader service temperatures than the best 100%-solids adhesives.[26]

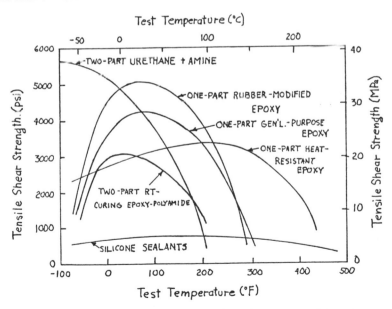

Figure 5-1: Typical tensile-shear strength data for paste and liquid adhesives. Modified after Reference 12.

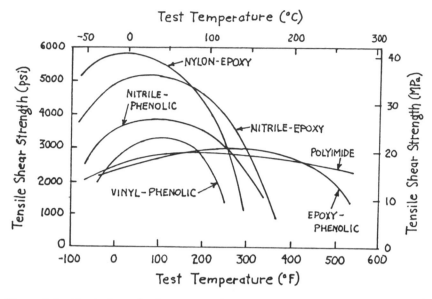

Figure 5-2: Typical tensile-shear strength data for tape-, film- and solvent-based adhesives. Modified after Reference 12.

The handling and reliability advantages of tape and film adhesive stem from the fact that they come ready to use, with no need for mixing, no degassing, and no possibility for error in adding catalyst. Tapes permit a variety of lay-up techniques which facilitate the production of a virtually defect-free structure. The use of a mesh support helps control the bond-line thickness with tape adhesives, avoiding thin, adhesive-starved areas where curvature or external pressure is the greatest.[12]

Tape and film adhesives are generally composed of three components.[12] (Exceptions to this generalization are the epoxy-phenolic adhesives, which are composed of two thermosetting adhesives.)

- A high-molecular-weight "backbone" polymer providing the elongation, toughness, and peel. (This is the thermoplastic or elastomeric component).

- A low-molecular-weight cross-linking resin, invariably either an epoxy or phenolic (thermosetting types).

- A curing agent for the cross-link resin.

Film and tape adhesives are also frequently called "two-polymer" or "alloyed adhesives." With few exceptions, all successful film and tape adhesive are, or have been, one of the types shown in Tables 5.3 and 5.4. The adhesive types based on phenolic cross-linking resins liberate volatiles during cure, while the types based on epoxies need only enough pressure to maintain alignment and compensate for cure shrinkage.[12]

Table 5-3: Most Important Tape and Film Adhesives[12]

Adhesive Type	Backbone Polymer	Cross-Linking Resin	Catalyst	High-Pressure Cure
Nylon-epoxy	Soluble nylon	Liquid epoxy	DICY-type	No
Elastomer-epoxy	Nitrile rubber	Liquid epoxy	DICY-type	No
Nitrile-phenolic	Nitrile rubber	Phenolic novalac	Hexa, sulfur	Yes
Vinyl-phenolic	PVB or PVF	Resol phenolic	Acid	Yes
Epoxy-phenolic	Solid epoxy	Resol phenolic	Acid	Yes

Table 5-4: Range of Bond Strengths of Tape and Film Adhesives at Room Temperature[12]

Adhesive Type	Tensile-Shear Strength (psi)	(MPa)	......T-Peel Strength...... (lb/in)	(N/m)
Nylon-epoxy	5,500-7,200	34-49	80-130	14,000-22,750
Elastomer-epoxy	3,700-6,000	26-41	22-90	3,850-15,750
Nitrile-phenolic	3,000-4,500	21-31	15-60	2,625-10,500
Vinyl-phenolic	3,000-4,500	21-31	15-35	2,625-6,065
Epoxy-phenolic	2,000-3,200	14-22	6-12	1,050-2,100

Furane Adhesives

These are dark-colored synthetic thermosetting resins containing the chemical group known as the furane ring:

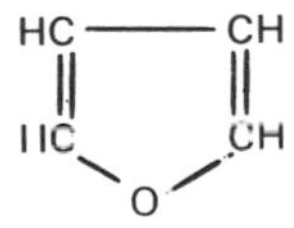

These compounds include the condensation polymers of furfuraldehyde (furfural) and furfuryl alcohol. Upon addition of an acid, these furane compounds polymerize, passing through a liquid resinous state, and have adhesive properties. Volatile loss during cure is low, so bonding pressure need not be high. Resistance to boiling water, organic solvents, oils, and weak acids and alkalies is good. However, strong oxidizing agents attack these materials. High-temperature resistance depends on the type and quantity of catalyst. For continuous exposure, service temperatures up to $302°F$ ($150°C$) are acceptable. Furane resin adhesives are used as bonding agents or modifiers of other adhesive materials. Applications include: surfacing and bonding agents for flooring compositions and acid-resistant tiles, chemically resistant cements for tank linings, phenolic laminates (shear strengths up to 5800 psi or 40 MPa), binder resins for explosives and ablative materials used in rockets and missiles at $2282°F$ ($1250°C$) service temperatures, foundry core boxes, binder resins for carbon and graphite products.[4,7]

Furane adhesives are suitable for gap-filling applications because their strength is maintained with thick glue lines. For this reason the resins are used as modifiers for urea-formaldehyde adhesive to improve gap-filling and craze resistance. Since furanes are compatible with a variety of other resins, they are used in mixtures with silicates and carbonaceous materials for chemically resistant grouting compositions.[7]

Hot-Melt Adhesives

Hot-melt adhesives are thermoplastic bonding materials applied as melts which achieve a solid state and resultant strength on cooling. These thermoplastic 100%-solids materials melt at a temperature range from 140–350°F (65–180°C). Theoretically any thermoplastic can be a hot-melt adhesive, but the ten or so better materials usually used are ideally solid up to 175°F (79.4°C) or higher, then *melt sharply* to give a low-viscosity fluid that is easily applied and capable of wetting the substrate to be bonded, followed by rapid setting upon cooling. When hot-melt adhesives are used, such factors as softening point, melt viscosity, melt index, crystallinity, tack, heat capacity, and heat stability must be considered, in addition to the usual physical and strength properties.[3,7]

The materials used in hot-melt applications are generally not newly developed materials. However, the combination of *properly formulated resins* and of *application equipment* to handle these resins has had much to do with the success of hot-melt technology.[3] While currently used hot-melt adhesives usually melt at about 175°F (79.4°C), they are usually applied at much higher temperatures, from 300–550°F (149–288°C). In addition to the thermoplastic polymer used, other ingredients are added to improve processing charactertistics, bonding characteristics, or service properties. *Stabilizers* are used to retard oxidation, *tackifiers* to improve bond strength, *wax* to reduce viscosity and to alter surface characteristics, and various *fillers* to increase viscosity, melting point, and bond strength. Hot melts are sold only by manufacturer's number or name designation, with no generic identification, as is common in most other adhesives. For this reason, comparisons between competing brands of similar hot-melt adhesives is not easy.[27]

One of the most important characteristics of hot-melt adhesive is *service temperature*. Since hot melts have low melting temperatures, service temperatures are necessarily low, which is a disadvantage. These materials also *creep* under load and with time. Thermoplastics have some characteristics of viscous liquids and, with few exceptions, are not dimensionally stable under load. For this reason, hot melts are recommended primarily for hold-in-place operations with negligable load requirements. The main disadvantages of hot melts are limited strength and heat resistance. Unlike other adhesives, the set-up process is reversible and, at about 170°F (77°C) most hot melts begin to lose strength. The maximum shear load capacity is usually about 500 psi (3.4 MPa).[28] Lap-shear strengths up to 630 psi (4.3 MPa) have very recently been reached with hot-melt adhesives used to bond untreated high-density polyethylene to high-density polyethylene.[29]

Foamable hot-melt adhesives—These materials were introduced in September 1981. The process involves introducing a gas, normally N_2 or CO_2, into the hot-melt adhesive in a volumetrically metered fashion. A patented two-stage gear pump is used. Ordinarily an increase in adhesive volume of 20-70% is obtained. Although all adhesives foam under these conditions, the quality of the foam varies with an individual adhesive. Foamed hot-melt adhesives can be used on the same substrates on which standard hot melts are used. A superior bond can often be achieved on metal, plastics, and paper products, as well as on heat-sensitive and porous substrates. This is because of the foaming characteristics of increased spreading ability, large open time, shorter set time, increased penetration, and reduced thermal distortion over traditional hot melts. *Polyethylene*, in particular, gives excellent results. Applications include gasketing and sealants.[30]

Ethylene-vinyl acetate (EVA) and polyolefin resins—These are the cheapest resin materials used in hot melts. Applications are for bonding paper, cardboard, wood, fabric, etc. for use at −30 to 120°F (−34 to 49°C). Compounded versions can be used for non-loadbearing applications up to about 160°F (71°C). The EVA's and polyethylenes represent the highest volume of hot-melt adhesives used, primarily in packaging and wood-assembly applications.[27]

Polyamide (nylon) and polyester resins—These compounds are the next step up in strength and general service in hot-melt adhesives. These so-called "high-performance" hot melts are being used to assemble products made from glass, hardboard, wood, fabric, foam, leather, hard rubber, and some metals. Service temperatures range from −40°F (−40°C) to about 180°F (82°C). A number of formulations are available that can be used at over 200°F (93°C). Some are capable of being used in no-load applications at over 300°F (149°C).[27]

Polyester-based hot melts are generally stronger and more rigid than the *nylon* compounds. The polyesters have sharp melting points because of their high crystallinity, a decided advantage in high-speed hot-melt bonding. Frequently they have a combination of high tensile strength and elongation. Both nylon and polyester types are sensitive to moisture during application. The nylons combine good strength with flexibility. If nylon compounds are not stored in a dry area, they may pick up moisture, which may cause foaming in the heated adhesive. This problem, in turn, may produce voids in the applied adhesive layer, reducing bond strength. Moisture affects polyesters in a somewhat different way, entering into the molecular structure of the resin, thereby lowering the molecular weight and viscosity. For this reason, polyester hot melts should not be used in a reservoir-type application system.[22]

Other hot-melt adhesives—Other materials, not yet widely used, are the *polyester-amides* and those formulated from *thermoplastic elastomers*. The former are said to have the desirable properties of polyesters, but with improved application characteristics. The principal base polymers in thermoplastic elastomers are currently being used mostly in pressure-sensitive applications, replacing other adhesives, such as contact cements, to eliminate the solvent-pollution problems. These materials are used for applications such as tape products and labels, which require relatively low strength.[27] One particular thermoplastic rubber formulation provides paper tear in the range from −10 to 140°F (−23 to 60°C). This adhesive may also be applied by guns for attaching items such as plastic molding to wooden cabinet doors.[31]

Thermoplastic elastomer hot melts are not as strong as the polyesters, but are stronger than conventional thermosetting rubbers. They provide good flexibility and toughness for applications requiring endurance and vibration resistance, and have good wetting properties. These compounds are very viscous, even at 450°F (232°C), because of their high molecular weights. This makes them more difficult to apply than the nonelastomeric materials, unless they are formulated with other ingredients.[31]

An example of 100%-solids, nonflammable, heat-reactivatable hot-melt adhesive recommended for *structural bonding* of aluminum, steel, copper, brass, titanium, fabric, and some plastics is 3M Company's Scotch-Weld Thermoplastic Adhesive Film 4060. Strength data is shown in Table 5.5.[32] Bonding with this clear amber unsupported film adhesive takes place very rapidly. The speed of bonding is limited only by the time needed to reach the optimum bonding temperature of 300°F (149°C) at a pressure sufficient to maintain contact between the surfaces to be bonded. The adhesive can also be preapplied to parts days or months in advance of the actual bonding operation. When parts are ready to be bonded, heat is applied to the previously applied adhesive to quickly reactivate the material for bonding. This adhesive is available in roll form in thicknesses of 3 and 5 mils (0.076 and 0.13 mm). Typical applications include non-loadbearing honeycomb panels, application of decorative trim, and installation of electronic parts.[32,33]

Table 5-5: Strength Characteristics of Scotch-Weld Thermoplastic Adhesive Film Used to Bond 2024-T3 Clad Aluminum, FPL-Etched*[32]

Temperature		Overlap Shear Strength		T-Peel Strength	
(°F)	(°C)	(psi)	(MPa)	(lb/in)	(N/m)
-67	-55	4,100	28.2	4	700
0	-18	–	–	5	875
75	24	2,200	15.2	30	5,250
180	82	400	2.8	20	3,500

*Bonding cycle 5 minutes at 300°F (149°C) in platen press at 100 psi (0.69 MPa). For additional conditions consult 3M Company Technical Bulletin on this film.[33]

Inorganic Adhesives (Cements)

These materials are widely used because they are durable, fire-resistant, and inexpensive when compared with organic materials. Inorganic adhesives are based on compounds such as sodium silicate, magnesium oxychloride, lead oxide (litharge), sulfur, and various metallic phosphates. The characteristics of some of the more important commercial materials are summarized below.

Soluble silicates (potassium and sodium silicate)—*Sodium silicate* is the most important of the soluble silicates. This material is often called "water glass" and is ordinarily supplied as a colorless, viscous water solution displaying little tack. Positive pressure must be used to hold the substrates together. This material will withstand temperatures up to 2012°F (1100°C). The main applications of sodium silicate adhesives are in bonding paper and making corrugated boxboard, boxes and cartons. They are also used in wood bonding and in bonding metal sheets to various substrates; in bonding glass to glass; porcelain; leather; textiles; stoneware, etc; bonding glass-fiber assemblies; optical glass applications; manufacture of shatter-proof glass; bonding insulation materials; refractory cements for tanks, boilers, ovens, furnaces; acid-proof cements; fabrication of foundry molds; briquettes and abrasive polishing wheel cements. Soluble silicates may also be reacted with silicofluorides or silica to produce acid-resistant cements with low shrinkage and a thermal expansion approaching that of steel.[7,34]

Phosphate cements—These cements are based on the reaction product of phosphoric acid with other materials, such as sodium silicate, metal oxides and hydroxides, and the salts of the basic elements. *Zinc phosphate* is the most important phosphate cement and is widely used as a "permanent" dental cement. It is also modified with silicones to produce dental filling materials. Compressive strengths up to 29,000 psi (200 MPa) are typical of these materials, which are formulated to have good resistance to water (saliva). *Copper phosphates* are used for the same applications, but have a shorter use-life and are used primarily for their antiseptic qualities. Magnesium, aluminum, chromium and zirconium phosphates are also used.[7]

Basic salts (Sorel cements)—There are basic salts of heavy metals, usually magnesite or magnesium cement or magnesium oxychloride cement. They are suitable for dry locations where 2–8 hours hardening will permit their immediate use for bonding many refractory materials, ceramics and glass. The final strength will be in the range of 7000–10,000 psi (48-069 MPa). *Magnesium oxychloride* is an inorganic adhesive notable for its heat and chemical resistance. It is usually supplied as a two-part product (magnesium oxide and magnesium chloride) which is mixed at the time of use. Copper is added to overcome the tendency to dissolve in water. These cements resist damage by cooking fats and greases, repel vermin, and prevent the growth of molds and bacteria. They also conduct static electricity from flooring and similar materials.[7,34]

Litharge cements—Mixtures of glycerine and litharge (lead oxide or PbO) are used as adhesives in the repair of tubs and sinks, pipe valves, glass, stoneware and common gas conduits. A mixture of one part slightly diluted glycerine with two to three parts of lead oxide requires approximately one day to form a crystalline compound. The resulting cement resists weak acids and nitric acid, but reacts with sulfuric acid. These materials have also been used as ceramic seals in potting electronic equipment.[7,34]

Sulfur cements—Liquid sulfur (M.Pt. 730°F or 388°C) can really be considered as an inorganic hot-melt adhesive. This material should not be exposed to temperatures higher than 199°F (93°C) because of a marked change in the coefficient of expansion at 205°F (96°C) as a result of a phase change. The addition of carbon black and polysulfides improves its physical properties. Tensile strengths of about 580 psi (4.0 MPa), falling to 435 psi (3.0 MPa) have been reported after 2 years exposure in water at 158°F (70°C). The principal use of sulfur cements is for acid tank construction, where high resistance to oxidizing acids, such as nitric and hydrofluoric acid mixtures at 158°F (70°C) is required. Resistance to oleic acid, oxidizing agents and strong bases or lime is poor. Adhesion to metals, particularly copper, is good.[7]

Melamine-Formaldehyde Adhesives (Melamines)

These synethetic thermosetting resins are condensation products of un-substituted melamines and formaldehyde. They are equivalent in durability and water resistance to phenolics and resorcinols. Melamines are often com-bined with ureas to lower costs. The service temperatures are higher than the ureas.[1,4,7]

Savia[35] has discussed amino resins, including melamines, in considerable detail. Another very recent comprehensive discussion is by Pizzi.[36]

Microencapsulated Adhesives

Microencapsulation is a method for separating a reactivating solvent or re-active catalyst from the adhesive base. The materials, whether solids or liquids, are packaged in very small "microscopic" capsules. When adhesion is desired, the encapsulated solvent is released by breaking the capsules through heat or pres-sure, and a tacky adhesive with instant "grab" is produced. In addition to sol-vents, small quantities of plasticizers or tackifiers may be contained in the capsules. The capsules are made of gelatin and are insoluble in water and non-reactive to the solvents. *Heat-activated adhesives* are another form of micro-encapsulation. A blowing agent is mixed with the solvent in the capsule. Upon application of heat, the blowing agent vaporizes and ruptures the capsule, releas-ing either the entire adhesive or the solvent needed to make the adhesive tacky. A third form of encapsulation involves *two-part adhesives*, such as epoxy or polyurethanes. In this type both the base resin and the catalyst are stored in the same container. The catalyst can be released by pressure, or by other means, to cure the adhesive.[26]

Natural Glues

These adhesives include vegetable and animal-based materials and are usually limited to use with paper, paperboard, wood and metal foil. Shear strengths range from 5-1000 psi (0.034-6.9 MPa). Few natural glues retain their strengths at temperatures over $212°F$ ($100°C$). Most of these materials have poor resistance to moisture, vermin and fungus, but they do have good resistance to organic solvents. Typical natural glues are discussed below.

Vegetable glues—These adhesives are soluble or dispersible in water and are produced or extracted from natural sources. Other adhesives, such as rubber cements, nitrocellulose and ethyl cellulose lacquer cements, are also produced from plant sources, but are not water-soluble or water-dispersible and are there-fore not classified as vegetable glues.[37]

Starch—Starch adhesives are derived primarily from the cassava plant, al-though other starch sources may be used. The starch is usually heated in alkaline solutions, such as NaOH, then cooled to room temperature to prepare the dis-persions. After cooling, they are applied as cold-press adhesives. They develop their strength by loss of water into the wood substrate. Tack is developed rapidly and normal wood processing is for 1-2 days at room temperature and 72-105 psi (0.5 - 0.7 MPa). Starch adhesives are also used for paper cartons, bottle labeling, and stationery applications. Joint strengths are low compared to other vegetable adhesive types. These adhesives are resistant to water and biodeterioration, but their resistance to these environments is improved by adding preserva-tives.[7;38] An example of starch adhesive application is found in Military Speci-fication MIL-A-17682E, "Adhesive, Starch" (see entry under Chapter 12), for use in mounting paper targets to target cloth. In this specification the starch source must be wheat. Perhaps some readers will remember using "flour and water" to make a simple paste for school and home use. It should be noted that this source of starch is not subjected to heating in alkaline solution, and, there-fore, does not have the strength of the commercial material.

Dextrins—These are degradation products of starch produced by heating in the presence or absence of hydrolytic agents. Usually basic or acid-producing substances are used. Other catalytic agents used include enzymes, alkali, and oxidizing agents. These adhesives can be used in formations using many differ-ent substrates. Their applications are primarily for paper and paperboard. Laminating adhesives are usually made from highly soluble white dextrins and contain fillers such as clay, as well as alkalies or borax. Blends of white dextrins and gums are common. Urea-formaldehyde is often added to produce water resistance.[37] Military Specification MIL-A-13374D, "Adhesive, Dextrin, for Use in Ammunition Containers", covers four classes of dextrin adhesives for use in making spirally wound containers and chipboard spacers (see entry under Chap-ter 12).

Soy(a)bean glue—Nitrogenous protein soybean is the most common plant protein adhesive derived from seeds and nuts, hemp and zein. These adhesives are cheap and can be used for making semi-water-resistant plywood and for coat-ing some types of paper. The protein from the soybean is separated out mechani-cally and used much like casein protein, with the addition of calcium salts to improve water resistance. The soybean glues are used as room-temperature-set-ting glues to produce interior-type softwood plywood, where only limited moisture resistance is needed. Soybean glues have been largely replaced by protein-blend glues, such as combinations of soybean and blood proteins, and by phenolics for plywood bonding. Cold-press bonding of plywood with soybean adhesives requires 4-12 hours at 102-145 psi (0.70-1.0 MPa). Hot-press bonding requires 3-10 minutes at 212-284°F (100-140°C) and 145-218 psi (1.0-1.5 MPa) pressure. Water resistance of soybean glues is limited. However, like casein glues, they become stronger on drying. These materials will biodeteriorate under humid conditions unless inhibitors are used. Heat resistance is poor and they are restricted to indoor applications because of their poor durability. Fillers are used for paper and paperboard laminating, cardboard and box fabrication, and particle binders.[7,38] Lambuth[39] has discussed these glues at length.

Rosin—The most common form of rosin adhesive is *colophony*, a hard amorphous substance derived from the oleoresin of the pine tree. This material is used in solvent solution as a hot-melt mastic. It has poor resistance to water, is subject to oxidation, and has poor aging properties. Plasticizers are usually added to reduce its brittleness. Bond strengths are moderate and develop rapidly. These materials are used as temporary adhesives in bonding paper and as label varnishes. They are also used as components of pressure-sensitive adhesives based on styrene-butadiene copolymers and in hot-melt adhesives and tackifiers. These materials have been largely replaced by synthetic-resin adhesives.[7] A specialized form of rosin adhesive is Canada Balsam, covered by the recently cancelled Military Specification, MIL-C-3469C, "Canada Balsam." This material was intended for cementing optical elements.

Glues of animal origin—These glues include glues derived directly and indirectly from animals, including mammals, insects and fish, as well as milk products. The category should not be called "Animal Glues", since that term is a specialized form.

Casein glue—Casein is the protein of (skim) milk, from which it is obtained by precipitation. Dry-mix casein glues are simply mixed with water before use. Casein glues are used at room temperature and set or hardened by loss of water to the wood substrate, and by a certain degree of chemical conversion of the protein to the more insoluble calcium derivative. Applications are in packaging, where the adhesives are used to apply paper labels to glass bottles. In woodworking they are used in laminating large structural timbers for interior applications. They are also used for general interior woodworking applications, including furniture. They cannot be used outdoors, although they are more resistant to temperature changes and moisture than other water-based adhesives. Casein adhesives will tolerate dry heat up to 158°F (70°C), but, under damp conditions, the adhesives lose strength and are prone to biodeterioration. Chlorinated phenols can be used to reduce this tendency to deteriorate, however. These glues are often compounded with materials such as latex and dialdehyde starch to improve durability. Resistance to organic solvents is generally good.[7,38] Salzburg[40] had described the use of these adhesives in some detail.

Blood albumen (Blood glues)—These glues are used in much the same manner as casein glues. The proteins from animal blood in slaughtering are precipitated out, dried, and sold as powders, which are then mixed with water, hydrated lime, or sodium hydroxide. The blood proteins undergo some heat coagulation so that they can be hardened by hot-pressing, as well as by loss of water. Processing is usually for 10–30 minutes at 158–248°F (70–230°C) for plywood, with bonding pressures of 72–102 psi (0.5–0.7 MPa). Porous materials require only several minutes at 176°F (80°C). Cold-press applications are also possible. Blood glues are used to a limited degree in making softwood plywood, sometimes in combination with casein or soybean proteins. They have also been used as extenders for phenolic-resin glues for interior-type softwood plywood. Another application is for bonding porous materials, such as cork, leather, textiles and paper, and for packaging, such as in bonding cork to metal in bottle caps.[7,38] Lambuth[39] has described blood glues in some detail.

Animal glues (bone and hide glues)—The term "animal glues" is normally used for glues prepared from mammalian collagen, the principal constituent of skin, bone and sinew. (What reader has not heard of old or injured horses being useless for any other purpose and therefore being headed for the "glue factory?") Other types of glues obtained from animal sources are usually referred to according to the material from which they are derived (casein, blood, fish, shellac, which see). *Bone glues* are made from animal bones, while *hide glues* are made from tannery waste. These glues are supplied as liquids, jellies, or solids in the form of flakes, cubes, granules, powders, cakes, slabs, etc. for reconstitution with water. They are used primarily for furniture woodworking, but are also used for leather, paper and textiles, and as adhesive binders for abrasive paper and wheels. Liquid hide glues are normally supplied with a gel depressant added to the molten-glue mixture. The purpose is to assure that the dispersion remains liquid when cooled to room temperature so that it can be bottled. These glues harden only by loss of water to the adherend, which must be relatively porous.[7,41] Barker[42] has discussed the current market for animal glues in a very recent article.

The processing conditions for animal glues are dependent on the type of glue. These glues set at temperatures in the range of 176-194°F (80-90°C), or they may set at room temperature. Bonding pressures range from contact pressure to 145-203 psi (1-1.4 MPa) for hardwoods and 51-102 psi (0.35-0.70 MPa) for softwoods. Application periods range from 5 minutes to several hours. Hide glues are stronger than bone glues. The bond strengths of these glues usually exceed the strength of wood and fibrous adherends. High-strength joints are achieved where the bonds are kept under dry conditions. Structural applications are limited to interior use, however. These glues are gap-filling, which makes them useful where close-fit joints are not feasible and filler products are not required.[7]

Animal glue is the primary adhesive component in gummed tapes used in sealing commercial solid fiber and corrugated shipping uses, as well as the more common lightweight types used in retail packaging. Hubbard[43] has described the uses and properties of animal glues in considerable detail.

Fish glues—These are by-products of desalted fish skins, usually cod, and have properties similar to the animal skin and hide glues, which have largely replaced them in woodworking applications. Fish glues were the forerunners of all household glues. Many of the original industrial applications developed because fish glue was liquid and had an advantage over animal glues, which required a heated glue pot. Fish glue has been in use well over 100 years. Even with the many synthetic adhesives available today, there are applications which require the unique properties of fish glue.[7,44]

Fish glues are available in cold-setting liquid form which does not gel at room temperature. Solvents such as ethanol, acetone, or dimethylformamide may be added to facilitate the penetration of the glue into substrates which may be coated or finished (e.g. paper, leather and fabrics). These glues may be exposed to repeated freezing and thawing cycles without adverse effects. Initial tack is excellent on remoistening dry fish-glue films with cold water. Water resistance of dried glue films can be improved by exposure to formaldehyde vapors, which insolubilize the fish gelatin component. Fish glues bond well to glass, ceramics, metals, wood, cork, paper and leather. The main uses are in the preparation of gummed tapes with animal/fish glue compositions, and in the bonding of stationery materials. Latex, animal glues, dextrins and polyvinyl acetate adhesives are sometimes modified with fish glues to improve wet-tack properties. High-purity fish glues are important photoengraving reagents. The service temperature range is from 30 to 500°F (−1 to 260°C). Shear strength (ASTM D 905) is 3200 psi (22 MPa) with 50% wood failure.[7,44]

Norland[44] has discussed fish glues in considerable detail.

Shellac—Shellacs are thermoplastic resins derived from insects. They are used in alcoholic solutions or as hot-melt mastics. They have good electrical insulating properties, but are brittle unless compounded with other materials. Shellacs are resistant to water, oils, and grease. Bond strengths are moderate.

Shellacs are used to bond porous materials, metals, ceramics, cork and mica. They are also used as adhesive primers for metal and mica, for insulating sealing waxes, and as components of hot-melt adhesives. Shellacs are the basic components of de Khotinsky cement. Their use has decreased in recent years because of their high cost.[7]

Shellac is used as an alternative to alkyd resins in binding mica splittings to produce mica board. This is pressed into shapes used as insulation in electric motors, generators, and transformers. Mica tape, used as insulators in motors and generator coil slots, is fabricated by bonding mica flakes to glass cloth and tissue paper with shellac or silicones.[45]

Neoprene (Polychloroprene) Adhesives

This type of synthetic rubber is used extensively to bond aluminum. Its characteristics are summarized in Table 5.2 under Elastomeric Adhesives. Neoprene is ordinarily used in organic solvents for convenient application. Although neoprene is similar in properties to natural rubber, it generally is stronger and has better aging and high-temperature-resistance properties. Neoprene solvent-based cements are also used extensively as shoe adhesives. For structural applications neoprene is usually combined with a phenolic resin plus a number of other additives for curing and stabilizing the mixture (neoprene-phenolic, which see). Both cold-setting and heat-curing formulations can be prepared.[3]

Neoprenes are general-purpose adhesives used for a wide range of materials. Gap-filling properties are good. Neoprene joints may require several weeks' conditioning to achieve optimum strength. The unalloyed adhesives should not be used for structural applications requiring shear strengths over 290 psi (2 MPa) because they are liable to cold flow under relatively light loads. Tack retention is generally inferior to natural rubber. Loads of 29–101.5 psi (0.2–0.7 MPa) can be sustained for extended periods soon after bonding.[7]

Steinfink[46] has discussed neoprene adhesives in considerable detail.

Neoprene-Phenolic Adhesives

These alloy adhesives are thermosetting phenolic resins blended with neoprene (polychloroprene) rubber. They are available in solvent solutions in toluene, ketones, or solvent mixtures, or on unsupported or supported films. The supporting medium may be glass or nylon cloth. Neoprene-phenolic adhesive may be used to bond a variety of substrates, such as aluminum, magnesium, stainless steel, metal honeycombs and facings, and plastic laminates, glass and ceramics. Wood-to-metal bonds are often primed with neoprene-phenolic adhesives. Compounding with neoprene rubber increases flexibility and peel strength of phenolic resins and extends the high-temperature resistance. The film form is preferred where solvent removal is a problem. The higher curing temperatures provide the highest strengths. Curing is under heat and pressure. Film is ordinarily cured at 302–500°F (150–260°C) for 15–30 minutes at 51–261 psi (0.35 - 1.8 MPa) bonding pressure. The liquid adhesives are ordinarily dried at 176°F (80°C) and then cured for 15–30 minutes at 194°F (90°C) and contact to 102 psi (0.7 MPa) pressure. The bond may be removed from the hot press while still hot. The liquid adhesive may be used as a metal primer for film types.[1,7]

The normal service temperatures of these adhesives are from -70 to $200°$F (-57 to $93°$C). Because of their high resistance to creep and most severe environments, neoprene-phenolic joints can withstand prolonged stress. Fatigue and impact strengths are excellent. Shear strength, however, is lower than that of other modified phenolic adhesives.[1]

Nitrile-Epoxy (Elastomer-Epoxy) Adhesives

The first term listed in the heading is frequently used to signify elastomer-epoxy, even though this is not the only elastomer-epoxy available. Tables 5.3 and 5.4 summarize some of the important properties of this important adhesive. The maximum bond strengths of these adhesives are generally below the maximum attainable with nylon-epoxies at room temperature. A major advantage of these adhesives, however, is that their peel strength does not decrease as abruptly at subzero temperatures as do the peel values of the nylon-epoxies. The bond durability of these new high-peel elastomer-epoxies is satisfactory, as measured by most long-term moisture tests, but it does not match the durability of the older vinyl-phenolic or nitrile-phenolic types.[12] Nitrile-epoxies should not be used for exposure to marine environments or under continuous immersion in water.[47]

Nitrile-Phenolic Adhesives

These adhesives are usually made by blending a nitrile rubber (see below) with a phenolic novalac resin, along with other compounding ingredients. Usage in tape, film, or solution form is very high. The major uses are in the automotive industry in bonding brake shoes and clutch disks. They are also used in aircraft assembly and in many other smaller applications because of their low cost, high bond strengths at temperatures up to $250°$F ($121°$C) and exceptional bond durability on steel and aluminum. Nitrile-phenolics have exceptionally high durability after extended salt spray, water immersion, and other corrosive environments. They constitute the most important tape adhesives (see Tables 5.3 and 5.4 under Film and Tape Adhesives). Their major disadvantages are the need for a high-pressure cure (up to 200 psi or 1.38 MPa), while the trend is toward cure at lower pressures, and the need for high-temperature ($300°$F or $149°$C), long-term cures, while the trend is toward adhesives which cure rapidly at or below $250°$F ($121°$C).[1]

The liquid nitrile-phenolic adhesives are dried at $176°$F ($80°$C) and cured for 15–30 minutes at $194°$F ($90°$C) at contact to 102 psi (0.70 MPa).[7]

Nitrile Rubber Adhesive

This is one of the most important synthetic thermoplastic elastomers. Nitrile rubber is a copolymer of butadiene and acrylonitrile. Usually the copolymer contains enough acrylonitrile (over 25%) so that good resistance to oil and grease is obtained. The adhesive properties also increase with increasing nitrile content. These adhesives are used to bond vinyls, other elastomers, and fabrics where good wear, oil and water resistance are important. Compatibility with additives, fillers and other resins is another advantage of this material.[3] Table 5.2 under Elastomeric Adhesives above summarizes the properties of nitrile rubber. Morrill and Marguglio[48] have discussed nitrile rubber adhesives in some detail.

Nylon Adhesives

Nylons are synthetic thermoplastic polyamides of relatively high molecular weight which have been used as the basis for several types of adhesive systems. They are used as solution adhesives, as hot-melt adhesives, and as components of other adhesive-alloy types (nylon-epoxy and phenolic-nylon). The high-molecular-weight products are generally referred to as *modified nylons.* Low-and intermediate-molecular-weight materials are also available. The latter two are more commonly used in hot-melt formulations (see Hot-Melt Adhesives), and the modified nylons are often blended with small amounts of a phenolic resin to improve surface wetting (hence nylon-phenolic). Solution systems of low-and intermediate-molecular weight nylon resins can be coated on paper, metal foil, or plastics, and when heat-activated, will act as adhesives for these substrates. The modified nylons have fair adhesion to metals, and good low- and high-temperature properties, good resistance to oils and greases, but poor resistance to solvents.[3,7]

Certain specialty nylon resins with low melting temperatures have been used quite successfully with extrusion techniques. Both nylon 11 and high-molecular-weight polyamide resins chemically related to dimer acid-based polyamides are used in high-strength metal-to-metal adhesives and are applied by extrusion.[3]

Nylon (polyamide) use in hot-melt adhesives is discussed briefly under Hot-Melt Adhesives above.

Nylon-Epoxy Adhesives

These are quite likely the best film-and-tape structural adhesives available. Their tensile strength of over 7000 psi (48 MPa) and climbing-drum peel strengths of over 150 lb/in (26,265 N/m) are the highest available in structural adhesives. These adhesives also have exceptional fatigue and impact strengths. Low-temperature performance is good down to the cryogenic range, except that brittleness occurs at cryogenic temperatures ($-400°F$ or $-240°C$). Other disadvantages are: poor creep resistance and extreme sensitivity to moisture.[3,12] Property data on these adhesives are shown in Tables 5.3 and 5.4 under Film and Tape Adhesives.

Nylon-epoxy film adhesives not only have the unfortunate tendency of picking up substantial amounts of water *before* use, but also tend to lose bond strength rapidly *after* use upon exposure to water or moist air. After 18 months exposure to 95% RH, conventional nitrile-phenolic adhesive loses only a fraction of its initial strength, going from 3000 to 2500 psi (21 to 18 MPa) in tensile shear. On the other hand, one of the best nylon-epoxy adhesives available degraded from about 5000 to 1000 psi (34 to 6.8 MPa) in just two months under the same test conditions.[49] Considerable effort has been made to solve this moisture problem, but there has been little success to date. Nitrile-epoxy or acetal-toughened epoxy film adhesives are still superior in durability.[12]

Phenolic Adhesives

These adhesives, more properly called *phenol-formaldehyde adhesives*, are condensation products of formaldehyde and a monohydric phenol.[4] They dominate the wood adhesives (plywood) field and represent one of the largest volumes of any synthetic adhesive. They are also among the lowest-cost adhesives. Phenolics may be formulated as water dispersions, where penetration into the cell structure of wood is important for the formation of permanent bonds. Beyond the wood and wood products area, *unmodified phenolics* are used mainly as *primers*, to prepare metal surfaces for bonding, and as *binders*, for such varied products as glass wool insulation mats, foundry sand, abrasive wheels, and brake lining composites. Phenolics are provided either as one-component, heat-curable liquid solutions or powders, or as liquid solutions to which catalysts must be added. The curing mechanism is different in these two types.[3]

Acid-catalyzed phenolics—The *acid-catalyzed* type forms wood joints requiring from 1-7 days conditioning, depending on the end use. Metals bonded with these adhesives require priming with a vinyl-phenolic or rubber-resin adhesive before bonding. These adhesives have good gap-filling properties, but they are not recommended as structural adhesives unless their glue-line pH is higher than 2.5. Because of the acidic nature of the adhesive, glass or plastic mixing vessels are required. The mixed adhesive is exothermic (gives off heat) and temperature-sensitive. These adhesives are cured under the following conditions:[7]

General-purpose	3-6 hrs @ 68°F (20°C)
Timber (hardwood)	15 hrs @ 59°F (15°C) and 174 psi (1.2 MPa)
Timber (softwood)	15 hrs @ 59°F (15°C) and 102 psi (0.7 MPa)

The curing time is reduced by increasing the curing temperature. Resistance to weather, boiling water, and biodeterioration is good. Resistance to elevated temperatures is good, but inferior to that of heat-cured phenolic and resorcinol adhesives. Excess acidity due to poor control of the acid catalyst content often leads to wood damage on exposure to warm, humid air. The durability of joints at high and low temperatures for extended periods is usually good. These adhesives are used for woodwork assemblies, where the service temperature range does not exceed 101°F (40°C). Applications are for furniture construction and, to a minor extent, plywood fabrication. This type is also used to join metal to wood for exterior use.[7]

Hot-setting phenolics—The hot-setting form of phenol-formaldehyde adhesive is supplied in spray-dried powder to be mixed with water, as alcohol, acetone or water-solvent solutions, or as glue films. It may be compounded with fillers and extenders. The gap-filling properties of this type of phenolic adhesive are poor and inferior to the acid-catalyzed phenolic adhesives described above. Joints require conditioning up to 2 days. Although durable and resistant to many solvents, the bonds are brittle and prone to fracture under vibration and sudden impact. These adhesive types are used as additives to other materials to form adhesives for glass and metals, or modifying agents for thermoplastic elastomer adhesives, or as components of thermoplastic resin-elastomer adhesives for metal bonding.[7]

Hot-setting phenolic adhesives are processed up to 15 minutes at 212-301°F (100-150°C) and 102-247 psi (0.7-1.7 MPa) bonding pressure. The film type is processed for up to 15 minutes at 248-302°F (120-150°C) and 102-203 psi (0.7-1.4 MPa). This type of phenolic is resistant to weather, boiling water, and biodeterioration. The temperature stability is superior to that of the acid-catalyzed type. Applications are for fabrication of exterior-grade weather- and boil-proof plywood and for bonding glass to metal for electric light bulbs.[7]

Barth[50] has discussed phenolic-resin adhesives in considerable detail. Another very recent source is a comprehensive discussion by Pizzi.[51]

Phenoxy Adhesives

These materials are synthetic thermoplastics in the form of polyhydroxy ethers. They are one-component materials supplied in powder, pellet or film form. They may be dissolved in solvents or supplied as special shapes. Phenoxies act as hot melts and set on cooling. The liquid forms require removal of solvent by drying before bonding. Time and temperature are important in obtaining optimum strength, but bonding pressure is not critical. Typical conditions are: Bond for 30 minutes at 378°F (192°C) or 2–3 minutes at 500°F (260°C), or 10 seconds at 572–662°F (300–350°C). Pressure is contact to 25 psi (0.17 MPa). Phenoxies are used as structural adhesives for rapid assembly of metals and rigid materials, for continuous lamination of metal to metal (cladding) or wood, and flexible substrates, paper, cloth, metal foil and plastic laminations. Other applications are in pipe jointing (with fiber type), assembly of automotive components, bonding of polymeric materials such as polyester film, polyurethane foam, acrylics, and phenolic composites. They are also used as components of hot-melt adhesives for conventional applications.[7]

Phenoxy adhesives withstand weathering and resist biodeterioration. They have excellent resistance to inorganic acids, alkalies, alcohols, salt spray, cold water, and aliphatic hydrocarbons. They swell in aromatic solvents and ketones. Thermal stability is good, with a service temperature range of −52 to 180°F (−62 to 82°C). Resistance to cold flow and creep is high, even at 176°F (80°C). These adhesives provide rigid, tough glue lines with high adhesive strength. Shear strengths are similar to epoxies, and for metals generally exceed 2465 psi (17 MPa), and may approach 3988 psi (27.5 MPa). Film thickness is not critical and can be as little as 0.47 mils (0.012 mm). Liquid adhesives do not usually provide optimum bond strengths, since complete solvent removal may be a problem. Hot-melt adhesive systems can also present difficulties. Thermal degradation may occur before the resin is completely melted, unless plasticizers are used.[7] Plasticizers used are diphenyl phthalate (DPP), tricresyl phosphate (TCP) and dicyclohexyl phosphate (DCHP), which are used in hot-melt formulations. Unplasticized phenoxies give peel strengths of 18-20 lb/in (3152-5253 N/m) in bonding Neolite to Neolite. Formulations with 60% DCHP raise the peel strength to 36–41 lb/in (5078-5213 N/m).[26,52,53] Good adhesion has been obtained with substrates such as copper, brass, steel, aluminum, wood, and many other non-metallic substances.[3]

Polybenzimidazole (PBI) Adhesives

For a general discussion of these high-temperature adhesives, compared with polyimides, see *Aromatic Polymers* above. These adhesives are supplied in film form on glass cloth. Normally filler (usually aluminum) and antioxidants are also components. Polybenzimidazoles are thermoplastics, although their thermoplastic nature is not evident below 700°F (371°C). These materials were developed specifically for use as high-temperature materials. They are relatively stable in air to 550°F (288°C) for short-term applications. Polyimides are superior for long-term strength retention.[3] Polybenzimidazoles are very expensive and are limited to the bonding of high-temperature metals (stainless steel, beryllium and titanium). Polybenzimidazoles are of greatest interest to the aerospace engineer for use in the adhesive assembly of lightweight honeycomb structures for supersonic aircrafts, missiles and other space systems. They are somewhat sensitive to moisture at room temeprature, The lap-shear strength drops gradually on heating to 600°F (316°C), then more rapidly at higher temperatures. Figure 5.3 shows the effect of heat aging of 700°F (371°C), compared with polyimides.[3]

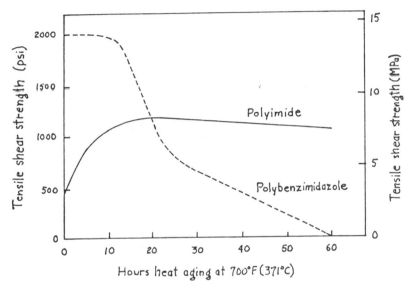

Figure 5-3: Performance of high-temperature adhesives (PBI and PI) at 700°F (371°C). Modified after Reference 3.

Processing is normally carried out in a pre-heated press at 698°F (370°C) with pressure maintained at 4.4 psi (0.03 MPa) for 30 seconds. The pressure is then increased to 87–203 psi (0.6–1.4 MPa) and the glue-line temperature maintained at 698°F (370°C) for 3 hours. The temperature is then reduced to 500°F (260°C) or less and the assembly removed from the press. Autoclave techniques can also be used. For improved mechanical properties, post-curing in an inert atmosphere (nitrogen, helium, or vacuum oven) is recommended. Recommended conditions are 24 hours each at 601°, 653°, 698° and 752°F (316°, 345°, 370° and 400°C) followed by 8 hours at 800°F (427°C) in air for optimum properties.[7] Obviously these "literature" recommendations should be checked against manufacturers' recommendations, but they do give an idea of the complexity of PBI processing.

PBI adhesives have good resistance to salt spray, 100% humidity, aromatic fuels, hydrocarbons, and hydraulic oils. About 30% loss of strength occurs on exposure to boiling water for two hours. Electrical properties are fairly constant throughout the temperature range up to 392°F (200°C). Thermal stability at high temperatures for short periods is good (1022°F or 540°C) for 10 minutes or 500°F (260°C) for 1000 hours. The useful service-temperature range as adhesives is −418°F (−250°C) to 572°F (300°C).[7] Note that this includes the cryogenic temperature range.

Polyester Adhesives

Polyester adhesives may be divided into two distinct groups, *saturated* (thermoplastic) and *unsaturated* (thermosetting). The saturated polyesters are reaction products of difunctional acids plus difunctional alcohols or glycols. Their adhesive uses are minor, except in hot melts (high performance). The unsaturated (thermosetting) polyesters which require a catalytic cure have a few uses as adhesives. These usually involve the bonding of polyester substrates. Polyester adhesives are also used in patching kits for repair of fiberglass boats, automobile bodies, and concrete flooring. Other minor uses are in bonding polyester laminates to polyester or to metal, and as adhesives for optical equipment. CR-39 allyl diglycol carbonate is an example of the latter. This material in the cured condition exhibits improved abrasion and chemical resistance over other transparent adhesive resins and displays the good heat resistance and dimensional stability associated with thermosetting systems. These properties are retained on prolonged exposure to severe environmental conditions, CR-39, meaning Columbia Resin 39, is an allyl resin, a special type of unsaturated polyester.[3,7,27]

Polyimide (PI) Adhesives

For a general discussion of these high-temperature adhesives, compared with polybenzimidazole, see *Aromatic Polymers* above. These adhesives are synthetic thermosetting resins formed by reaction of a diamine and a dianhydride. As with polybenzimidazoles, they were developed specifically for high-temperature aerospace applications. They are superior to PBI's for long-term strength retention, as shown in Figure 5.3.[7]

Polyimide adhesives are supplied as solvent solutions of the polyimide prepolymer, or as film, usually containing a filler, such as aluminum powder, with glass-cloth interliner. Processing is as follows: liquid types—removal of solvent by heat or under reduced pressure, and by precuring the resin to the desired degree (B-staging), usually at 212-302°F (100-150°C). The volatile content may range from 8-18% w/w after B-staging. Final cure (C-staging) is carried out by stages over the range 302-572°F (150-300°C), or higher. The film types may require B-staging. Their typical cure schedule involves heating to 482°F (250°C) over a 90-minute period and maintenance at this temperature for 90 minutes. Postcuring at higher temperatures up to 572°F (300°C) and beyond is recommended where maximum mechanical properties are required. Bonding pressures should be 38-137 psi (0.26-0.65 MPa).[7] Like PBI's the adhesives require tedious preocessing, compared to other adhesives.

Polyimide adhesives have good hydrolytic stability and salt-spray resistance and excellent resistance to organic solvents, fuels and oils. They are resistant to strong acids, but are attacked slowly by weak alkalies. Ozone causes bond deterioration. The service temperature range is −321 to 500°F (−196 to 260°C) for long-term exposure, but these materials will withstand short exposures (200 hours) up to 662°F (250°C) and 10 minutes at 711°F (377°C). Polyimides are exceptionally good high-temperature electrical insulating materials and also have exceptional resistance to atomic radiation (electrons and neutrons). These materials are used as structural adhesives for high and low-temperature applications, including cryogenic, for bonding metals such as stainless steel, titanium, and aluminum, and in aircraft applications generally. They are also used in preparing glass-cloth-reinforced composites for electrical insulation, and in bonding ceramics.[7]

Polyimide adhesives require higher cure temperatures than epoxy-phenolic adhesives (which see). Curing at 500°F (250°C) is usually adequate where service temperatures do not exceed that temperature. Volatiles are released during the cure of PI adhesives, and for this reason the best results are obtained when the volatiles are free to escape (e.g. honeycomb or perforated-core structures). For long-term aging at temperatures in the range of 400-600°F (204-316°C) polyimides are superior to PBI and epoxy-phenolic adhesives.[7] (See Figure 5.3 for comparison with PBI).

According to Edson[54] polyimide adhesives are capable of withstanding temperatures up to 600°F (316°C) for hundreds of hours, and up to 400°F (204°C) for thousands of hours. Thermal "spikes" of 1000-1500°F (538-816°C) can also be accommodated. Polyimides are about 5 times as expensive as epoxies (in 1983). According to Alvarez[55] polyimide adhesives can be processed at 350°F (177°C) and postcured at 450°F (232°C) to produce bonds capable of 600°F (316°C) service. Other new developments include new "exchange" polymers with a processing range of 350°F (177°C) to 550°F (288°C) at a pressure of 15 psi (0.10 MPa). These materials will withstand 600°F (316°C) with normal 450°F (232°C) postcuring. According to Steger[56] polyimides are useful for bonding high-temperature metals like titanium and graphite/polyimide composite for use at 500-600°F (260-316°C).

Serlin et al.[57] have discussed polyimides in considerable detail.

Polyisobutylene Adhesives

These thermoplastic elastomers are covered briefly in Table 5.2 in the section on Elastomeric Adhesives. Polyisobutylene is a homopolymer. Stucker and Higgins[15] have discussed these adhesives in some detail.

Polystyrene Adhesives

Polystyrene is a transparent, colorless thermoplastic resin available in solvent-solution or aqueous-emulsion form. In both forms applications are limited to conditions where at least one of the adherends is porous. An example is sticking polystyrene tiles onto a plaster wall. Polystyrene adheres well to wood, but not to plastics, except polystyrene itself. For bonding polystyrene, a low-molecular-weight styrene polymer with a perioxide catalyst is used. This adhesive polymerizes in the glue line.[4]

With some woods, shear strengths up to 1885 psi (13 MPa) may be obtained Polystyrene is used as a modifier for other adhesives such as unsaturated polyesters, hot-melt materials, and in optical cements. Resistance to high temperatures is limited. The heat-distortion temperature is about 171°F (77°C). Electrical insulating properties are very good. Polyester adhesives have good resistance to water, nuclear radiation, and biodeterioration. However, they generally have poor resistance to chemicals. Other undesirable properties include high flammability and a tendency to brittleness and crazing. Copolymers of styrene and butadiene (SBR), also discussed in Table 5.2, are much less brittle and more valuable as adhesives. These materials are commonly used in footwear for bonding leather and rubber soles.[4,7]

Polysulfides (Thiokols)

Polysulfides are flexible materials belonging to the synthetic rubber family. Some of the more important characteristics of polysulfide adhesive/sealants are tabulated in Table 5.2. Although polysulfides are primarily used as sealants for automotive, construction and marine uses, they are used to some extent as flexibilizing hardeners for epoxy adhesives. Their sulfur linkages combine good strength with the ability to rotate freely, resulting in a strong, flexible polymer. Polysulfides utilize atmospheric moisture to accelerate cures. A two-component system is usually used, consisting of formulated polysulfide and formulated lead dioxide catalyst. Moisture converts a position of the lead dioxide catalyst to a faster-reacting form.[3]

Polysulfides cure at room temperature and reach maximum strength in 3–7 days. Polysulfides and epoxies are mutually soluble in all proportions (See epoxy-polysulfides above). Polysulfides are also alloyed with phenolics.[7]

Curing agents may be furnished in powder, paste, or liquid form. The activity of the metallic curing agents is a function of surface area, so that particle size becomes important. Since it is necessary to effect a fairly complete dispersion throughout the polymer in order to obtain a complete cure, it is generally more effective to combine the lead oxide with a plasticizing agent into the form of a paste. A finished polysulfide adhesive/sealant will generally contain the following ingredients as a minimum:

(a) liquid polymer

(b) reinforcing filler to increase strength and reduce cost

(c) plasticizer to modify modulus and hardness

(d) retarder to control set time

(e) oxidizing agent

Heat, humidity and sulfur will accelerate the cure.[58]

Cook[58] has discussed sealants at length in a 1967 book, Panek[59] has also covered polysulfides in detail in a more recent publication.

Polysulfone Adhesives

These are temperature-resistant thermoplastic adhesives which require fairly high temperatures for heat activation after solvents have been removed.[3] Polysulfones are a family of tough, high-strength thermoplastics which maintain their properties over a temperature range from −150°F (−101°C) to above 300°F (149°C). Bakelite's UDEL Polysulfone P-1700 has the following properties: Tensile strength 10,200 psi (70 MPa); flexural strength 15,400 psi (106 MPa); heat-distortion temperature 345°F (174°C); second-order glass transition temperature 375°F (191°C). The flexural modulus is maintained over a wide temperature range. At 300°F (149°C) over 80% of the room-temperature stiffness is retained. Resistance to creep is excellent. Polysulfone adhesives are resistant to strong acids and alkalis, but attacked and/or dissolved by polar organic solvents and aromatic hydrocarbons.[60]

These adhesives maintain their structural integrity up to 375°F (191°C). More than 60% of their room-temperature shear strength, as well as excellent creep resistance, is retained at 300°F (149°C). Cure cycles need only be long enough to introduce enough heat to wet the substrate with the P-1700 polysulfone. For unprimed aluminum a temperature of 700°F (371°C) should be used after drying the adhesive film 2–4 hours at 250°F (121°C) to remove the equilibrium moisture. With a platen temperature of 700°F (371°C) and a pressure of 80 psi (0.55 MPa) joints with tensile lap-shear strengths avove 3000 psi (21 MPa) are developed in 5 minutes. Higher temperatures at shorter dwell times may be used whenever the metal will tolerate such temperatures. Tensile-shear strengths of over 4000 psi (27.5 MPa) have been obtained with stainless steel after pressing at 700°F (371°C).[60]

Polysulfone adhesives have good gap-filling properties. In general, polysulfone adhesives combine the high strength, heat resistance and creep resistance of a thermosetting-type adhesive with the processing characteristics and toughness of a high-molecular-weight thermoplastic.[60]

Polyurethane Adhesives

Urethane polymers were developed during World War II and have since been used frequently in flexible and rigid foams, cryogenic sealants, and abrasion-resistant coatings. They have not been used to any great extent as adhesives, although this application is expanding. Their principle use is in bonding plastics that are difficult to bond, usually to a dissimilar material or to metals. Cured urethanes may be considered as very lightly crosslinked thermoset resins, almost thermoplastic. This gives them a flexible rubbery characteristic. A brief description of their characteristics is given in Table 5.2. Their flexibility combined with good adhesion, insures good bonding to flexible plastics, where peel strength is important. The outstanding feature of urethanes is strength at cryogenic temperatures. Table 5.6 compares the strength of urethane, epoxy-nylons, and epoxy-polyamides at −400°F (−240°C).[3]

Table 5-6: Comparison of Typical Urethane Adhesive with Other Adhesives on Aluminum at −400°F (−240°C)[3]

Adhesive	Lap Shear Strength		Peel Strength	
	(psi)	(MPa)	(lb/in)	(N/m)
Urethane	8,000	55.2	26	4,550
Epoxy-Nylon	4,600	31.7	(brittle)	(brittle)
Epoxy-Polyamide	1,600	11.0	(brittle)	(brittle)

Polyurethanes are one-component thermoplastic systems in solvents (ketones, hydrocarbons) often containing catalysts in small amounts to introduce a degree of thermosetting properties. They are also available as two-part thermosetting products in liquid form, with or without solvents. The second part is a catalyst. The one-part solvent type is used for contact bonding of tacky adherends following solvent release or heat-solvent reactivation of dried adhesive coating. The two-part thermosetting products are mixed and fully cured at 68°F (20°C) in 6 days. They may also be heat-cured in 3 hours at 194°F (90°C) or in 1 hour at 356°F (180°C). Bonding pressures range from contact to 51 psi (0.35 MPa).[7]

A new one-component urethane prepolymer adhesive, Accuthane UR-1100 (H. B. Fuller Co.), is designed for bonding various substrates, including plastic to plastic, plastic to metal, and metal to metal.[61] This adhesive can be used to bond imperfectly matched substrates and can be used for tack welding. No priming of the substrate surface is said to be required, except for a solvent wipe. Average tensile strength and elongation (ASTM D 638, a *plastics* method, not an *adhesive* method) after 30 minutes cure at 260°F (127°C) is as follows:

−40°F (−40°C)	7,836 psi (54 MPa)	(8.7% elongation)
72°F (22°C)	2,410 psi (16.6 MPa)	(32% elongation)
180°F (82°C)	582 psi (4 MPa)	(22% elongation)
260°F (127°C)	318 psi (2.2 MPa)	(16% elongation)

Another urethane one-part adhesive (Urethane Bond) developed by Dow Corning is cured by moisture in the air at room temperature. This material requires a thin glue line and clamping to produce the strongest joints. The resultant bonds are moisture resistant and are claimed to work well on polystyrene, polyvinyl chloride, and acrylics, and fair on polyethylene.[62]

Schollenberger[63] has discussed polyurethanes and isocyanate-based adhesives in some detail. Frisch et al. have covered diisocyanates as wood adhesives in a very recent comprehensive discussion.[64]

Polyvinyl Acetal Adhesives

Polyvinyl acetal is the generic name of a group of polymers that are products of the reaction between polyvinyl alcohol and an aldehyde. In preparing these acetals, polyvinyl acetate is used, which is then partially hydrolyzed to the alcohol. As adhesives, the most important acetals are those from formaldehyde, namely, the *formal*, and from butyraldehyde, the *butyral*. The properties of these polymers are largely dependent on the molecular weights and on the degree of hydrolysis of the acetate. Considered as an adhesive, the butyral (polyvinyl butyral) is much more imporant than the formal (polyvinyl formal) because it is more readily soluble, has a lower melt viscosity, and is a softer and more flexible polymer providing better peel strength, and therefore, higher apparent adhesion with thin adhesives. In the two-polymer adhesive system the formal is at least as important as the butyral[4] (See vinyl-phenolic adhesive below for a detailed discussion of these materials).

Polyvinyl butyral is commonly used in safety-glass laminates. Polyvinyl acetals are used in flexibilizing thermoset resins to obtain structural adhesives for metals. The vinyl-phenolics (which see) and vinyl-epoxies are examples.[3]

Lavin and Snelgrove[65] have discussed polyvinyl acetal adhesives in some detail.

Polyvinyl Acetate Adhesives

By far the most widely used .resin in water-dispersion form is polyvinyl acetate, both the homopolymer and the copolymer. Polyvinyl acetate latex is the basis for the common household "white glue", of which Elmer's® is probably the most well-known. Products of this type are good adhesives for paper, plastics, metal foil, leather, and cloth. The major use is as a packaging adhesive for flexible substrates. This material is also used as a lagging adhesive to bond insulating fabric to pipe and duct work in steam plants and ships. It is also used in frozen-food packaging where low-temperature flexibility is important. Polyvinyl acetates are also used in hot-melt adhesive formulations. Other uses include bookbinding and the lamination of foils. Organic solvent solutions are also used, in addition to water dispersions.

For wood bonding, 10 minutes to 3 hours at 68°F (20°C) and contact to 145 psi (1 MPa) pressure is recommended. These adhesives have low resistance to weather and moisture. Reistance to most solvents is poor, although these adhesives will withstand contact with grease, oils, and petroleum fluids and are not subject to biodeterioration. The cured films are light-stable, but tend to soften at temperatures approaching 113°F (45°C). These are low-cost adhesives with high initial-tack properties. They set quickly to provide almost invisible glue lines. One to 7 days conditioning is recommended before handling bonded assemblies. Maximum bond strength up to 2030 psi (14 MPa) can be reached by baking the adhesive films, followed by solvent reactivation and assembly. Polyvinyl acetates tend to creep under substantial load. Gap-filling properties are good.[7]

Polyvinyl acetate adhesives are used in the construction of mobile homes. The purpose is to provide temporary bonds during construction until the units are supported on foundations. They provide strong initial bonds that develop strength quickly. Immediate strength and stiffness are needed to resist stress induced by flexing and racking of long mobile homes as they are moved within the factory and during hauling and lifting at the construction site.[66]

Polyvinyl acetate glues should be applied at 60-90°F (16-32°C) working temperatures. They soften on sanding.[67]

Corey et al. have discussed polyvinyl acetate emulsion glues in some detail.[68] Another very recent discussion is by Goulding of General Chemical Corporation, Johannesburg, South Africa.[69]

Polyvinyl Alcohol Adhesives

This is a water-soluble thermoplastic synthetic resin with limited application as an adhesive. The chief uses are in bonding porous materials such as leather, cork and paper in food packaging, and as a remoistenable adhesive. It is available as a water solution with good wet-tack properties. It sets by losing water to give a flexible transparent bond with good resistance to oils, solvents and mold growth, but poor resistance to water. It is nontoxic and odorless. Cured films are impermeable to most gases. The maximum service temperature is about 151°F (66°C). Polyvinyl alcohol is also used as a modifier for other aqueous adhesive systems to improve film-forming properties, or to promote adhesion. These materials are used with dextrins and starches to provide low-cost laminating adhesives. They are also used for envelopes and stamps.[7]

Corey et al have discussed these resins in some detail.[68]

Polyvinyl Butyral Adhesives

See the discussion of these adhesives under Polyvinyl Acetal Adhesives.

Premixed Frozen Adhesives

Ablestik Laboratories in Gardena, California has available frozen reactive adhesives, such as epoxies, in disposable tubes or syringes ranging upward in size from 1 cc. These adhesives are packed in dry ice and shipped in insulated cartoons. Included in each carton is a safety indicator which is formulted to melt and lose shape when exposed to temperatures unsafe for adhesive storage. Storage life at −40°F (−40°C) before use is usually from two to six months. In use the frozen adhesive is thawed to room temperature and applied within two hours after thawing. These adhesives preclude production-line delays necessitated by on-the-job mixing of messy two-part adhesives, saving valuable assembly time. They also guarantee accurate formulation of components. Another advantage is the reduction in the possibility of workers contracting dermatitis from handling irritating amine curing agents.[70]

Pressure-Sensitive Adhesives (PSAs)

These materials are frequently used on tapes. Pressure-sensitive adhesives, in the dry state, are aggressively and *permanently* tacky at room temperature and firmly adhere to a variety of dissimilar surfaces without the need for more than finger or hand pressure. They require no activation by water, solvent, or heat in order to exert a strong adhesive holding force.[7]

Most pressure-sensitive adhesives are based on natural rubber. Since rubber by itself has very low tack and adhesion to surfaces, tackifying resins based on rosins, petroleum or terpenes are added, along with hydrogenated resins to help in long-term aging. In recent years adhesives based on *acrylic polymers* have become commercially important and are now probably second only to natural rubber. These acrylics have good ultraviolet stability, are resistant to hydrolysis, and are water-white with good resistance to yellowing or aging. Unfortunately they do have poor creep properties, compared to natural rubber. Blends of *natural rubber* and *styrene-butadiene rubber (SBR)* also provide excellent PSAs. Other less desirable adhesives include *polyisobutylene* (which see) and *butyl rubber* (which see). The adhesives discussed above are all applied in solution form. However, hot-melt adhesives are also applied. These are pressure-sensitive at room temperature. These materials may be based on *ethylene/vinyl acetate copolymers* tackified with various resins and with softeners. These produce rather soft adhesives with poor cohesive strength. Their use is small, mostly on label stock. Of wider interest are hot-melt adhesives based on the block copolymers of styrene with butadiene or isoprene. *Vinyl ether polymers* are also used, particularly in medicinal self-adhering plasters or dressings.[71]

Silicone adhesives are also used to a small degree in pressure-sensitive adhesives. These products are based on silicone rubber and synthetic silicone resins. They have excellent chemical and solvent resistance, excellent elevated-temperature resistance, excellent cold-temperature performance, and high resistance to thermal and oxidative degradation. Their disadvantages are that they lack agressive tack and cost 3-5 times as much as acrylic systems.[72]

Pressure-sensitive adhesives are almost always supplied to the final consumer coated onto some substrate, such as cellophane tape or insulating tapes based on plasticized PVC film. These consist of the backing film, a primer or key coat, and the adhesive. If the product is to be rolled up in tape form, a release coat may be applied to the back of the film to reduce unwind tension when the tape is used. Otherwise it is omitted. The adhesive, generally of the types discussed above, is usually applied from an organic solvent. However, aqueous dispersions, and more recently, hot-melt forms may be used. The coating weights range from about 10 g/m^2 upward, but are generally around 20-50 g/m^2. The primer is applied at a coating weight of about 2-5 g/m^2 from solvent or aqueous dispersion. Currently *nitrile rubber, chlorinated rubbers*, and *acrylates* are used as primers. A graft copolymer of methyl methacrylate and natural rubber can be used as a primer coat for plasticized PVC. The release coat is also applied to a lightweight coating at about 1-5 g/m^2. *Acrylic acid esters* of long-chain fatty alcohols, *polyurethanes* incorporating long aliphatic chains, and *cellulose esters* have also been used as release coats. As for backing or support for the adhesive, almost any material which can be put through a coating process can be used.[71]

The reader is probably aware of most applications of pressure-sensitive adhesives, and space will not permit covering this subject.

Bemmels has also discussed pressure-sensitive adhesives in some detail.[73] In 1982 an English translation of a comprehensive book on pressure-sensitive adhesives was published. This book was written by 27 contributors.[74]

Resorcinol-Formaldehyde Adhesives

These adhesives cure on addition of formaldehyde, compared to phenolics, which cure on addition of strong acids.[1] Commercially, these adhesives are supplied as two-part systems. A liquid portion, the "A" part, is the resinous constitutent. It is generally a solution of the preformed formaldehyde-deficient resin in a mixture of alcohol and water, with a solids content of about 50%. This resin is stable if kept in closed containers at or below room temperature. The pH at which the liquid is buffered controls the reactivity of the glue. The solid portion, or "B" part, is a solid, powdered mixture of paraformaldehyde, or "para", and fillers. The para is selected for control of glue-mix working life and curing efficiency. Once the "A" and "B" portions are mixed, the life of the mixture is limited. Since many of these glue mixtures produce exotherms (give off heat on mixing), the temperature of the mixes increases, and the crosslinking reaction proceeds more rapidly, reducing the pot life considerably. In these cases it is important to remove the heat by stirring and cooling as rapidly as the heat is generated. Actual gluing may take place anywhere in the range of 70-110°F (21-43°C), with clamping at moderate pressures.[75]

These adhesives are suitable for exterior use and are unaffected by water (even boiling water), molds, grease, oil, and most solvents. They are used primarily on wood, plywood, plastics, paper and fiberboard.[1] Resorcinol-formaldehyde adhesives are excellent marine-plywood adhesives. Curing at room temperature normally takes 8-12 hours, while phenolic wood adhesives require a high-temperature cure. The adhesives are also used for indoor applications because of their high reliability.[3,75]

Pizzi[51] had discussed these adhesives in some depth in a very recent source.

Rubber-Base Adhesives

See Elastomeric Adhesives.

Silicone Adhesives

Silicones are semi-inorganic polymers (polyorganosiloxanes) which may be fluid, elastomeric, or resinous, depending on the types or organic groups on the silicone atoms and the extent of cross-linkage between polymer chains.[7]

An example of silicone resin structure is seen in the following structure:

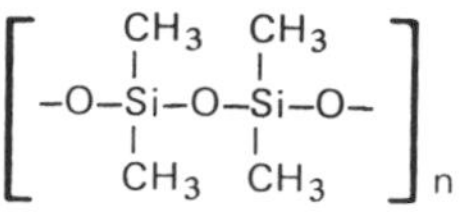

$$\left[\begin{array}{cc} CH_3 & CH_3 \\ | & | \\ -O-Si-O-Si-O- \\ | & | \\ CH_3 & CH_3 \end{array} \right]_n$$

Polydimethylsiloxane

The silicone resins owe their high heat stability to the strong silicon-oxygen-silicon bonds. The resin systems vary significantly in their physical properties as a result of the degree of cross-linkage and the type of radical (R) within the monomer molecule. In this regard the chief radicals are methyl, phenyl, or vinyl groups.[76]

The polymers have unusual properties and are used both to promote and to prevent adhesion. Silicones have good heat stability, chemical inertness, and surface-active properties. Silicone adhesive uses fall into four types:[3]

- *Primers*, or coupling agents
- *Adhesives* and *sealants* (adhesive/sealants)
- *Pressure-sensitive adhesives*
- *Heat-cured adhesives*

The use of silicones as adhesives has not been extensive, primarily because of their high cost. Applications are numerous, however, and varied. Silicones are used where organic materials (based on carbon) cannot withstand harsh environmental conditions, where superior reliability is required, or where their durability gives them economic advantages. As *coupling agents*, silicones are widely used for surface treatment of fiberglass fabric for glass-reinforced laminates. Epoxy or polyester adhesion to glass cloth is improved both in strength and in moisture resistance of the cured bond by the use of silicone coupling agents. The retention of flexibility and some degree of strength at a temperature range from cryogenic to over 500°F (260°C) is unusual in polymers. Generally, however, the room-temperature strength properties of silicone adhesives are quite low compared to more typical polymers.[3]

The excellent peel strength properties of silicones are more important in joint designs than the tensile or lap-shear properties. Some examples of peel and lap-shear strengths with silicones are as follows:[3]

Peel strength
    Rubber to aluminum   17-20 lb/in (2975-3500 N/m)
    Urethane sealant to aluminum
      —without primer   3.5 lb/in (612 N/m)
      —with silicone coupling agent 14 lb/in (2450 N/m)
Lap-shear strength
    Metal to metal   250 psi (1.7-3.4 MPa)

Silicone use in adhesives includes:[3]

- Two-part adhesive for bonding insulating tape to magnet wire (Class M performance).

- One- or two-part adhesive to prepare pressure-sensitive tapes useful for −80°F (−62°C) to 500°F (260°C). These tapes are used in electronic and aerospace applications.

Silicone use in primers includes:[3]

- Bond promoters with phenolic binders for foundry sand on abrasive wheels.

- Filler treatment in filled polyester or epoxy coatings (epoxy concrete patching formulations).

- Improved bonding of polysulfide or urethane sealants to metal substrates or glass.

In some cases the silicone is equally effective when blended into an adhesive formulation as when it is applied separately as a primer. For silicone *coupling agents*, moisture adsorbed on the substrate plays an important role in attaching the silicone molecule through hydrolysis. The opposite end of the molecule contains a chemical group such as a vinyl or amine, which is reactive with the epoxy, polyester, or other resin that is to be adhered to the substrate. In this manner a single layer of silicone molecules "couples" the resin to the substrate. In addition to bond strength, improved moisture resistance usually results.[3]

Silicone adhesives cure without the application of heat or pressure to form permanently flexible silicone rubber. The rubber remains flexible despite exposure to high or low temperatures, weather, moisture, oxygen, ozone, or UV-radiation. This makes them useful for joining and sealing joints in which considerable movement can be expected, such as between plastics and other materials of construction (as acrylic glazing). Several type of silicone adhesive/sealants are available. These include one-part and two-part systems. *One-part silicone systems* are ready to use. They require no mixing, present no pot-life problem, and are generally the least expensive. Conventional one-part adhesive/ sealants are available with two different types of cure systems; an *acid cure* and a *non-acid cure.* Both require moisture from the atmosphere to cure. The acid-curing type has the greatest unprimed adhesion and the longest shelf life. The non-acid-curing type is useful when the acetic acid released by the cure reaction may cause corrosion (see MIL-A-46146 in Chapter 12), or be otherwise objectionable.[77]

The *two-part silicone adhesive/sealants* do not require moisture to cure and produce a superior deep-section cure. Two types are available, addition-cure and condensation-cure. *Addition curing* produces no by-products, can be heat-accelerated, produces negligible shrinkage, and provides the best high-temperature resistance of all the silicone adhesives. *Condensation-cure* silicones are not easily inhibited and can be used on a greater variety of materials.[77]

In 1982 Dow Corning announced the availability of a new silicone adhesive/ sealant for high-temperature use. This is a one-part, non-slumping paste that cures to a tough, rubber solid at room temperature on exposure to water vapor in the air. This material is claimed to perform at temperatures ranging from −85 to 500°F (−65 to 260°C) for continuous operation, and to 600°F (316°C) for intermittent exposure. This material will meet the requirements of MIL-A-46106A(2), Type 1 (discussed in Chapter 12). The adhesive/sealant is acid-cured and acetic acid is evolved during cure.[78]

Table 5-2 summarizes some of the characteristics of silicone adhesives. Beers[79] has discussed silicone adhesive/sealants in some detail. Silicone (silane) coupling agents have been discussed by Marsden and Sterman.[80] An excellent general discussion of room-temperature-vulcanizing (RTV) silicone adhesive sealants has been written by Romig and Bush.[81]

Solvent-Based Systems

Natural and synthetic rubber and synthetic resins are capable of being dissolved in organic solvents, resulting in *cements, resin solutions,* or *lacquers.* In addition, there are many cellulose derivatives, such as nitrocellulose, ethyl cellulose, and cellulose acetate butyrate used in preparing solvent-based adhesives. Solvent-based adhesives are also prepared from cyclized rubber, polyamide, and polyisobutylene. Low-molecular-weight polyurethane and epoxy compounds can be used with or without solvent. On the other hand, high-molecular-weight types or prepolymers require solvent to make application possible.[14]

Solvents, or solvents containing small amounts of bodying resin, are used for bonding types of thermoplastic resins and films. An example is *toluol,* which can be used to soften and dissolve polystyrene molded articles so that the softened pieces can be joined. Ketones can be used to bond polyvinyl chloride films in a similar manner. A small amount of resin can be used to thicken the solvent so that a sufficient amount will stay in place to dissolve the substrate. It should be noted, however, that solvent welding of molded plastics can cause stress cracking and weakening of the structure as the parts age.[14]

Another class of solvent-based dispersion is the *organosols.* In this case, vinyl chloride copolymer resins are dispersed in suitable nonvolatile plasticizers and solvent. The solvent is evaporated and the remaining film is heated to approximately 350°F (177°C). The heat helps dissolve the resin in the plasticizer, and a tough, flexible film is obtained on cooling to room temperature.[14]

The major polymers used for solvent-based adhesives are listed in Table 5.7.[14]

Table 5-7: Principal Polymers Used for Solvent-Based Adhesives and Solvent- or Water-Base Adhesives[14]

Solvent-base
  Nitrocellulose
  Cellulose acetate butyrate

  Cyclized rubber
  Polyisobutylene

Solvent- or water-based
  Natural rubber
  SBR rubbers
  Butyl rubbers
  Neoprene rubbers
  Nitrile rubbers
  Reclaim rubbers
  Polyvinyl acetate and copolymer
  Polyvinyl chloride copolymer

  Polyvinyl ether
  Polyvinylidene chloride and copolymers
  Polyacrylate and polymethacrylate
  Polyamide
  Asphalt
  Urea-formaldehyde
  Phenol-formaldehyde
  Resorcinol-formaldehyde

Resin esters

Solvent-based adhesives are more expensive than water-based. They usually make bonds which are more water-resistant and have higher tack and higher early strength than water-based adhesives. Solvent-based adhesives also wet oily surfaces and some plastics considerably better than water-based. With organic solvents, explosion-proof equipment must be used and other precautions taken for handling and application. In addition, ventilation to remove toxic hazards must be provided for personnel exposed to solvent vapors.[14] Similar problems exist when organic solvents are used in surface preparation (see Chapter 4).

Thermoplastic Resin Adhesives

A thermoplastic resin adhesive is one which melts or softens on heating and rehardens on cooling without (within certain temperature limits) undergoing chemical change. At temperatures above the melting point, an irreversible chemical change such as depolymerization or oxidative degradation may take place. When used as adhesives, thermoplastic resins are used in the form of solutions, dispersions in water, or solids. They are usually set by purely physical means. When applied as solutions or dispersions, adhesion follows evaporation or absorption of the liquid phase, as in solvent activation. When applied by melting and cooling the solids, the terms "hot-melt" or "melt-freeze" are used to describe the method of application (See hot-melt adhesives). Although the terms "setting" and "curing" are frequently used synonomously for both thermoplastic and thermosetting adhesives, the term "setting" is more common with thermoplastic adhesives, unless a chemical process such as polymerization is involved, when the term "curing" is more appropriate. Although thermoplastic adhesives fall into many different chemical classes, they all are comprised predominantly of linear macromolecules. Most thermoplastic resins are capable of bonding a wide variety of substrates, such as paper, wood and leather. Some are capable of bonding rubbers, metals, and some plastics, without special surface treatment. The most notable exceptions are the silicone and fluorocarbon plastics.[4]

Thermoplastic Rubber (for Use in Adhesives)

Thermoplastic rubber is a relatively new class of polymer. It has the solubility and thermoplasticity of polystyrene, while at ambient temperatures it has the toughness and resilience of vulcanized natural rubber or polybutadiene. These rubbers are really block copolymers. The simplest form consist of a rubbery midblock with two plastic endblocks (A-B-A), as shown in Figure 5-4. Examples of commercial products are Kraton (Shell Chemical Co.) and Solprene (Phillips Chemical Co.). These materials are often compounded with plasticizers to decrease hardness and modulus, eliminate drawing, enhance pressure-sensitive tack, improve low-temperature flexibility, reduce melt and solution viscosity, decrease cohesive strength or increase plasticity if desired, and substantially lower material costs. Low levels of thermoplastic rubber are sometimes added to other rubber adhesives. These materials are used as components in the following applications: Pressure-sensitive adhesives, hot-melt adhesives, heat-activated-assembly adhesives, contact adhesives, reactive contact adhesives, building construction adhesives, sealants, and binders. Two specific varieties of thermoplastic rubber adhesives used are styrene-butadiene-styrene (S-B-S) and styrene-isoprene-styrene (S-I-S).[13]

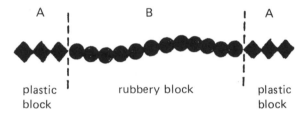

Figure 5-4: Simplified representation of a thermoplastic rubber molecule. Modified after Reference 13.

Thermosetting Resin Adhesives

A thermosetting synthetic resin is one that, on curing, undergoes an irreversible chemical and physical change to become substantially infusible and insoluble. The term *thermosetting* is applied to the resin both before and after curing. Some thermosetting adhesives are *condensation polymers* and some are *addition polymers*. The important thermosetting resin adhesives are urea-formaldehydes, melamine-formaldehydes, phenol-formaldehydes, resorcinol-formaldehydes, epoxies, polyisocyanates and polyesters (see the discussions under these headings).[4]

Ultraviolet-Curing Adhesives

Ultraviolet curing of adhesives is relatively new and has followed the development of UV-sensitive inks and varnishes. The adhesives used are fast-curing, clear liquids which have stabilizers that are destabilized at 365 nm wave length at an intensity of 10,000 μ watts/cm^2. This is approximately the same amount of energy in sunlight without the heat or visible portions. Ultraviolet-light curing is very rapid and bond strengths are high, usually exceeding the strength of glass.[82]

It has been known for a long time that certain synthetic resins, such as vinyls, acrylics and olefins, will cure rapidly when irradiated by suitable light. Only recently has this effect been applied to the development of one-component adhesives that cure within very short periods of time (5–15 seconds) when exposed to ultraviolet light. The cured polymers display good adhesion to glass, metals, and certain thermoset plastics. These adhesives have obvious industrial applications. One possibility under consideration is the joining of glass to metal in automotive applications, such as windows. The formation of comparatively stress-free joints based on glass-to-glass or metal as a result of cold curing is another attractive aspect in glass laminate or optical component production.[83]

The Loctite Corporation has two UV-curing adhesives on the market, #353 and #354, for rapid glass bonding, tacking and coating. These products have very high light transmittance. (>95%) and refractive indices similar to glass. Cure is obtained on exposure to ultraviolet light with wavelengths between 2700Å to 4100Å, but light centered at 3600Å is most effective. Adhesive 353 is recommended for bonding glass components such as stemware, lens components, or quartz crystals. Bondline stresses are minimal, since the adhesive cures without heat, which could distort parts. Adhesive 353 is an anaerobic adhesive. Adhesive 354 cures only under ultraviolet light. When cured with high-intensity mercury vapor lamps, the adhesive cures rapidly to a hard surface. When cured in this manner Adhesive 354 is excellent for tacking wires and potting small components. The fixture time under low-intensity light is 30 seconds for Adhesive 353 and 15 seconds for 354. The fixture time for high-intensity light is 15 seconds for Adhesive 353 and 5 seconds for 354. Both these adhesives can be partially cured under UV light and allowed to post-cure to ultimate strength without further light exposure. Their strengths will increase over a 48-hour period. XX15 or XX40 lamps are recommended for slow fixture of Adhesives 353 and 354 (available from Ultra-Violet Products, Inc. San Gabriel, CA 91778). Porta Cure 1000 curing lamps are high-intensity lamps recommended for very fast curing or surface curing of Adhesive 354. These lamps are available from American Ultraviolet Company, Chatham, NJ 07928.[84]

Bluestein[85] has recently reviewed radiant energy-curable adhesives. An excellent and comprehensive review article on ultraviolet adhesive bonding was very recently written by Marino of Loctite Corporation. This article was published too late to be used in preparing this brief discussion.[86]

Urea-Formaldehyde Adhesives (Ureas)

These adhesives commonly called *urea glues*, are the condensation products of unsubstituted urea and formaldehyde. They are usually two-part systems, consisting of the resin and hardening agent (liquid or powder). They are also available as spray-dried powders with incorporated hardener. The latter is activated by mixing with water. Fillers are also added. Curing is normally accomplished under pressure without heat. For general purposes curing is carried out for 2–4 hours at 68°F (20°C) and 51–102 psi (0.35–0.70 MPa) bonding pressure. For plywood manufacture heat is used. Conditions recommended are 5–10 minutes at 248°F (120°C) and up to 232 psi (1.6 MPa) pressure. Timber (hardwood) is cured 15-24 hours at 68°F (20°C) and 203 psi (0.14 MPa). Softwood curing conditions are 15-24 hours at 68°F (20°C) and 102 psi (0.70 MPa). The bonding pressure depends on the type of wood, shape of parts, and similar factors. The most common application of urea-formaldehyde adhesives is in plywood. Urea glues are not as durable as other types, but are suitable for a fair range of service applications. When glue-line thicknesses range from 0.002-0.004 inch (0.05-0.10 mm) the bond strengths usually exceed the strength of the wood. However, when glue-line thicknesses exceed 0.0146 inch (0.37 mm) gap-filling properties are poor. Thick glue lines will craze and weaken the joints unless special modifiers, such as furfural alcohol resin, are used. These adhesives are not suitable for outdoor applications or extreme temperatures.[1,4,7]

Savia[35] has discussed amino resins, including ureas, in considerable detail. Pizzi[36] has covered amino resins in wood adhesives in a very recent discussion.

Vinyl-Epoxy Adhesives

These structural adhesive alloys are polyvinyl acetals (which see).

Vinyl-Phenolic Adhesives

These structural adhesive alloys are also polyvinyl acetals (which see). They may be phenolic-vinyl butyral or phenolic-vinyl formal.[7] The "vinyl" in vinyl-phenolic adhesive is a somewhat misleading term referring either to polyvinyl formal or polyvinyl butyral. Vinyl phenolics generally have excellent durability, both in water and in other adverse environments. Cures are at 350°F (177°C) for the polyvinyl formal-phenolic and 302°F (150°C) for the polyvinyl butyral-phenolic. Although these adhesives provide excellent performance, primarily as film adhesives, the trend is to the use of the newer types which have been developed to provide cures at lower temperatures and pressures, and with higher hot strength, higher peel strength, and other performance advantages.[7,12] These adhesives are discussed briefly under Alloyed Adhesives and more extensively under Film and Tape Adhesives. Tables 5-3 and 5-4 in the latter section, provide useful information on these adhesives and their strength properties.

Cure conditions for the polyvinyl formal-phenolic film are 350°F (177°C) for 5 minutes or 302°F (150°C) for 30 minutes at 51-510 psi (0.35-3.5 MPa) bonding pressure. For the polyvinyl butyral-phenolic film curing conditions are 302°F (150°C) for 14.5-29.0 psi (0.10-0.20 MPa). The polyvinyl formal-phenolic film, which is the most commonly used, retains adequate strength when exposed to weather, mold growth, salt spray, humidity and chemical agents, such as water, oils, and aromatic fuels. These adhesives generally have good resistance to creep, although temperatures up to 194°F (90°C) produce creep and softening of some formulations. Fatigue resistance is excellent, with failure generally occurring in the adherends, rather than in the adhesive. The useful service temperature range is −76 to 212°F (−60 to 100°C).[7]

Polyvinyl formal-phenolics—These are structural adhesives in bonding metal to metal in aircraft assemblies; metal honeycomb panels and wood-to-metal sandwich construction; cyclized rubber and, in some cases, vulcanized and unvulcanized rubbers; primer for metal-to-wood bonding with resorcinol or phenolic adhesives, copper foil to plastic laminates for printed circuits. Polyvinyl formal-phenolics are among the best thermoset adhesives for metal-honeycomb and wood-metal structures. These adhesives are generally equivalent to nitrile-phenolics for strength, but have slightly better self-filleting properties for honeycomb assembly. They are superior to epoxy types where strength is desirable in sandwich construction.[7]

Polyvinyl butyral-phenolics—These are used in metal or reinforced plastic facings to paper (resin impregnated) honeycomb structures; cork and rubber compositions; cyclized and unvulcanized rubbers; steel to vulcanized rubber; electrical applications; primer for metals to be bonded to wood with phenolics. Polyvinyl butyral-phenolics lack the shear strength and toughness of the polyvinyl formal-phenolic type.[7]

Vinyl-Resin Adhesives

Several vinyl monomers are used to prepare thermoplastics which are useful in certain adhesive applications. The most important vinyl resins for adhesives are *polyvinyl acetate* (which are), the *polyvinyl acetals* (butyral and formal) (See Polyvinyl Acetal Adhesives and Vinyl-Phenolic Adhesives), and *polyvinyl alkyl ethers*. Polyvinyl chloride and copolymers of both vinyl chloride and vinyl acetate with other monomers, such as maleic acid esters, alkyl acrylates, maleic anhydride, and ethylene, are also used for solvent-based adhesives.[3]

Water-Based Adhesives

These adhesives are made from materials that can be dispersed or dissolved in water alone. Some of these materials are the basis for solvent-based adhesives (which see) and are the principal materials used for liquid-adhesive formulations given below in Table 5.8

Table 5-8: Principal Polymers Used Exclusively for Water-Based Adhesives[14]

Starch and dextrin	Casein
Gums	Sodium carboxymethylcellulose
Glue (animal)	Lignin
Albumen	Polyvinyl alcohol
Sodium silicate	

Table 5.7 lists polymers used for both water-based and solvent-based adhesives. Water-based adhesives are lower in cost than the equivalent solvent-based compounds. Even inexpensive organic solvents are costly when compared to water. The use of water eliminates problems of flammability and toxicity associated with organic solvents. However, in most cases water-based adhesives must be kept from freezing during shipment and storage because of possible permanent damage to both the container and contents.[14]

Because of government legislation on the use of organic solvents the use of solvent-based adhesives will undoubtedly decline in the future. Replacement products will most likely be hot melts, 100%-reactive adhesives, and *water-based adhesives*. According to one prediction, solvent-based adhesives will become almost obsolete by the year 2000.[87]

There are two general types of water-based adhesives—*solutions* and *latices*.

Solutions are made from materials that are soluble in water alone, or in alkaline water. Examples of materials that are soluble in water alone are:[87] Animal glue, starch, dextrin, blood albumen, methyl cellulose, and polyvinyl alcohol. Examples of materials that are soluble in alkaline water are:[87] Casein, rosin, shellac, copolymers of vinyl acetate or acrylates containing carboxyl groups, and carboxymethyl cellulose.

A *latex* (plural *latices*) is a stable dispersion of a polymeric material in an essentially aqueous medium. An *emulsion* is a stable dispersion of two or more immiscible liquids held in suspension by small percentages of substances called *emulsifiers.* In the adhesives industry the terms *latex* and *emulsion* are sometimes used interchangeably. There are three types of latices: Natural, synthetic, and artificial. *Natural latex* refers to the material obtained primarily from the rubber tree. *Synthetic latices* are aqueous dispersions of polymers obtained by emulsion polymerization. These include polymers of chloroprene, butadiene-styrene, butadiene-acrylonitrile, vinyl acetate, acrylate, methacrylate, vinyl chloride, styrene, and vinylidene chloride. *Artificial latices* are made by dispersing solid polymers. These include dispersions of reclaimed rubber, butyl rubber, rosin, rosin derivatives, asphalt, coal tar, and a large number of synthetic resins derived from coal tar and petroleum.[86]

Latex adhesives are more likely than solution adhesives to replace solvent-based adhesives. Most latex adhesives are produced from polymers that were not designed for use as adhesives. For this reason they require extensive formulation in order to obtain the proper application and performance properties. Application methods for latex adhesives are: Brush, spray, roll coat, curtain coat, flow, and knife coat. The bonding techniques used for latex adhesives are similar to those used for solvent adhesives. The following techniques are used:[87]

- *Wet bonding.* Used when at least one of the bonded materials is porous. The adhesive is usually applied to only one surface. Bonding takes place while the adhesive is still wet or tacky.

- *Open-time bonding.* In this method the adhesive is applied to both surfaces and allowed to stand "open" until suitable tack is achieved. At least one of the adherends should be porous.

- *Contact bonding.* In this method both surfaces are coated and the adhesive is permitted to become dry to the touch. Within a given time these surfaces are pressed together and near ultimate bond strength is immediately achieved. In this method two non-porous surfaces are usually used. Neoprene latices are commonly utilized in contact bonding.

- *Solvent reactivation.* In this method the adhesive is applied to the surface of the part and allowed to dry. To prepare for bonding, the adhesive is reactivated by wiping with solvent or placing the part in a solvent-impregnated pad. The surface of the adhesive tackifies and the parts to be bonded are pressed together. This method is suitable only for relatively small-size parts.

- *Heat reactivation.* In this method a thermoplastic adhesive is applied to one or both surfaces and allowed to dry. To bond, the part is heated until the adhesive is soft and tacky. The bond is made under pressure while hot. On cooling, a strong bond results. This method is used with non-porous heat-resistant materials. This method can also be used in a continuous in-line operation. The adhesive is applied in liquid form to a film or sheet, force-dried with heat to remove the water, and than laminated to a second surface while still hot. Temperatures are usually in the range of 250-350°F (121-177°C).

Solids contents of latex adhesives are in the 40-50% range compared to about 20-30% for solvent-based adhesives. The main disadvantage of latex adhesives is the longer drying time required before tack or strength develops. On the other hand, latex adhesives have good brushability and usually require less pressure to pump or spray than solvent-based adhesives. Prior to drying they can be cleaned up with water. In general, the state of the art of latex adhesives is considerably behind that of solvent based adhesives.[86]

Applications of water-based adhesives are mentioned briefly in the above sections on: Natural glues, Elastomeric adhesives (natural rubber), polyvinyl acetate, Neoprene and Styrene-butadiene rubber (SBR).

REFERENCES

1. Petrie, E.M., Chapter 10, Plastics and Elastomers as Adhesives, *Handbook of Plastics and Adhesives*, (C.A. Harper, ed.), McGraw-Hill, NY (1977).
2. Merriam, J.C., Adhesive Bonding, *Materials in Design Engineering*, 50(3): 113-128 (September 1959).
3. Chapter 1, Types of Adhesives, *Adhesives in Modern Manufacturing*, published by the Society of Manufacturing Engineers (SME), (E.J. Bruno, ed.), (1970).
4. Rayner, C.A., Chapter 4, Synthetic Organic Adhesives, *Adhesion and Adhesives*, 2nd Edition, Vol. 1 — Adhesives, R. Houwink and G. Salomon, eds., Elsevier Publishing Co. (1965).
5. Miska, K.M., Second Generation Acrylic Adhesives at Lower Cost, *Materials Engineering*, 84(5): 40-42 (November 1976).
6. Second-Generation Acrylic Adhesives, *Adhesives Age*, 19(a): 21-24 (September 1976).
7. Shields, J., *Adhesives Handbook*, 2nd Edition, Newnes-Butterworth, London (1976). (The 3rd Edition was published by Butterworths, London, 1984.)
8. Murray, B.D., Anaerobic Adhesive Technology; Preprint Booklet, Symposium on Durability of Adhesive Bonded Structures, sponsored by U.S. Army Research and Development Command, held at Picatinny Aresenal, Dover, NY, October 27-29, 1976, pp. 599-610. [Also published in *Journal of Applied Polymer Science*, Applied Polymer Symposia 32, (M.J. Bodnar, ed.), Wiley-Interscience, 1977, pp. 411-420].
9. Pearce, M.B., How to Use Anaerobics Successfully, Applied Polymer Symposium No. 19, *Processing for Adhesive Bonded Structures*, presented at symposium held at Stevens Institute of Technology, Hoboken, NJ and sponsored by Picatinny Arsenal and Stevens Institute of Technology, August 23-25, 1972. (M.J. Bodnar, ed.), Wiley-Interscience (1966), pp. 207-230.

10. Karnolt, C.L., Anaerobic Adhesives for Sheet Metal Assembly, presented at SAE Automotive Engineering Congress and Exposition, Detroit, Michigan, February 24-28, 1975. SAE Paper No. 750140.

11. Burnham, B.M. et al., Chapter 6, Anaerobic Adhesives, *Handbook of Adhesive Bonding*, (C.V., Cagle, H. Lee, and K. Neville, eds.), McGraw-Hill, NY (1973).

12. Bolger, J.C., Chapter 1, Structural Adhesives for Metal Bonding, *Treatise on Adhesion and Adhesives*, Vol. 3, (R.L. Patrick, ed.), Marcel Dekker (1973).

13. Harlan, J.T. and Petershagen, L.A., Chapter 19, Thermoplastic Rubber (A-B-A Block Copolymers) in Adhesives, *Handbook of Adhesives*, 2nd Edition, (I. Skeist, ed.), Van Nostrand Reinhold, NY 1977).

14. Lichman, J., Chapter 44, Water-based and Solvent-Based Adhesives, *Handbook of Adhesives*, 2nd Edition, (I. Skeist, ed.), Van Nostrand Reinhold, NY (1977).

15. Stucker, N.E. and Higgins, J.J. Chapter 16, Butyl Rubber and Polyisobutylene, *Handbook of Adhesives*, 2nd Edition, (I. Skeist, ed.), Van Nostrand Reinhold, NY (1977).

16. Landrock, A.H., Effects of Varying Processing Parameters on the Fabrication of Adhesive-Bonded Structures, Part VII, *Electrically and Thermally-Conductive Adhesives—Literature Search and Discussion*. Picatinny Arsenal Technical Report 4179 (March 1971).

17. De Lollis, N.J. *Adhesives, Adherends, Adhesion*, Robert E. Krieger Publishing Co., Huntington, NY (1980).

18. Sharpe, L.M., et al., Chapter 1, Development of a One-Part Electrically Conductive Adhesive System, *Adhesion 7*, (K.W. Allen, ed.), Applied Science Publishers, London (1983).

19. Wake, W.C., *Adhesion and the Formulation of Adhesives*, 2nd Edition, Applied Science Publishers, London (1982).

20. Brumit, T.M., Cyanoacrylate Adhesives—When Should You Use Them? *Adhesives Age*, 18(2): 17-22 (February 1975).

21. Lee Pharmaceuticals, Technical *Staff, Cyanoacrylate Resins—The Instant Adhesives*, (H. Lee, ed.), Pasadena Technology Press, Pasadena, CA (1981).

22. Thermal Cycling Makes Strong Adhesive Stronger, *News Trends, Machine Design*, 56(6):10 (March 22, 1984).

23. Eisenträger, K. and Druschke, W., Chapter 32, Acrylic Adhesives and Sealants *Handbook of Adhesives*, 2nd Edition, (I. Skeist, ed.), Van Nostrand Reinhold, NY (1977).

24. Dunn, B. and Cianciarulo, A., Epoxies Used in Liquid Gas Containment, Proceedings, 1974 Spring Seminar, *Designing with Today's Engineering Adhesives*, sponsored by The Adhesive and Sealant Council, Cherry Hill, NJ, pp. 141-153 (March 11-14, 1979).

25. Chastain, C.E. and Cagle, C.V., Chapter 3, Epoxy Adhesives, *Handbook of Adhesive Bonding*, McGraw-Hill, NY (1973).

26. Twiss, S.B., Adhesives of the Future, Applied Polymer Symposium No. 3, *Structural Adhesive Bonding*, presented at symposium held at Stevens Institute of Technology, Hoboken, NJ and sponsored by Picatinny Arsenal, September 14-16, 1965, (M.J. Bodnar, ed.), pp. 455-488, Wiley-Interscience (1972).

27. Dreger, D.R., Hot Melt Adhesives Put it All Together, *Machine Design*, 51(3): 54-60 (February 8, 1979).

28. Aronson, R.B., Adhesives Cure Getting Stronger in Many Ways, *Machine Design*, 51(3):54-60 (February 8, 1979).

29. Gulf Oil Chemical Company, Olefins and Derivatives Div., *Numel® Engineering Adhesives*, product brochure (1982).
30. Hughes, F.T., Foamed Hot Melt Adhesives, *Adhesives Age*, 25(9): 25–29 (September 1982).
31. Bell, J.J. and Robertson, W.J., Hot Melt Bonding with High Strength Thermoplastic Rubber Polymers, presented at SAE Automotive Engineering Congress Exposition, Detroit, MI, Feb. 25–March 1, 1974. SAE Paper No. 740261.
32. Miska, K.M., Hot Melt Can Be Reactivated, *Materials Engineering*, 83(4):32 (April, 1976).
33. 3M Company, Adhesives, Coatings and Sealers Division, Product Specification 4060, *Scotch-Weld Thermoplastic Adhesive Film 4060*, September 15, 1971, Rev. December 1, 1975.
34. Wills, J.M., Chapter 8, Inorganic Adhesives and Cements, *Adhesion and Adhesives*, 2nd Edition, Volume 1—Adhesives, (R. Houwink and G. Salomon, eds.), Elsevier Publishing Co. (1965).
35. Savia, M., Chapter 25, Amino Resin Adhesives, *Handbook of Adhesives*, 2nd Edition, (I. Skeist, ed.), Van Nostrand Reinhold Co., NY (1977)
36. Pizzi, A., Chapter 2, Amino Resin Wood Adhesives, *Wood Adhesives—Chemistry and Technology*, (A. Pizzi, ed.), Marcel Dekker, NY (1983).
37. Kirby, K.W., Chapter 3, Vegetable Adhesives, *Adhesion and Adhesives*, 2nd Edition, Volume 1—Adhesives, (R. Houwink and G. Salomon, eds.), Elsevier Publishing Co. (1965).
38. Blomquist, R.F., Chapter 10, Soybean Glues, *Handbook of Adhesives*, 2nd Edition, (I. Skeist, ed.) Van Nostrand Reinhold, NY (1977).
39. Lambuth, A.L., Chapter 10, Soybean Glues, *Handbook of Adhesives*, 2nd Edition, (J. Skeist, ed.), Van Nostrand Reinhold, NY (1977).
40. Salzberg, H.K., Chapter 9, Casein Glues and Adhesives, *Handbook of Adhesives*, 2nd Edition, (I. Skeist, ed.), Van Nostrand Reinhold, NY (1977).
41. Krogh, A.M. and Wooton, J., Chapter 2, Animal Glues and Related Protein Adhesives, *Adhesion and Adhesives*, 2nd Edition, Volume 1—Adhesives, (R. Houwink and G. Salomon, eds.), Elsevier Publishing Co. (1965).
42. Barker, A., Animal Glues Holding On, *Adhesives Age*, 27(5): 16–17 (May 1984).
43. Hubbard, J.R., Chapter 7, Animal Glues, *Handbook of Adhesives*, 2nd Edition, (I. Skeist, ed.), Van Nostrand Reinhold, NY (1977).
44. Norland, R.E., Chapter 8, Fish Glue, *Handbook of Adhesives*, 2nd Edition, (I. Skeist, ed.), Van Nostrand Reinhold, NY (1977).
45. Buchoff, L.S., Chapter 50, Adhesives in the Electrical Industry, *Handbook of Adhesives*, 2nd Edition (I. Skeist, ed.), Van Nostrand Reinhold, NY (1977).
46. Steinfink, M., Chapter 21, Neoprene Adhesives: Solvent and Latex, *Handbook of Adhesives*, 2nd Edition, (I. Skeist ed.), Van Nostrand Reinhold, NY (1977).
47. De Lollis, N.J., Durability of Adhesive Bonds (A Review), Proceedings, 22nd National SAMPE Symposium, Vol. 22, *Diversity-Technology Explosion*, San Diego, CA, pp. 673–698 (April 26–28, 1977).
48. Morrill, J.R. and Marguglio, L.A., Chapter 17, Nitrile Rubber Adhesives, *Handbook of Adhesives*, 2nd Edition, (I. Skeist, ed.), Van Nostrand Reinhold, NY (1977).
49. De Lollis, N., Theory of Adhesion—Part 2—Proposed Mechanism for Bond Failure, *Adhesives Age*, 13(1): 25–29 (January 1969).
50. Barth, B.P., Chapter 23, Phenolic Resin Adhesives, *Handbook of Adhesives*, 2nd Edition, (I. Skeist, ed.), Van Nostrand Reinhold, NY (1977).

51. Pizzi, A., Chapter 3, Phenolic Resin Wood Adhesives, *Wood Adhesives—Chemistry and Technology*, (A. Pizzi, ed.), Marcel Dekker, NY (1983).
52. Union Carbide Corporation. Technical Literature (J-2421A 106-3) on Phenoxy Adhesives (PRDA-8080), undated.
53. Mayer, W.P. and Young, R.M. Formulating Hot Melts with EEA Copolymers, *Adhesive Age*, 19(5): 31–36 (August 1976).
54. Edson, D.V., Adhesives Take the Heat, *Design News*, pp. 45–48, December 19, 1983.
55. Alvarez, R.T., 600°F Thermoplastic Polyimide Adhesive, 29th National SAMPE Symposium and Exposition, (Vol. 29), Reno, Nevada, April 3-5, 1984, *Technology Vectors*, pp. 68–72
56. Steger, V.Y., Structural Adhesive Bonding Using Polyimide Resins, proceedings, 12th National SAMPE Technical Conference, (Vol 12), Seattle, WA, October 7-9, 1980, *Materials 1980*, pp. 1054-1059.
57. Serlin, I. et al., Chapter 37, Aromatic Polyimide Adhesives and Bonding Agents, *Handbook of Adhesives*, 2nd Edition, (I. Skeist, ed.), Van Nostrand Reinhold, NY (1977).
58. Cook, J.P., Chapter 6, Polysulfide Sealants, *Sealants*, (A. Damusis, ed.), Reinhold, NY (1967).
59. Panek, J.R., Chapter 22, Polysulfide Sealants and Adhesives, *Handbook of Adhesives*, 2nd Edition, (I. Skeist, ed.), Van Nostrand Reinhold, NY (1977).
60. Udel® Polysulfone for High-Temperature Structural Adhesive Applications, Union Carbide Engineering Polymers, *Udel® Polysulfone Product Data*, Union Carbide Corp. Brochure F-43410B, 5 pp. (November 1977).
61. H.B. Fuller Co., St. Paul, MN, Technical Data Bulletin on ACCUTHANE™ UR-1100 Adhesive, Rev. (May 1983).
62. Day, R., An Epoxy-Tough Urethane Glue, Personal-Use Report, *Popular Science*, (1975). Reprint Supplied by Dow Corning Co., Midland, MI.
63. Schollenberger, C.S., Chapter 27, Polyurethane and Isocyanate-Based Adhesives, *Handbook of Adhesives*, 2nd Edition, (I. Skeist, ed.), Van Nostrand Reinhold, NY (1977).
64. Frisch, K.C., et al., Chapter 6, Diisocyanates as Wood Adhesives, *Wood Adhesives-Chemistry and Technology*, (R. Pizzi, ed.), Marcel Dekker, NY (1983).
65. Lavin, E. and Snelgrove, J.A., Chapter 31, Polyvinyl Acetal Adhesives, *Handbook of Adhesives*, 2nd Edition, (I. Skeist, ed.), Van Nostrand Reinhold, NY (1977).
66. Blomquist, R.F. and Vick, C.B., Chapter 49, Adhesives in Building Construction, *Handbook of Adhesives*, 2nd Edition, (I. Skeist, ed.), Van Nostrand Reinhold, NY (1977).
67. Miller, R.S., *Home Construction Projects with Adhesives and Glues*, Franklin Chemical Industries, Inc., Columbus, Ohio (1983). (This book was published in 1980 in almost identical form under the title *Adhesives and Glues—How to Choose and Use them*).
68. Corey, A.E., et al, Chapter 28, Polyvinyl Acetate Emulsions and Polyvinyl Alcohol for Adhesives, *Handbook of Adhesives*, 2nd Edition, (I. Skeist, ed.), Van Nostrand Reinhold, NY (1977).
69. Goulding, T.M., Chapter 7, Polyvinyl Acetate Wood Adhesives, *Wood Adhesives-Chemistry and Technology*, (A. Pizzi, ed.), Marcel Dekker, NY (1983).
70. Ablestik Laboratories, Gardena, CA, *Ablestik Solves the Quality Control Problem with Preserved Frozen Adhesives*, Company literature (1 sheet), undated.

71. Hodgson, M.E., Chapter 13, Pressure Sensitive Adhesives and Their Applications, *Adhesion* 3, (K.W. Allen, ed.), Applied Science Publishers, London (1978).

72. Abraham, W.A., Pressure Sensitive Adhesives—The Solventless Solutions, proceedings, papers presented at conference *Adhesives for Industry*, Technology Conference in Conjunction with So. California, Section, SPE, El Segundo, CA, pp. 140-150, (June 24-25, 1980).

73. Bemmels, C.W., Pressure-Sensitive Tapes and Labels, *Handbook of Adhesives*, 2nd Edition, (I. Skeist, ed.), Van Nostrand Reinhold (1977).

74. Satas, D., (editor), *Handbook of Pressure-Sensitive Adhesive Technology.* (This book has 30 chapters, 620 pages, and was written by 27 contributors), Van Nostrand Reinhold, NY (1982).

75. Moult, R.H., Chapter 24, Resorcinolic Adhesives, *Handbook of Adhesives*, 2nd Edition, (I. Skeist, ed.), Van Nostrand Reinhold, NY (1977).

76. Fenner, O.H., Chapter 4, Chemical and Environmental Properties of Plastics and Elastomers, *Handbook of Plastics and Elastomers*, (C.A. Harper, ed.), McGraw-Hill, NY (1975).

77. Smith, J.S., Silicone Adhesives for Joining Plastics, *Adhesives Age*, 17(6): 27-31 (June 1974).

78. Dow Corning Literature, *Silastic 736 RTV High-Temperature Adhesive/ Sealant*, 2 pp. (1982).

79. Beers, M.D., Chapter 39, Silicone Adhesive Sealants, *Handbook of Adhesives*, 2nd Edition, (I. Skeist, ed.), Van Nostrand Reinhold, NY (1977).

80. Marsden, J.G. and Sterman, S., Chapter 40, Organofunctional Silane Coupling Agents, *Handbook of Adhesives*, 2nd Edition, (I. Skeist, ed.), Van Nostrand Reinhold, NY (1977).

81. Romig, C.A. and Bush, S.M., Room Temperature Vulcanizing (RTV) Silicone Adhesive/Sealants, in Proceedings, 1979 Spring Seminar, *Designing with Today's Engineering Adhesives*, sponsored by The Adhesive and Sealant Council, Cherry Hill, NJ, pp 75-80 (March 11-14, 1979).

82. Haviland, G.S., Manufacturing Aspects of Six Adhesive Cure Systems, pp. 175-189, *High-Performance Adhesive Bonding*. Society of Automotive Engineers (SAE) (1983).

83. Shields, J., Chapter 11, Evaluation of an Ultraviolet-Curing Structural Adhesive, *Adhesion* 1, (K.W. Allen, ed.), Applied Science Publishers, London (1977).

84. Loctite Product Data Sheet *Ultraviolet Curing Adhesive PDS 353, 354*, Loctite Corp. Newington, CT, 06111 (Dec. 1978).

85. Bluestein, C., Radiant Energy Curable Adhesives, *Adhesives Age*, 25(12): 19-22 (December 1982).

86. Marino, F., Ultraviolet Adhesives for Quick, Easy Bonding, *Machine Design*, 56(18): 50-54 (August 9, 1984).

87. Yaroch, E.J., Water-Based Adhesives, pp 138-152, *High Performance Adhesive Bonding*, Society of Automotive Engineers (SAE) (1983).

88. Bolger, J.C. and Morano, S.L. Conductive Adhesives: How and Where they Work, *Adhesives Age*, 27(7):17-20 (June 1984).

6

Adhesives for Specific Adherends

INTRODUCTION

Chapter 4 outlined recommended surface preparation methods for all types of adherends, and Chapter 8 will offer specific recommendations for solvent cementing of thermoplastics. This chapter suggests adhesives for specific adherend types. Tables are occasionally published listing large numbers of adhesive types recommended for specific adherends. Such tables can be misleading in supplying information needed to provide strong durable bonds, because the user tends to forget that some combinations of adhesives and adherends are superior to others, particularly in durability, although resistance to other environments is also important. The emphasis in the following material will be on a listing of the adhesive types believed to provide strong lasting bonds. Chapter 5 discussed the adhesive types in detail.

Metals

Aluminum and alloys—Adhesives recommended include modified epoxies, modified phenolics, epoxy-phenolics, neoprene-phenolics, second-generation acrylics, cyanoacrylates, silicone rubbers, and vinyl plastisols. Sell[1] has ranked a number of adhesives in the order of decreasing durability with aluminum adherends as follows:

- nitrile-phenolics
- high-temperature epoxies
- 250°F (121°C)-curing epoxies
- 250°F (121°C)-curing rubber-modified epoxies
- vinyl epoxies
- two-part room-temperature-curing epoxy paste with amine cure
- two-part polyurethanes

190

Brewis[2] has recently discussed the nature of adhesives used for aluminum. The two major aluminum manufacturers, Aluminum Company of America (ALCOA) and Reynolds Metals, have published small useful volumes on all aspects of aluminum bonding, although these volumes are not recent.[3,4] Another excellent detailed discussion of aluminum adhesives, particularly from the viewpoint of durability, is given by Minford of ALCOA.[5]

Beryllium—Adhesives recommended include epoxy-phenolics, nitrile-phenolics, epoxies (RT cure, contact pressure), epoxy-nylon, polyimide, polybenzimidazole, epoxy-nitrile, and polyurethane. Since beryllium retains significant strength at temperatures up to 1000°F (538°C), the high-temperature application area is significant for this somewhat exotic metal. Polybenzimidazoles (PBI's) are relatively stable in air at temperature up to 550°F (288°C), as indicated in Chapter 5, for short periods of time. Polyimides (PI's) can be used at somewhat lower temperatures for longer periods. The more conventional adhesives listed above are much more temperature-sensitive than PBI and PI, but are considerably stronger at room temperature, and have equivalent, or even slightly higher strength, at 250°F (121°C).[6,7] Bond strengths of >4350 psi (30 MPa) in shear and tension can be obtained by adhesive bonding beryllium, with fracture being due to cohesive failure within the adhesive.[8]

Brass—Adhesives used with copper and copper alloys (see below) can also be used with brass, although the surface preparation methods may be different.

Bronze—Adhesives used with copper and copper alloys (see below) can also be used with bronze.

Cadmium (plated on steel)—Adhesives recommended include nitrile-phenolic and anaerobics.

Copper and Copper Alloys—Adhesives recommended include epoxies, polyurethane, silicone, nylon-epoxy, nitrile-phenolic, neoprene-phenolic, acrylic, cyanoacrylate, anaerobics, and partially hydrogenated polybutadiene (for bonding copper to polyethylene).

Only heat-cured epoxies containing dicyandiamide (DICY) or melamine should be used. DICY has been shown to be beneficial, either when used as the sole curing agent with epoxy resins, when mixed with other curing agents, or when used to pretreat the copper surface before bonding. Even when simply added to coatings (e.g. phenolic-cured epoxies, which cure by a different mechanism), DICY and a melamine compound both increased time to adhesive failure significantly on either bare or alkaline permanganate-treated copper.[9]

Gold—Adhesives recommended include epoxies, epoxy-phenolic, polyvinyl alkyl ether, and anaerobics (need primer to activate system).

Lead—Adhesives recommended include epoxies, vinyl alcohol-vinyl acetate copolymer, polyvinyl alkyl ether, polyacrylate (carboxylic), polyurethane (two-part), epoxy-phenolics, silicones, and cyanoacrylates. The high-strength thermoset and alloy adhesives are rarely justified for bonding lead. Even where other properties recommend these adhesives, the designer should check to see whether some lower-cost or easier-to-use adhesive is also suitable. An exception is terne (lead-coated steel). This is a much stronger metal than lead, and lap-shear strengths exceeding 300 psi (21 MPa) have been reported for adhesive joints with terne.[10]

Magnesium and magnesium alloys—Adhesives recommended include epoxies, epoxy-phenolics, polyurethanes, silicones, cyanoacrylates, polyvinyl acetate, vinyl chloride-vinyl acetate copolymer, vinyl-phenolic, nitrile-phenolic, neoprene-phenolic, and nylon-epoxy. A wide variety of adhesives can be used for bonding magnesium so long as proper corrosion protection is maintained in keeping with joint design and end-use requirements. Because of magnesium's sensitivity to moisture and the galvanic couple, water-based adhesives would be expected to cause problems. Surface preparation should always be carried out to assure that the adhesive itself does not react with the alloy to create a corrosive condition. Another important observation is that high-modulus adhesives tend to provide lower bond strengths than lower-modulus adhesives.[11]

Nickel and nickel alloys—Nickel is usually used in alloy form. Relatively little work has been carried out on adhesive bonding of nickel-base alloys because most of these alloys are used at temperatures above the service temperature of organic adhesives, or under corrosive conditions. Inorganic adhesives of sufficient ductility and low-enough maturing temperatures have not been developed to compete effectively with brazing and welding for joining high-temperature structures.[12] To date, *epoxy* adhesives are the most common adhesives used to bond nickel and its alloys. In all likelihood, PBI and PI adhesives can also be used for high-temperature applications. Other adhesives used include epoxy-nylon, polyamides, nitrile-phenolic, vinyl-phenolic, polyisocyanates, melamines and neoprenes.[13]

Plated metals—See Cadmium and Zinc.

Silver—Adhesives recommended include epoxies, polyvinyl alkyl ether, polyhydroxy ether and neoprene rubber.

Steel, mild, carbon, (iron)—Adhesives recommended include acrylics, epoxies, nitrile-phenolic (high moderate-temperature strength, but drops off rapidly at higher temperatures), polybenzimidazole (high strength over a wide temperature range), polyimide, and epoxy-phenolic for high-strength applications. For lesser-strength applications use thermoplastics and rubber-base materials such as chlorinated natural rubber, reclaim rubber, styrene-butadiene rubber, butadiene-acrylonitrile rubber, neoprene, butyl rubber, polyisobutylene, polyurethane rubber, polysulfide and silicone rubber.[14]

Bitumen and soluble silicates are also used for some applications.

Stainless steel—Although surface preparation methods are usually different, the adhesives used for mild steel can generally be used for stainless steel.

Tin—Adhesives recommended include caesin glue, epoxies, polyvinyl alkyl ether, polyacrylate (carboxylic), styrene-butadiene rubber, and polyisobutylene.

Titanium and titanium alloys—Adhesives recommended include epoxies, nitrile-epoxy, nitrile-phenolic, polyimide, and epoxy-phenolic. *Polyimide* adhesives provide strengths of 1600-1800 psi (11.0–12.4 MPa) at 600°F (316°C). These adhesives are not used for skin-to-core bonds because the temperature environment is not high enough to make them attractive, and because of the inherent problems caused by high volatile release during cure. *Epoxy-phenolics* (novalacs) and *nitrile-epoxies* are normally tested at 350°F (177°C). *Nitrile-phenolics*, because of their high peel strengths, are recommended for use in metal-to-metal bonds at the sacrifice of lap-shear strength at temperatures above 350°F (177°C), provided the application will permit a reduction in shear strength. *Nitrile-epoxies* are recommended for skin-to-core applications because less volatiles are released during cure than with epoxy-phenolics. The volatiles released during cure by the latter adhesive and by polyimides create internal pressure, which can result in core-node bond and skin-to-core bond failure.[15,16]

The use of titanium adhesive-bonded structures for high-temperature (392-572°F or 200-300°C) applications has been limited because of the rapid degradation of the adhesive at these temperatures. Recently polyimide adhesives have been developed with terminal acetylenic groups. These adhesives have been found to retain 45-50% of their original strength after 1000 hours of thermal aging at 500°F (260°C). In another approach, the introduction of perfluoroalkylene groups into aromatic polyimides has resulted in a high degree of strength retention after 5000 hours at 572°F (300°C). To improve the oxidation resistance at elevated temperatures many formulations are pigmented with fine alumina powder. The only really high-temperature adhesive not based on polyimide resin is *polyphenylquinoxaline*. An adhesive based on this heteroaromatic polymer showed a decrease of only 25% of its original strength after 500 hours at 698°F (370°C).[17]

Keith[18] has covered all aspects of titanium adhesive bonding, including adhesive selection, in a 1973 discussion.

Tungsten and tungsten alloys—Little information has been found on recommended adhesives for tungsten, although nitrile rubber and epoxies have been used.[13]

Uranium—Epoxies have been used to bond this exotic material.[13]

Zinc and zinc alloys—Adhesives recommended include nitrile-epoxies, epoxies, silicones, cyanoacrylates, and rubber-based adhesives.[19]

Thermoplastics

With these materials solvent cementing or thermal-welding methods are often preferable alternatives to adhesive bonding. However, where dissimilar materials are being bonded, or where the thermoplastic is relatively inert to solvents, adhesive bonding is recommended.

Acetal copolymer (Celcon®)—Although thermal welding is ordinarily used for bonding this material to obtain optimum bond strength, adhesives are used under certain conditions. Three types of adhesives are used: *solvent, structural,* and *non-structural*. Hexafluoroacetone sesquihydrate is used for solvent cementing (see Chapter 8). Structural adhesives are generally thermosets. Many of these adhesives can be used continuously at temperatures up to 350°F (177°C), which is higher than the recommended continuous-use temperature of 220°F (104°C) of the copolymer. Structural adhesive types recommended by the manufacturer (Celanese) are: epoxy (to 160°F or 71°C), polyester with isocyanate curing agent (to 200°F or 121°C), and cyanoacrylate (to 181°F or 82°C). Structural adhesives for bonding acetal copolymer to itself have yielded shear strengths of 600-800 psi (4.1-5.5 MPa). Non-structural adhesives are usually one-component, room-temperature curing systems based on either thermoplastic resins or elastomeric materials dispersed in solvents. They are normally used in applications which will not have to sustain heavy and/or continuous loading and will not reach temperatures above 180°F (82°C). Neoprene rubber adhesives have been used to provide shear strengths of 325 psi (2.24 MPa) to sanded surfaces and 300 psi (2.1 MPa) to unsanded surfaces. As in structural adhesives, a reduction in strength can be expected under peeling load.[20]

Acetal homopolymer (Delrin®)—Adhesives used to bond acetal homopolymer to itself and to other materials, such as aluminum, steel, natural rubber, neoprene rubber, and Buna rubber, include polyester with isocyanate curing agent, rubber-base adhesives, phenolics, epoxies, modified epoxies, and vinyls. Solvent cementing cannot be used unless the surfaces are specially roughened, because of the high solvent resistance of this material.[21] Other adhesive types sometimes used are resorcinol, vinyl-phenolic, ethylene vinyl acetate, cyanoacrylates and polyurethane.

Acrylonitrile-Butadiene-Styrene (ABS)—Bodied solvent cements are usually used to bond ABS. Adhesives recommended include epoxies, urethanes, second-generation acrylics, vinyls, nitrile-phenolics and cyanoacrylates.[22,23]

Cellulosics—These plastics (cellulose acetate, cellulose acetate butyrate or CAB, cellulose nitrate, cellulose propionate, and ethyl cellulose) are ordinarily solvent cemented, but for bonding to non-solvent-cementable materials, conventional adhesives must be used. Adhesives commonly used are polyurethanes, epoxies, and cyanoacrylates. Cellulosic plastics may contain plasticizers that are not compatible with the adhesive selected. The extent of plasticizer migration should be determined before an adhesive is selected.[22] Recommendations for conventional adhesives for specific cellulosic types are as follows:

- *cellulose acetate*—natural rubber (latex), polyisobutylene rubber, neoprene rubber, polyvinyl acetate, ethylene vinyl acetate, polyacrylate (carboxylic), cyanoacrylate, polyamide (versamid), phenoxy, polyester + isocyanate, nitrile-phenolic, polyurethane, and resorcinol-formaldehyde.

- *cellulose acetate butyrate (CAB)*—natural rubber (latex), polyisobutylene rubber, nitrile rubber, neoprene rubber, polyvinyl acetate, cyanoacrylate, polyamide (versamid), polyester + isocyanate, nitrile-phenolic, resorcinol-formaldehyde, and modified acrylics.

- *cellulose nitrate*—same as CAB above.

- *ethyl cellulose*—cellulose nitrate in solution (or general-purpose household cement), epoxy, nitrile-phenolic, synthetic rubber or thermoplastic resin combined with thermosetting resin, and resorcinol-formaldehyde.

Ethylene-chlorotrifluoroethylene (E-CTFE)—See Fluoroplastics.

Fluorinated-ethylene propylene (FEP Teflon)—See Fluoroplastics.

Fluoroplastics—Epoxies and polyurethanes give good bond strengths with properly treated fluoroplastic surfaces.[22]

Ionomer (Surlyn®)—Adhesives recommended are epoxies and polyurethanes.

Nylons (polyamides)—There are a number types, based on their chemical structure, but the most important and most widely used is nylon 6,6. The best adhesives for bonding nylon to nylon are generally solvents, as discussed below in Chapter 8. Various commercial adhesives, especially those based on phenol-formaldehyde (phenolics) and epoxy resins, are sometimes used for bonding nylon to nylon, although they are usually considered inferior to the solvent type because they result in a brittle joint. Adhesives recommended include nylon-phenolic, nitrile-phenolic, nitriles, neoprene, modified epoxy, cyanoacrylate, modified phenolic, resorcinol-formaldehyde, and polyurethane. Bonds in the range of 250–1,000 psi (1.7–6.9 MPa), depending on the thickness of the adherends, have been obtained.[22,23]

Perfluoroalkoxy Resins (PFA)—See Fluoroplastics.

Phenylene-oxide based resins (Noryl®)—Although solvent cementing is the usual method of bonding these resins, conventional adhesive bonding can be used. *Epoxy* and *acrylic* adhesives are generally recommended because of the versatile product lines and cure-rate schedules. Other adhesives recommended include cyanoacrylates, polysulfide-epoxy, RTV silicones, synthetic rubber, and hot melts. The manufacturer, General Electric, has recommended specific commercial designations of these types. The cure temperatures of the adhesives selected must not exceed the heat-deflection temperature of the Noryl resin, which ranges from 185 to 317°F (85 to 158°C), depending on the formulation. Adhesives not tested for compatibility with Noryl resins should be avoided or tested. Such testing should consider operational conditions of temperature and stress.[24]

Polyaryl ether (Arylon T)—This material is normally joined by solvent cementing.

Polyaryl sulfone (Astrel 360) (3M Co)—Hysol EA 9614 (modified epoxy on a nylon carrier) has been used to give good bonds with this plastic in a steel-plastic-steel bond.[25] Curing is at 160°F (71°C) for 4 hours, or 1 hour at 200–250°F (93–121°C) at 30 psi (0.21 MPa) pressure. Bonds of strengths up to 2023 psi (14 MPa) have been obtained with solvent-cleaned surfaces.

Polycarbonate—Polycarbonate is usually solvent cemented, but it can be bonded to other plastics, glass, aluminum, brass, steel, wood and other materials using a wide variety of adhesives. Silane primers may be used when joining polycarbonates with adhesives to promote adhesion and ensure a dry surface for bonding.[23] Adhesives recommended include *epoxies, urethanes, silicones, cyanoacrylates,* and *hot melts.* Generally the best results are obtained with solventless materials, such as epoxies and urethanes. Polycarbonates are very likely to stress crack in the presence of solvents. When cementing polycarbonate parts to metal parts a non-temperature-curing adhesive should be used to avoid creating strains in the adhesive caused by the differences in the coefficients of thermal expansion. This differential causes adherend cracking and considerably decreases expected bond strengths. Under no conditions should curing temperatures exceed 270°F (132°C), the heat-distortion temperature of standard polycarbonate resins.[26]

Polychlorotrifluoroethylene (PCTFE) (KEL-F) (Aclar)—Epoxy-polyamide and epoxy-polysulfide adhesives have been used successfully for bonding properly treated PCTFE. An epoxy-polyamide adhesive (Epon 828/Versamid 125 (60:40 ratio) cured 16 hours at room temperature and followed by 4 hours at 165°F (74°C) has given tensile-shear strengths of 2840–3010 psi (19.6–20.8 MPa) for various KEL-F resins treated with sodium naphthalene etch solutions and also abraded.[27,28]

Polyester (Thermoplastic Polyester)—Solvent cementing is usually used with these materials. Conventional adhesives recommended include single-and two-component polyurethanes, cyanoacrylates (Loctite 430 Superbonder), epoxies and silicone rubbers.

Polyetheretherketone (PEEK)—Epoxy adhesives, such as Ciba-Geigy's Araldite AW 134 with HY 994 hardener [cured 15 minutes at 248°F (120°C)] and Araldite AV 1566 GB [cured 1 hour at 248°F (230°C)] give the best results with this new engineering resin, according to ICI, the manufacturer. Other adhesives usable are cyanoacrylate (Loctite 414 with AC primer), anaerobics (Loctite 638 with N primer), and silicone sealant (Loctite Superflex). The highest lap-shear strength was obtained with the Araldite AW 134. This adhesive has balanced properties, good resistance to mechanical shock, thermal resistance to 212°F (100°C), and reasonable stability in the presence of aliphatic and aromatic solvents. Some solvents, particularly chlorinated hydrocarbons, will deteriorate the bond.[29]

Polyetherimide (Ultem®)—Adhesives for this new engineering plastic are *polyurethane* [cure at RT to 302°F (150°C)], *RTV silicones, hot-melts* (polyamide types) curing at 401°F (205°C), and *epoxies* (non-amine type, two-part).[30]

Polyethersulfone—Polyethersulfone may be solvent cemented (see Chapter 8). Conventional adhesives recommended by the manufacturer, ICI Ltd., are *epoxies* (CIBA-Geigy's Araldite AV 138 with HV 998 hardener) Araldite AW 134B with HY 994 hardener. Hysol 9340 two-part epoxy paste, and Silcoset 153 RTV *silicone sealant* supplied by ICI Ltd. with primer OP, also supplied by ICI Ltd., and Silcoset RTV 2 with Superflex primer, the latter supplied by an English source (Douglas Kane Group, Herts, UK). The highest lap-shear strength was obtained with the Araldite AW 134B.[31]

Other adhesives recommended by ICI are 3 M Company's Scotch Weld 2216 two-part epoxy, Amicon's Uniset A-359 one-part aluminum-filled epoxy, American Cyanamid's BR-89 one-part epoxy, Bostik's 7026 synthetic rubber and 598-45 two-part adhesive, GE's Silgrip SR-573, and Goodyear's Vitel polyester with isocyanate curing agent.[32]

Polyethylene—Acceptable bonds have been obtained between polyethylene surfaces with polar adhesives such as epoxies (anhydride- and amine-cured and two-component modified epoxies) and solvent cements containing synthetic rubber or phenolic resin. Other adhesives recommended include styrene-unsaturated polyester and solvent-type nitrile-phenolic.[33]

Polymethylmethacrylate (PMMA)—Ordinarily solvent cementing or thermal welding is used with PMMA. These methods provide stronger joints than adhesive bonding. Adhesives used are cyanoacrylates, second-generation acrylics, and epoxies, each of which provide good adhesion, but poor resistance to thermal aging.[22]

Polymethyl pentene (TPX)—No information has been found on adhesives for bonding TPX, but it is likely that the adhesives used for polyethylene will prove satisfactory for this polyolefin.

Polyphenylene sulfide (PPS) (Ryton®)—Adhesives recommended by the manufacturer. Phillips Chemical Company, include anaerobics (Loctite 306), liquid two-part epoxies (Hughson's Chemlok 305), and a two-part paste epoxy (Emerson & Cuming's Eccobond 104). Also recommended are USM's BOSTIK

7087 two-part epoxy and 3M Company's liquid two-part polyurethane EC-3532.[34]

Polypropylene—In general, adhesives recommended are similar to those used for polyethylene. Candidate adhesives include epoxies, polyamides, polysulfide epoxies, nitrile-phenolics, polyurethanes, and hot melts.[23]

Polystyrene—This material is ordinarily bonded by solvent cementing. Polystyrene can be bonded with vinyl acetate/vinyl chloride solution adhesives, acrylics, polyurethanes, unsaturated polyesters, epoxies, urea-formaldehyde, rubber-base adhesives, polyamide (Versamid-base), PMMA, and cyanoacrylates.[22,23,35] Monsanto Plastics and Resins Company has published an excellent bulletin recommending particular cements for both non-porous and porous surfaces. Cements are recommended for the fast-, medium-, and slow-setting ranges.[36]

Polysulfone—Adhesives recommended by the manufacturer, Union Carbide, include:[37,38]

3M Company
   Scotch-Grip 880 one-part solvent-based chloroprene
   Scotch-Weld 1838 two-part epoxy
   Scotch-Weld 2214 one-part epoxy
   Scotch-Weld 2216 two-part epoxy

American Cyanamid
   BR-89 one-part epoxy
   BR-92 two-part epoxy with either DICY curing agent
   or curing agent Z.

M&T Chemicals
   Uralane 5738 two-part polyurethane
   Uralane 8615 two-part polyurethane

The Scotch-Grip 880 elastomeric adhesive is recommended for bonding polysulfone to canvas, and the Uralane 8615 for bonding polysulfone to polyethylene.

Polytetrafluoroethylene (PTFE) (TFE Teflon)—See Fluoroplastics. Other adhesives used include nitrile-phenolics, polyisobutylene and silicones, the last two mentioned as pressure-sensitive adhesives.[35]

Polyvinyl chloride (PVC)—Solvent cementing is usually used for PVC. Since plasticizer migration from vinyls to the adhesive bond line can cause problems, adhesives selected must be tested for their compatibility with the plasticizer. *Nitrile rubber* adhesives are particularly good in this respect, although *polyurethanes* and *neoprenes* are also useful. 3M Company's Scotch-Grip 2262 adhesive (synthetic resin in solvent) is claimed to be exceptionally resistant to plasticizer migration in vinyls. A number of different plasticizers can be used with PVC's, so an adhesive that works with one plasticizer may not work with another.[22] Even rigid PVC contains up to 5% plasticizer, making it difficult to bond with epoxy and other non-rubber type adhesives. Most vinyls are fairly easy to bond with elastomeric adhesives after proper surface preparation. Cyanoacrylates can be used with rigid PVC. The highest bond strengths with semi-rigid or rigid PVC are obtained with two-component room-temperature-curing epoxies. Other adhesives used with rigid PVC include polyurethanes, modified acrylics, silicone elastomers, anaerobics, polyester-polyisocyanates, PMMA, nitrile-phenolics, polyisobutyl rubber, neoprene rubber, epoxy-polyamide and polyvinyl acetate.

Polyvinyl fluoride (PVC) (Tedlar®)—Adhesives recommended include acrylics, polyesters, epoxies, elastomers and pressure-sensitives.

Polyvinylidene fluoride (PVDF) (Kynar®)—See Fluoroplastics.

Styrene-acrylonitrile (SAN) (Lustran®)—Solvent cements are frequently used for SAN. Commercial cements include cyanoacrylate, epoxy, and the following 3M Company elastomeric adhesives:

Scotch-Grip 847 nitrile rubber
Scotch-Grip 1357 neoprene rubber
Scotch-Grip 2262 synthetic rubber

Several other commercial adhesives not specified as to type can be found in Reference 36.

Thermosetting Plastics (Thermosets)

Most thermosetting plastics are not particularly difficult to bond. Since these materials are not soluble, solvent cementing cannot be used. In some cases, however, solvent solutions can be used to join thermosets to thermoplastics. In general, adhesive bonding is the only practical way to join a thermoset to a thermoplastic, or to another thermoset. *Epoxies* or *modified epoxies* are the best adhesives for this purpose.

Diallyl phthalate (DAP)—Suggested adhesives include urea-formaldehyde, epoxy-polyamine, neoprene, nitrile-phenolic, styrene-butadiene, phenolic-polyvinyl butyral, polysulfides, furans, polyesters and polyurethanes.

Epoxies—Suggested adhesives include modified acrylics, epoxies, polyesters, resorcinol-formaldehyde, furane, phenol-formaldehyde, polyvinyl formalphenolic, polyvinyl butyral, nitrile rubber-phenolic, polyisobutylene rubber, polyurethane rubber, reclaimed rubber, melamine-formaldehyde, epoxy-phenolic, and cyanoacrylates. For maximum adhesion primers should be used. Nitrile-phenolics give excellent bonds if cured under pressure at temperatures of 300°F (149°C). Lower-strength bonds are obtained with most rubber-base adhesives.

Melamine-formaldehyde (melamines)—Adhesives recommended are epoxies, phenolic-polyvinyl butyral, epoxy-phenolic, nitrile-phenolic, polyurethane, neoprene, butadiene-nitrile rubber, cyanoacrylates, resorcinol-polyvinyl butyral, furane, and urea-formaldehyde.

Phenol-formaldehyde (phenolics)—Adhesives recommended are neoprene and urethane elastomers, epoxies and modified epoxies, phenolic-polyvinyl butyral, nitrile-phenolic, polyester, cyanoacrylates, resorcinol-formaldehyde, phenolics, polyacrylates, modified acrylics, PVC, and urea-formaldehyde. Phenolic adhesives give good results, but require higher cure temperatures and are less water-resistant than resorcinol-based adhesives.

Polyester (thermosetting polyester)—These materials may be bonded with neoprene or nitrile-phenolic elastomer, epoxy, epoxy-polyamide, epoxy-phenolic, phenolic, polyester, modified acrylic, cyanoacrylates, phenolic-polyvinyl butyral, polyurethane, butyl rubber, polyisobutylene, and PMMA.

Polyimide—Little published information is available on adhesives for bonding polyimides. A recent NASA study,[39,40] evaluated six adhesives for this purpose. Those recommended were:

American Cyanamid's FM-34 polyimide-tape

American Cyanamid's FM-34 B-18 polyimide tape (arsenic-free)

LARC-13, developed at NASA Langley Research Center

NRO56X, an adhesive resin of the NR 150 polymer family (polyimide) developed by Du Pont under a NASA contract.

Polyurethane—Elastomeric adhesives are prime candidates for polyurethanes, and polyurethane elastomer adhesives are particularly recommended.[23] Other suitable adhesives include epoxies, modified epoxies, polyamide-epoxy, neoprene, and resorcinol-formaldehyde. The latter offers excellent adhesion, but is somewhat brittle and can fail at relatively low loads.[22]

Silicone resins—These are generally bonded with silicone adhesives, either silicone rubber or silicone resin. Primers should be used before bonding.

Urea-formaldehyde—Adhesives recommended are epoxies, nitrile-phenolic, phenol-formaldehyde, urea-formaldehyde, resorcinol-formaldehyde, furane, polyester, butadiene-nitrile rubber, neoprene, cyanoacrylates and phenolic-polyvinyl butyral.

Reinforced Plastics/Composites

Adhesives which bond well to the base resin can be used to bond plastics reinforced with such materials as glass fibers or synthetic high-strength fibers. *Reinforced thermoplastics* can also be solvent cemented to themselves or joined to other thermoplastics by using a compatible solvent cement. For *reinforced thermosets*, in general, the adhesives recommended above for thermosetting plastics apply.

Plastic Foams

Solvent cements are usually preferable to conventional adhesives for thermoplastic structural foams. Some solvent cements and solvent-containing pressure-sensitive adhesives will collapse thermoplastic foams. Water-based adhesives based on SBR, polyvinyl acetate, or neoprene are frequently used. Solvent cementing is not effective on polyethylene foams because of their inertness. Recommendations for adhesives for *thermoplastic foams* are:

- *Phenylene oxide-based resins (Noryl®)*—epoxy, polyisocyanate, polyvinyl butyral, nitrile rubber, neoprene rubber, polyurethane rubber, polyvinylidene chloride, and acrylic.

- *Polyethylene*—nitrile rubber, polyisobutylene rubber, flexible epoxy, nitrile-phenolic, and water-based (emulsion) adhesives.

- *Polystyrene*—for these foams (expanded polystyrene or EPS, etc.)

aromatic solvent adhesives (toluol, etc.) can cause collapse of the foam cell walls. For this reason it is advisable to use either 100%-solids adhesives or water-based adhesives based on SBR or polyvinyl acetate.[22] Specific adhesives recommended include urea-formaldehyde, epoxy, polyester-isocyanate, polyvinyl acetate, vinyl chloride-vinyl acetate copolymer, and reclaim rubber. Polystyrene foam can be bonded satisfactorily with any of the following *general* adhesive types:

> *Water-based (emulsion)*—best for bonding polystyrene foam to porous surfaces.

> *Contact-bond*—for optimum initial strength. Both the water-based and solvent-based types may need auxiliary heating systems for further drying. Solvent types are recommended for adhering to metal, baked enamel, and painted surfaces.

> *Pressure-sensitive adhesives*—these will bond to almost any substrate. Both water-base and solvent-base types are used. However, they are not usable in applications requiring long-term resistance to stress or resistance to high heat levels.

> *100%-solids adhesives*—these are two-part epoxies and polyurethanes. They form an extremely strong heat- and environmental-resistant bond.

- *Polyvinyl chloride*—epoxy, polyester-isocyanate, unsaturated polyester, vinyl chloride-acetate copolymer, polyvinyl acetate, polyvinyl alkyl ether, ethylene-vinyl acetate, nitrile rubber-phenolic, neoprene rubber, polyisobutylene rubber, polyurethane rubber, and polysulfide rubber. See discussion under Thermoplastics above concerning migration of plasticizers in polyvinyl chloride.

- *Polycarbonate*—urethane, epoxy, rubber-based adhesives.

- *Thermoplastic polyester*—urethane, epoxy.

Recommendations for *thermosetting foams* are:

- *Epoxy*—(including syntactic foams), heat-cured epoxies (one-part).

- *Phenolic*—epoxy, polyester-isocyanate, polyvinyl acetate, vinyl chloride-acetate copolymer, polyvinyl formal-phenolic, nitrile rubber, nitrile rubber-phenolic, reclaim rubber, neoprene rubber, polyurethane rubber, butyl rubber, melamine-formaldehyde, neoprene-phenolic, and polyvinyl formal-phenolic.

- *Polyurethane*—epoxy, polyester, polyacrylate, polyhydroxyether, nitrile rubber, butyl rubber, water-based (emulsion), polyurethane rubber, neoprene, SBR, melamine-formaldehyde and resorcinol-formaldehyde are specific types. Generally, a *flexible* adhesive should be used for flexible polyurethane foams. Synthetic elastomer adhesives with fast-tack characteristics are available in spray cans. Solvent-base neoprenes are recommended for resistance to stress, water and weathering. Solvent-base nitriles are recommended for resistance to heat, solvents and oil. Water-base adhesives generally dry too slowly for most industrial applications, unless accelerated equipment is used. For immediate stress resistance, *contact bonding* is preferred. In this method, the adhesive is applied to the foam and to the other substrate by spraying or brushing. *Wet bonding* can be used where the adhesive is applied to the other surface. This reduces "soak-in" on the highly absorbent and porous foam.[41]

- *Silicone*—silicone rubber.

- *Urea-formaldehyde*—urea-formaldehyde, resorcinol-formaldehyde.

Rubbers (Elastomers)

Bonding of vulcanized elastomers to themselves and to other materials is generally accomplished by using a pressure-sensitive adhesive derived from an elastomer similar to the one being bonded. Adhesives used include the following rubber-base materials: natural-, chlorinated-, reclaim-, butyl-, nitrile-, butadiene-styrene-, polyurethane-, polysulfide-, and neoprene-rubber; also, acrylics, cyanoacrylates, polyester-isocyanates, resorcinol-formaldehyde, phenolic-resorcinol-formaldehyde, silicone resin, epoxies, polyisocyanates, furanes, nitrile-phenolics, neoprene-phenolic, polyvinyl formal-phenolic, and flexible epoxy-polyamides.[13,22] *Neoprene* and *nitrile rubber* adhesives are particularly recommended for bonding rubber. Neoprene adhesives are good all-around adhesives for rubber. Nitrile adhesives are particularly recommended for gaskets formulated with nitrile or polysulfide rubber.[41]

Wood

Adhesives (glues) used for wood include animal or hide glues, starch, casein, soybean, blood glues, fish glues, and synthetic resin adhesives. *Fish glues* have been used for bonding wood for many years. *Resorcinol-formaldehyde* resins are cold-setting adhesives used for wood structures. *Urea-formaldehyde* adhesives, often modified with *melamine-formaldehydes*, are used in the production of plywood and in wood veneering for interior use. *Phenol-formaldehyde* and *resorcinol-formaldehyde* adhesive systems have the best heat and weather resistance of wood adhesives. *Polyvinyl acetates* are quick-drying, water-based adhesives commonly used for furniture assembly. These adhesives produce very strong bonds, but are not resistant to moisture or high temperatures.[42]

Epoxies have been used for certain specialized applications (wood-to-metal bonds). *Rubber-based contact adhesives* have also been used. *Mastic* adhesives have been used in construction work and are usually applied in caulking guns. These materials are based on elastomers, including reclaimed rubber, neoprene, butadiene-styrene, polyurethane, and butyl rubber.[42]

Excellent references on wood bonding adhesives include those by Pizzi,[43] Millett et al.,[44] Selbo,[45] and Miller.[46] The latter is an easy-to-read book prepared by Franklin Chemical Industries, a major wood glue manufacturer.

Glass and Ceramics

Adhesives for bonding glass are generally transparent, heat-setting resins that are water-resistant to meet the requirements of outdoor application. They include polyvinyl butyral, phenolic butyral, nitrile-phenolic, neoprene, polysulfide, silicone, vinyl acetate, and epoxies.[22] Adhesives used for bonding glass and glazed ceramics should have minimum volatile content, since these materials are non-porous. *Epoxy* adhesives, being 100%-solids materials, are probably the best for these applications. Second-generation *acrylics* are good for bonding glass and ceramics to thermoplastic polymers. Where large areas and materials with greatly differing coeffcents of thermal expansion are being bonded, polysulfides should be considered, since these adhesives have very high elongation.[47]

Optical adhesives used for bonding glass lenses are usually styrene-modified polyesters and styrene-monomer-based adhesives. Optically clear epoxies are also available.[22,47,48]

REFERENCES

1. Sell, W.D., Some Analytical Techniques for Durability Testing of Structural Adhesives, proceedings, 19th National SAMPE Symposium and Exhibition, Vol. 19, *New Industries and Applications for Advanced Materials Technology*, p. 196 (abstract only) Buena Park, CA (April 23-25, 1974).
2. Brewis, D.H., Chapter 5, Aluminum Adherends, *Durability of Structural Adhesives*, (A.J. Kinloch, ed.), Applied Science Publishers, London (1983).
3. Reynolds Metals Company, Richmond, VA, *Adhesive Bonding Aluminum*, (1966).
4. Aluminum Company of America, Pittsburgh, PA, *Adhesive Bonding ALCOA Aluminum*, (1967).
5. Minford, J.D., Chapter 3, Aluminum Adherend Bond Permanence, *Treatise on Adhesion and Adherends*, Vol. 5, (R.L. Patrick, ed.), Marcel Dekker, NY (1981).
6. Cagle, C.V. Chapter 21, Surface Preparation of Beryllium and Other Adherends, *Handbook of Adhesive Bonding*, (C.V. Cagle, ed.), McGraw-Hill, NY (1973).
7. St. Cyr, M., Adhesive Bonding of Beryllium, proceedings, 15th National SAMPE Symposium and Exhibition, Vol. 15, *Materials and Processes for the 70's*, Los Angeles, CA, pp. 719-731 (April 29-May 1, 1969).
8. Cooper, R.E., Chapter 7, Adhesive Bonding of Beryllium, *Adhesion* 3 (K. W. Allen, ed.), Applied Science Publishers, London (1979).
9. Bolger, J.C. et al., A New Theory for Improving the Adhesion of Polymers to Copper. Final Report, INCRA Project No. 172, International Copper Research Association (INCRA), 708 Third Avenue, NY 10017 (August 16, 1971).
10. Fader, B., Chapter 13, Adhesive Bonding of Lead, *Handbook of Adhesive Bonding* (C.V. Cagle, ed.), McGraw-Hill, NY (1973).
11. Jackson, L.C., Chapter 15, Principles of Magnesium Adhesive-Bonding Technology, *Handbook of Adhesive Bonding*, McGraw-Hill, NY (1973).
12. Keith, R.E. et al., Adhesive Bonding of Nickel and Nickel-Base Alloys, NASA Technical Memorandum NASA TM X-53428 (October, 1965).

13. Shields, J., *Adhesives Handbook*, 2nd Edition, Newnes-Butterworth, London (1976). (The 3rd edition was published by Butterworths, London, in 1984).
14. Cagle, C.V., Chapter 14, Bonding Steels, *Handbook of Adhesive Bonding* (C.V. Cagle, ed.), McGraw-Hill, NY (1973).
15. Lively, G.W. and Hohman, A.E., Development of a Mechanical-Chemical Surface Treatment for Titanium Alloys and Adhesive Bonding, proceedings, 5th National SAMPE Technical Conference, Vol. 5, *Materials and Processes for the 70's*, pp. 145-159, Kiamesha Lake, NY (October 9–11, 1973).
16. Walter, R.E. et al. Structural Bonding of Titanium for Advanced Aircraft, proceedings, 2nd National SAMPE Technical Conference, Vol. 2, *Aerospace Adhesive and Elastomers*, pp. 321-330, Dallas, TX (October 6-8, 1970).
17. Mahoon, A., Chapter 6, Titanium Adherends, *Durability of Structural Adhesives* (A.J. Kinloch, ed.), Applied Science Publishers, London (1983).
18. Keith, R.E., Chapter 12, Adhesive Bonding of Titanium and its Alloys, *Handbook of Adhesive Bonding* (C.V. Cagle, ed.), McGraw-Hill, NY (1973).
19. McIntyre, R.T., Adhesive Bonding to Cadmium and Zinc Plated Steel Substrates, *Applied Polymer Symposia No. 19* (M.J. Bodnar, ed.), pp. 309-327, Wiley-Interscience (1972).
20. Celanese Plastics Company, Chatham, NJ, *The Celcon Acetal Copolymer Design Manual*, undated.
21. E. I. du Pont de Nemours & Co., *Delrin Acetal Resins Design Handbook*, A-67041, (1967).
22. Petrie, E. M., Chapter 10, Plastics and Elastomers as Adhesives, *Handbook of Plastics and Elastomers*, (C.A. Harper, ed.), McGraw-Hill, NY (1975).
23. Cagle, C.V., Chapter 19, Bonding Plastic Materials, *Handbook of Adhesive Bonding* (C.V. Cagle, ed.), McGraw-Hill, NY (1973).
24. General Electric Co., Plastics Dep't, Selkirk, NY, *Noryl® Resin Design Guide*, CDX-83.
25. Landrock, A.H , Processing Handbook on Surface Preparations for Adhesive Bonding, Picatinny Arsenal Technical Report 4883 (December 1974).
26. General Electric Company, Plastics Dept., Pittsfield, MA, *Designing with Lexan Resin*, CDC-536B, (April 1982).
27. DeLollis, N.J. and Montoya, O., Surface Treatment for Difficult to Bond Plastics, *Adhesives Age*, 6(1): 32-3 (January 1963).
28. St. Cyr, M., State of the Art—Methods of Bonding Fluorocarbon Plastics to Structural Materials, *PLASTEC* Rept. 6, May 1961.
29. ICI Petrochemicals and Plastics Div., Victrex PEEK Aromatic Polymer, *Adhesive Bonding*, Provisional Data Sheet PK PD 11 (February 1982).
30. General Electric Co., Pittsfield, MA, *The Comprehensive Guide to Material Properties, Design, Processing, and Secondary Operations*, ULTEM® *Polyetherimide Resin*, ULT-201. (undated).
31. ICI Americas, "Victrex" PES Aromatic Polymer—Adhesives for Use with "Victrex" Polyethersulfone, data sheets VX TS 1.78 (August 1978).
32. ICI Americas, Adhesives for Bonding Victrex® Polyethersulfone Systems and Cures, undated, received March 1984.
33. Anonymous, 3 Prime Factors in Adhesive Bonding of Plastics, *Plastics Design and Processing*, 8(6): 10-22 (June 1968).
34. Phillips Chemical Company, Bartlesville, OK, *Adhesives for Bonding Polyphenylene Sulfide*, Technical Service Memorandum (TSM) 283 (January 1979).

35. Miron, J. and Skeist, I., Chapter 41, Bonding Plastics, *Handbook of Adhesives*, 2nd Edition, Van Nostrand Reinhold, NY (1977).

36. Monsanto Plastics and Resins Co., *Fabrication Techniques for Lustrex and Lustran Styrene Plastics*, Technical Bulletin 6422, undated.

37. Union Carbide Plastics, Polysulfone News No. 10, *Adhesives for Polysulfone*, BA-107-17.

38. Union Carbide Plastics, Udel® Design Engineering Data Handbook, Section 6 *Fabrication*, p. 32, Adhesive Bonding.

39. Pragar, D., Polyimide Adhesive Bonding, pp. 123-138 in *Graphite/Polyimide Composites*, NASA Conference Publication 2079 (Feb. 28-Mar. 1, 1979).

40. Blatz, P.S., NR 150B2 Adhesive Development, *NASA CR 3017* (July 1978).

41. 3M Company, *Adhesives Answer Book for Product Assembly*, Z-CPSS(731)R.

42. Blomquist, R.F., Chapter 17, Adhesive Bonding of Wood, *Handbook of Adhesive Bonding*, (C.V. Cagle, ed.) McGraw-Hill, NY (1973).

43. Pizzi, A., (editor), *Wood Adhesives—Chemistry and Technology*, Marcel Dekker, NY (1983).

44. Millett, M.A. et al., Evaluating Wood Adhesives and Adhesive Bonds: Performance Requirements, Bonding Variables, Bond Evaluation, Procedural Recommendations, prepared by USDA Forest Products Laboratory for US Dept. of Housing and Urban Development (Feb. 1977). (Available from NTIS as PB 265646).

45. Selbo, M.L., *Adhesive Bonding of Wood*, USDA Forest Service Technical Bulletin No. 1512 (August 1975). (Available from the Supt. of Documents, U.S. Government Property Office, No. 001-000-03382).

46. Miller, R.S., *Home Construction Projects With Adhesives and Glues*, Franklin Chemical Industries, Columbus, Ohio, (1983). (This book was published in 1980 in almost identical form under the title *Adhesives and Glues—How to Choose and Use Them*).

47. DeLollis, N.J., *Adhesives, Adherends, Adhesion*, Robert E. Krieger Publishing Co., Huntington, NY (1980).

48. Cagle, C.V., Chapter 18, Bonding Glass, Optics, Ceramics, and Related Substrates, *Handbook of Adhesive Bonding*, (C.V. Cagle, ed.), McGraw-Hill, NY (1973).

7

Adhesive Bonding Process

INTRODUCTION

In assembling components by adhesive bonding, the availability of the multiplicity of adhesive-bonding methods can be used to advantage. This multiplicity can also be a problem to the bonder, however. His choice of method of application can, for example, restrict the degree of freedom to be exercised in designing the end product. The method of application can also affect the selection of materials that can be used in manufacturing the product, the quality or performance of the product, and, very often, assembly cost. The following factors must be considered in selecting the bonding method:[1]

- The size and shape of the parts to be bonded.
- The specific areas to which the adhesive is to be applied.
- The number of assemblies to be produced.
- The required production speed.
- The viscosity or other characteristics of the adhesive.
- The form of the adhesive (liquid, paste, powder, film, hot melt).

ADHESIVE STORAGE

Many adhesives must be stored in the dark or in opaque containers, while others should be stored at low temperatures (e.g. 41°F or 5°C) to prolong shelf life.[2] The manufacturer's directions, usually found in technical bulletins on the particular adhesive, frequently provide information on storage requirements. For example, the epoxy-phenolic film adhesive HT-424 (Bloomingdale) has the following storage schedule for newly prepared adhesive film that has been properly refrigerated during storage:[3]

. Storage Temperature. .		Useful Life
°F	°C	(days)
-10	-23	180
0	-18	150
30	-1	75
75	24	12
85	29	3
100	38	1

Basic resins and curing agents for thermosetting adhesives should be kept apart so that accidental container breakage will not lead to contamination problems. Containers for solvent-based adhesives should generally be sealed immediately after use to prevent solvent loss or the escape of toxic or flammable vapors.[2]

ADHESIVE PREPARATION

Adhesive preparation requires very careful attention. When removed from refrigerated storage the adhesive must be brought to application temperature. Usually this is at room temperature, but in some cases, such as in hot-melt adhesives, much higher temperatures are used. When separate components are mixed together (two-part adhesives) it is usually important to measure the proportions correctly to obtain optimum properties. This is particularly true with catalytic reactions, such as with amine curing agents for epoxy resins, where insufficient catalyst prevents complete polymerization of the base resin, while too much catalyst may lead to brittleness in the cured material. Excess unreacted curing agent may also cause corrosion of metallic adherends. Some two-component adhesives (fatty polyamides used with epoxies) have less critical mixing ratios and component volumes may often be measured by eye without too adverse an effect on the ultimate bond strength.[2]

The weighed-out components of multiple-part adhesives must be mixed together thoroughly until no color streaks or density stratifications are noticeable. Air should not be permitted to be absorbed into the mix as a result of over-agitation. The introduction of air can cause foaming of the adhesive during heat cure, resulting in porous bonds. If air presents a problem it may be necessary to remove it by vacuum degassing prior to application. Only enough adhesive should be mixed to work with before the adhesive begins to cure. The working life (pot life) of the adhesive is decreased as the ambient temperature increases and the batch size becomes larger. One-part and some heat-curing two-part adhesives have very long working lives at room temperature, and application and assembly speed or batch size are not critical. For a large-scale bonding operation, hand mixing is costly, messy and slow. Repeatability is entirely dependent on the operator. Equipment is available that can meter, mix, and dispense multi-component adhesives on a continuous or short basis.[4]

Small-Portion Mixer Dispensers

A number of packaging systems currently available will store small amounts of liquid adhesive components (thermosetting) and provide means for convenient mixing and dispensing, all within the package. These may take the form of flexible plastic pouches with removable dividers (clamps) separating the components (for two-or three-part resin systems with mixing ratios from 1:1 to 1:100). The pouches contain resin amounts varying from 2 grams up to as much as 100 grams. Other forms of container dispensers are also available in the form of two-barrel hypodermic syringes fused together. Polyethylene cartridge assemblies have an advantage over the pouches in that they make it easier to apply the adhesive to localized areas. The components in the pouches, which are frequently colored, are kneaded by hand until completly mixed, as evidenced by the resulting uniformity of color. A corner of the pouch is then snipped off with a scissors and functions as a dispenser when the adhesive is squeezed out. A large number of adhesive systems are available in these units, many for the home craftsman. The useful life of these units is extended considerably by refrigerated storage.

METHODS OF ADHESIVE APPLICATION

The selection of a method of application depends on the adhesive form, whether liquid, paste, powder, film or hot melt. Other factors influencing the choice of application method are the size and shape of parts to be bonded, the areas where the adhesive is to be applied, and production volume and rate.[4]

Liquid Adhesives

Adhesives in liquid form may be applied by the methods described below.

Brushing[2,5]—Brushing is often used when the adherends have complex shapes, or when it is desired to apply the adhesive to selected areas of a surface without the use of masks. With brushing the control of film thickness is limited and the resultant adhesive films are often uneven and blobbed. Brushing requires very little equipment and results in a minimum of wastage. Brushing is generally not suited to rapid assembly work. Stiff brushes provide the best results. Good brushes may be used repeatedly.

Flowing[2,5]—Flowing is particularly useful for applying liquid adhesives to flat surfaces having irregular shapes. This method is superior to brushing in that it provides a more uniform adhesive film thickness and a higher production rate. In the method the adhesive is fed under pressure through either a simple nozzle or a hollow brush. When a nozzle is used, the device is known as a "flow gun". With a brush it is called a "flow brush". The adhesive should be brushed on in a single smooth sweep. A second sweep over the same area is not practical with most adhesive types.

Spraying[2,5]—Spraying is primarily used for covering large areas with uniform contours. This method provides a higher production rate than flowing and also offers a more uniform film thickness. The solids content and consistency of the adhesive, however, must be rigidly controlled to insure an even coating. The equipment used is quite similar to that used for spraying parts. The exact equipment design and operating conditions may be varied to permit the production of almost any desired type and thickness of coverage. The presence of a possible health hazard coming from the solvent-spray mists must be considered and adequate ventilation provided.

Roll coating[2,5]—This technique is based on the transfer of adhesive material from a trough by means of a pick-up roller partially immersed in it, to a contacting transfer roller sheet. Material is continuously coated with adhesive when fed between the transfer roller an a pressure roller which is adjusted to determine the thickness required. Roll coating is most suitable for applying adhesives to flat sheets and film and may be used for parts as wide as 6 feet (1.83 meters). Where feasible, it provides the highest possible production rate and the most uniform coverage possible. On many occasions, when small stampings must be coated, it is economical to roll-coat large sheets, then cut the parts from the coated sheets. The wastage of the coating on the uncured portion is rarely as costly as the coating of small individual parts.

When multiple coats of adhesive must be applied by any process, the most uniform film thickness is achieved by applying the second coat perpendicular to the first. In addition, the time between successive coats must be carefully regulated. Too short a drying time may result in sagging, bubbling, or blistering, while too long a time may lend to the lifting of earlier coats.

Knife coating—This method employs an adjustable knife blade, bar, or rod to control the deposition of adhesive flowing to a sheet moving under the blade. The distance between the blade edge and the adherend surface determines the adhesive coating thickness.[2]

Silk screening[1,2]—This manual technique is used to insure that when selected areas or patterned spaces are adhesive coated, low-viscosity adhesives work well because they pass readily through the cloth onto the work. However, only relatively thin films can be applied by this method. By filling the pores in a cloth over areas to remain uncoated and leaving other portions of the cloth unfilled, effective masking is set up. The adhesive is then poured on top of the cloth and a rubber squeegee is used to facilitate flow of the adhesive down through the cloth. The squeegee, alone is another manual method of applying adhesive. As a general rule, fast-drying or tacky adhesives cannot be satisfactorily silk-screened.

Oil can and squeeze bottle—The hand-pump oil can and polyethylene squeeze bottle are often used for spot application of adhesives. Where it is necessary to apply adhesive inside a blind hole, or some other point with limited access, the oil can with a stem of appropriate length can be a simple solution. Tips at the end of the stem can be devised to deliver multiple spots of adhesive. The squeeze bottle can sometimes be used for the same purpose, and also to apply a continuous bond. Some liquid adhesives are applied in squeeze-bottle containers. When using polyethylene squeeze bottles, care should be taken that the adhesive will not affect the polyethylene itself.[1]

Hand dipping—The dipping process is capable of speeding up the coating of relatively large surfaces. Automatic versions are the most satisfactory, but hand dipping can be used to speed up the application of adhesive. The devices used to facilitate immersion of parts into the adhesive bath or trough range from baskets to screens to perforated drums.[1]

Pastes

Bulk adhesives such as pastes or mastics are the simplest and most reproducible adhesives to apply. These systems can be trowelled on or extruded through a caulking gun. Little operator skill is required. Since the thixotropic nature of the paste prevents it from flowing excessively, application is usually clean and little waste results.[4] A typical use is the application of adhesives in paste or mastic form to wall panels instead of using nails, which would be visible.

Spatulas, knives, trowels—A simple application tool particularly effective for hard-to-spread adhesive is the spatula. Knives and trowels, both of which may have notches cut into their applying edges, are variations of the spatula. The depth of notches and the spacing between them help regulate the amount of adhesive applied. The blade should be held firmly and at a right angle to the surface. The preferred shape for trowel notches is square. Triangular notches present sharp teeth which wear out faster than the wide contact surfaces of the square notches. Shallow, rounded, and closely spaced notches are often used with adhesives of higher liquid content, because such notches permit the adhesive to flow together and form a continuous unbroken film.[1]

Powders

Powder adhesives can be applied in three ways:

(1) They may be sifted onto a preheated substrate. The powder which falls onto the substrate melts and adheres to it. The assembly is then mated and cured according to recommended processes.

(2) A preheated substrate can also be dipped into the powder and then removed with an attached coating of adhesive (as in fluidized-bed coating). This method helps assure even powder distribution.

(3) The powder can be melted into a paste or liquid form and applied by conventional means.

Powder adhesives are generally one-part epoxy-based systems that require heat and pressure to cure. They do not require metering and mixing, but must often be refrigerated to obtain maximum shelf life. Because coating uniformity is poor, large variations in joint strength may result with these adhesives.[4]

Films

The use of dry adhesive films (see *Film and Tape Adhesives* in Chapter 5) is expanding more rapidly than other forms because of the following advantages:

● High repeatability—no mixing or metering; constant thickness

● Ease of handling—low equipment cost; relatively hazard-free; clean operating

● Very little waste—preforms can be cut to size

● Excellent physical properties—wide variety of adhesive types available.

Film adhesives may be used only on flat surfaces or simple curves. Application requires a relatively high degree of cure to ensure nonwrinkling and removal of separator sheets. Characteristics of available film adhesives vary widely, depending on the type of adhesives used. Film adhesives are supplied in both *unsupported* and *supported* types. The carrier for supported films is generally fibrous fabric or mat. Film adhesives are supplied in heat-activated, pressure-sensitive, or solvent-activated forms. Solvent-activated forms are made tacky and pressure-sensitive by wiping with solvent. This type of adhesive is not as strong as other types, but is well suited for contoured, curved, or irregularly shaped parts. Manual solvent-reactivation methods should be closely monitored so that excessive solvent is not used. Chemical formations available in solvent-activated forms include neoprene, nitrile, and butyral phenolics. Decorative trim and nameplates are usually fastened onto a product with solvent-activated adhesives.[4]

Hot Melts

Although there are a number of variations in equipment, two basic types of systems are used to heat and apply hot-melt adhesives: *melt-reservoir* and *pressure-feed systems.*[2,7]

Melt-reservoir systems (Tank-type applications)—This type consists of a melting pot or reservoir, a pump, feed hose, and an extrusion gun or application wheel to apply the melted adhesive to the product. Spray or jet guns are also used in packaging applications. Metering pumps are used in automated systems. Extrusion guns are triggered manually for assembly products that do not lend themselves to automation. Hot-melt adhesive is loaded into the reservoir in granular, block, or chip form, where it is heated and maintained at the desired delivery temperature. A typical reservoir system delivers about 10 pounds (4.54 kg) per hour, depending on the adhesive used. Operation temperatures are below $400°F$ ($204°C$). These systems are excellent for handling low-performance, low-viscosity adhesives (1,000–50,000 cps.) (1-50 Pa·s).

While the reservoir-type system has the advantage of holding a large amount of fluid adhesive in readiness, it has some drawbacks. It is sometimes difficult to maintain a uniform temperature in so large a volume of adhesive, since fresh solid adhesive must be added at periodic intervals to replace the adhesive used. If the temperature of the fluid adhesive is too low at the nozzle, the bond may be degraded, or carbonized deposits may form and plug the nozzle. The higher-performance hot melts are especially susceptible to these problems. These formulations have higher molecular weights and high viscosities and they degrade more rapidly when heated in the presence of oxygen than the packaging hot melts. These materials require very close temperature control and, for optimum results, elimination of oxygen from the dispensing system.

Progressive-feed systems—These systems heat only a small amount of adhesive at a time, thereby eliminating some of the problems inherent in a reservoir system. Several versions of the "first-in, first-out" equipment are in use. In one type, granules or pellets of adhesive are loaded into a hopper. The adhesive melts on a heated grid, then flows to a gear pump and is immediately transferred under pressure, through an electrically heated hose, to a heated gun. This equipment can feed hot melts at the rate of 20 pounds (9.1 kg) or more per hour, even though less than a pound (454 grams) of adhesive is maintained in the molten state.

Another type of progressive-feed system is a self-contained applicator gun (portable hot-melt gun). This system requires no insulated hose, because the adhesive is melted at the gun, from adhesive cartridges, pellets (slugs) or coiled cord. In guns using cartridge or pellet adhesives, air pressure moves the fluid adhesive to the nozzle; in the coil-fed system, the coil is mechanically driven, which forces the adhesive through the gun onto the workpiece.

Progressive-feed systems benefit from the first-in, first-out principle of application, which minimizes the possibility of adhesive degradation and permits the use of higher-performance (usually more heat-sensitive) materials. Operating temperatures range from 350–600°F (176–315°C). Another advantage of self-contained applicators is their portability. These guns handle high-performance adhesives with viscosities up to 500,000 cps. (5000 Pa·s) of the type that are used for structural applications. A disadvantage of the systems for some adhesives (principally polyamides) is that there is no holding period to allow time for moisture flash-off. Material cost is higher than for systems that use the granular form. The lower delivery rates (3–4 pounds) (1.4–1.8 kg) pr hour of cartridge and coil-type progressive-feed systems may be a limitation, but only for applicators requiring very long adhesive beads. These applicators cannot be used for tacky or low-durometer adhesives without special arrangements.

Hand-held applicator guns of the type described above can be used for applications such as:[6]

- Plastic to wood or metal
- Metal parts assemblies
- Fiberglass to wood laminates
- Foam padding to metal and wood
- Decorative panels and reliefs to wood and metal
- Chipboard to wood or metal
- Doweling and tongue-and-grooving
- Welt and gimp attachment
- Moldings to cabinets
- Potting of electronic parts
- Replacement for jigs and fixtures

JOINT-ASSEMBLY METHODS

A number of methods have been developed for assembling bonded joints; all have certain basic points in common:[5]

- The surface of the adhesive coating must become liquid at some point in the process to insure wetting of the adherend surface and promote contact
- Foreign materials, such as solvents and moisture must be expelled from the joint to prevent the formation of voids, vapor locks, and faults in the glue line

- Pressure must be applied to the joint until the adhesive sets sufficiently to hold the assembly together without accidental misalign-
ment of the adherends.

While various specific methods of joint assembly may differ in detail, most
of these methods may be classified as one of the four basic methods described
below: wet assembly, pressure-sensitive bonding, solvent activation, and heat
activation.

Wet Assembly

This is probably the commonest of the joint-assembly methods. With porous
adherends it may be used with any type of adhesive that will wet the adherend
surface without being heated. With nonporous adherends, however, wet assembly is practicable only when the adhesive used contains no volatile solvents, since
otherwise the sealed-in solvent would produce voids. The process consists essentially in aligning the parts, pressing them together while the adhesive coating is
still wet from application, and maintaining the pressure until the bond is strong
enough to hold the assembly together without pressure. To minimize the length
of time that this pressure must be maintained, the adhesive coating is usually
permitted to partially dry before assembly. The permissable predrying time is
known as the "open assembly time" of the adhesive.

Pressure-Sensitive and Contact Bonding

These methods are usuable only when the adhesive retains some tack when
dry. The joint assembly processes used for each are identical; the only difference
is that with pressure-sensitive bonding only one of the adherends is coated, while
with contact bonding both are. For both processes, therefore, joint assembly
consists in permitting the adhesive coating to dry completely, then aligning the
parts and pressing them together to form the bond.

Solvent Activation

This is essentially a postponed wet-assembly process and, as such, is usually
unsuitable for nonporous adherends. It can be used only with adhesives that can
be reactivated by solvents, unlike most two-part adhesives. Solvent activation is
primarily desirable when it is convenient to precoat parts that are to be bonded
at a later date, or when the adhesive used has an exceptionally slow drying rate.
Solvent activation consists in permitting the adhesive coating to dry completely,
dampening the surface of the coating with a fast-drying solvent, such as methyl
ethyl ketone, quickly aligning the parts, and promptly applying pressure until
the adhesive sets sufficiently to hold the assembly together. Water is used as
the solvent for "gummed" adhesives such as are used in gummed-tape labels.
Water-activated adhesives are usually made from animal glues or dextrin. Adhesives activated with organic solvents have not received wide acceptance
because of the fire and health hazard involved and the inconvenience of activating with solvents. They are used in certain applications where a pressure-sensitive

adhesive tape cannot supply the bond strength, solvent—or heat resistance required.[5],[8]

Heat Activation

This method may be used only with adherends that can tolerate heat. Since no volatile solvents are employed, it is especially useful with non-porous adherends. However, it is suitable only for adhesives that can be reactivated by heating. Two-part adhesives are therefore excluded. Heat activation consists in permitting the adhesive coating to dry completely, aligning the parts, heating them, applying pressure so that the adhesive flows together, then allowing the adhesive to cool sufficiently to form the bond. Occasionally the parts are heated before their bonding surfaces are mated.[5] Heat-activated adhesives are used on tapes which can be applied with heat and pressure. They can be made from a wide variety of *thermoplastic* materials such as waxes, polyethylene, cellulose esters and ethers, nitrocellulose, polyvinyl acetate, polyvinyl chloride and many rubber-resin combinations. Typical tape applications are in fabric-mending tape and paperboard box manufacture (stay-tapes). Heat-activated adhesives provide a very high irreversible bond. No fire or health hazards are involved in their application. One disadvantage is the need for special high-temperature presses to apply this type of adhesive.[8]

Curing

A number of high-strength structural adhesives must be cured to develop joint strength. These materials exist in film and liquid forms, the latter usually relying on catalytic reaction to effect a cure. Application of a pre-mixed liquid adhesive-catalyst system must be completed within its working life if spreading and wetting are to be adequate. Following its application, an adequate time allowance must be made for the adhesive to cure. Room-temperature-curing adhesives often require a number of hours to set, during which period the assembly must be jig-supported. Heat-curing adhesives, which set within minutes, are also available. Sometimes an inert atmosphere must be used for certain metallic adherends. Some adhesives contain volatile components to improve their consistency. In these cases processing may involve an intermediate liquid-removal stage before bonding (i.e. evaporation by heat, or even force-drying). Other forms are supplied as films, which may or may not be supported on carrier cloths. Structural adhesive films generally require high bonding pressures to be sustained during hot-curing schedules. Another recent approach to rapid curing at room temperature is ultra-violet radiation.[2]

The exact conditions required for curing the adhesive joints depend on the properties of the specific adhesive used. The manufacturer invariably recommends the optimum procedure. In most instances curing is accomplished through the application of heat or pressure, or both. Depending on the properties of the adhesive, curing pressures may range from contact pressure, 1–5 psi (6.9–34.5 kPa) to 500 psi (3447 kPa), while curing temperatures may range from room temperatures up to 662°F (350°C), although the maximum temperature usually used is 350°F (177°C). With ceramic-based adhesives dependent upon a sintering action for adhesion, the processing temperature can reach 982°F (1800°C).[2],[5]

BONDING EQUIPMENT

After application of the adhesive the assembly must be mated as quickly as possible to prevent contamination of the adhesive surface. The substrates are held together under pressure and heated, if necessary, until cure is achieved. The equipment required to perform these functions must provide adequate heat and pressure, maintain constant pressure during the entire cure cycle, and distribute pressure uniformly over the bond area. For adhesives curing with simple contact pressure at room temperature, extensive bonding equipment is not necessary.[4]

Pressure Equipment

Pressure devices should be designed to maintain constant pressure in the bond during the entire cure cycle. These devices must compensate for thickness reduction from adhesive flow-out and thermal expansion of assembly parts. Screw-activated devices, such as C-clamps, and bolted fixtures are not acceptable when constant pressure is important. Spring pressure can often be used to supplement clamps and compensate for thickness variation. Dead-weight loading may be applied in many instances. Such loading takes the form of bags of sand or shot, or similar materials. Dead loads may be used only on relatively flat adherends to provide relatively low pressures. This method is impractical for heat-cure conditions.[4]

Pneumatic and hydraulic presses are excellent tools for applying constant pressure. Steam—or electrically heated platen presses with hydraulic rams are often used for adhesive bonding. Some units have multiple platens, thereby permitting the bonding of several assemblies at one time. Large bonded areas, such as on aircraft parts, are usually cured in an *autoclave*. The parts are first joined, then covered with a rubber blanket to provide uniform pressure distribution. The entire assembly is then placed in an autoclave which can be pressurized and heated. This method, which is widely used in the aerospace industry, requires having capital-equipment investment.[4]

In autoclave bonding pressure in the assembly is applied as fluid pressure obtained principally from compressed air, and in older installations, from steam. The actual pressure on the parts is exerted by the pressure differential over the autoclave blanket or membrane, which is connected with the bonding table by means of an airtight seal. The differential can be produced by placing the assembly on a sealed table in a pressure vessel, while the assembly is connected with the atmosphere or a vacuum installation. The autoclave itself can be either cylindrical or of the clam-shell (watch-case) type. The latter is limited to relatively small panels. Heat and pressure can be applied by injection of live steam. Subsequent injections of cold air will allow for relatively rapid cooling. The advantages of live steam are simple operation and short curing cycles. The disadvantages are many. Independent variations of temperature and pressure are impossible to achieve. The high moisture content in the autoclave leads to several practical problems, of which the entry of moisture into the bonded parts before the final cure is an important one. Pressurizing is usually accomplished by application of *compressed air* from an accumulator next to the autoclave.[9]

Autoclave bonding generally involves high pressures which can create assembly-distortion problems not significant in other bonding equipment. For example, variations in glue-line pressure as a result of non-conformity of assembly parts is often a problem. The concentration of bonding pressure at adherend edges causes thinning of the glue-lines at the edge, which may lead to distortion of the structure.[2]

Vacuum bags are effective not only in applying pressure, but also in withdrawing any volatiles produced in curing. This type of assembly is usually accomplished with flat panels, or panels of simple contours, where light pressures suffice to hold mating surfaces in contact. Such an assembly can be laid out on a steam or electrically heated, water-cooled table. A rubber blanket or plastic sheet, such as polyvinyl alcohol, is placed over the assembly and sealed to the table with a bead-type seal. The table is provided with a pipe or hose outlet to a vacuum pump. When the vacuum is applied, atmospheric pressure holds the assembly together. The table is then heated and the adhesive cured. This type of equipment is considerably less cumbersome than a press, but it is limited to assemblies that need no more than atmospheric pressure (14.7 psi or 101.3 kPa).[10] Greater pressures may be achieved by using "pressure bars" between the wall of the bag and the enclosed joint. For example, 60 psi (412 kPa) may be achieved on a 1-inch (2.54 cm) joint by using a T-bar with a 1-inch (2.54 cm) face against the wall of the bag.[5]

Heating Equipment

Many structural adhesives require heat as well as pressure. Most often the strongest bonds are achieved by an elevated-temperature cure. With many adhesives trade-offs between cure time and temperature are permissable. Generally, however, the manufacturer will recommend a certain curing schedule for optimum properties. If a cure of 60 minutes at 300°F (149°C) is recommended, this does not mean that the assembly should be placed in a 300°F (149°C) oven for 60 minutes. Such conditions would result in an under-cure. It is the bond line that should maintain these conditions. Total oven time should be 60 minutes *plus* whatever time is required to bring the adhesive up to 300°F (149°C). Large parts may act as heat sinks and may require substantial time for an adhesive in the bond line to reach the required temperature. Bond-line temperatures are best measured by thermocouples placed very close to the adhesive. In some cases it may be desirable to place a thermocouple directly in the adhesive joint for the first few assemblies being cured.[4]

Direct heating curing—Oven heating—This is the most common source of heat for bonded parts, even though it requires long curing cycles because of the heat-sink action of large assemblies. Ovens may be heated with gas, oil, electricity, or infrared units. Good air circulation within the oven is essential to prevent nonuniform heating.[4] Recent oven models realize temperatures of 842 ± 1.8°F (450 ± 1°C).[2]

Liquid baths—Various liquids are used to provide rapid heat transfer by conduction. Water is commonly used, but for higher curing temperatures mineral or silicone oils are required. The silicone oils are useful non-toxic heating media for temperatures up to 572°F (300°C). Direct contact between the bond and the heating medium should be avoided. This method depends on heat conduction through the adherend to cure the adhesive.[2]

Hot presses or platens—This equipment relies on electrical resistance heaters or steam to provide heat to the platens compressing the bonded assembly. The highest temperatures are obtained with electrical heating elements. These elements can be controlled by relay mechanisms where the curing cycle involves various temperature-time stages. Steam heating is a faster process, and it is often advantageous to circulate cooling water through the piping after curing. This technique is particularly effective in bonding assemblies that must be cooled under pressure.[2,5]

Radiation curing—This technique, involving infrared radiation heaters, produces an increase in the heat-transfer rate exceeding that of oven heaters. Infrared lamps provide a useful way of removing solvents from contact adhesives prior to bonding, and are useful in the rapid heating of localized areas of a substrate. The rate of heat transfer is dependent, to some extent, on the color or the workpiece. The darker the part, the more rapid the heating.[2]

Electric resistance heaters—In this method a conductive strip of metal is embedded in the adhesive to act as an internal resistance heater. The heating of the bond is accomplished by passing electric current through a metallic adherend or conductor within, or adjacent to, the glue line for non-metallic adherends. Recently graphite has been used as an internal electric resistance heater for curing structural adhesives. The graphite is available in various physical forms, such as felts, yarns, woven fabrics and tapes. It can be utilized as a heating element over a wide range of temperatures, up to 680°F (360°C) in air and beyond 5072°F (2800°C) in an inert atmosphere. A negative coefficient of resistance with temperature prevents current surges during heating. Rapid heating and cooling of the cloth results from the low thermal mass and high emissivity of radiation per unit area of graphite fabric. This internal-heat-surface method with graphite resistance elements provides bond strengths of joints which compare favorably with similar joints prepared by oven curing. Advantages of this method over conventional external heating methods are:[2]

- Rapid attainment of curing temperatures, since the adhesive is heated directly; provision is usually made for heat loss to adherends and the environment

- Easy application of heat to localized areas of an assembly

- Fabrication of assemblies with high-temperature-curing adhesives without the risk of distortion; uniform heating of fabric eliminates hot spots

- Closer control of glue-line temperatures with consequent realization of maximum adhesion performance. The fabric acts as a glue-line spacer and insures uniform thickness of the adhesive layer

- Restriction of heating to the glue line, thereby avoiding unnecessary heating of adherends; reduced expenditure on large assemblies following a lower power consumption

- Simplicity of process, obviating the need for ovens or platens; ready on-site repair of damaged assemblies with transportable power equipment

- Realization of improvements in design and on-site modifications to structural units in a room-temperature environment

The graphite-fabric technique may be used to advantage with hot-melt adhesives to achieve easier processing. In conventional usage assemblies are heated up to the melting point of the hot melt (applied as a film or powder between the adherends) and then cooled. By impregnating graphite fabric with the hot melt and passing a current for a short period to liquify the adhesive, the need to heat up and cool down the entire assembly is eliminated and processing times considerably reduced.[2]

Other methods for electrical heating of bonded assemblies utilize wrap-around electrical heating tapes or resistance elements within the jig supporting the jointed structure. It is generally difficult to attain uniform heating with these methods.[2]

High-frequency dielectric (radio frequency) heating—The curing of glue lines by heat conduction from hot plates is inefficient where thick, non-conductive adherends are involved. High-frequency dielectric heating has been developed as a curing means for bonds based on organic polar materials, or materials that behave as polar materials through the water they contain, which are poor conductors (unlike metals, for which inductive heating is preferred) or insulators (e.g. polystyrene). This process is particularly effective with the thermosetting resins used for woodworking applications, and to lesser extent, animal and casein glues. The process is based on the absorption of energy by the adherend material (or dielectric) when it is placed in an alternating electric field. At high frequencies, from 10–15×10^6 Hz, molecular vibration (resonance) occurs, which leads to heat generation within the material, provided the material has an appropriate loss factor at that frequency. High loss factors favor rapid heating. (NOTE: loss factor = dielectric constant (permittivity) $\times$ power factor).[2,11]

Induction heating—This is similar to dielectric heating. Electric power is used to generate heat in a conducting material. The treatment is therefore applicable to metal adherends, or to adhesive materials filled with metal powder. If one adherend is conductive and the other non-conductive, either dielectric or induction heating may be applied. The possibility of heat charring the adhesive, where rapid heat curing is involved, should be recognized and care taken to provide adequate control of heat input. Rapid heating, in general, should be avoided.[2]

Low-voltage heating—This method utilizes the principle of resistance heating in a simple and straightforward manner. The use of low voltage permits an inexpensive heating element to be used and allows relatively safe handling. For efficient operation the heating element should be in direct contact with the work being glued, so that heat transfer is by conduction. The only apparatus required is a step-down transformer with a capability of providing low voltages (4, 6, or 12V with currents of 500–1000 Å) which are applied to a metal platen, usually made from galvanized or tinned mild steel or stainless steel. Low-voltage heating is used for a variety of wood-bonding applications (e.g. scarf-joint manufacture, boat-building hull work). This method is inexpensive in comparison with dielectric heating, but is not as efficient for gluing wood if the distance to the glueline is more than a few millimeters. The temperature of LVH elements may vary between $167°F$ ($75°C$) and $392°F$ ($200°C$).[2,11]

Ultrasonic activation—This method of curing the adhesive is based on the transmission of mechanical vibrations from an ultrasonic transducer to the adhesive at the interface between the mating parts. It is used most efficiently where a bead or film adhesive can be incorporated. Energy concentrations of 700 ft lb/sec/in^2 (1.085 $\times$ 10^6 W/m^2/sec) are sufficient to melt, flow and cure many thermoplastic and thermosetting adhesives. The equipment used is the same as that made for ultrasonic welding of thermoplastics.[12]

A wide variety of adhesives can be used with ultrasonic activation, but efficiency may vary with viscoelastic and curing characteristics. Thermoplastics of value include most hot-melt and heat-reactivated adhesives, particularly those with some elastomeric qualities. Thermoset adhesives may be used if a proper balance of activation and cure rate can be accomplished. Epoxy, nitrile and phenolic adhesives have been used with some success. A fast-curing thermoset (B-stage) gives good results, since the resin can be quickly melted and flowed to the adherends. The curing reaction is initiated by the heat remaining in the adherends. With slow-curing thermosets flow can be accomplished, but continued ultrasonic activation causes a rapid temperature rise. This may initiate chemical or mechanical decomposition of the adhesive before the cross-linking reactions have been completed.[12]

The adhesive can be used in the form of a film, scrim or coating preapplied to one or both adherends, or a ribbon applied to one of the adherends. Liquid adhesives are generally unsatisfactory. Viscoelastic adhesives are particularly suitable. Ultrasonic activation usually increases the bond strength and reduces curing time where it can be employed as an alternative to conventional thermal or drying processes.[12]

Adhesive thickness—It is highly desirable to have a uniformly thin (2-10 mil or 0.05-0.25 mm) adhesive bond line. Starved adhesive joints (where some areas have no adhesive), however, will result in poor bonds. Three basic methods are used to control adhesive (bondline) thickness:[4]

- Use mechanical shims or stops, which can be removed after the curing operation. Sometimes it is possible to design stops into the joints.

- Employ a film adhesive that becomes highly viscous during the cure cycle, preventing excessive adhesive flow-out. With supported films, the adhesive carrier itself can act as the "shim". Generally the cured bondline thickness will be determined by the original thickness of the adhesive film.

- Use trial and error to determine the correct pressure-adhesive viscosity factors that will yield the desired bond thickness.

WELDBONDING

Weldbonding, also called spot-weld adhesive bonding, is a method of fabricating hardware that uses both welding and adhesive bonding techniques. A layer of adhesive, either in paste or film form, is applied to one of the metal members to be joined. The other metal member is placed on top, forming a lap-type joint, and the assembly is clamped or resistance tack-welded to maintain part alignment. The two members are then joined by resistance welding through the adhesive, using a conventional spot or seam welder. The welds are commonly spaced 1 to 2 inches (2.54 to 5.08 cm) apart, center to center, as shown in Figure 7-1. After curing at ambient or elevated temperature, the adhesive forms a gas-tight seal. Hardware fabricated by this method has higher tensile-shear strength, and improved corrosion resistance, compared to equivalent resistance-welded structures. Both cost and weight savings are significant compared to riveted or mechanically fastened structures.[13]

Weldbonding was initially developed in the Soviet Union and used in the fabrication of transport aircraft (AN 22, AN 24 and YAK 40). In recent years several American aircraft manufacturers (Lockheed-Georgia, Northrop, Martin-Marietta, TRW) have studied the process and published reports, some in-house and some sponsored by NASA and the Air Force. Annotated bibliographies have been prepared covering developments in weldbonding.[14,15]

When properly applied, weldbonding will provide a joint structurally comparable to an adhesive-bonded joint, while eliminating the complex and expensive tooling generally required for adhesive bonding. Weldbonding may be accomplished by applying the adhesive to the faying surfaces prior to welding (weld-thru), or a low-viscosity adhesive may be used to infiltrate the faying surfaces by capillary action after the welding has been accomplished. Both techniques have been used in Russia.[16,17]

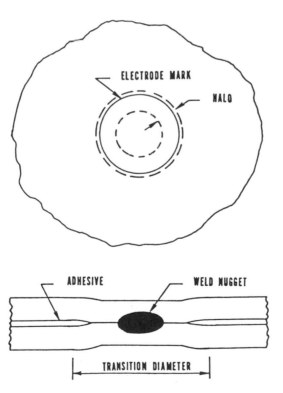

Figure 7-1: Schematic description of a spotweld in a single-lap weldbond joint.[19]

Weldbond Configuration

Figure 7-1 is a schematic (not to scale) detail of a spotweld and the surrounding adhesive bonded region of a single-lap weldbonded joint. In the most-used process, a paste adhesive is applied to the metal sheet material and the metal is then spotwelded through the uncured adhesive. The spotwelding pressure and heat result in displacement of the adhesive and fusion of the metal to form a solid weld nugget. In the figure the solid circle represents the visible mark at the edge of the surface of contact between the spotwelding electrode and the metal sheet. The inner dashed circle outlines the weld nugget. The area between the two dashed circles, or "halo", is effectively unbonded due to the displacement and heating of the adhesive during the spotwelding process. Beyond the halo is a region of transition to full adhesive thickness. The exact shape and dimensions of the features are functions of several variables. These include stiffness and thickness of the metal sheet and uncured adhesive and such welding parameters as pressure, current, resistance, time and electrode shape.[18,19] Figure 7-2 shows a single-lap and comparable double-lap weldbond joint.

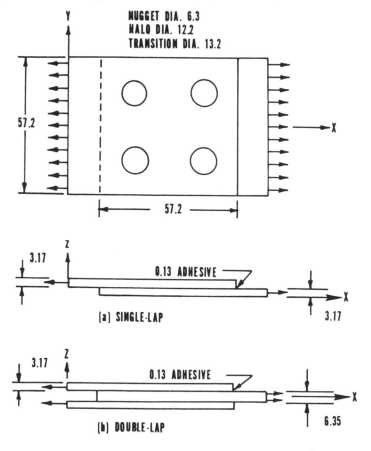

Figure 7-2: Comparable single- and double-lap weldbond joints.[19] Dimensions are in mm; 1 inch = 25.4 mm.

High-quality welds can be made in high-strength aluminum alloys up to 72 hours after layup of the parts with adhesive at the parts interface. Because the weldbond process is highly adaptable to mechanization and automation, a substantial production cost savings is realized when compared to other methods of joining.[20]

Advantages and Limitations

The advantages of a weldbond structure over a similar structure joined by rivets or mechanical fasteners are as follows:[21]

- Increased tensile shear strength (static strength)
- Increased fatigue life
- Gas-tight structure suitable for gas or liquid containers
- Increased rigidity
- Improved corrosion resistance of the lap-type joint
- Inexpensive tooling requirements for fabrication
- Weight savings
- Smooth hermetically sealed inner and outer surfaces improve aerodynamics and hardware attachments
- Complete interface bond improves load transfer between inner and outer members

Weldbond is competitive in static strength with titanium fasteners in sheet thicknesses up to 0.156 inch (0.40 cm). Beyond this point the spot-welding limitations and static-strength capacity make weldbond less desirable.

Surface Preparation

Chapter 4 provides details on surface preparation methods for aluminum and titanium. Chemical cleaning of surfaces for weldbonding is necessary for high-strength joints. For non-critical weldbonds used in automotive applications involving aluminum alloys and steel, no cleaning or special surface preparation is necessary.[21,22]

The low-voltage phosphoric acid/sodium dichromate (P/SD) anodize is a versatile process, since it is capable of producing the required boehmite oxide layer ($\alpha Al_2O_3 \cdot H_2O$), which is tenaciously bonded to the aluminum alloy substrate, and a variety of alloys. However, it is essential to remove the "as received" oxide layer (bayerite) before anodization to assure maximum corrosion-resistance and bond durability. Since the boehmite layer thickness is controllable, Class A spot welds can be produced consistently, even with corrosion-inhibited weldbond paste adhesive in place. With the addition of chromates to adhesives improved corrosion resistance under salt and humid environments has been possible.

Adhesive Choice

Epoxy, modified epoxy, and urethane adhesives are commonly used in weldbonding aluminum. Epoxy and polyimide adhesives are used for titanium. Polyimides are particularly suitable for titanium because of their very high temperature resistance (matching the resistance of titanium). Epoxy and modified-epoxy adhesives are available in one- or two-part liquid, paste, capillary, or unsupported-film form.[21]

The most suitable adhesive for general weldbond applications is the modified-epoxy one-part paste type containing metal filler. Most adhesives used for weldbonding are arbitrarily selected from adhesives that were developed for other purposes, resulting in a compromise when used for weldbonding. Adhesives specifically developed for weldbonding should show improvements over those presently available.[21]

Fillers for epoxy adhesives used in weldbonding include silica (Cab-O-Sil, 7% by wt.) to prevent run-off, 3% strontium chromate to provide corrosion inhibition, and conductive metal powders. Viscosity of adhesives in paste form has an important effect on the weld and bonded joint. The viscosity must be low enough to allow the force of the electrodes to push the adhesive out of the interface contact area, yet sufficiently high or thixotropic so that it will not flow out of the joint during the cure cycle.[23]

Film adhesives have been found by some workers to be impractical because spot welds could not be made consistently, due to the carrier.[24] On the other hand film adhesives have been found to provide substantial benefits because they improve workmanship in the production facility. The first requirement for any adhesive for the weld-through process of weldbonding is that the adhesive have the capability of being moved under pressure of the welding electrodes in order for metal-to-metal contact to occur at the joint interface. The second requirement is that the heat resulting from the spot weld cause only limited detrimental effect on the strength of the bond.[25]

Tooling for Weldbonding

Tooling for weldbonding falls into four general categories (1) part handling, (2) adhesive application, (3) tacking, and (4) welding. Since three of these categories are commonly used in industry, only tooling for adhesive application will be discussed. Tooling for paste-adhesive application must fixture the parts and control location, width and thickness of the adhesive stripe. The tooling should have a platen to locate the part, and a moveable head consisting of gridded shim stock to control width and thickness of the adhesive stripe. The adhesive can be applied with an air-activated caulking gun and spread to a uniform thickness (0.006–0.016 inch) (0.15–0.41 mm) and width by drawing a plastic spreader along the shim stock. The open position of the fixture should provide accessibility for solvent cleaning of the shim stock prior to the next adhesive application. Automatic adhesive application and spreading can be incorporated into the fixture if high production rates are required.[21]

Weldbonding Techniques

Class A resistance spot or seam welds can be made through the adhesive in weldbond structures when recommended spot-weld cleaning or surface-preparation techniques are used. When a modified-epoxy paste adhesive containing a conductive powder filler is used, the welding parameters are nearly the same as those used for welding without adhesive. Welding through an adhesive causes a higher percentage of irregularly shaped nuggets, but the strength of the weld-bond joint is not adversely affected. When welding through adhesive on material prepared by the recommended surface preparation, there is a tendency for a high percentage of the spot welds to exhibit expulsion (production of a black patch of mixed aluminum oxide and charred adhesive for aluminum substrates), resulting in lower-quality and lower-strength welds.[24]

Weldbond process specifications have been prepared by several contractors for use by government agencies.[25-27] These process specifications give detailed steps to be taken to provide optimum weldbonds. The actual joining of parts by weldbonding is relatively straightforward. Most of the processes involved in resistance spot welding are applicable to weldbonding. The parts are chemically cleaned as for spot welding, wrapped and stored for up to 36 hours, if required, and removed for welding. Paste adhesive of the consistency of room-temperature honey is applied to the parts by laying a small bead of adhesive on the part surface and spreading it, using a nylon spatula. The parts are then brought together and temporarily clamped. The parts are then placed between the electrodes of a conventional three-phase, variable pressure-type spot welder and welded together. The weld set-up used to join the parts is only slightly modified from a conventional set-up. After welding, the structure is placed in a low-temperature oven and the adhesive cured for approximately one hour. Time and temperature are dependent on the type of adhesive and method involved.[20,28]

Quality Control

This subject will be discussed briefly in Chapter 10 on Quality Control.

REFERENCES

1. Society of Manufacturing Engineers, Chapter 5, Methods of Bonding, *Adhesives in Modern Manufacturing*, (E.J. Bruno, ed.) (1970).
2. Shields, J., *Adhesives Handbook*, 2nd Edition, Newnes-Butterworth, London (1976). (The 3rd edition was published by Butterworths, London, in 1984).
3. Bloomingdale Dept., American Cyanamid Co., Technical Literature.
4. Petrie, E.M., Chapter 10, Plastics and Elastomers as Adhesives, *Handbook of Plastics and Elastomers* (C.A. Harper ed.), McGraw-Hill, N.Y. (1975).
5. U.S. Dept. of Defense, Military Handbook 691A, *Adhesives*.
6. Dreger, D.R., Hot Melt Adhesives Put it All Together, *Machine Design*, 47 (1): 88-94 (January 9, 1975).
7. Nordson Corporation, Amherst, Ohio, Technical Bulletins.
8. Bemmels, C.W., Chapter 47, Pressure-Sensitive Tapes and Labels, *Handbook of Adhesives*, 2nd Ed., (I. Skeist, ed.), Van Nostrand Reinhold, N.Y. (1977).
9. Schliekelmann, R.J., Chapter 15, Adhesive-Bonded Metal Structures, *Adhesion and Adhesives*, Vol. 2 - Applications, 2nd Ed. (R. Houwink and G. Salomon, eds.) Elsevier Publishing Co. (1967).
10. De Lollis, N.J., *Adhesives, Adherends, Adhesion*, Robert E. Krieger Publishing Co., Huntington, N.Y. (1980).
11. Rayner, C.A.A., Chapter 13, Adhesive Bonding Process, *Adhesion and Adhesives*, Vol. 2 - Applications, 2nd Edition, (R. Houwink and G. Salomon, eds.) Elsevier Publishing Co. (1967).

12. Hauser, R.L., Ultrasonic Bonding Techniques, *Applied Polymer Symposia No. 19, Processing for Adhesives Bonded Structures*, Wiley-Interscience, pp. 257-265 (1972).

13. Hall, R.C., Environmental Resistance of a Weldbond Joint, *SAMPE Journal, 10* (14) : 10-14 (Aug/Sept 1974).

14. Kizer, J.A. and Kopkin, T.J., Weldbond Joining Process Bibliography, Lockheed-Georgia Company (September 1970).

15. Winans, R.W. et al., Weldbonding in the United States: an Annotated Bibliography and History, *PLASTEC* Note N26. AD A008048 (December 1974).

16. Kizer, J.A., Development of Weld-Thru Weldbond Process Surface Preparation, Proceedings, 5th National *SAMPE Technical Conference*, Vol. 5, Kiamesha Lake, N.Y., pp. 124-130 (Oct. 9-11, 1973).

17. Russell, W.J. and Tanner, W.C., *Component Parts Assembly with Joints: Adhesive-Mechanical*, Part 1. Surface Preparation of Aluminum for Weldbonding, Picatinny Arsenal Technical Report (PATR) 4610, AD 919 514L (February 1974).

18. Mitchell, R.A. et al. *Finite Element Analysis of Spotwelded Bonded and Weldbonded Lap Joints*, National Bureau of Standards, NBSIR 75-957 (December 1975).

19. Mitchell, R.A. et al., *Component Parts Assembly with Joints: Adhesive-Mechanical*, Part 4. Analysis and Test of Bonded and Weldbonded Lap Joints, Picatinny Arsenal Technical Report (PATR) 4965. AD A029426 (Prepared by NBS). (March 1976).

20. Fields, D., Summary of the Weldbonding Process, *Adhesives Age, 16* (9) : 41-44 (September 1973).

21. Beemer, R.D., Introduction to Weld Bonding, *SAMPE Quarterly, 5* (1) : 37-41 (October 1973).

22. Wu, K.C. and Bowen, B.B., Northrop Corp., Aircraft Div., *Advanced Aluminum Weldbond Manufacturing* Methods, AFML-TR-76-131, Sept. 1976. AF Contract F33615-71-C-5083. AD CAB 016865. Final Report, (June 1975-March 1976). Also in Preprint Book, 22nd National SAMPE Symposium, Vol. 22, San Diego, CA pp. 536-554 (Apr. 26-28, 1972).

23. Bower, B.B., *Preparation of Aluminum Alloy Surfaces for Spot Weld Bonding*, Proceedings, 7th National SAMPE Technical Conference, Albuquerque, New Mexico. Vol. 7, pp. 374-385.

24. Salkind, M.J., Spot-Weld Bonding in the Blackhawk Helicopter, Proceedings, Symposium on *Welding, Bonding, and Fastening*, sponsored by NASA, George Washington University and ASM, May 1972.

25. Szabo, R.L., General Dynamics Convair Div., *Feedback-Controlled Spotwelding.* AFML-TR-76-35 (April 1976). Final Report, June 15, 1975-April 15, 1976. AF Contract F33615-75-C-5229. AD B014690L.

26. Bowen, B.B. et al., Northrup Corp., Aircraft Div., *Improved Surface Treatments for Weldbonding Aluminum*, AFML-TR-76-159 (October 1976). Final Report April 1975-June 1976. AF Contract F33615-74-C-5027. AD CAB017811L.

27. Grosko, J.J. and Kizer, J.A. Lockheed-Georgia Co., *Weldbond Flight Component Design/Manufacturing Program*, AFML-7R-74-179, AFFDL-7R-74-106, December 1974. Final Report AF Contract F33615-71-C-1716. AD B002822L. (July 15, 1971-July 15, 1974).

28. Miller, F.R., Advanced Fusion Welding Processes, Solid State Joining and a Successful Marriage, Proceedings, Symposium on *Welding, Bonding, and Fastening*, sponsored by NASA, George Washington University and ASM, May 1972.

8

Solvent Cementing of Plastics

INTRODUCTION

Solvent cementing is a process in which thermoplastics, usually amorphous, are softened by the application of a suitable solvent, or mixture of solvents, and then pressed together to effect a bond. The resin itself, after evaporation of the solvent, acts as the adhesive. Many thermoplastic resins are easier to join effectively by solvent cementing than by conventional adhesive bonding. Often mixtures of solvents give better results than individual solvents. Frequently small amounts of the plastic to be cemented are dissolved in the solvents to form "bodied" cements. These additions of polymer aid in gap filling and accelerate setting. They also reduce shrinkage and internal stresses. If the evaporative rates of the solvents used are too high due to excessive volatility of the solvent, crazing or blushing often results.

It is possible to solvent cement different plastic types to each other, but in this case the solvent must be compatible with both plastics. Usually a mixture of fast-evaporating solvent is combined with a high-boiling solvent, often with resin addition (up to 25% by weight). Upon softening the plastic adherends, they are allowed to become tacky. At this point they are pressed together and held under pressure until dry. As thin a coat of solvent as possible should be used. Recommended solvents and solvent mixtures will be given below for each plastic type. Bonding should be carried out in a warm, dry situation to avoid condensation due to solvent evaporative cooling of the part. The solvents may be brushed, sprayed, applied by dipping, or with a syringe. Caution should be taken in applying the solvent, because excess may run into unwanted areas and result in damage to the appearance of the part. Heating the part is not always recommended because it can cause stress cracking as the solvent leaves the surface. Heating can also result in bubbling in the solvent layer.

Solvent cementing is the simplest and most economical method of joining thermoplastics. Solvent-cemented joints are less sensitive to thermal cycling than joints bonded with conventional adhesives and are as resistant to degrading environments as the parent plastic. This is true because the final joint consists solely of the parent plastic, with no other adhesive or solvent material present. Bond strengths 85-100% of the strength of the parent plastic can be obtained.[1-3]

Solvents used to cement plastics should be chosen with approximately the same solubility parameter (δ) as the plastic to be bonded. The solubility parameter is the square root of the cohesive energy density (CED) of the liquid solvent or polymer. It is defined as follows:

$$\delta = (\Delta E/V)^{1/2} \text{ (measured in hildebrands)}$$

where ΔE = energy of vaporization

V = molar volume

$\Delta E/V$ = cohesive energy density (or internal pressure)

A non-polar molecule, such as methane, evaporates readily and is a gas at ordinary temperatures. It has a low CED, hence a low δ ($\sim$6). By contrast, a highly polar, associated (hydrogen-bonded) molecule of the same size, such as water, requires high heat input to evaporate it, and consequently has a very high δ of 23.4. Literature sources provide data on δ's of a number of plastics and resins.[3,4] Much of this data is shown in Tables 8.1 and 8.2. The solubility parameters help explain why polystyrene (δ = 9.1) is soluble in butane (δ = 9.3), but not in acetone (δ = 10.0), while cellulose acetate (δ = 10.9) dissolves in ethyl acetate (δ = 9.1), but not in butyl acetate (δ = 8..5). The concept of δ also explains why a plastic will sometimes dissolve in a mixture of two liquids, neither one of which is itself a solvent. The classic example is the solubility of nitrocellulose (δ = 11.0) in the non-solvents ethyl alcohol (δ = 12.7) and diethyl ether (δ = 7.4). The mixture of nonsolvents, one with too high a δ and the other with too low a δ has its δ within the right range. Similarly, the solvent acetone (δ = 10.0), mixed with a small percentage of the nonsolvent ethanol (δ = 12.7), is a better solvent for cellulose acetate (δ = 10.9) than acetone alone. Generally, the more polar plastics require more polar solvents. Figure 8.1 shows good correlation between solubility parameter and critical surface tension for polymers at the low end of both scales. With increasing values, however, anomalies become apparent. The discrepancy can be attributed, at least in part, to differences in crystallinity, the presence of compounding ingredients, and differences in chemical composition of the bulk polymer from the surface. Surface treatment of polyethylene may have a strong effect in that polymer. Where solubility parameters and contact-angle measurements disagree, the latter provides the better direction for choosing adhesives, provided they have been carried out on the materials as they are actually prepared for bonding.[3,4]

A very recent 1983 handbook, not yet seen by the author, has been published with much information on the solubility parameters of a wide range of materials.[5]

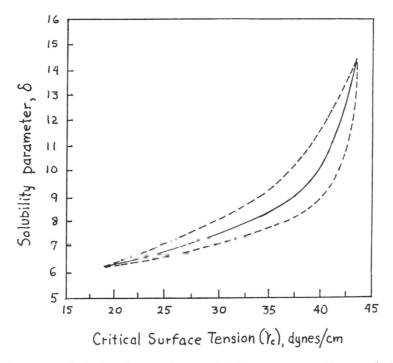

Figure 8-1: Critical surface tension vs solubility parameter of thermoplastics. Modified after Reference 4.

RECOMMENDATIONS FOR SPECIFIC ADHERENDS

Acetal Copolymer (Celcon®)

A room-temperature solvent, *hexafluroacetone sesquihydrate*, when used at full strength, is a very effective bonding agent for adhering Celcon to Celcon, Celcon to nylon, and Celcon to ABS. This solvent is available from the Allied Corporation. Celcon to Celcon and Celcon to nylon bond strengths in shear greater than 850 psi (5.86 MPa) have been obtained on "as molded" surfaces. Bond strengths of 350 psi (2.41 MPa) were obtained on Celcon to ABS. Hexafluoroacetone sesquihydrate is a severe eye and skin irritant and care should be taken in its use.[6]

Acetal Homopolymer (Delrin®)

Because of the high solvent resistance of Delrin, molded surfaces cannot be joined by the use of cements unless they have been specially roughened.[7]

Acrylonitrile-Butadiene-Styrene (ABS)

See Acetal Copolymer above for discussion of use *hexafluoroacetone sesquihydrate* in bonding ABS to Celcon. Solvents used in cementing ABS should be quick-drying to prevent moisture absorption, yet slow enough to allow assemblage of parts. The recommended cure time is 12–24 hours at room temperature, 73°F (23°C). The required time can be reduced by setting at 130–150°F (55–70°C). The solvents recommended by Borg-Warner Corp., a major ABS manufacturer, are methyl ethyl ketone (MEK), methyl isobutyl ketone (MIBK), tetrahydrofuran (THF), and methylene chloride. These solvents can be "bodied" with up to 25% ABS resin.[8] ASTM D 2235, a specification on solvent cement for ABS pipe and fittings, calls for a solution of ABS (min. 15%) in MEK.[9] ASTM D 3138 covers solvent cements for ABS-PVC transition joints for plastic pipe. A minimum of 10% PVC must be used in tetrahydrofuran in combination with cyclohexanone, or MEK, or both.[10]

Cellulosics

These plastics include cellulose acetate, cellulose acetate butyrate (CAB), cellulose nitrate, cellulose propionate, and ethyl cellulose. These materials are most commonly bonded by solvent cementing. Adhesive bonding is used to a minor extent.

Cellulose acetate—The following solvents may be used alone in cementing cellulose acetate.[11]

acetone	dioxane	ethyl lactate
methyl acetate	nitromethane	Cellosolve acetate
ethyl acetate	methyl Cellosolve*	diacetone alcohol
methyl ethyl ketone	methyl Cellosolve acetate	

   *Cellosolve is an ethylene glycol monoethyl ether (2-ethoxy ethanol) made by Union Carbide Corporation.

Solvent mixtures recommended include:

   acetone/ethyl lactate (70/30)
   ethyl acetate/acetone/ethyl lactate (30/40/30)

Dope-type formulations may be made by including a 10–20% solution of cellulose acetate in any of the appropriate solvent mixtures shown above. A typical example is:

   cellulose acetate/acetone/methyl Cellosolve/methyl
   Cellosolve acetate (18/55/20/7)

Cellulose acetate butyrate (CAB)—CAB may be cemented with the solvents listed for cellulose acetate above and, in addition, with the following solvents.[11]

methylene chloride	isopropyl alcohol	cyclohexanone
chloroform	nitromethane	butyl lactate
ethylene dichloride	butyl acetate	

Solvent mixtures suitable for CAB and CAB to CP are:

acetone/Ektasolve* EM acetate (70/30)
acetone/methyl Cellosolve acetate (70/30)
butyl acetate/butyl lactate (80/20)
acetone/butyl acetate/Ektasolve EM acetate (30/50/20)
acetone/butyl acetate/methyl Cellosolve acetate (30/50/20)
acetone/ethyl lactate (90/10)
acetone/methoxyethyl acetate (80/20)
acetone/Ektasolve EM acetate (80/20)
butyl acetate/acetone/methyl acetate (50/30/20)

*Ektasolve is a solvent series of glycol ethers and glycol-ether esters available from Eastman Chemical Products, Inc. Kingsport, TN.

Formulas for dope or "bodied" cements are as follows:

cellulose acetate butyrate/acetone/ethyl acetate (20/40/40)
cellulose propionate/acetone/methyl Cellosolve/methyl-
Cellosolve acetate) (18/55/207)

Cellulose acetate butyrate should *not* be solvent cemented to cellulose acetate. A nitrocellulose-base adhesive should be used for such joints. ASTM D 2560, a specification for CAB pipe, tubing and fittings, calls for a solution of CAB in one of the three thinners as follows:[12]

Thinner A—acetone/butyl acetate/methyl Cellosolve acetate (30/50/20)

Thinner B—butyl acetate

Thinner C—acetone/toluene/methyl Cellosolve acetate (equal parts)

Cellulose nitrate—This material may be readily joined to itself with acetone. To obtain high optical clarity, use medium-boiling ketones and esters, or mesityl oxide (isopropylidene acetone). Ethyl acetate, methyl acetate, butyl acetate, ethyl lactate, diacetone alcohol, and methyl ethyl ketone are also used. A dope-type cement made with 10% by weight of cellulose nitrate and 90% by weight of diacetone alcohol has also been suggested.[11]

Cellulose propionate (CP)—Cellulose propionate may be cemented with the solvents listed for cellulose acetate, and, in addition, with the following solvents.[11]

methylene chloride	butyl acetate
chloroform	mesityl oxide (isopropylidene acetate)
ethylene dichloride	Cellosolve
isopropyl acetate	cyclohexanone
nitromethane	butyl lactate

Suitable mixtures of solvents for cellulose propionate are:

acetone/Ektasolve EM acetate (70/30)
acetone/methyl Cellosolve acetate (70/30)
butyl acetate/butyl lactate (80/20)
acetone/butyl acetate/Ektasolve EM acetate (30/50/20)
acetone/methyl acetate (70/30)

A "bodied" or dope cement is as follows:

cellulose propionate/acetone/methyl Cellosolve/methyl-Cellosolve acetate (18/55/20/7).

Ethyl cellulose—Solvents recommended are ethylene dichloride, butyl alcohol, and ethyl acetate. Mixtures suggested are:[11]

toluol/ethanol (80/20) or (90/10)
benzol *(highly toxic)*/methanol (2:1) (67/33)
xylol/butanol (80/20)
ethyl acetate/ethanol (60/40) or (80/20)
butyl acetate/toluol/ethanol (equal parts)
toluol/methanol (80/20)

For a bodied cement a suggested formulation is to use solvent solutions of the polymer ethyl cellulose in ethyl acetate/ethyl cellulose (80/20).

Nylons (polyamides)

At room temperature, conventional solvents will not provide effective bonds on nylon. Generally conventional adhesives are used. Three non-conventional solvent cements are sometimes used. These are the following:

aqueous phenol cement

resorcinol-ethanol solvent cement

nylon-bodied calcium chloride-ethanol

Bonds produced with these cements are nonembrittling, tough, and quick-setting. Detailed directions for their preparation and use are given in a duPont design handbook.[14]

- Aqueous phenol—This cement, containing 10–15% water, is useful in bonding nylon 6,6 to itself. It can be purchased in this form from chemical supply houses.

- Resorcinol-ethanol—Equal parts of resorcinol and ethanol are stirred together or shaken together at room temperature for 15–20 minutes to dissolve the resorcinol. The concentration is not critical.

- Nylon-bodied calcium chloride-ethanol.

10 parts Zytel 101 NC-10
22.5 parts calcium chloride
67.5 parts ethanol

For nylon 6, 15% by weight of nylon 6 is mixed with 85% formic acid (20 grams nylon 6/100 ml formic acid). This mixture is brushed onto both surfaces to be joined. A one-minute period should be used to permit some of the formic acid to evaporate, leaving a tacky film. The surfaces should then be joined and allowed to set for 5 minutes at 100 psi (0.69 MPa) at RT or 200°F (93°C). The higher temperature will double the peel strength. Shear strengths of 1200-2500 psi (8.28-17.2 MPa) are obtained with this cement. Unfortunately, the cement is very *toxic* and *highly corrosive* because of the formic acid. Rubber gloves and proper ventilation are required when working with this material.[11]

Polycarbonate

Solvent cementing is the most common method of bonding polycarbonate. It can be carried out with specific solvents, mixtures of solvents, and mixtures of polycarbonate and solvents. Solvents used include:

methylene chloride   tetrachloroethane
ethylene dichloride   1,1,2-trichloroethane
methyl methacrylate monomer (used with methylene chloride)

Methylene chloride, when used by itself, has an extremely fast evaporation rate. This solvent is recommended for most temperate climate zones and for small areas. A solution of 1-5% of polycarbonate in methylene chloride can be used in extreme cases where perfectly mated bonding areas are impossible to obtain. A mixture of methylene chloride with a maximum of 40% ethylene dichloride, may be used where it is difficult to join parts quickly enough to prevent complete evaporation of methylene chloride. The evaporative rate of methylene chloride is 6.7 times faster than that of ethylene dichloride. Bonds made with the mixture have strengths of 9000-10,000 psi (62.1-69.0 MPa). Tensile shear strengths of 4500-6500 psi (31.0-44.8 MPa) have been obtained by General Electric after 48 hours of setting at room temperature for both methylene chloride and ethylene chloride solvent bonds. This is superior to conventional adhesive bond tensile shear strengths of 350-2900 psi (2.41-20.0 MPa).[11]

Polystyrene

Polystyrene may be bonded to itself by solvent cementing, conventional adhesive bonding, thermal, spin, and ultrasonic welding, or electromagnetic bonding. However, solvent cementing is the most effective approach. In many applications, solvent cementing can be used to bond polystyrene to a variety of dissimilar materials. A wide variety of solvent types are available, and the selection of the specific solvent to be used is determined by the time required to set the joint, which, in turn, is governed by the evaporation rate of the solvent. Fast evaporation rates result in a quick-setting joint, usually with crazing. Slower-drying solvents are often mixed with fast-drying cements for optimum results. A 50:50 mixture of ethyl acetate and toluol bodied with polystyrene is an excellent general-purpose adhesive. Perchloroethylene can be added to reduce flammability. Bond strengths up to 100% of the strength of the parent material are common.[1]

Table 8.1 is a list of some of the solvents recommended for polystyrene, along with notations on crazing and joint strength. As noted, the fast-drying solvents tend to cause crazing in the relatively low-elongation polystyrene. The less soluble impact grades contain polybutadiene. Solvents attack this additive and cause subsequent stress cracking. Impact-grade polystyrene should be bonded with medium-to-slow-drying solvents. Quantities of polystyrene from 5–15% by weight are often added to provide gap filling.[11] Polystyrene resin ground to a powder and dissolved in an appropriate solvent provides excellent gap filling when poor-fitting parts are to be joined. One formula suggested is 90% toluene and 10% PS. A polystyrene cement recommended for transparent joints consists of the following mixture:[14]

solvent boiling below 212°F (100°C)
solvent boiling at 212°F–392°F (100°C–200°C)
high-boiling solvent (boiling above 392°F or 200°C)
dissolved PS (raised to 15% for airtight or watertight seals)

In general, where optical clarity or maximum mechanical properties are not mandatory, solvents in the 170°–250°F (77°–121°C) range provide satisfactory drying time, good sealing, and high bond strength.[15]

Table 8-1: Solvents Recommended for Cementing Polystyrene[16,17]

Solvent	. . . Boiling Point . . .		Crazing	Tensile Strength . . of Joint. . .	
	(°F)	(°C)		(psi)	(MPa)
Fast-drying*					
Methylene chloride	104	40	yes	1800	12.4
Ethyl acetate	171	77.2	yes	1500	10.3
Methyl ethyl ketone	175	79.4	yes	1600	11.0
Ethylene dichloride	182	83.3	slight	1800	12.4
Trichloroethylene	187	86.1	slight	1800	12.4
Medium-drying					
Toluene (toluol)	232	111	slight	1700	11.7
Perchloroethylene (tetrachloro-ethylene)	249	120.6	very slight	1700	11.7
Xylene (xylol)	271–289	133–143	very slight	1450	10.7
Diethyl benzene	365	185	very slight	1400	9.7
Slow-drying					
Monoamyl benzene	396	202.2	very slight	1300	9.0
Ethyl naphthalene	495	257.2	very slight	1300	9.0

*20 seconds or less.

Styrene-Acrylonitrile (SAN)

The techniques used for solvent cementing polystyrene are applicable to SAN, but the list of solvents is more restricted. Solvent cements recommended are:

acetone   methylene chloride
methyl ethyl ketone   ethylene dichloride
tetrahydrofuran

Solutions of approximately 5% SAN in methyl ethyl ketone may be used effectively as bodied cements.[17]

Polysulfone

Solvent cementing of polysulfone can be carried out with chlorinated hydrocarbons. A solution of 5% polysulfone resin in methylene chloride can be used to bond polysulfone to itself. High pressures (500 psi or 3.45 MPa) for 5 minutes are required. A minimum amount of solvent should be applied to the mating surfaces. The strength of a properly prepared joint will exceed the strength of the polysulfone parts. Polysulfone can be solvent-cemented to other plastics by using a solvent compatible with both plastics.[11,17]

Polybutylene Terephthalate (PBT) (Valox®)

Solvents recommended for this solvent-resistant plastic are hexafluoro-isopropanol and hexafluoroacctone sesquihydrate, used separately or in combination. The solvent is brushed on the mating surface and dried under pressure. These solvents are toxic and should be applied only in areas of positive ventilation.[18] A recent design guide by the manufacturer (General Electric) omits any mention of solvent cementing.[19]

Polymethylmethacrylate (PMMA)

Acrylics such as PMMA should be annealed before solvent cementing to minimize the formation of internal stresses that can cause crazing. Acrylic sheet can be annealed by heating in a forced-air oven at about $10°F$ ($5°C$) below the temperature which will cause the part to distort (heat distortion temperature). Thin sections of acrylics are ordinarily heated for 2 hours at $140°F$ ($60°C$) for easy-flow formulations, while hard flows will require temperatures of $170°F$ ($77°C$). Thicker sections will require much longer periods. Solvent cements recommended for acrylics include ethylene dichloride, methylene chloride, and methylene chloride/diacetone alcohol (90/10) for medium joint strength, and a blend of methylene chloride/methylmethacrylate monomer (60/40) with 0.2 parts benzoyl perioxide catalyst and sufficient acrylic resin for body for high joint strength. This type of monomer-polymer cement sets by conversion of the liquid monomer into solid polymer. These cements have the advantage of fast initial set, and the cemented joints are usually sufficiently hard and strong for machining within four hours after assembly. The pot life of monomer-polymer mixtures is very short (1 hour), however, because of the non-reversible polymerization reaction. Joint strengths of these mixtures is excellent and weathering resistance is very good.[20]

Phenylene-Oxide Based Resins (Noryl®)

This material may be solvent cemented to itself, or to certain dissimilar plastics, using a number of commercially available solvents, solvent mixtures, and solvent solutions containing 1–7% of the resin. The addition of 5–20% of the resin will reduce the evaporation rate and fill minor imperfections on the surface of the bonded joints. Recommended solvents are shown in Table 8.2. The solvents and solvent combinations shown for Noryl to Noryl are especially designed to control the evaporation rate.[21]

Table 8-2: Solvent Combinations for Cementing Phenylene-Oxide-
Based Resins (Noryl)[21]

Noryl to Noryl
    For surface areas $<$1 ft^2 and/or open time $<$60 sec
        Trichloroethylene/Methylene chloride (1/1)*
        Trichloroethylene/1,2-dichloroethylene (1/1) (15 sec)
        Trichloroethylene (30 sec)
        Trichloroethylene/monochlorobenzene (4/1)** (45 sec)
    For surface areas $>$1 ft^2 and/or open time $>$60 sec
        Trichloroethylene/monochlorobenzene (1/1)
        Trichloroethylene/toluene (1/1)
        Trichloroethylene/monochlorobenzene (4/1) + 5–25%
            Noryl weight/vol. If more open time is needed, in-
            crease monochlorobenzene by about 10 parts at
            a time up to a maximum of 60 parts
Noryl to ABS/PVC alloy
    Trichloroethylene/monochlorobenzene/tetrahydrofuran (1/1/2)
Noryl to ABS
    Trichloroethylene/methyl ethyl ketone (4/1)
    Trichloroethylene/xylene (1/1)
Noryl to PVC or CPVC
    Xylene/methyl ethyl ketone (1/1)
    Tetrahydrofuran
    Tetrahydrofuran/trichloroethylene (1/1)

 *Signifies equal parts on a volume basis.
**Signifies 4 parts to 1 part, respectively, on a volume basis.

To attain maximum bond strength with solvent-cemented Noryl joints the manufacturer of the resin recommends the following steps:[21]

(1)   Remove all surface contaminants, such as grease, oil and dust with an isopropyl alcohol wipe. Avoid use of mold release agents if possible, either directly or in the vicinity of the molding or extrusion operation.

(2)   Abrade the bond surface lightly with fine sandpaper or treat with chromic acid etchant (E-20 etchant, Marbon Co.). When etching, best results are obtained by immersing the areas to be bonded in an 80°C (176°F) bath for 50–60 seconds.

(3)   Wipe the bond surfaces again with a cloth dampened in isopropyl alcohol.

(4)   Apply the solvent to be used for cementing to both surfaces and quickly join the two parts. Rapid connection of the bonding surfaces will prevent excessive solvent evaporation.

(5)   Clamp the parts together as soon as they are joined. The amount of pressure required will generally depend on the part geometry. Moderate pressure will usually suffice. Clamping pressure should be sufficient to insure good interfacial contact, but not so high that the parts are deformed, or that the solvent is forced from the joint.

(6)   Maintain uniform clamping pressure for 30–50 secs., or as long as the particular part requires. Bonded parts may be handled safely after the original hold time, although maximum bond strength is usually reached at a later time (see Table 8.2).

Polyvinyl Chloride (PVC)

The homopolymer of PVC is not readily soluble, and is therefore difficult to bond by solvent-cementing techniques, although a number of solvents and solvent mixtures have been used with varying degrees of success.[14,15] A large number of solvents have been suggested for solvent cementing PVC. These are:[11]

- Ketones

    acetone
    methyl ethyl ketone
    methyl isobutyl ketone
    isophorone
    cyclohexanone

- Alcohols

    methanol
    ethanol
    isopropanol

- Other Solvents

    propylene oxide   trichloroethylene
    toluol   petroleum ether (low-boiling fraction)
    xylol   methylene chloride
    tetrahydrofuran   ethyl acetate
    dimethyl formamide   dichlorobenzene

When smooth, rigid PVC surfaces are to be joined, the preferred method of using solvent cements is to apply the cement to the two edges of the pieces while they are clamped closely together, thus permitting the solvent to flow between them by capillary action.[14] Ketones are often used and propylene oxide (boiling point 35°C) is usually included as an ingredient, since it contributes to very rapid attack on the plastic. The propylene oxide should be blended with high-boiling ketones, such as methyl ethyl ketone and methyl isobutyl ketone. A moderate percentage of an aromatic hydrocarbon is sometimes used to hasten softening of the PVC. Methyl ethyl ketone and methyl isobutyl ketone are better solvents for the low- and medium-molecular-weight copolymers, and the homopolymers usually require the more powerful cyclohexanone, or 5% dioctylphthalate (DOP) may be added to improve the flexibility at the joint and to reduce stresses. Acetic acid is sometimes added to increase the "bite" of the cement. A mixture of solvents and non-solvents is frequently used. One such mixture is as follows:[11]

dioxane	20 pbw
methanol	12 pbw
MEK	60 pbw
DOP	3 pbw
glacial acetic acid	2 pbw
isophorone	3 pbw

Dissolved chips or shavings of PVC will increase the viscosity of the solution and make the solvent more effective in joining mating surfaces that are not perfectly smooth. Another formulation that will work with either flexible or rigid PVC is as follows:[11]

PVC resin, med. mol. wt (22.4% by wt.)	100 pbw
tetrahydrofuran	100 pbw
methyl ethyl ketone	200 pbw
methyl isobutyl ketone	25 pbw
dioctyl phthalate	20 pbw
organic tin stabilizer	1.5 pbw

Care must be used in handling this formulation because of the slightly toxic nature of the tetrahydrofuran. Good ventilation is required.[11]

ASTM D 2564 covers a solvent cement used for PVC pipe and fittings. No particular solvent is recommended, but a minimum of 10% PVC resin must be used for bodying. Solvent systems consisting of blends of tetrahydrofuan and cyclohexanone are suggested.[22]

Chlorinated Polyvinyl Chloride (CPVC)

An ASTM specification, ASTM F 493, covers solvent cements for CPVC pipe and fittings. No particular solvent system is specified, but solvent systems consisting of blends of tetrahydrofuran and cyclohexanone are suggested. A minimum of 10% CPVC resin is required for bodying.[23]

Polyetherimide (Ultem®)

Methylene chloride, with or without a 1-5% solution of Ultem resin, is recommended by General Electric, the manufacturer. Moderate pressures of 100-600 psi (6.89-41.3 MPa) for 5 minutes are required.[24]

REFERENCES

1. Petrie, E.M., Chapter 10, Plastics and Elastomers as Adhesives, *Handbook of Plastics and Elastomers*, (C.A. Harper, ed.), McGraw-Hill, N.Y. (1975).
2. Gentle, D.F., Bonding Systems for Plastics, pp. 142-170 in *Aspects of Adhesion* — 5, Proceedings of Conferences held at the City University, 5-6 April 1967 and 9-10 April 1968. (D.J. Alner, ed.), CRC Press (1969).
3. Skeist, I., Choosing Adhesives for Plastics, *Modern Plastics 33*(9): 121-130, 236 (May 1956).
4. Miron, J. and Skeist, I., Chapter 41, Bonding Plastics, *Handbook of Adhesives*, 2nd Edition. (I. Skeist, ed.), Van Nostrand Reinhold, N.Y. (1977).
5. Barton, A.F., *CRC Handbook of Solubility Parameters and Other Cohesion Parameters*, CRC Press, Boca Raton, FL, (1983).
6. Celanese Plastics Co., Chatham, NJ, *The Celcon® Acetal Copolymer Design Manual*, undated.
7. E.I. duPont de Nemours & Co., Plastics Dept., Wilmington, DE, *Delrin® Acetal Resins Design Handbook*, A-67041 (1967).

8. Borg-Warner Chemicals, Parkersburg, W. Va., *Fastening, Bonding and Thermal Welding Assembly Techniques for Cycolac® Brand ABS*, Technical Publication P-404, undated.

9. American Society for Testing and Materials (ASTM), ASTM D 2235-81, Standard Specification for Solvent Cement for Acrylonitrile-Butadiene-Styrene (ABS) Plastic Pipe and Fittings, *Annual Book of ASTM Standards*, Vol. 08.04.

10. American Society for Testing and Materials (ASTM), ASTM D 3138, Standard Specification for Solvent Cement for Transition Joints Between Acrylonitrile-Butadiene-Styrene (ABS) and Poly(Vinyl Chloride) (PVC) Non-Pressure Piping Components, *Annual Book of ASTM Standards*, Vol. 08.04.

11. Landrock, A.H., Effects of Varying Processing Parameters in the Fabrication of Adhesive-Bonded Joints, Part XVIII, Adhesive Bonding and Related Joining Methods for Structural Plastics – Literature Review, Picatinny Arsenal Technical Report 4424 (November 1972).

12. American Society for Testing and Materials (ASTM), ASTM D 2560-80, Standard Specification for Solvent Cements for Cellulose Acetate Butyrate (CAB) Plastic Pipe, Tubing, and Fittings, *Annual Book of ASTM Standards*, Vol. 08.04.

13. E.I. duPont de Nemours & Co., Polymer Products Dept., Wilmington, DE. *DuPont ZYTEL® Resin Design Handbook*, E-21920 (undated).

14. Been, J.L., Bonding, article under Adhesion and Bonding, *Encyclopedia of Polymer Science and Technology*, Vol. 1, pp 503–308, Wiley-Interscience, N.Y. (1964).

15. Riley, M.W., Joining and Fastening Plastics, *Materials in Design Engineering*, 47(1): 129–144. Manual No. 145. (January 1958).

16. Chapter 27, "Joining and Assembling Plastics" *Plastics Engineering Handbook of the Society of the Plastics Industry*, Inc., 4th Edition, (J. Frados, ed.), Van Nostrand Reinhold, N.Y., (1976).

17. Monsanto Plastics and Resins Co., St. Louis, Mo., *Fabrication Techniques for Lustrex® and Lustran® Styrene Plastics*, Technical Bulletin 6422, 8-940-0277-6. (undated).

18. General Electric Company, Plastics Dept., Pittsfield, MA, *Valox® Thermoplastic Polyester*, Technical Booklet, VAL-5A, undated.

19. General Electric Company, Plastic Operations, Pittsfield, MA, *Valox® Resin Design Guide*, VAL-50A, (undated).

20. Rohm and Haas Co., Philadelphia, PA, *Plexiglas® Design & Fabrication Data*, PL – 7N, *Cementing Plexiglas® Brand Acrylic Sheet* (1983).

21. General Electric Company, Noryl Products Div. Selkirk, N.Y., *Noryl® Resin Design Guide*, CDX – 83B (June 1984).

22. American Society for Testing and Materials (ASTM), ASTM D 2564, Standard Specification for Solvent Cements for Poly(Vinyl Chloride) (PVC) Plastic Pipe and Fittings, *Annual Book of ASTM Standards*, Vol. 08.04.

23. American Society for Testing and Materials (ASTM), ASTM F 493-80, Standard Specification for Solvent Cements for Chlorinated Poly(Vinyl Chloride) (CPVC) Plastic Pipe and Fittings, *Annual Book of ASTM Standards*, Vol. 08.04.

24. General Electric Co., Pittsfield, MA, *The Comprehensive Guide to Material Properties, Design, Processing, and Secondary Operations, Ultem® Polyetherimide Resin*, ULT 201 (undated).

9

Effects of Environment on Durability of Adhesive Joints

INTRODUCTION

Adhesive bonds must withstand the mechanical forces acting on them, but must also resist the service environment. Adhesive strength is affected by many common environments, including temperature, moisture, chemical fluids, and outdoor weathering.[1]

In applications where possible degrading elements are present, candidate adhesives should be tested under simulated service conditions. Common ASTM environmental test methods often reported on in the literature include the following, described in Chapter 12:

ASTM D 896	ASTM D 2295
ASTM D 1151	ASTM D 2557
ASTM D 1828	ASTM D 4299
ASTM D 1879	ASTM D 4300

The service environments to which adhesive-bonded assembies are exposed vary from highly protected sealed systems to exterior unprotected systems. The latter may include semi-arid sites, such as New Mexico or Arizona, and highly corrosive tropical jungle or marine locations. Durability must always be linked with the service environment. The durability of an adhesive bond to extreme environmental conditions is very much dependent on the surface treatment of the adherends. Another factor which must be considered is the durability of bonds under stress. This is of particular concern in primary structures in aircraft applications. Most other types of bonds are not subject to high loads for prolonged periods. In laboratory ambient storage no degradation was found in bond strengths of epoxy/aluminum bonds stored for periods as long as 11 years, nor for elastomeric adhesive sealant bonds to aluminum and cold-rolled steel stored for up to 5 years.[1]

Applied stress will, however, cause an adhesive bond to degrade at a faster rate than an unstressed bond, although stress-relief primers, such as vinyl-phenolic, or stress-relief adhesives, such as nitrile-modified phenolics, will minimize bond degradation due to stress.[2]

The ranking of adhesives on the basis of durability is very much dependent on the type of exposure conditions employed. Table 9.1 illustrates a comparison of five common adhesives to (1) typical laboratory accelerated environments and (2) typical weathering environments. Artificial aging (accelerated testing) is considered by some workers to rank adhesives with respect to their resistance to penetration of water into the adhesive, with corresponding influence on the cohesive strength of the adhesive. The normal outdoor weathering tests, however, rank the adhesives with respect to their resistance to penetration of corrosion of the metal along the interface.[3]

Table 9-1: Environmental Failure Resistance of Different Adhesive-Aluminum Joints[3]

Adhesive Type	Laboratory Exposure* (Picatinny Arsenal Tests)	Outdoor Exposure** (Hockney, UK)
Nitrile-phenolic (with primer)	Excellent	Excellent
Epoxy-phenolic	Good	Excellent
Epoxy-polyamide	Good	Very poor
Vinyl-phenolic	—	Good
Modified epoxy paste	Average	Good

*Chromic acid solution- or paste-treated. Lap-shear 1 year storage at 160°F (66°C) and MIL-STD-304 conditioning. MIL-STD-304 conditioning consists of 30-day cycles under the following conditions: –65°F (–54°C), 160°F (71°C) dry heat, 160°F (71°C) and 95% RH-heat and humidity.
**Chromic acid-etched Al adherends. Lap-shear, 90° peel and honeycomb joint geometries. Two years exposure in temperate, hot-dry, and hot-wet regions.

Wangsness[4] has made the following general statements about durability of adhesive systems:

- Heat-curing systems possess greater durability than RT-curing systems.

- Surface preparation is an important factor, and it is important in making comparisons between adhesive systems to use the same surface preparation techniques.

- All systems do have an endurance limit. The RT-cure systems have a low endurance limit, $\sim$ 100 psi (0.7 MPa).

- The chemistry of adhesive systems does have an effect on durability, i.e., highly crosslinked systems such as aromatic amine-cured systems and phenolic systems generally possess superior durability.

- Materials such as nylon that show hydrogen bonding tend to have lower durability.

- Materials such as vinyl polymers which can break down to form HCl are detrimental to durability.

- The use of chromate-pigmented primer systems greatly enhances durability.

Because of the large number of factors that influence the durability of adhesive-bonded joints a durability test should be conducted on all systems before they are selected for any particular application. This test should include the adherends, surface preparation, adhesive, and cure parameters needed for each application.[4]

HIGH TEMPERATURE

All polymers are degraded to some extent by exposure to high temperatures. Physical properties are lowered after exposure to testing at high temperatures, but they also degrade during thermal aging. Some recently developed polymeric adhesives are capable of withstanding temperatures up to 500–600°F (260–316°C) continuously. Unfortunately, to use these adhesives the designer must pay a premium in adhesive cost and must also have facilities for carrying out long, high-temperature cures.[1]

For an adhesive to withstand high-temperature exposure it must have a high melting point or softening point and must be resistant to oxidation. Thermoplastic adhesives may provide excellent bonds at room temperature. However, once the service temperature approaches the glass-transition temperature of the adhesive, plastic flow results in deformation of the bond and degradation in cohesive strength. Thermosetting adhesives have no melting point and consist of highly cross-linked networks of macromolecules. Many of these materials are suitable for high-temperature applications. The critical factor in thermosets is the rate of strength reduction due to thermal oxidation and pyrolysis.[1]

Adhesives that are resistant to high temperatures usually have rigid polymeric structures, high softening temperatures, and stable chemical groups. These same factors make the adhesives very difficult to process. Few thermosetting adhesives can withstand long-term service temperatures over 350°F (177°C).[1]

Tape and film adhesives provide different high-temperature properties than paste adhesives, as shown in Figures 5.1 and 5.2 above. The distinguishing feature of tape and film adhesives is that they contain a high proportion of a high-molecular-weight polymer. On the other hand, typical 100%-solids paste adhesives or liquid adhesives must, to remain fluid and usable, contain only low-molecular-weight resins. Tape and film adhesives (discussed above in Chapter 5) frequently contain polymers with a degree of polymerization of 600 or more and molecular weights of 20,000 or higher. Network polymers made from these high-molecular-weight linear polymers can be very much tougher and more resilient and can provide more recoverable elongation than the highly branched networks formed by curing the low-molecular-weight resins used in paste adhesives. Figures 5.1 and 5.2 illustrate this point by comparing typical tensile-shear data reported by manufacturers of a variety of adhesive types. The best of the tape and film adhesives have higher peak values and broader service-temperature ranges than the best of the 100%-solids types.[5]

Epoxies

Epoxy adhesives are generally limited to applications below 250°F (121°C). Some epoxy adhesives have been able to tolerate short-term service at 500°F (260°C) and long-term service at 300 to 500°F (149 to 260°C). These systems were formulated especially for thermal environments by incorporation of stable epoxy coreactants or high-temperature curing agents into the adhesive. One of the most successful epoxy coreactants is an epoxy-phenolic alloy. The excellent thermal stability of the phenolic resins is coupled with the adhesion properties of epoxies to provide an adhesive capable of 700°F (371°C) short-term operation and continuous use at 350°F (177°C).[1]

Anhydride curing agents give unmodified epoxy adhesives greater thermal stability than most other epoxy curing agents. Phthalic anhydride, pyromellitic dianhydide, and chloriendic anhydride allow greater crosslinking and result in short-term heat resistance to 450°F (232°C). Long-term thermal endurance, however, is limited to 300°F (149°C).[1]

Epoxy-based adhesive systems offer the advantages of relatively low cure temperatures, no volatiles formed during cure, low cost, and a variety of formulating and application possibilities. The higher-temperature-resistant adhesives lose these advantages in favor of improved thermal-aging characteristics.[1]

Modified Phenolics

Nitrile-phenolics—Of the common modified phenolic adhesives, the nitrile-phenolic blend has the best resistance to elevated temperatures. Nitrile-phenolics have high shear strengths up to 250–350°F (121–177°C), and their strength retention on aging at these temperatures is very good. Nitrile-phenolic adhesives are extremely tough and provide high peel strength.[1] These materials are available in solvent solutions and unsupported and supported film.

Epoxy-phenolics—These adhesives are mostly used for the military market and are designed for service between 300 and 500°F (149 and 260°C). Epoxy-phenolics, however, do not withstand exposure to 350°F (177°C) as well as nitrile-phenolics. Since 350°F (177°C) cures tend to cause outgassing and foaming, 200°F (93°C) cures for 24 hours are recommended.

Polysulfone

This is a high-temperature hot-melt thermoplastic that has been used as an adhesive. Polysulfone is capable of adhering to hot metals, has a high softening point (190°F or 88°C) and outstanding heat stability. It has a 345°F (174°C) heat-distortion temperature and a 375°F (193°C) second-order glass-transition temperature. Flexural modulus is retained fairly constant over a wide temperature range. Polysulfone adhesive is supplied as dry pellets. Metal-to-metal joints exhibit high peel and shear strength. Polysulfone maintains its structural integrity up to 375°F (193°C). Over 60% of its room-temperature shear strength, as well as excellent creep resistance, is retained at 300°F (149°C). Since this is a hot melt, long cure cycles are not needed. The cycle need only be long enough to introduce sufficient heat for adequate wetting of the substrate by the polysulfone. Polysulfone hot melt has been used successfully with clad aluminum alloy, stainless steel, and cold-rolled steel. Adequate surface preparation is important before bonding. Table 9.2 shows bond strength from −67°F (−55°C) to 400°F (204°C) on aluminum alloy.[6]

Table 9-2: Effect of Temperature Variation on Tensile Lap-Shear Strength of Hot-Melt Polysulfone Adhesive (UDEL P-1700) on 2024-T3 Clad Aluminum Alloy (0.002–0.003") (0.05–0.076 mm) Glue Line[6]

Test Temperature		Tensile Lap-Shear .... Strength	
(°F)	(°C)	(psi)	(MPa)
−67	−54	3,300	22.8
77	25	3,500	24.1
180	82	2,700	18.6
300	149	2,200	15.2
350	177	1,950	13.4
400	204	520	3.6

For unprimed aluminum, a temperature of 700°F (371°C) should be used to permit the polysulfone to flow sufficiently to wet the substrate. With this temperature and a pressure of 80 psi (0.6 MPa), joints with tensile lap-shear strengths above 3000 psi (20.7 MPa) are developed in 5 minutes. Stainless steel has also been bonded with good results at 700°F (371°C). Shear strengths above 4000 psi (27.6 MPa) are obtained at this temperature. With both carbon steel and aluminum, a bonding temperature of 500°F (260°C) can be used with satisfactory results if the metal is first primed with a dilute solution (5–10%) of polysulfone applied by spray or flow-coated, and baked for 10 minutes at 500°F (260°C). The primed metal surfaces can then be bonded by pressing for 1 minute at 500°F.[6]

Silicones

Silicone adhesives have very good thermal stability, but low strength. Their primary application is in nonstructural uses, such as high-temperature pressure-sensitive tapes. Attempts have been made to incorporate silicones in other adhesives, such as epoxies and phenolics, but long cure times and low strength have limited their use.[1] The maximum service temperature for silicone adhesive/sealants is 500°F (260°C) for continuous operation and up to 600°F (316°C) for intermittent exposure, depending on the type used.

Polyaromatics

The polyaromatic resins, polyimide and polybenzimidazole, offer greater thermal resistance than any other commercially available adhesive. The rigidity of their molecular chains decreases the possibility of chain scission caused by thermally agitated chemical bonds. The aromaticity of the structure provides high bond dissociation energy and acts as an "energy sink" to the thermal environments.[1]

Polyimides (PI)—The strength retention of polyimide adhesives for short exposures to 1000°F (538°C) is slightly better than that of an epoxy-phenolic alloy. The thermal endurance of polyimides at temperatures greater than 500°F

(260°C), however, is unmatched by other commercially available adhesives. Polyimide adhesives are usually supplied as glass-fabric-reinforced film with a limited shelf life. A cure of 90 minutes at 500–600°F (260–316°C) and 15–200 psi (0.10–1.4 MPa) pressure is necessary for optimum results. High-boiling volatile constituents are released during cure, resulting in a somewhat porous adhesive layer. Because of the inherent rigidity of polyimides, peel strength is low.[1]

Polybenzimidazoles (PBI)—These adhesives offer the best short-term performance of any adhesives at elevated temperatures. PBI adhesives, however, oxidize very rapidly and are not recommended for continuous use at temperatures above 450°F (232°C). PBI adhesives require a cure at 600°F (316°C). As with polyimide adhesives release of volatiles during cure contributes to a porous adhesive bond. These adhesives are supplied as very stiff, glass-fabric-reinforced films and are very expensive. Their applications are limited by a long, high-temperature curing cycle.[1]

LOW AND CRYOGENIC TEMPERATURES

Cryogenic adhesives have been defined as those capable of retaining shear strengths above 1000 psi (6.89 MPa) at temperatures varying from room temperature to −423°F (20K) (−253°C). With space vehicles carrying cryogenic fluids and traveling through outer space and re-entering the earth's atmosphere at speeds greater than Mach 3, adhesives encounter temperatures varying from −423°F (−253°C) to 1500°F (816°C).[7]

The major use of adhesives for cryogenic applications is for bonding external insulation for both metallic and nonmetallic substrates. Other applications are listed in Reference 7. Adhesives are also capable of acting as sealants. Many wing structures utilize adhesive-sealed tanks and pressure-type bulkheads. Room-temperature-vulcanizing (RTV) silicones have been evaluated as sealants and adhesives for cryogenic applications. The adhesive strengths obtained with methyl-phenyl RTV silicones were only 1/4 to 1/10 the values for the better structural cryogenic adhesives. These values are adequate, however, for nonstructural bonding applications where low tensile and shear forces are anticipated. The RTV silicones may prove useful where high-temperature extremes up to 600°F (316°C) are encountered for short periods. The better cryogenic adhesives will not tolerate these high temperatures.[7]

Many problems associated with bonded joints at cryogenic temperatures are the result of stress concentrations and gradients developed within the bond. There are a number of causes of stress concentrations in adhesive joints, and a number of these causes are aggravated by cryogenic temperatures. Some of the principal causes are as follows:[7]

- Difference in thermal coefficients of expansion between adhesive and adherends
- Shrinkage of adhesive in curing
- Trapped gases or volatiles evolved during bonding
- Difference in modulus of elasticity and shear strengths of adhesive and adherends

- Residual stresses in adherends as a result of the release of bonding pressure
- Inelasticity in the adhesive or adherend
- Plasticity in the adhesive or adherend

At room temperature a low-modulus adhesive may readily relieve stress concentration by deformation. At cryogenic temperatures, however, the modulus of elasticity may increase to a point where the adhesive can no longer effectively release the concentrated stresses. The modulus of elasticity generally increases with decreasing temperature. More constant properties are usually obtained when attempts are made to match the coefficient of expansion of the adhesive to that of the adherends. The thermal conductivity is important in minimizing transient stresses during cooling. These stresses are minimized by thinner glue lines and higher thermal conductivities.

A PLASTEC report issued in 1965 by the author provides data on a number of cryogenic adhesives.[7]

Polyurethanes are among the better adhesives for cryogenic applications. Room-temperature-curing polyurethane adhesives in current use provide higher ultimate shear and tensile stress and higher peel and shock properties at −423°F (−253°C) than the earlier polyurethanes. This situation is the inverse of what happens to most structural adhesives. The polyurethane adhesives increase in strength at −423°F, but become weaker at ambient and higher temperatures, as shown in Table 9.3.[7]

Table 9-3: Comparison of "Tough" Cryogenic Adhesives at Liquid-Hydrogen and Ambient Temperatures (Modified after Roseland[8])

	. . Tensile Shear Strength. . .		. . . . . . .T-Peel Strength					
	−423°F (−253°C)		. Ambient .		−423°F (−253°C)		. . . Ambient. . . .	
Adhesive Type	(psi)	(MPa)	(psi)	(MPa)	lbf/in	N/m	lbf/in	N/m
Polyurethane*	7500	51.7	1500	10.3	70	12,260	20-40	3502~7004
Nylon-Epoxy Film**	3600	24.8	5000	34.5	4	700	104	18,200

*RT-curing paste.
**300° to 350°F (149° to 177°C) cure.

The shear-strength properties of several classes of adhesives suitable for cryogenic applications are shown in Figure 9.1. Although bond strengths are reasonably good, the *unmodified epoxy resins* suffer from brittleness and corresponding low peel and impact strength at cryogenic temperatures. The *epoxy-phenolics* have excellent cryogenic-temperature strengths, as well as good high-temperature properties. The *epoxy-nylons* give consistently high strength at cryogenic temperatures. The flexiblity of the nylons imparts greater peel strength to the epoxies and produces systems with unusual low-temperature properties. *Epoxy-polyamides* are readily mixed, easily applied, have good pot life, and can be cured at room temperature to yield a flexible system. Their low-temperature performance, however, is not as good as that of the epoxy-nylons.

Vinyl acetal-phenolic adhesives are available as supported and unsupported films, solutions, and solutions with powder, and show reasonably good low-temperature strength. Their strength falls off, however, with decreasing temperature due to the increasing modulus of elasticity characteristic of thermoplastics.[9]

Rubber phenolics (elastomer-phenolics) are of value because of their high peel strengths, but their shear strengths are relatively low. Nitrile-phenolics are examples of this type. Polyurethane adhesives have excellent adhesion to a number of substrates, along with inherently good low-temperature flexibility. Peel strength is excellent at cryogenic temperatures. Epoxy-nylon adhesives, mentioned above have higher strength in the low temperature (−100°F or −73°C) range, as seen in Figure 9.1, than any other cryogenic adhesive. At liquid-nitrogen temperature (−321°F or 77K) there is little difference in the shear strengths of the *polyurethane* and *epoxy-nylon* types. At liquid hydrogen temperature (−423°F or 20.4K), however, the newer *polyurethane* adhesives surpass the *epoxy-nylons*.[9]

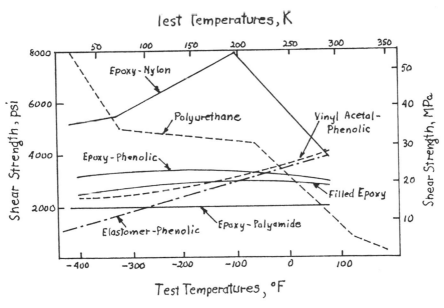

Figure 9-1: Comparison of cryogenic and low-temperature adhesive types. (Modified after Reference 9)

The National Bureau of Standards (NBS) at Boulder, Colorado has prepared an excellent survey for NASA on available reports and publications on cryogenic adhesives and sealants.[10]

A 1974 contractor report[11] describes work performed on the development of a structural adhesive system for use over a −423°F (−253°C) to 600°F (316°C) range. Two *polyimide* adhesives, BR 34/FM 34 (American Cyanamid) and P4/A5F or P4/A5FA (TRW Systems), were found to provide excellent results with both titanium alloy and stainless steel substrates.

HUMIDITY AND WATER IMMERSION

Moisture can affect adhesive strength in two significant ways. Some polymeric materials, particularly ester-based polyurethanes, will "revert", i.e. lose hardness, strength, and in the worst cases, liquefy during exposure to warm, humid air. Water can also permeate the adhesive and preferentially displace the adhesive at the bond interface. This mechanism is the most common cause of adhesive-strength reduction in moist environments.[1]

The rate of hydrolytic reversion depends on the chemical structure of the base adhesive, the type and amount of catalyst used, and the flexibility of the adhesive. Certain chemical linkages, such as ester, urethane, amide and urea, can be hydrolyzed. The rate of such attack is fastest for ester-based linkages. Such linkages are present in certain types of *polyurethane* and *anhydride-cured epoxies*. In most cases amine-cured epoxies provide better hyrolytic stability than anhydride-cured types. The reversion rate of hydrolytic materials is also dependent on the amount of catalyst used in the formulation. The best hydrolytic properties are obtained when the proper stoichiometric ratio of base material to catalyst is used. Reversion is usually much faster in flexible materials because water permeates them more readily.[1]

Structural adhesives not susceptible to the reversion phenomenom are also likely to lose strength when exposed to moisture, particularly at high temperatures. The mode of failure in the initial stages of exposure under the conditions is *cohesive*. After 5 to 7 days the failure becomes one of adhesion (*adhesive* failure). Water vapor apparently permeates the adhesive through its exposed edges and concentrates in weak boundary layers at the interface. This effect is very much dependent on the type of adhesive. *Nitrile-phenolic* adhesives do not fail as a result of the mechanism of preferential displacement at the interface. Failures occur cohesively within the adhesive, even when tested after 24 hours immersion in water. A *nylon-epoxy* adhesive degrades rapidly under the same conditions because of permeability and preferential displacement by moisture. Adhesive strength deteriorates more rapidly in an aqueous-vapor environment than in liquid water because of the more rapid permeation of the water vapor. Because of the importance of the interface, primers and surface treatments tend to hinder adhesive strength degradation in moist environments. A fluid primer that easily wets the interface presumably tends to fill in minor discontinuities on the surface. Chemical etching, which removes surface flaws, also improves resistance to high-humidity environments.[1]

Environmental effects on adhesive joints are accelerated by stress. Few data are available on this phenomenon, however, because of the time and expense involved with stress aging tests. It is recognized, however, that the moisture environment significantly decreases the ability of an adhesive to bear prolonged stress.[1]

Effects of Surface Preparation on Moisture Exposure

The Aluminum Company of America (ALCOA) has conducted studies on *unstressed* joint durability after room-temperature water immersion and after 100%-RH exposure at 125°F (52°C). Table 9.4 gives results on the room-temperature tests in water. Direct comparison of durability for all joints can only be made after one-year soaking exposure. The two-year tests had not been completed for the chromic acid anodized and phosphoric acid anodized surface joints at the time of publication of the report. After one year in this moderately accelerated laboratory weathering exposure the relative bond retention averages were similar for all but the vapor-degreased surface joints, and ranged between 70–80% of initial bond strengths. The vapor-degreased surface joints averaged only 46.2% of their initial bond strength values. Table 9.5 gives the results after exposure to condensing humidity (100% RH) at 125°F (52°C). After 12 months exposure all of the anodized surface joints showed significantly high retained bond strength, ranging from a high of 73% for the sulfuric acid-anodized surface joint to a low of 54.8% for the phosphoric acid-anodized surface joints. The average bond strength retention for the acid-etched surface joints in high humidity ranged from 31% for the inhibited alkaline cleaned (Ridolene 53) and chromic-sulfuric acid-etched surface joint to 16.1% for the alcohol-phosphoric acid-etched surface joints.[12] ALCOA has also studied the effects of stress with the system described above.

Stressed Temperature/Humidity Test

One of the earlier methods used by the Army for evaluating the durability of adhesive-bonded joints is the stressed temperature/humidity test described in ASTM D 2919-71 (1981) (See Chapter 12). Army ARDC workers at Picatinny Arsenal have used an environmental condition not listed in the standard test environments of this standard, 60°C (140°F) and 95–100% RH. These conditions were selected during an evaluation of adhesive-bonding processes used in helicopter manufacture. This stressed method is very time-consuming and often expensive, since it requires environmental test chambers.[13] Figure 9.2 shows the stressing jig. This jig can be used with multiple units fitted with automatic failure-recording devices.

Figures 9.3 and 9.4 show the degradation of the epoxy adhesive bond due to the effect of temperature on the anodic aluminum joint at 95% RH.[14] The plots show that those joints exposed at 73°F (23°C) and 95% RH degraded rapidly when the stress level exceeded 880 lbs force (3914 N), which would be equal to approximately 40% of the room-temperature tensile-shear strength of the joint. At 120°F (49°C) rapid deterioration occurred at a stress level of 440 lbs force (1957N), or approximately 20% of the room-temperature strength of the joint. The durability or stress vs log time-to-failure curves are shown in Figure 9.4 for the various temperatures at 95% RH. The resultant curves are close to being parallel, indicating that the effect of temperature at a constant RH is constant, and as the temperature increases, the durability decreases as a function of the strength change caused by the change in temperature. This data is for aluminum, but similar results are obtained with titanium. Thus, an increase in % RH will result in a decrease in joint durability, and appears to be a significant factor in joint failure.[14] This ASTM method has some drawbacks, such as the large scatter of data, the inability to check stress level at environmental conditions, and the inability to run very low loads because of the high k-factor of the springs. Springs with lower k-factors could be used, however.[15]

Table 9-4: Effects of Surface Treatment on the Durability of 6061-T6 Aluminum Alloy Joints Exposed to Immersion in the Unstressed Condition (Nitrile-Modified Epoxy Paste Adhesive) (Modified after Minford[12])

Surface Treatment	Initial Shear . .Strength . .		Average Percent Retained Bond Strength After Indicated . . Exposure Time (months). . .			
	psi	MPa	3	6	12	24
Vapor degreased	4330	29.9	70.9	59.4	46.2	27.0
Deoxidine 526* (5 min, 25% concentration at RT)	4970	34.3	83.1	84.5	73.8	56.9
Chromic-sulfuric [5 min, 180°F (82°C)]	5330	36.8	83.9	82.6	78.2	66.2
Chromic acid anodize	5513	38.0	85.6	83.6	70.9	NA
Phosphoric acid anodize (Boeing procedure, 5–10 volts, 20 min)	6480	44.7	89.8	80.2	74.4	NA
Sulfuric acid anodize (12 asf, 60 min boiling-water seal)	3550	24.5	69.6	66.8	67.6	68.5

*Amchem Corp.

Table 9-5: Effects of Surface Treatment on the Durability of 6061-T6 Aluminum Alloy Joints Exposed to 100% RH (Condensing Humidity) at 125°F (52°C) in the Unstressed Condition (Nitrile-Modified Epoxy Paste Adhesive) (Modified after Minford[12])

Surface Treatment	Initial Shear . .Strength . .		Average Percent Retained Bond Strength After Indicated . . Exposure Time (months). . .			
	psi	MPa	3	6	12	24
Deoxidine 526* (5 min, 25% concentration at RT)	4970	34.3	46.1	30.2	16.1	10.9
Chromic-sulfuric [5 min, 180°F (82°C)]	5330	36.8	39.9	23.8	16.6	12.4
Ridoline 53 [3 min, 180°F (82°C)] Chromic-sulfuric [5 min 180°F (82°C)]	5430	37.4	51.4	36.3	31.3	26.3
Chromic acid anodize	5513	38.0	70.7	69.8	60.9	NA
Phosphoric acid anodize (Boeing procedure, 5–10 volts, 20 min)	6480	44.7	70.7	62.8	54.8	NA
Sulfuric acid anodize (12 asf, 60 min boiling-water seal)	3550	24.5	74.9	74.1	72.7	67.6

*Amchem Corp.

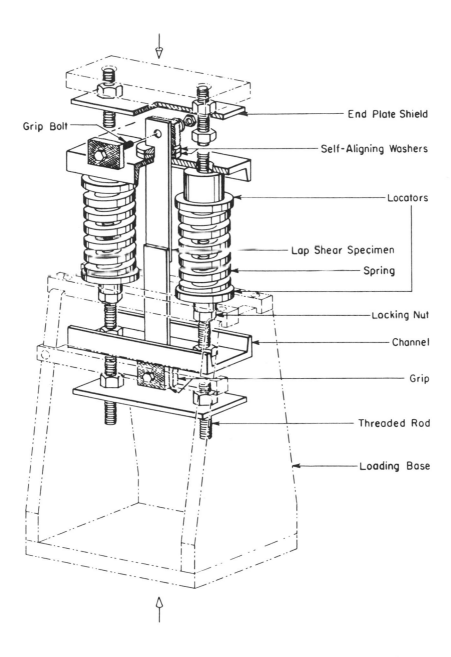

End Plate Shield

Grip Bolt

Self-Aligning Washers

Locators

Lap Shear Specimen

Spring

Locking Nut

Channel

Grip

Threaded Rod

Loading Base

Figure 9-2: Stressing jig used in loading test fixture. (Reprinted, with permission from the Annual Book of ASTM Standards, volume 15.06. Copyright ASTM, 1916 Race Street, Philadelphia, PA 19103.)

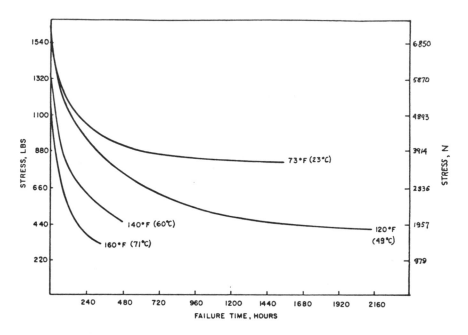

Figure 9-3: Effect of temperature on the durability of adhesive-bonded ano-dized aluminum at 95% RH. (Modified epoxy film adhesive[14])

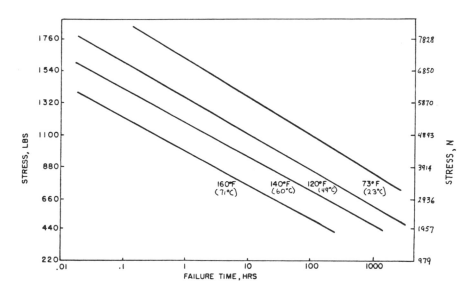

Figure 9.4: Effect of temperature on the durability of adhesive-bonded ano-dized aluminum at 95% RH (log plot). (Modified epoxy film adhesive[14])

Hot-Water-Soak Test

With this newer method developed by Army ARDC workers at Picatinny Arsenal, the durability curves for a large number of adhesive/adherend combinations can be predicted simultaneously. The method involves the soaking of test specimens in a tank of deionized water at 60°C (140°F). The specimens are allowed to soak for periods up to 100-1000 hours in the 60°C (140°F) water. At the end of the soak period the specimens are removed and placed in a container of water at 60°C (140°F). This container is then placed into the test chamber of a universal tensile-testing machine. The chamber is also kept at 60°C (140°F). The test specimens are then removed, one at a time, from the water and placed in the grips of the test machine. A thermocouple is attached to the specimen and the temperature monitored. When the temperature of the specimen reaches 60°C (140°F) the test is started. The load is supplied at a rate of 120-1400 lbf/min (270-315 N/min) until the specimens fail. This test is carried out on at least four specimens and the results are plotted on semi-logarithmic paper.[13,15] When the data is plotted as a function of the residual strength vs log of exposure time, the resulting plot is a straight line similar to the stressed durability curves shown in Figure 9.4.[15]

Figure 9.5 shows the data obtained on an epoxy-nitrile film adhesive on 5052-H34 aluminum alloy after immersion in hot water for 50, 100, 300, 500, and 1000 hours. This test is very useful because it permits a large number of adhesive-bonded specimens with different adhesives, adherends, and surface pretreatments to be tested at the same time with a relatively small investment in man hours and equipment. Figure 9.6 shows a comparison of the stressed-durability data and unstressed hot-water-soak data on the same epoxy-nitrile film adhesive, using 2024-T3 aluminum alloy. Note the parallelism of the plots. The curve in the lower left was obtained when lap-shear specimens were subjected to various levels of stress and then exposed to an environment of 60°C (140°F) and 95% RH until failure. The failure time is plotted as a log function. The curve in the upper right portion is a plot of the data when the same types of lap-shear specimens were subjected to 60°C (140°F) water for specified periods of time and then tested for their residual strength. In the first case *failure time* was recorded. In the latter case, *residual strength* was determined. The same type of data is obtained with both curves.[15]

Fatigue-Life Data

Bonded joints in helicopters are subjected to constant fatigue. The operation of a helicopter subjects the aircraft to a constant state of vibration. The mission of the Army's helicopter fleet subjects them to conditions of elevated temperature and frequently high humidity. Environmental exposure is believed to have a strong detrimental effect upon the fatigue endurance of bonded joints in these aircraft.[16]

A vibration frequency of 1000 cycles per minute is considered representative of the vibration of Army aircraft during flight. Accordingly, Army ARDC workers have designed a test for determination of fatigue life data in the tension stress cycle from zero to a specific maximum stress of 2800 psi (19.3 MPa). Endurance curves were determined at room temperature, 60°C (140°F) with 95% RH.[16]

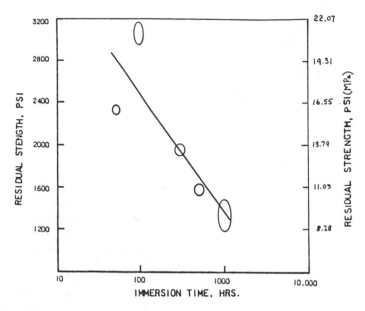

Figure 9-5: Hot-water-soak data with epoxy-nitrile film adhesive on 5052-H34 aluminum joints.[15]

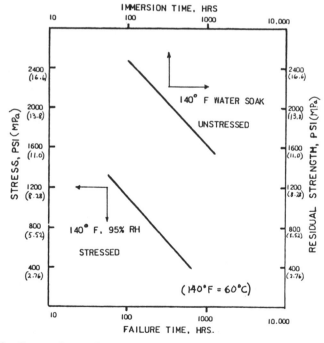

Figure 9-6: Comparison of stressed durability data and unstressed hot-water-soak data with epoxy-nitrile film adhesive on anodized 2024-T3 aluminum joints.[15]

Figure 9.7 compares the stress vs. log of cycles to failure (S/Log N) curves for an epoxy-nitrile film adhesive used for bonding sulfuric-dichromate etched 2024-T3 aluminum at a frequency of 1000 cycles per minute at RT, 60°C (140°F) (dry) and at 60°C (140°F) with 95% RH.[16] In general, the fatigue curves obtained at these environments follow the trend shown in Figures 9.3 and 9.4, where a noticeable drop is experienced when the environment is changed from 73°F (23°C) to 140°F (60°C), and again when the environment becomes more severe with the addition of 95% RH to the 60°C (140°F) environment.[17]

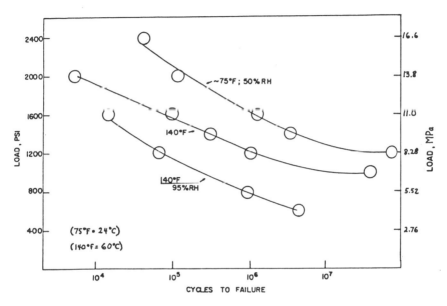

Figure 9-7: Environmental S-Log N curves for nitrile-epoxy film adhesive-bonded 2024-T3 aluminum joints at 1000 cycles/minute.[16]

SALT WATER AND SALT SPRAY

Salt water and especially salt spray are known to have a deleterious effect on adhesive joints. Testing for the effects of salt spray (salt fog) is usually carried out using ASTM B 117, "Standard Method of Salt Spray (Fog) Testing". This method has been approved for use by DOD agencies to replace Method 811.1 of Federal Test Method Standard No. 151B.

Bond durability results on an aluminum alloy are shown in Table 9.6 for chromic acid-anodized and four variations of phosphoric acid-anodized surfaces, plus two variations of sulfuric acid-anodized surfaces, i.e. sealed and unsealed. The excellent resistance to surface corrosion in salt-water exposure shown by unbonded anodized surfaces is also shown in the bonded joints. All chromic acid and phosphoric-acid anodized surface joints were highly resistant to bond failures, which ordinarily occur by undercutting corrosion of the bondline in this environment. The essential absence of such undercutting was noted visually in joints failed deliberately after 12 months exposure, and was confirmed by the high percentage of retained bond strength in the range of 79–96%. Although thicker sulfuric acid-anodized coatings would be expected to offer the highest degree of corrosion resistance to salt spray in the sealed and unbonded state, the only bond failure encountered when tested weekly to 50% of initial bond strength was with a sealed sulfuric acid-anodized pretreated joint. This is not surprising, since it is known that an unsealed sulfuric acid-anodized surface can be bonded with higher initial bonding strength and better bond durability. The good bond durability in salt-water exposure of anodized surface pretreated joints is a good reason for selecting this type of pretreatment over the acid-etching procedures for marine applications.[12]

Table 9-6: Effect of Surface Treatment and Exposure to 3½% Salt Water Intermittent Spray on the Durability of 6061-T6 Aluminum Alloy Joints Exposed in the Unstressed Condition (Nitrile-Modified Epoxy Paste Adhesive) (Modified After Minford[12])

Surface Treatment	Initial Shear . .Strength . .		Average Percent Retained Bond Strength After Indicated . . Exposure Time (months)* . .			
	psi	MPa	3	6	9	12
Chromic acid anodize	5513	38.0	>50	>50	>50	82.0
Phosphoric acid anodize (30 volts, 30 min)	5700	39.3	>50	>50	>50	89.8
Phosphoric acid anodize (60 volts, 18.5 min)	6030	41.6	>50	>50	>50	92.9
Phosphoric acid anodize (110 volts, 6 min)	6070	41.9	>50	>50	>50	95.9
Phosphoric acid anodize (Boeing procedure)	6480	44.7	>50	>50	>50	79.3
Sulfuric acid anodize (12 asf, 60 min unsealed)	3940	27.2	>50	>50	>50	75.1
Sulfuric acid anodize (12 asf, 60 min, boilwater seal)	3550	24.5	>50	>50	>50	49.6

*In this procedure the specimens were stressed weekly to 50% of initial shear strength and then returned to the exposure conditions, providing no bond failure occurred. After 52 weeks testing the joints were deliberately failed for quantitative determination of the actual bond strength as shown.

Seacoast Weathering Environment

The results of studies with vapor-degreased, acid-etched, and sulfuric acid anodized surface joints exposed to a seacoast environment are shown in Table 9.7. These tests were carried out for periods as long as 8 years. This type of natural weathering test environment is highly discriminating between the various surface pretreatments. The surprisingly high 62% bond strength retention for

the sealed sulfuric acid-anodized surface joints after 8 years is especially significant when compared to the less-than-one-year survival time for vapor-degreased
surface joints and the approximately 700-to 1440-day survival time for alcohol-
phosphoric acid and chromic-sulfuric etched surface joints.[12] Minford[18] showed
that the seacoast atmosphere is the most deteriorating to *heat-cured epoxies* as
a group, many failing completely after the end of four years exposure. *Anhydride-cured epoxies* give better results and retain about half their initial shear
strength after four years in this aggressive marine enviornment. *Nitrile-modified
epoxies* give better results than non-modified epoxies, as is the case with
phenolics and *nitrile-phenolics*.

Table 9-7: Effect of Surface Treatment and Exposure to Seacoast Environment
on the Durability of 6061-T6 Aluminum Alloy Joints Exposed in the Unstressed
Condition (Nitrile-Modified Epoxy Paste Adhesives) (Modified after Minford[12])

Surface Treatment	Initial Shear . Strength. .		Average Percent Retained Bond Strength After Indicated . . . Exposure Time (years)* . . .			
	psi	MPa	1	2	4	8
Vapor degreased	4330	29.9	0***	—	—	—
Deoxidine 526** (5 min 25% concentration at RT)	4970	34.3	72.4	10.9	0†	—
Chromic-sulfuric [5 min 180°F (82°C)]	5330	36.8	91.2	63.2	0††	—
Chromic acid anodize	5513	38.0	82.5	—	—	—
Sulfuric acid anodize (12 asf, 60 min boiling-water seal)	3270	22.6	69.4	68.2	85.4	62.3

　*Exposed at Point Judith, Rhode Island.
　**Amchem Corp.
　***Time span of bond failure 71 to 270 days.
　†Time span of bond failure 720 to 1440 days.
　††Time span of bond failure 760 to 1440 days.

Two-part epoxy adhesives (RT-curing) give poor results in seacoast atmospheres unless a compatible organic sealer is placed over the edge of the bond-
line. In the case of *tape and film adhesives*, nylon or nylon-modified epoxy
adhesive bonds either failed to survive four years exposure, or lost 73% of initial
strength. Excellent performances were shown by all *nitrile-phenolic* and
phenolic-type adhesives. As a group, all joints fabricated from ten of twelve tape-
and-film adhesives tested in a seacoast atmosphere survived for the total test
period of 48 months exposure. By contrast, no two-part epoxy joints lasted
longer than 30 months, and only one heat-cured, one-part-epoxy survived 48
months exposure.[18]

Salt Water Immersion

Minford also studied the effects of four different phosphoric acid processing
conditions under stress and intermittent salt-water immersion testing of 6061-T6

aluminum alloys. None of the joints pretreated by varying phosphoric acid anodizing conditions failed after 480 days exposure, even after 2268 psi (15.6 MPa) stress. A few of the stressed joints pretreated by chromic acid anodizing failed during the 480 days of exposure, but only at a stress level of approximately 2000 psi (13.8 MPa), or approximately 35% of the initial bond strength. Because of the lower initial bond strength of the sulfuric acid anodized surface joints, the highest stress levels imposed were 1379 psi (9.5 MPa) (35% of initial bond strength) for the unsealed and 1242 psi (8.6 MPs) (35% of initial bond strength) for the sealed joints. After about 100 days, the sealed sulfuric acid anodized joints failed in exposure, while the corresponding unsealed joints survived after 482 days exposure.[12]

Nitrile-Phenolic Adhesives

Minford has shown the exceptional strength retention of nitrile-phenolics, such as FM-61, on aluminum after extended salt spray, water immersion, and other long-term exposure tests. It is probably true that no other adhesive type exceeds the ability the nitrile-phenolics to maintain good strength on steel or aluminum after extended exposure to water, salts, or other corrosive media, and to prevent undercutting through corrosion of the metal substrate.[5]

Boeing/Air Force Studies on Salt-Spray Effects

A 1976 Air Force-sponsored study reported on the effects of corrosive salt-spray environment on bondlines of different bonded systems.[19] The system variations included clad and bare alloys, surface treatments, adhesive primers, and adhesives. Five specimens were fabricated for each of the bonded systems. The specimens were then placed in a salt-spray environment of 5% NaCl at 95°F (35°C). The change in wedge-test crack length of each specimen was recorded periodically. At the end of one month, one specimen was randomly selected from each bonded system and opened for visual inspection of the bondline condition, both in the stressed zone (crack-tip zone) and in the unstressed zone. The same procedure was carried out after 2, 3, 6, and 12 months, when the last specimen was removed from test. The conclusions were as follows:[19]

- The phosphoric acid-anodize process provides markedly improved stressed-bond joint durability and retards bond-line crevice corrosion (started at an edge) in severely corrosive environments when compared to chromic acid-anodize and FPL etch.

- Stressed-bond joint durability is markedly affected by the adherend prebond surface treatment and the adhesive/primer system in contact with it. This is evidenced by the poor performance of FM 123-L/BR 123 (non-CIAP) adhesive/primer system on FPL-etched and chromic acid-anodized 2024-T3 aluminum alloy, clad and bare, and the superior performance of the same systems when BR 127 (CIAP, Corrosion-Inhibiting Adhesive Primer) is substituted for BR 123 (non-CIAP).

- The wedge test method is discriminatory and provides a relative ranking for many of the parameters that affect bond-joint durability.

- Clad aluminum in bondlines is undesirable under severely corrosive salt-spray environments.

WEATHERING

By far the most detrimental factors influencing adhesives aged outdoors are *heat* and *humidity*. Thermal cycling, ultraviolet radiation, and low temperatures are relatively minor factors. When exposed to weather, structural adhesives rapidly lose strength during the first 6 months to 1 year. After 2 to 3 years the rate of decline usually levels off at 25 to 30% of the initial joint strength, depending on the climate zone, adherend, adhesive, and stress level. The following generalizations are important in designing a joint for outdoor services:[1]

- The most severe locations are those with high humidity and warm temperatures.

- Stressed panels deteriorate more rapidly than unstressed.

- Stainless-sreel panels are more resistant than aluminum panels because of the corrosion in the latter.

- Heat-cured adhesive systems are generally more resistant to severe outdoor weathering than room-temperature-cured systems.

- Using the better adhesives, unstressed bonds are relatively resistant to severe outdoor weathering, although all joints will eventually exhibit some strength loss.

Simulated Weathering/Accelerated Testing

Army workers at Picatinny Arsenal have carried out a number of experiments in the laboratory (accelerated testing) using the MIL-STD-304 conditioning, and in actual weathering sites throughout the world.[20] MIL-STD-304 has now been replaced by MIL-STD-331, but the MIL-STD-304 conditions were: exposure to alternating cycles of cold ($-65°$F) ($-54°$C), dry heat ($160°$F) ($71°$C), and heat and humidity (95% RH) for 30 days. After the exposure period, the aluminum alloy panels used in the studies (2024-T3) were cut into individual specimens and tested at $-65°$F ($-54°$C), $73°$F ($23°$C) and $160°$F ($71°$C). Eleven types of adhesives were tested. Only one adhesive actually failed. The results are shown in Table 9.8. Virtually all the adhesives showed a loss in strength when tested at $160°$F ($71°$C), some being more affected than others. One adhesive, No. 7, the epoxy-anhydride RT-cured system, lost approximately 70% of its joint strength at $73°$F ($23°$C) after cycling. The RT-cured epoxy-polyamide systems (No. 1) seemed to be the least affected of the adhesive types tested. Picatinny Arsenal carried out the tests in conjunction with actual field weathering tests.

Table 9-8: Effect of MIL-STD-304* Conditioning (JAN cycle) on Strength
of Bonded Aluminum Alloy 2024-T3 Joints (Modified after Tanner[24])

| | Test Temperatures | | . . Average Shear Strength. . . | | | |
| | | | . Control . . | | MIL-STD-304* | |
Adhesive Type	(°F)	(°C)	(psi)	(MPa)	(psi)	(MPa)
1. Epoxy-polyamide, RT-cured	−65	−54	1700	11.7	2100	14.5
	73	23	1800	12.4	2100	14.5
	160	71	2700	18.6	1800	12.4
2. Epoxy-polyamide w/mica filler, RT-cured	−65	−54	2200	15.2	3080	21.2
	73	23	2500	17.2	3140	21.7
	160	71	2200	15.2	1120	7.7
3. Resorcinol epoxy-polyamide, RT-cured	−65	−54	2600	17.9	2440	16.8
	73	23	3500	24.1	3120	21.5
	160	71	3300	22.8	2720	18.8
4. Epoxy aromatic amine, RT-cured	−65	−54	1700	11.7	**	**
	73	23	2000	13.8	**	**
	160	71	720	5.0	**	**
5. Epoxy-polysulfide, RT-cured	−65	−54	1800	12.4	1940	13.4
	73	23	1900	13.1	1640	11.3
	160	71	1700	11.7	1070	7.4
6. Nylon-epoxy, RT-cured	−65	−54	2400	16.6	3020	20.8
	73	23	2600	17.9	1730	11.9
	160	71	220	1.5	80	0.6
7. Epoxy-anhydride, RT-cured	−65	−54	2400	16.6	1900	13.1
	73	23	3000	20.7	920	6.3
	160	71	3300	22.8	1330	9.2
8. Modified epoxy, cured 1 hr at 350°F (177°C)	−65	−54	3700	25.5	2700	18.6
	73	23	4900	33.8	3400	23.4
	160	71	4100	28.3	3200	22.1
9. Epoxy-phenolic, cured 45 min at 330°F (166°C)	−65	−54	2800	19.3	2610	18.0
	73	23	2900	20.0	2350	16.2
	160	71	2900	20.0	2190	15.1
10. Nitrile-phenolic, cured 1 hr at 350°F (177°C)	−65	−54	4700	32.4	5300	36.6
	73	23	4600	31.7	3900	26.9
	160	71	3070	21.2	2900	20.0
11. Polyurethane, RT-cured	−65	−54	3500	24.1	4200	29.0
	73	23	2600	17.9	1970	13.6
	160	71	1600	11.0	1560	10.8

*Alternating cycles of cold (−65°F) (−54°C), dry heat 160°F (71°C), and heat and
humidity (160°F) (71°C) (95% RH) for 30 days. MIL-STD-304 has been superseded by MIL-STD-331.
**Panels fell apart.

Outdoor Weathering (Picatinny Arsenal Studies)

Weathering studies after exposures up to one year were also made by Picatinny Arsenal on the adhesives covered in Table 9.8, along with several additional adhesives.[21] The results are given in Table 9.9 as percent retention of
original joint strength. In addition to controls, the following climates were used:

- Hot, dry (Yuma, Arizona)
- Hot, humid (Panama Canal Zone)
- Temperate (Picatinny Arsenal, Dover, N.J.)

Results of JAN cycling (MIL-STD-304), as described above, were also given. The results in Table 9.9 show that, in general, most of the adhesive joints, when stored in the laboratory (controls), retain most of their original joint strength for one year. The joints that were stored in the hot, dry area (Yuma) generally retained most of their original strength. Two adhesives, epoxy-resorcinol and epoxy-phenolic, show a trend towards decreased joint strength. Where climatic conditions subject the bond to humidity and precipitation, i.e., at Picatinny and in Panama, most of the joints show a decrease in joint strength. Four adhesive joints, those with filled epoxy-nylon, unfilled epoxy-nylon, nitrile-phenolic, and silicone, do not appear to be affected to any large extent by weathering, regardless of the site and climatic conditions. The MIL-STD-304 temperature and humidity cycle (now MIL-STD-331) appears to be useful in predicting the changes that occur in panels exposed to high humidity.[21]

Picatinny Arsenal workers also carried out a three-year weathering program on aluminum joints, using seventeen different adhesives.[22] Of the seventeen adhesives, thirteen were epoxies or modified epoxies, since epoxy types are most widely used in adhesive-bonding applications. In the study, only five of the original seventeen adhesives retained a minimum of 50% of their original joint strength and approximately 2000 psi (13.8 MPa) at all test temperatures after two years of weathering, regardless of the test site. After three years, only two adhesives, the phenolic-epoxy film (#13) and the nitrile-phenolic film (#14) met these requirements. Table 9.10 shows the percentage of retention of the original adhesive-joint strength for the seventeen adhesives after three years of weathering at the three test sites and after MIL-STD-304 cycling. It is apparent that the joints stored in the laboratory (controls) retained most of their original strength for the three years. So did most of the joints weathered at Yuma. Where humidity and moisture were prevelant, as at Picatinny and Panama, most of the joints showed a decrease in joint strength.

The joints made with an aluminum-filled one-component epoxy paste, 350°F (177°C) curing adhesive (#1) showed very good durability at Yuma. However, these joints showed a marked decrease in joint strength in the high humidity of Panama. At the end of three years there was some evidence of corrosion of the aluminum beneath the bondline. The joints made with another filled, one-component modified-epoxy paste, 350°F (177°C) curing adhesive (#2) showed good retention of joint strength during the three-year exposure at all sites other than Panama. After two years, the Panama joints had lost more than 50% of their original strength due to corrosion of the aluminum. After three years they were so badly corroded that they fell apart in the racks. In general, the joints formed with a (121°C) 250°F curing, one-component epoxy-paste adhesive (#3) showed a good retention of the original bond strength. However, here again, the higher-humidity sites appeared to have an adverse effect after three years. This effect was also noted on the joints made with the polyamide-modified epoxy (#4).

Table 9-9: Percent Retention of Original Adhesive Joint Strength* After Weathering One Year (2024-T3 Aluminum Alloy)[25]

Weathering	Epoxy 350°F (177°C)	Epoxy 250°F (121°C)	Poly-amide Epoxy	Epoxy Anhy-dride	Poly-Epox-ide	Epoxy Powder	Filled Epoxy Nylon	Epoxy Resor-cinol	Epoxy Nylon	Epoxy Phen-olic	Nitrile Phenolic	Silicone (RTV)	Polyure-thene	Poly-ester
Tested at 73°F (23°C)														
Control	92	115	115	91	107	82	99	87	100	89	103	200	168	80
Picatinny	76	84	88	84	76	71	93	80	93	80	90	184	98	95
Panama	60	90	90	34	74	104	96	74	98	76	90	208	24	0
Yuma	91	112	137	90	100	117	96	81	102	78	100	300	183	61
MIL-STD-304	58	89	135	32	83	68	105	92	69	72	92	107	135	47
Tested at 160°F (71°C)														
Control	96	124	113	100	139	72	158	98	115	85	98	230	130	138
Picatinny	88	103	77	71	122	81	82	86	101	82	98	140	57	102
Panama	55	83	76	30	90	117	65	65	95	73	88	220	87	0
Yuma	97	127	100	91	174	99	120	89	90	80	103	220	155	162
MIL-STD-304	62	86	84	46	181	69	90	77	22	72	100	170	130	40
Tested at -65°F (-54°C)														
Control	95	98	123	83	110	68	112	89	74	94	107	171	102	81
Picatinny	67	62	119	90	108	68	92	90	78	83	104	172	45	89
Panama	67	74	114	85	77	84	107	78	119	78	104	145	27	0
Yuma	67	90	130	98	140	71	97	86	85	85	93	172	64	109
MIL-STD-304	64	83	143	93	145	73	110	76	115	82	119	109	119	121

*Based on each testing temperature.

Table 9-10: Percent Retention of Original Adhesive Joint Strength After 3 Years of Weathering (2024-T3 Aluminum Alloy)[26]
(Column headings explained on the following page)

Weathering	1	2	3	4	5	6	7	8	9	10	11	12	13	14	15	16	17
							Tested at 73°F (23°C)										
Control	79	94	108	102	109	100	102	ND	92	93	36	110	81	103	250	177	159
Picatinny	70	68	77	78	58	35	64	63	54	58	65	110	67	95	102	24	22
Panama	22	0	53	39	0	0	0	87	90	64	100	0	68	127	250	165	0
Yuma	90	80	103	91	94	103	102	90	95	88	85	77	77	147	183	70	74
MIL-STD-304																	
Control	96	90	128	115	105	96	105	80	85	104	103	—	89	108	208	180	93
Cycled	58	51	89	135	32	83	0	68	105	92	69	—	72	92	167	135	47
							Tested at 160°F (71°C)										
Control	84	100	107	83	100	256	274	ND	142	94	108	84	80	106	276	195	147
Picatinny	75	75	89	65	65	95	100	61	70	51	97	109	76	107	115	30	70
Panama	39	0	46	43	0	0	0	84	54	59	94	0	70	88	204	71	0
Yuma	103	93	105	85	87	324	154	83	71	84	129	120	85	108	200	88	80
MIL-STD-304																	
Control	93	94	129	125	114	289	89	95	176	93	56	—	95	107	250	133	140
Cycled	62	56	86	84	46	181	0	69	90	77	22	—	72	100	170	130	40
							Tested at -65°F (-54°C)										
Control	79	110	108	118	97	139	97	ND	102	74	81	154	81	128	153	121	92
Picatinny	73	78	76	111	91	52	59	58	71	64	94	83	68	111	31	44	66
Panama	22	0	72	75	0	0	0	69	113	73	98	0	63	121	136	47	0
Yuma	76	73	72	130	111	150	76	70	98	84	95	98	88	137	115	64	86
MIL-STD-304																	
Control	97	88	109	115	118	122	119	73	85	81	91	—	88	106	208	99	85
Cycled	64	63	83	143	93	145	0	73	119	76	115	—	82	119	109	119	120

Note: ND means No Data.

Table 9-10 (continued)

Key to Column Headings

Column Heading	Polymer Type	Trade Name	Cure Temperature °F	°C
1	Epoxy paste, Al-filled*	EC-2086	350	177
2	Epoxy paste, filled	EC-2186	350	177
3	Epoxy paste, filled*	EC-2214	250	121
4	Polyamide-epoxy	Epon-828/V-140	RT	
5	Epoxy anhydride	Epon-31-59	RT	
6	Polysulfide-epoxy	C-14	RT	
7	Modified epoxy	Epon-913	RT	
8	Epoxy powder	Epon-917	300	149
9	Nylon-epoxy/polyamide	N-159	RT	
10	Resorcinol-epoxy/polyamide*	K-159	RT	
11	Nylon-epoxy paste	7133/3170	RT	
12	Nylon-epoxy film	FM-1000	375	190
13	Phenolic-epoxy film**	HT-424	330	166
14	Nitrile-phenolic film**	AF-30	350	177
15	RTV silicone rubber	RTV-102	RT	
16	Polyether-based polyurethane	PR-1538	180	82
17	Styrene-modified polyester	Laminac 4116/4134	RT	

Note: RT is room temperature.

*Adhesive joint retained a minimum of 50% of its original bond strength and approximately 2000 psi shear strength at all test temperatures after 2 years of weathering at all test sites.

**Adhesive joint retained a minimum of 50% of its original bond strength and approximately 2000 psi shear strength at all test temperatures after 3 years of weathering at all test sites.

The joints with the two-part epoxy-anhydride adhesive (#5), the two-part aliphatic amine-cured polysulfide-modified epoxy (#6), and the two-part mixed amine-cured, filled epoxy (#7) fell apart at the Panama site. Numbers 5 and 6 both fell apart during the second year of the program. The #6 joints fell apart during the third year, and also showed a sharp decrease in strength after three years at the Picatinny site. Joints #8, 9, 10 and 11 retained better than 50% of their original joint strength after three years at all test temperatures and sites.

The joints made using the epoxy-nylon film adhesive (#12) fell apart on the racks at Panama due to crevice and exfoliation corrosion. The epoxy-phenolic film adhesive (#13) and the nitrile-phenolic film adhesive (#14) retained better than 50% of their original joint strength with average joint strengths of approximately 3000 psi (13.8 MPa) under all test conditions after exposure to the environments of all the test sites. The joints made with the RTV silicone rubber (#15) showed a general increase in bond strength in the early stages of exposure, probably due to further cure. These joints also demonstrated a general retention of the initial bond strength throughout the three-year period. The joints made with polyurethane (#16) showed signs of degradation at all the outdoor sites. Those made with the polystyrene-modified unsaturated polyester (#17) fell apart in Panama after two years exposure. The panels at the other sites generally retained a fair percentage of their original strength.[22]

In summary, the joints that retained better than 50% of their original bond strengths at all test temperatures after exposure to three years at any of the test sites were #'s, 8, 10, 13 and 14. Of these, only #13 and #14 retained approximately 2000 psi (13.8 MPa) shear strength at all test temperatures after three years at any of the test sites. Five of the joints retained better than 50% of their original bond strength and approximately 2000 psi (13.8 MPa) shear strength at all test temperatures after exposure to MIL-STD-304 cycling. These were #'s 1, 2, 3, 13, and 14. Subjection of bonded panels to the temperature and humidity aging of MIL-STD-304 does, in general, tend to show up those adhesive systems which will not form joints that will give satisfactory performance in highly humid atmospheres.[22]

CHEMICALS AND SOLVENTS

Most organic adhesives tend to be susceptible to chemicals and solvents, especially at elevated temperatures. Among the standard test fluids and immersion conditions (other than water, high humidity and salt spray) are the following:

- 7 days in JP-4 jet engine fuel
- 7 days in anti-icing fluid (isopropyl alcohol)
- 7 days in hydraulic oil (MIL-H-5606)
- 7 days in HC test fluid (70/30 v/v isooctane/toluene)

Unfortunately, exposure tests lasting less than 30 days are not applicable to many service-life requirements. Practically all adhesives are resistant to these fluids over a short time period and at room temperature. Some epoxy adhesives even show an *increase* in strength during aging in fuel or oil.[1] Hysol Division (Dexter Corporation) reported studies on their Aerospace Adhesive EA 929, a fast-curing, one-part thixotropic epoxy paste adhesive. With gasoline at $75°F$ ($24°C$) and gear oil at $250°F$ ($121°C$) there was definite increase in $250°F$ ($121°C$) tensile-shear strength in etched 2024-T3 Al clad cured 20 minutes at $400°F$ ($204°C$). This increase tended to level off after 4–6 months immersion.[23] This effect may be due to postcuring or plasticizing of the epoxy by the oil.[1]

Epoxy adhesives are generally more resistant to a wide variety of liquid environments than other structural adhesives. However, the resistance to a specific environment is greatly dependent on the type of epoxy curing agent used. Aromatic amines, such as m-phenylenediamine, are frequently preferred for long-term chemical resistance.[1]

Urethane adhesives generally show good resistance to most chemicals, solvents, oils and greases.

There is no one adhesive that is optimum for all chemical enviornments. As an example, maximum resistance to bases almost axiomatically means poor resistance to acids. It is relatively easy to find an adhesive that is resistant to one particular chemical environment. Generally, adhesives which are most resistant to high temperature have the best resistance to chemicals and solvents.[1]

The temperature of the immersion medium is a significant factor in the aging properties of adhesives. As the temperature increases, more fluid is generally adsorbed by the adhesive and the degradation rate increases. In summary:[1]

- Chemical resistance tests are not uniform in concentrations, temperature, time or properties measured.

- Generally, chlorinated solvents and ketones are severe environments.

- High-boiling solvents, such as dimethylformamide, dimethyl sulfoxide, and Skydrol (Monsanto Corp.) are severe environments.

- Acetic acid is a severe environment.

- Amine curing agents for epoxies are poor in contact with oxidizing acids.

- Anhydride curing agents are poor in contact with caustics.

ASTM D 896-84, "Standard Test Method for Resistance of Adhesive Bonds to Chemical Reagents" (see Chapter 12) covers the testing of all types of adhesives for resistance to chemical reagents. The standard chemical reagents are those listed in ASTM D 543 and the standard oils and fuels are given in ASTM D 471. Additional supplementary reagents, for which the formulations are given, are: Hydrocarbon Mixture No. 1, Standard Test Fuel No. 2, and Silicone Fluid (Polydimethylsiloxane).

VACUUM

The ability of an adhesive to withstand long periods of exposure to a vacuum is of primary importance for materials used in space travel. The degree of adhesive evaporation is a function of its vapor pressure at a given temperature. Loss of low-molecular-weight constituents such as plasticizers or diluents could result in hardening and porosity of adhesives or sealants. Most structural adhesives are relatively high-molecular-weight polymers, and for this reason exposure to pressures as low as 10^{-9} torr (1.33×10^{-7} Pa) is not harmful. However, high temperatures, nuclear radiation, or other degrading environments may cause the formation of low-molecular-weight fragments which tend to bleed out of the adhesive in a vacuum.[1]

The space vacuum is one of the more important components of the space environment. Although volatility of materials at high vacuum is certainly important, volatility of the polymer is usually not high enough to be significant. Polymers, including adhesives, will not volatilize as a result of vacuum alone. Incomplete polymerization, often the result of poor manufacturing processes not detected by quality-control sytems, frequently results in the presence of residual lower-molecular-weight species, which, in return, are responsible for observed outgassing of polymeric materials. The vacuum is no real problem in itself when the molecular weight of the polymer is reasonably high and the polymer is free of low-molecular-weight components. The effect of vacuum on polymers is not one of evaporation or sublimation, but is a degradation caused by the breaking down of the long-chain polymers into smaller, more volatile fragments. Chain length (molecular weight), extent of branching, and cross-linking have a direct effect upon the rate of decomposition. Polymers which show high decomposition rates in vacuum near room temperature are nylon, polysulfides and neoprene.[24]

Douglas Aircraft reported on a study[25] of outgassing of commercially available (in 1966) structural adhesives, sealants and seal materials at 10^{-9} torr (1.33×10^{-7} Pa). Disks of 1-inch diameter were punched from nineteen adhesives, sealants, and seal materials. The disks were dried in a desiccator over phosphorus pentoxide, weighed on an automatic balance, and placed in a vacuum under the above-mentioned conditions for seven days. At the conclusion of the exposure period each specimen was immediately placed in a desiccator and reweighed for determination of any weight change. After this second weighing the specimens were exposed to the atmosphere for one week and again weighed to determine any additional weight change. Table 9.11 shows a few results of this study and indicates that under *ambient* conditions a high vacuum does not cause significant weight loss in the materials.[25]

Table 9-11: Effect of 10^{-7} Torr (1.33×10^{-5} Pa) on Commercial
Adhesives/Sealants[25]

Adhesive	Type	Weight Change (%)	Moisture Change (%)
Lefkoweld 109	Modified epoxy	−0.03	+0.60
EC2216 B/A	Flexibilized epoxy	−0.06	+0.61
Adiprene L-100 + MOCA	Polyurethane	+0.01	+0.38
PR 1535	Polyurethane	+0.01	+0.44
EC 1605	Polysulfide	−0.23	+0.39

To show the significant effect of temperature on the rate of decomposition and volatilization under vacuum, the same experiment described above was conducted at $225°F$ ($107°C$) on the two polyurethane materials. The Adiprene L-100 and MOCA formulation now showed a weight change of −0.75% and the PR 1535 showed a weight change of −1.45%. The $225°F$ ($107°C$) temperature was considerably higher than intended for urethane formulas available at the time of testing.[25]

RADIATION

High-energy particulate and electromagnetic radiation, including neutron, electron, and gamma radiation, have similar effects on organic adhesives. Radiation causes molecular-chain scission of polymers used in structural adhesives, which results in weakening and embrittlement of the bond. This condition is worsened when the adhesive is simultaneously exposed to elevated temperatures. Figure 9.8 shows the effect of radiation dosage on the tensile-shear strength of structural adhesives (1957 data). Generally, heat-resistant adhesives have been found to resist radiation better than less thermally stable systems. Fibrous reinforcements, fillers, curing agents, and reactive diluents affect the radiation resistance of adhesive systems. In epoxy-based systems, aromatic curing agents offer greater radiation resistance than aliphatic types.[1,26]

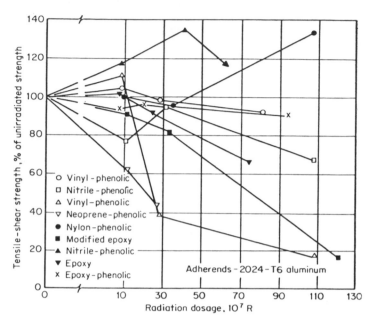

Figure 9-8: Effect of nuclear radiation (gamma rays) dosage on structural adhesives.[26]

ASTM D 1879-70 (1981), "Standard Practice for Exposure of Adhesive Specimens to High-Energy Radiation", is the test method currently in use. This method was originally adopted in 1961. It is described briefly in Chapter 12. Polyester resins and cured anaerobic products have high radiation resistance based on a radiation spectrum for electrical insulation and materials. Anaerobic resins are classed in a radiation-exposure category with all dose rates based on 100 hours up to 1000 Mrad with a dose rate of 10^6 to 10^7 rads/hour, or 10^{11} to 10^{12} neutrons/hour. Thread-locking grades of anaerobic adhesives have sustained 2×10^7 rads without molecular change or loss in locking torque. Anaerobic threaded connections have been exposed to radiation in a reactor for several years with no apparent loss in holding strength.[27]

Adhesives generally react to radiation in much the same manner as the plastics or elastomers from which they are derived. Generally, those containing aromatic compounds show good resistance to radiation. Fillers and reinforcing materials improve the radiation stability of these products substantially, while also improving other properties.[28]

The following conclusions were made by Battelle workers in 1979 on sterilizing radiation effects in polymers that might be used in adhesives:[28]

Polysulfones—can withstand radiation doses greater than 1000 Mrads without significant effect. (General radiation resistance: excellent)

Phenol-formaldehyde or Urea-formaldehyde—these are usually filled or reinforced. The addition of several fillers increases the radiation stability significantly, by as much as 100-fold. Filled resins of this type usually show good radiation resistance up to 500 Mrads or higher (General radiation resistance: good).

Epoxies—these materials are above average in radiation resistance of polymers, although this resistance may be varied somewhat, depending upon the hardeners used. Resins using aromatic curing agents generally are more stable than those using aliphatic hardeners. These polymers are stable to radiation doses above 1000 Mrads. (General radiation resistance: excellent).

Unsaturated polyesters—these thermoset materials have quite good radiation resistance, especially if they contain mineral fillers or glass fibers. They can be expected to withstand greater than 1000 Mrads. (General radiation resistance: good).

Polyimides—these materials are well-known for their very high thermal and radiation resistance. They can be expected to withstand radiation doses of about 1000 Mrads at high temperatures (500°F) (260°C). (General radiation resistance: excellent).

Polyurethanes—properties vary from those of an elastomer to those of hard, rigid cross-linked polymers with mechanical properties showing no reduction after an exposure to 1000 Mrads. (General radiation resistance: excellent).

In 1962 a study was carried out on the effects of gamma radiation on the performance of several structural adhesive bonds.[29] The general conclusions were:

- Nitrile-phenolic adhesives are more resistant to radiation damage than epoxy-based adhesives.

- The peel strength of adhesives deteriorates more rapidly than other properties.

- Thick adhesives layers retain useful strength better than thin glue lines. Ten mils (0.01 inch or 0.25 mm) is recommended as the minimum glue-line thickness where radiation is a factor.

Radiation does not appear to have serious effects on the overlap-shear strength of highly cross-linked adhesives (EC 1469 modified epoxy, AF 31 elastomer-phenolic film, AF 32 elastomer-phenolic film, and EC 1639 modified phenolic). In this study by McCurdy and Rambosek in 1962[29] the adhesives seemed to benefit slightly from the additional cross-linking caused by the low orders of radiation (to 300 Mrads), but eventually began to degrade after 500-600 Mrads. The principal effect noted was embrittlement due to high (800–900 Mrads) amounts of radiation. Very probably there was a loss of cohesive strength due to considerable amount of chain scission, as well as cross-linking. McCurdy and Rambosek also studied the effects of high temperatures (250°F and 300°F) (121°C and 149°C) combined with radiation to 900 Mrads, using the same adhesives listed above, with the exception of the EC 1639 modified phenolic. In these studies the high-temperature performance appeared to fall off in a manner parallel to the room-temperature performance. The modified-epoxy adhesive maintained its properties up to about 600 Mrads and the modified phenolic was relatively unaffected by very high doses of radiation. The elastomer-phenolic films, as might be expected, were more greatly affected by radiation, since they have a greater flexibility and more sites for cross-linking. These films maintained their performance at room temperature up to 400 Mrads, but fell below the old MIL-A-4090D, type II requirements of 2500 psi (17.2 MPa) after about 100 Mrads of radiation.[29]

As long as an adhesive maintains its adhesion to the substrate, along with a certain amount of cohesive strength, its performance is relatively unaffected by radiation. The results obtained with a more complex property such as peel strength are shown in Table 9.12.

Table 9-12: Effect of Gamma Radiation on T-Peel Strength of Elastomer-Phenolic Film Adhesives (Modified after McCurdy and Rambosek[29])

| | . . . . Radiation Dose, Mrads . . . . | | | | | | | | |
| | . . . 0 . . . . | | . . . 100 . . . | | . . .300 . . . | | . . .600 . . . | | . . 900 . . |
Adhesive	lbf/in	N/m	lbf/in	N/m	lbf/in	N/m	lbf/in	N/m	lbf/in	N/m
AF-30	48.5	8490	34	5950	12	2100	5	876	5	876
AF-31*	30	5250	20	3500	8	1400	4.5	788	4	700
AF-32	34	5950	10	1750	4	700	3	525	3	525

*Most rigid.

An important question is, What is the probable radiation dose during exposure to the space environment? In outer space unshielded structures may be exposed to as much as 3500 Mrads per day during a solar storm. Organic structures with minimum shielding are exposed to perhaps 10 to 15 rads per hour. Even at the upper level of 50 rads per hour (Van Allen belt), it would take 2,000,000 hours (229 years) to reach the 100-Mrad dosage level, and most missions should be completed by that time.[29]

Most structural adhesives will perform well under radiation encountered in outer space over any reasonable time period. The rigid metal-to-metal adhesives will perform under fairly high radiation dosages, although it is not recommended that they be exposed directly to the space environment. The other major requirement for radiation resistance of adhesives is for *nuclear reactors* and related equipment with high radiation flux zones.[29]

BIOLOGICAL[30,31]

Adhesives in bonded joints may or may not be attacked and degraded by biological organisms (fungi, bacteria, insects and rodents), depending on how attractive the adhesives are to these organisms. Adhesives based on animal or plant materials (animal and fish glues, starch, dextrins) are much more likely to be affected than synthetic adhesives. Fungi and bacteria are classified as microorganisms, although the former may consist of forms readily visible to the naked eye (mycelia).

Polyurethane resins based on *polyether* polyols are moderately to highly resistant to fungal attack, while all *polyester* urethane resins are highly susceptible to such attack. This susceptibility is related to the number of adjacent unbranched methylene groups in the polymer chain. At least two and preferably three such groups are required for appreciable attack to occur. The presence of side chains on the diol moiety of the polyurethane reduces susceptibility to attack. With the polyethers, attack is dependent on the diol and the diisocyanate used. Adipic acid and diethylene glycol, used in making polyester, are capable of supporting mold growth. It is not always easy to determine whether deterioration of polyurethane is caused by hydrolytic reversion or by fungal attack. However, a magnifying lens can be used to detect channel-like lesions on the surface of the polyurethane (this is more difficult with adhesives), and below the surface, tubular formations branch out in different directions. This network of tunnels is frequently seen to radiate from a single point. The tunnels contain individual mold hyphae or mold strands of varying thicknesses. Over a long period, and more rapidly when exposed to high temperature, humidity, and light, the damaged material softens increasingly. Eventually, after several months, it becomes a gelatinous mixture of degradation products and fungal hyphae. Of the mold species attacking polyester polyurethanes, *Stemphylium* is one of the most active. Biodeterioration can be slowed down by the use of hydrolysis inhibitors (stabilizers) and/or certain fungicides, such as 8-hydroxyquinoline.

A very recent interim report by the Army Natick Research and Development Center (NRDC) describes a program of evaluation of commercial adhesive formulations and bases for microbial susceptibility or resistance. Work will continue on this program on the evaluation of thermoplastic and thermosetting resins and plasticizers used in adhesive formulations. The use of biocides will also be evaluated.[32]

TEST METHODS

ASTM Committee D-14 on Adhesives has five test methods published that are applicable to biological attack, as follows. Brief descriptions of these procedures are given in Chapter 12.

- ASTM D 1382-64 (1981)—Standard Test Method for *Susceptibility of Dry Adhesive Films to Attack by Roaches*

- ASTM D 1383-64 (1981)—Standard Test Method for *Susceptibility of Dry Adhesive Films to Attack by Laboratory Rats*

- ASTM D 1877-77—Standard Test Method for *Permanence of Adhesive-Bonded Joints in Plywood Under Mold Conditions*

- ASTM D 4299-84—Standard Test Methods for *Effect of Bacterial Contamination of Adhesive Preparations and Adhesive Films*

- ASTM D 4300-84—Standard Test Methods for *Effect of Mold Contamination on Permanence of Adhesive Preparations and Adhesive Films*

REFERENCES

1. Petrie, E.M. Chapter 10, Plastics and Elastomers as Adhesives, *Handbook of Plastics and Elastomers*, (C.A. Harper, ed.), McGraw-Hill, NY (1975).
2. DeLollis, N.J., *Adhesives, Adherends, Adhesion*, 2nd. Edition, Robert E. Krieger Publishing Co. Inc., Huntington, NY (1980). (First edition was published in 1970 as *Adhesives for Metals*).
3. Kinloch, A.J., Explosives Research and Development Establishment, UK, *Environmental Failure of Structural Adhesive Joints—A Literature Survey*, ERDL Technical Note No. 95, Unlimited Distribution (AD 784 890).
4. Wangsness, D.A., Sustained Load Durability of Structural Adhesives, *Journal of Applied Polymer Science, Applied Polymer Symposia 32*, 1977, pages 296–300, based on symposium on Durability of Adhesive Bonded Structures held at Picatinny Arsenal (Ocotber 27–29, 1976).
5. Bolger, J.C., "Chapter 1, Structural Adhesives for Metal Bonding, *Treatise on Adhesion and Adhesives*, Vol. 3, (R.L. Patrick, ed.) Marcel Dekker, NY (1973).
6. Union Carbide Corporation, Product Data Bulletin F-43410B 11/77, *Udel Polysulfone for High-Temperature Structural Adhesive Applications* (November 1977).
7. Landrock, A.H. *Properties of Plastics and Related Materials at Cryogenic Temperatures*, Plastics Technical Evaluation Center, PLASTEC Rept. 20, July 1965. (AD 469 126).
8. Roseland, L.M., *Materials for Cryogenic* Usage, Proceedings, 21rst Annual Meeting, Reinforced Plastics Division, SPI, Chicago, Il. Section 4C, (Feb. 8–10, 1966).
9. Kausen, R.C. High and Low-Temperature Adhesives—Where Do We Stand? Proceedings, 7th National SAMPE Symposium on Adhesives and Elastomers for Environmental Extremes, Section 1, Los Angeles, CA. (May 20–22, 1964).
10. Williamson, F.R. and Olien, N.A. Cryogenic Div. Institute for Basic Standards, National Bureau of Standards (NBS), Cryogenic Adhesives and Sealants - Abstracted Publications, NASA 5P-3101, prepared for the Aerospace Safety Research and Data Institute, NASA-Lewis Research Center (1977).
11. Vaughan, R.W. and Sheppard, C.M. TRW Systems Group, *Cryogenic/High Temperature Structural Adhesives*, NASA CR-134465, prepared for NASA on Contract NAS3-10780 (Jan. 1974).
12. Minford, J.D., Comparison of Aluminum Adhesive Joint Durability as Influenced by Etching and Anodizing Treatments of Bonded Surfaces, *Journal of Applied Polymer Science, Applied Polymer Symposia 32*, 1977:91–103, based on symposum on Durability of Adhesive Bonded Structures held at Picatinny Arsenal (October 27–29, 1976).

13. Wegman, R.F. *et al*, A New Technique for Assessing Durability of Structural Adhesives, *Adhesives Age, 21* (7):38:41 (July 1978). Based on a paper presented at the 9th National SAMPE Technical Conference, Atlanta, Ga. (Oct. 4-6, 1977).

14. Wegman, R.F. *et al.*, *Evaluation of the Adhesive Bonded Processes Used in Helicopter Manufacture, Part 1. Durability of Adhesive Bonds Obtained as a Result of Processes Used in the UH-1 Helicopter*, Picatinny Arsenal Technical Report 4186, AD 732 353 (Sept. 1971).

15. Wegman, R.F. et al., *Durability Studies of Adhesive Bonded Metallic Joints*, Picatinny Arsenal Technical Report 4707 (Dec. 1974).

16. Wegman, R.F. et al., *The Effect of Environmental Exposure on the Endurance of Bonded Joints in Army Helicopters*, Picatinny Arsenal Technical Report 4744, (May 1975).

17. Wegman, R.F. et al., Durability of Some Newer Structural Adhesives, *Journal of Applied Polymer Science, Applied Polymer Symposia 32*, 1977:1-10, based on symposium on Durability of Adhesive Bonded Structures held at Picatinny Arsenal (October 27-29, 1976).

18. Minford, J.D., Chapter 2, Durability of Adhesive Bonded Aluminum Joints, *Treatise on Adhesion and Adhesives*, Vo. 3, (R.L. Patrick, ed.), Marcel Dekker, NY (1973).

19. Marceau, J.A. and McMillan, J.C., Boeing Commercial Airplane Co., *Exploratory Development on Durability of Adhesive Bonded Joints*, AFML-TR-76-173 (Oct. 1976). Air Force Contract AF 33615-74-C-5065.

20. Tanner, W.C., Adhesives and Adhesion in Structural Bonding for Military Material, *Applied Polymer Symposia*, No. 3:1-25 (1966). Based on symposium on Structural Adhesive Bonding, held at Picatinny Arsenal (Sept. 14-16, 1965).

21. Wegman, R.F. et al., How Weathering and Aging Affect Bonded Aluminum, *Adhesives Age*, 10(10):22-26 (Oct. 1967).

22. Wegman, R.F. et al., *Effect of Outdoor Aging on Unstressed, Adhesive-Bonded, Aluminum-to-Aluminum Lap-Shear Joints. Three-Year Summary Report*, Picatinny Arsenal Technical Report 3689 (Mary 1968).

23. Hysol Division, Dexter Corporation, *Aerospace Adhesive EA 929*, Bulletin A5-129.

24. Landrock, A.H. *Effects of the Space Environment on Plastics: A Summary with Annotated Bibliography*, PLASTEC Rept. 12, AD 288 682 (July 1962).

25. L.M. Roseland, Structural Adhesives in Space Applications, *Applied Polymer Symposia* No. 3:361-367 (1966), based on symposium on Structural Adhesive Bonding held at Picatinny Arsenal (Sept. 14-16, 1965).

26. Arlook, R.S. and Harvey, D.G., *Effect of Nuclear Radiation on Structural Adhesive Bonds*, Wright Air Development Center Report WADC-TR-46-467, AD 118 063 (Feb. 1957).

27. Pearce, M.B., How to Use Anaerobics Successfully, *Applied Polymer Symposia No. 19*:207-230 (1972), based on symposium on Processing for Adhesives Bonded Structures held at Hoboken, NJ and sponsored by Picatinny Arsenal.

28. Skiens, W.E., Battelle Pacific Northwest Laboratories, Sterilizing Radiation Effects on Selected Polymers, PNL-SA-7640, Conf. 7403108-1. Paper presented at Symposium on Radiation Sterilization of Plastic Medical Products (28 March 1979). Cambridge, MA. Available from NTIS as ADD 432 161.

29. McCurdy, R.M. and Rambosek, G.M. *The Effect of Gamma Radiation on Structural Adhesive Joints*, Preprint Book, SAMPE National Symposium on The Effects of the Space Environment on Materials, held at St. Louis, MO (May 7-9, 1962).
30. Evans, D.M. and Levisohn, I., Biodeterioration of Polyester-Based Polyurethane, *International Biodeterioration Bulletin*, 4(2):89-92 (1968).
31. Darby, R.T. and Kaplan, A.M., Fungal Susceptibility of Polyurethanes, *Applied Microbiology*, *16*(6):900-905 (June 1968).
32. Wiley, B.J. et al., *Microbial Evaluation of Some Adhesive Formulations and Adhesive Bases*, U.S. Army Natick Research and Development Center, Natick, MA Tech. Rept. TR-84/023. Interim Report (Oct. 1982-Sept. 1983).

10

Quality Control

INTRODUCTION

Industrial processing of adhesives has made considerable progress from the crude processes of the past. Unfortunately one of the disadvantages of adhesive bonding as an assembly method is that a bond area cannot be inspected visually. Inspection must be carried out by two methods, *destructive* and *nondestructive*. Destructive inspection may be carried out on process-control test specimens prepared from the same adherend and adhesive materials as the production parts. The process-control specimen, as the name implies, accompanies the production parts throughout the stages of cleaning, assembly and cure. The adhesives and adherends are all assembled at the same time and cured in the same press or autoclave. As an additional control, each part may be designed with an expendable tab as an integral part of the assembly. After the cure, the tab is removed and subjected to the same tests as the control test specimens. The results are checked against the specification requirements and the part is accepted or rejected based on these results. The rejected parts may subsequently be inspected nondestructively for final acceptance or rejection. Final rejection would result in systematic destruction to learn how good or bad the parts really were. In initial production of critical parts, such as primary bonded structures for aircraft, where human lives are dependent on reliability, a sampling and destructive analysis of actual production parts may be included in the test program.[1]

A flow chart of a quality-control system for a major aircraft manufacturer is shown in Figure 10-1. This system is designed to detect substandard bonds before they are shipped, and to recommend methods of correcting the causes. It combines nondestructive Fokker tests of individual joints with rigid controls over process operations, and destructive tests of sample bonded parts and test specimens. The level of quality control applied to a particular bonded assembly depends upon its structural requirements. Critical joints are controlled by high sampling levels for destructive testing, and by tight acceptance requirements. Less critical bonds are controlled by less stringent procedures.[2]

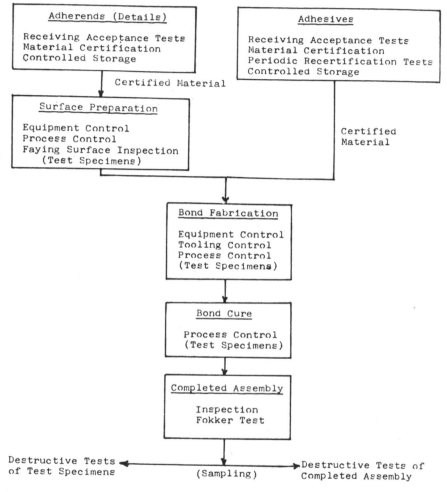

Figure 10-1: Flow chart of a quality-control system for adhesive bonding[2]

The first phase of the quality-control system outlined in Figure 10-1 controls the quality of adhesive material and adherend details making up the joint. Inspections and tests are performed upon incoming materials to assure their meeting acceptance requirements. Shortly before use, *destructive tests* (in which the test specimens are damaged) are conducted on specimens bonded with each batch of adhesive to be used, to insure their capability of developing bond strength after proper cure. The use of carefully controlled storage conditions insures that only certified adherend details and adhesive materials are used in each joint.[2]

Rigid process controls insure that each batch of bonded joints receives proper processing during the surface preparation, fabrication and cure processes. Surface preparation processes (see Chapter 4) are controlled with respect to temperature and composition of baths and immersion time of parts, which is followed by inspection of treated faying surfaces to determine wettability. Fabrication operations are regulated by process controls in conjunction with tests and controls over dimensions, alignment, and pressurization provided by the tooling. Cure conditions are generally controlled by incorporating thermocouples into the bondline to monitor actual cure temperature and time. Although these rigid controls do not completely assure proper processing, there is a high level of assurance that each batch of parts is processed to develop acceptable bond quality in the lot.[2]

Test coupons, or preferably extensions of the actual parts (i.e. tabs), pass through the entire bonding process with the particular lot of assemblies they represent. These specimens are destructively tested in shear, tension or peel (see Chapters 11 and 12 for test methods applicable), and the strength of each joint within the lot is assumed to be that of the accompanying test specimens. Test specimens with substandard bond strength cause rejection of the entire lot. In addition, destructive tests are conducted to qualify the *first* article produced, and subsequently, upon a sampling of assemblies produced from each piece of tooling during the production run, to insure that the process and tooling remain under control. These process-control and sampling methods are capable of detecting discrepancies affecting the entire lot of assemblies, but cannot evaluate factors affecting individual joints or specific areas of a particular joint.[2]

Incorporating the nondestructive Fokker Test method into the inspection and testing system makes it possible to evaluate many of the factors affecting the bond strength of individual joints. The major limitation of existing quality-control systems for adhesive bonding is lack of ability to detect weak bonds caused by local areas of poor adhesion. The major causes of such discrepancies are inadequate surface preparation in particular areas, nonhomogeneous adherend surfaces, or contamination of prepared adherend surfaces or adhesive material during processing. Process controls are incapable of controlling these factors, and existing nondestructive test methods are incapable of detecting weak bonds caused by such discrepancies. The incidence of substandard bonds can be decreased by rigid controls over materials and processes, and by particular care being taken by production personnel. These methods are not capable of providing complete assurance of high-quality bonds, however. The solution to this problem would be the development of a nondestructive test method capable of measuring the properties of adhesive-adherend interfaces and the adhesion of films to adherend surfaces. Until such a nondestructive test method is developed, the present combination of rigid process controls, destructive tests of specimens, and nondestructive tests of each completed joint will remain the most reliable means of assuring the quality of adhesive bonds and bonded structures.[2]

RAW MATERIAL INSPECTION AND PROCESS CONTROL

Incoming Material Control

Quality control begins upon the receipt of raw materials, such as adhesives and catalysts. The purchase order ordinarily defines the required quality properties of this material. This is accomplished by an actual statement of requirements, or by what is called out in the Material Specifications. The inspection requirements are normally specified in the Material Specifications as Quality Acceptance Tests, or as Receiving Inspection Requirements.[3] Military and Federal Specifications on adhesives are listed and described in Chapter 12.

Containers—The first inspection requirement is normally the condition of the container. The following items should be checked when inspecting the container:[3]

Damage. Physical damage to a container of film adhesive can rupture its sealed wrapper, allowing moisture, dirt, etc. to reach and contaminate the adhesive. Damage can render a pail of liquid measure unusable in automatic measuring equipment.

Leakage. Leakage of liquid adhesive components can change the ratio of the catalyst to the base resin if premeasured kits are involved. It can also result in the receipt of less material than the purchaser needs and is paying for.

Identification. Identification of a container should include:

- product number
- manufacturer's name
- date of manufacture
- batch or lot number

- shelf life
- recommended storage conditions
- manufacturer's instructions for use
- safety precautions

Adhesives—Incoming adhesive material control includes two types of tests, *physical properties*, such as percent flow, gel time, and *percent volatiles*, that are of interest to the process engineer in assuring the quality of the bond. An example is the test for percent flow. This test is of value in maintaining the bonding process so the adhesive flow won't be too high, which could cause an adhesive-starved bond. Too little flow, on the other hand, would cause a thick or inadequately filled bond (4). Test methods used for physical properties include the following, which are described in Chapter 12:

ASTM methods for physical properties—D816, D898, D899, D1084, D1337, D1448, D1489, D1490, D1579, D1582, D1583, D1584, D1875, D1916, D2183, D2556, D2979, D3121 and D3236.

Federal Test Method Standard No. 175B for physical properties—Methods 4032.1, 4041.1, 4051.1.

Adhesives—mechanical properties—The mechanical properties of incoming adhesive materials are of interest since they are indicative of the structural results to be obtained in the final bonded assembly. The various tests and requirements for mechanical strength properties of structural adhesives are described in various specifications and test methods described in Chapter 12. The test methods covering mechanical strength properties, including durability, flexibility and fatigue, are as follows:

ASTM methods for mechanical properties—ASTM D897, D903, D905, D906, D950, D1002, D1062, D1144, D1184, D1344, D1781, D1876, D2095, D2182, D2295, D2339, D2557, D2558, D2918, D2919, D3111, D3163, D3164,

D3165, D3166, D3167, D3527, D3528, D3568, D3702, D3807, D3808, D3931, D4027, E229.

Federal Test Method Standard No. 175B for mechanical properties—Method 1081.

Adhesives—Miscellaneous properties (including creep)—ASTM 896, D904, D1146, D1151, D1174, D1183, D1286, D1304, D1382, D1383, D1581, D1713, D1780, D1828, D1877, D1879, D2294, D2739, D3310, D3632, D3929.

Surface Preparation Control

The second step, after determining the quality of incoming materials, is adherend surface preparation. Surface preparation must be carefully controlled for reliable production of adhesive-bonded parts. Adherend surface preparation is discussed in considerable detail in Chapter 4 and in Chapter 7 for (Weldbonding).

If a chemical surface treatment is required, the process must be monitored for proper sequence, bath temperature, solution concentration, and contaminants. If sand- or grit-blasting is employed, the abrasive must be changed regularly. An adequate supply of clean wiping cloths for solvent cleaning is also required. Fresh solvents for cleaning should be on hand. Checks should be made to determine if cloths or solvent containers have become contaminated. The specific surface preparation used can be checked for effectiveness by the water-break-free test described in Chapter 4. After the final treatment step, the substrate surface is checked for its ability to form a continuous film of water when deionized water droplets are applied to the surface. After the surface treatment has been found to be adequate, precautions must be taken to assure that the substrates are kept clean and dry until the bonding operation. The adhesive or primer should be applied to the treated surface as quickly as possible.[5]

Process Control of Bonding

In addition to surface preparation of the adherends described above, production of adhesive-bonded parts involves (1) prefit, (2) adhesive application, (3) assembly, and (4) cure.

Prefit—All detail parts must be *dry-fitted* together to insure a close contact of the faying surfaces. If two or more detail parts do not fit prior to being bonded, they are not likely to fit well enough after being bonded to produce a good joint. If a high production rate exists where a reproducible fit accuracy can be established, the prefit can sometimes be omitted. *First article fits* can be checked by using tool-proofing films which produce an imprint or image of the joint fit. This can greatly reduce the risk factor of poor fit where expensive or critical components are involved. After prefit conditions are verified, each detail part fitted in that assembly should be identified as such to facilitate mating of those specific parts after adhesive application. Process control test panels or excess *tag-end* portions of the assembly should be included with the kit or prefitted details at this point and verified at the time of prefit inspection. These process-control test specimens must be processed through all operations simultaneously with the end product. They should be tested after curing to verify the adhesive batch, surface preparation and other processing conditions used on that end item.[3]

Adhesive application—Most structural film adhesives require a primer. Adhesive primers are usually sray-applied by air or by airless spray systems. Roller or brush application is sometimes used in small areas, or where spray equipment is not available. The primer coat must be air-dried and sometimes over-baked to remove solvents. The thickness of the prime coat will usually affect the adhesive bond strength and must be controlled and verified. This is usually accomplished by periodically certifying the primer applicator, and by monitoring primer thickness after drying.[3]

Film adhesives are applied by removing a paper or plastic separator/protective film and laying the adhesive on the faying surface smoothly, taking care not to allow wrinkles to develop, nor air to become entrapped between the adhesive and the substrate surface. A common workmanship error is failure to remove the separator film before assembly of the detail parts. Some bonders utilize special check-off points to insure its removal. The batch number, lot number, time and date of application, and adhesive type should be logged into the inspection record for traceability should a failure occur. The shop-life expiration date and time should also be logged to aid in controlling assembly and cure of the adhesive.[3]

Assembly—The adhesive-coated detail parts are usually joined in a tool or holding fixture. Cleanliness and proper preparation of the tools should be verified. Time limits on the surface preparation, shop life of the adhesive, and remaining time during which the adhesive must be cured need verification at the point of assembly.[3]

Assembly of detail parts in their proper sequence and fit should be verified. Maintenance of cleanliness and atmospheric control is important. The atmosphere to which the parts and the adhesive are exposed must be controlled from the time the detail parts are prepared for adhesive application until the cure is initiated. The atmosphere is usually controlled by the following steps: (1) keeping the temperature between 65–90°F (18–32°C), (2) keeping relative humidity between 20 and 65%, (3) filtering of all incoming air to preclude airborne contaminants, and (4) maintaining a slight positive pressure differential between the controlled environment area and all surrounding areas. Temperature and humidity indicators of the recording type should be used to verify the conditions.[3]

Curing—Curing an adhesive in any joint is usually a time-temperature-pressure function. No matter how these three variables are controlled, the documentation-verification means are essentially the same. Controlling the length of cure time can be by manual or automatic timing devices. Verification is usually documented on a cure chart taken from a temperature and/or pressure recorder. Recording of pressure and temperature are made in the same manner.[3]

The heat source must be certified for its basic capabilities and uniformity with respect to its intended use. The following factors must be considered: (1) heat-up rate, (2) maximum temperature limits, (3) temperature range or spread during heat-up and at cure temperatures, and (4) cool-down characteristics. The same degree of verification (namely certification) is required for the pressure characteristics of the facility, whether it be an autoclave, a vacuum system, or a press.[3]

Standard test specimen—It is very desirable to fabricate a standard test specimen in the same cycle as the part being bonded. This specimen should be designed for a test method that is indicative of the prime structural loading requirement. For example, if the critical item is normally loaded in tensile shear, the specimen should be of the lap-shear type.

Final Inspection

After the adhesive is cured, the joint area can be inspected to detect gross flaws or defects. This inspection procedure can be either destructive or nondestructive, as discussed in the introduction above. *Destructive testing* generally involves placing samples of the production run in simulated or accelerated service and determining if it has properties similar to a specimen that is known to have a good bond and adequate service performance. The causes and remedies for a number of faults revealed by such mechanical tests are described in Table 10-1. Most of the destructive (mechanical) tests that can be carried out on adhesive bonds are listed above (ASTM methods for mechanical properties). *Nondestructive tests* are far more economical, and every assembly can be tested, if desirable.[5]

Nondestructive tests—*Visual inspection*—Visual inspection, with the help of a strong light, can be used to detect gross flaws and defects. Table 10-2 lists the characteristics of faulty joints that can be detected visually. The most difficult to detect by any means are those defects related to improper curing and surface treatments. For this reason, great care and control must be given to surface preparation procedures and shop cleanliness.[5]

Sonic methods—

- *Tap test.* In this method a coin is used as a special tapping hammer. Tone differences indicate inconsistencies in the bonded joint. Sharp, clear tones indicate that the adhesive is present and adhering to the substrate to some degree. Dull, hollow tones indicate a void or unattached area. Some improvement in the tap test can be achieved by using a solenoid-operated hammer with a microphone pickup. The resulting electrical signal can be analyzed on the basis of amplitude and frequency.[3,5]

- *Sonic resonator.* This method uses a vibrating crystal to excite a structure acoustically at sonic frequencies (5-28 KHz). The elastic properties of the structure are changed by unbonds or other structural defects. Resulting changes in the crystal loading are processed electronically to obtain an electrical signal for display or recording. The technique can be used to test bonded honeycomb structure without regard to the material of either the facing sheet or the honeycomb core. The method requires comparison standards and a liquid for coupling the probe to the specimen. The apparatus used is capable of detecting unbonds, crushed core, and water content.[6]

- *Eddy-sonic test method.* This method is based on the principle that a mechanical force is inherently associated with flow of eddy currents. Since the eddy current field is time-variant, the mechanical force is also time-variant. Therefore, an acoustic vibration can be induced in the proper sample. To use this principle in nondestructive testing of honeycomb materials, some constitutents

Table 10-1: Faults in Adhesive-Bonded Joints Revealed by Mechanical Tests[5]

Fault	Cause	Remedy
Thick, uneven glue line	Clamping pressure too low	Increase pressure, check that clamps are seating properly
	No follow-up pressure	Modify clamps or check for freedom of moving parts
	Curing temperature too low	Use higher curing temperature; check that temperature is above the minimum specified throughout the curing cycle
	Adhesive exceeded its shelf life, resulting in increased viscosity	Use fresh adhesive
Adhesive residue has spongy appearance or contains bubbles	Excess air stirred into adhesive	Vacuum-degas adhesive before application
	Solvents not completely dried out before bonding	Increase drying time or temperature. Make sure drying area is properly ventilated
	Adhesive material contains volatile constituent	Seek advice from manufacturer
	A low-boiling constituent boiled away	Curing temperature is too high
Voids in bond (i.e., areas that are not bonded), clean bare metal exposed, adhesive failure at interface	Joint surfaces not properly treated	Check treating procedure; use clean solvents and wiping rags. Wiping rags must not be made from synthetic fiber. Make sure cleaned parts are not touched before bonding. Cover stored parts to prevent dust from settling on them.
	Resin may be contaminated	Replace resin. Check solids content. Clean resin tank
	Uneven clamping pressure	Check clamps for distortion
	Substrates distorted	Check for distortion; correct or discard distorted components. If distorted components must be used, try adhesive with better gap-filling ability
Adhesive can be softened by heating or wiping with solvent	Adhesive not properly cured	Use higher curing temperature or extend curing time. Temperature and time must be above the minimum specified throughout the curing cycle. Check mixing ratios and thoroughness of mixing. Large parts act as a heat sink, necessitating larger cure times

Table 10-2: Faults in Adhesive-Bonded Joints Revealed by Visual Inspection[5]

Fault	Cause	Remedy
No appearance of adhesive around edges of joint, or adhesive bond line too thick	Clamping pressure too low	Increase pressure. Check that clamps are seating properly
	Starved joint	Apply more adhesive
	Curing temperature too low	Use higher curing temperature. Check that temperature is above the minimum specified
Adhesive bond line too thin	Clamping pressure too high	Lessen pressure
	Curing temperature too high	Use lower curing temperature
	Starved joint	Apply more adhesive
Adhesive flash breaks easily away from substrate	Improper surface treatment	Check treating procedure; use clean solvents and wiping rags. Make sure cleaned parts are not touched before bonding
Adhesive flash is excessively porous	Excess air stirred into adhesive	Vacuum-degas adhesive before application
	Solvent not completely dried out before bonding	Increase drying time or temperature
	Adhesive material contains volatile constituent	Seek advice from manufacturers
Adhesive flash can be softened by heating or wiping solvent	Adhesive not properly cured	Use higher curing temperature or extend curing time. Temperature and time must be above minimum specified. Check mixing

must be electrically conductive. The major advantage of the method is that no liquid energy coupliant is needed, since air serves as a satisfactory coupling medium. The eddy-sonic method is useful for detecting both near-side and far-side unbonds in thin honeycomb structures. It can also be used to detect crushed core, fractured core, and voids in the adhesive.

• *Pulsed eddy-sonic test method/Shurtronic harmonic bond tester.* This method can detect both near-side and farside unbonds in many types of honeycomb and laminar structures. It can also detect crushed core, fractured core, and excessive build-up in repaired structures. At least one of the surfaces must be electrically conductive to some extent.[6]

• *Arvin acoustic analysis system.* This is an indicator system which produces and detects acoustical vibrations in metal surfaces. It is useful for bond inspection of aluminum honeycomb materials. No acoustic coupling is required.

Ultrasonic methods—These methods are based on the response of the bonded joint to loading by low-power ultrasonic energy.[1] Ultrasonic methods are especially useful in detecting unbonds of the following types:[6]

(1)   Unbonds between the facing sheet to adhesive interfaces in honeycomb structures

(2)   Unbonds between the adhesive-to-core interfaces in honeycomb structures

(3)   Unbonds between adherends in adhesive-bonded laminate structures

• *Ultrasonic pulse echo and contact impedance testing. The contact impedance technique* is based on the fact that when a vibrating crystal is placed in a composite structure, the characteristic impedance or elastic properties of the structure determine the manner in which it is loaded. Changes in loading will change the amplitude or phase of the crystal with respect to the applied voltage. These changes can be indicated by a suitable meter readout, or can be displayed on a cathode-ray tube. The *pulse-echo technique* can be evaluated by observing energy reflection from defects and from the back surface of the structure being inspected. Both these methods are useful in detecting unbonds in honeycomb and laminar structures. They are also capable of detecting crushed core, fractured core, and adhesive build-up in repair areas. Unfortunately the response of these methods to a completely unbonded area in a honeycomb panel is difficult to differentiate from some other anomoly. Water, for example, shows the same response as an unbond.[6]

• *Sweep-frequency resonance method.* This method, of which the well-known Fokker Bond Tester is an example, has the advantage of producing a quantitative estimate of bond strength in metal-to-metal and metal-to-core structures, as well as similar structures made from nonmetallic materials. The energy introduced into the structure is varied over a wide temperature range; the resonance set up by the probe, face sheet, adhesive, and the remainder of the structure is observed.[3] The Fokker Bond Tester is believed to be the only system currently capable of providing an accurate direct *quantitative* reading of the quality of an adhesive bond. The principle is as follows: when a crystal resonating at its natural frequency is placed on a composite structure, the characteristic impedance or elastic properties of the structure determine the manner in which it is loaded. Changes in loading are shown by the combination of the two instrument readings—*resonance frequency shift* and a *change in amplitude of the resonant frequency.* Such a change is indicated by a meter readout and displayed on a cathode-ray tube. A light oil is used as a coupliant. The Fokker Bond Tester has been used successfully in determining near-side unbonds in a wide variety of adhesively bonded structures. It does not give good results in detecting unbonds in honeycomb panels with laminated facing sheets. In addition to unbonds, it is capable of detecting crushed core, fractured core, and voids in the adhesives.[6] The Fokker Bond Tester is most sensitive to properties which physically affect adhesion, such as voids, porosity, and incomplete wetting. It is not capable of detecting incomplete cure, poor surface preparation, or contamination of the interface.[5]

Liquid crystals—Cholesteric liquid crystals are compounds that go through a transition phase in which they flow like a liquid, yet retain much of the molecular order of a crystalline solid. Liquid crystals are able to reflect iridescent colors, depending on the temperature of their environment. Because of this property they may be applied to the surfaces of bonded assemblies and used to project a visual color picture of minute thermal gradients associated with bond discontinuities. Cholesteric crystals are potentially a simple, reliable and economical method for evaluating bond defects in metallic composite structures.[6] Materials with poor heat-transfer properties are difficult to test by this method. The joint must also be accessible from both sides.[5]

Holography—Holography is a method of producing photographic images of flaws and voids by using coherent light such as that produced by a laser. The major advantages of this method is that it photographs successive "slices" through the scene volume, thereby making it possible to reconstruct a three-dimensional image of a defect or void.[5] It is possible, using stored-beam holographic techniques, to make real-time differential interferometric measurements to a precision of the order of one millionth of a centimeter on ordinary surfaces. A simple method of inspecting bonded panels is to place them horizontally and to apply a thin layer of sand in the top surface. Upon vibration of the panel any unbonded areas will be revealed by the pattern resulting from the movement of the sand particles. Bond quality can also be determined by making circular cuts through one adherend down to the bondline in a zone where the strength of the assembly will not be affected. The disks are then pried out to expose the adhesive to visual inspection. Plugs may be inserted later in the cutouts. In certain cases, test specimens are treated and bonded simultaneously with production parts under identical conditions. These specimens are then tested for strength.[7]

Holographic techniques are very useful in their ability to measure differential displacements. This property of holography makes it a useful tool for detecting nonbonds in laminar structures. If these structures are stressed by any of several means (heat, differential pressure, or mechanical), then the displacement of the surface can be related to the integrity of the bonded layers beneath the surface. A laminar material which is well bonded will have a uniform surface displacement which is a function of the physical properties of the material, the means of stressing, and the holographic technique. If the material has a non-bonded region somewhere in the different layers, then the surface above that region will displace in a different manner than the rest of the surface, due to the change in the boundary conditions. This change in the surface displacement is a microscopic differential change, and would not normally be visible. Because of holography's remarkable sensitivity, such microscopic changes are clearly visible. The means of acquiring the surface displacements and thereby the integrity of the bond is called *double exposure holography*.[8]

Thermal image inspection—In this method bond discontinuities are revealed through temperature differences on the assembly surface. Ultraviolet radiation is used to permit direct visual detection of these discontinuities as dark regions in an otherwise bright (fluorescent) surface. For practical purposes, to preclude thermal damages to the adhesive and/or heat-sensitive adherends, a *phosphor* is used that shows a large change in brightness with a small change in temperature in the near-room-temperature range (25–65°C). The coatings used for this purpose provide a stable (non-settling) suspension of the phosphor in the vehicle that can be applied by conventional paint spray equipment.[7]

Thermal infrared inspection (TIRI)—This technique has been used to detect internal voids and unbonded areas in solid-propellant rocket engines and in large, panel-shaped components.[7] The technique uses a dynamic heating principle with continuous injection of thermal energy from an induction generator into the exposed surface of the specimen along a line of scan. Continuous radiometric detection of the emission from contiguous surface regions along the line of scan, after a fixed time interval following heat injection, results in temperature gradients at the outer surface. The depth of the flaw below the surface of the material (interface level) is determined by comparison of inspection recordings taken at pre-established exposure times. Destructive sectioning of representative specimens following TIRI examination shows a correlation of 95% for first and second interface defects.[9]

Radiography—Radiographic inspection techniques have been used successfully for detecting defects in adhesive-bonded metal-to-metal joints and metal sandwich structures. In the case of metal-to-metal joints, the adhesive must contain some metal powder or other suitable radio-opaque filler to create sufficient contrasts to show up defects. The same procedure can be used with nonmetallic adhesive-bonded joints. An experienced inspector, using radiography, can often detect undesirable concentrations of adhesive, or evaluate the quality of adhesive-bonded structures. Damage can occur in handling, or may be the result of unequal pressure during the bonding cycle.[3] Radiography will not detect lack of bond areas where the adhesive is present, but not bonded to one or both adherends.[6]

• *X-ray techniques*—These methods may be used only if fillers are added to the adhesive. This results from the slight weakening of the rays in penetrating through the unfilled adhesive because of its low density. Excellent results may be obtained by adding lead oxide. In this case it is possible to detect even the smallest air and gas bubbles. Conventional X-ray equipment for flaw detection is used. Because of the thinness of the adhesive layer, very long rays must be used.[7]

• *Radioisotope methods*—For inspecting the toughness of combined bonded and spot-welded joints radioactive isotopes may be used to check the possibility of electrolyte penetrating to the bondline during subsequent anodizing in sulfuric acid. A radioactive sodium isotope, such as sodium-22, with a half-life of 2.5 years, is introduced into the most active electrolyte. If there are voids in the adhesive layer the electrolyte penetrates into the clearance between the adherends. The joint is then washed clean and examined with a radiometer. If voids are present radioactive substance is retained in them and the radiation intensity is higher. The application of this method in industry is limited, however, because of the danger from radiation.[7]

• *Neutron radiography.* If the adhesive used is not X-ray opaque, neutron radiography may be used. The hydrogen atoms in the adhesive absorb neutrons, making the adhesive radio-opaque.[4] The neutron radiographic technique detects within adhesive bondlines and predicts the lap-shear strength, usually within 5 to 10%. A portable system permits the method to be applied to aircraft with adhesively bonded parts. Although ultrasonic and X-ray techniques can determine void content and joint strength, neutron radiography appears to have more sensitivity. In addition, it seems to be more nearly independent of metal thickness than X-ray and less dependent on scattering and geometric complexity than the ultrasonic method.[9]

Penetrant inspection. This method is used for local examination of sections of seam joints. The surface of the specimens must first be cleaned and degreased. Then a penetrant solution is applied along the joint. Capillary action pulls the solution into any defect open to the surface. The penetrant on the surface is rinsed with a solvent, leaving the penetrant in the defects. A developer is then applied to draw back the penetrant to the surface. Because the penetrants are brilliantly colored, each defect is easy to see.[7]

WELDBONDING

The extent of process and quality control used in weldbonding (Chapter 7) must be based on the end use of the hardware being bonded. Methods in current use should be selected to fit a specific application. Consistent joint strength can be assured by evaluating cured weldbond tensile-shear specimens, cleaned with each batch of parts, for strength and bond quality. Consistent weld quality can be assured by hourly evaluation of uncured tensile shear and macro specimens for strength and weld quality. Even higher assurance can be obtained from the use of an in-process weld monitor that will detect unacceptable welds. The extent of the inspection of the production item must be determined from or based on, the end use of the part, and will also be affected by the size and complexity of the part. If the parts are small, visual inspections for surface defects and surface adhesive irregularities may be adequate. If the parts are large and complex and the end use is critical, radiographic inspections may be used for determination of weld quality and ultrasonic inspection for determination of bond quality.[10] X-ray radiography will reveal weld nugget defects such as cracking, expulsion and porosity.[11] Infrared nondestructive methods cannot be used for weldbonded structural assemblies.[12] Process specifications prepared by leading aerospace corporations under government contracts have sections covering quality assurance provisions and should be consulted.[12-14]

REFERENCES

1. De Lollis, N.J., *Adhesives, Adherends, Adhesion,* Robert E. Krieger Publishing Company, Huntington, N.Y. (1980).
2. Smith, D.F. and Cagle, C.V., A Quality-Control System for Adhesive Bonding Utilizing Ultrasonic Testing, *Applied Polymer Symposia No. 3,* pp. 411–434, (M.J. Bodnar, ed.) (1966).
3. Society of Manufacturing Engineers, *Adhesives in Modern Manufacturing,* (E.J. Bruno, ed.) (1970).
4. Bandaruk, W., Process Control Considerations for Adhesive Bonding in Production, *Journal of Applied Polymer Science,* 6(20:217–220 (March/April 1962).
5. Petrie, E.M., Chapter 10, Plastics and Elastomers as Adhesives, *Handbook of Plastics and Elastomers,* (C.A. Harper, ed.) McGraw-Hill, N.Y. (1975).
6. Kraska, I.R. and Kamm, H.W., General American Transportation Corporation, *Evaluation of Sonic Methods for Inspecting Adhesive Bonded Honeycomb Structures,* AFML-TR-69-283, AD 876 977 (August 1970).

7. Semerdjiev, S., *Metal-to-Metal Adhesive Bonding*, Business Books Ltd., London, (1970).

8. Barbarisi, M.J. et al, *Evaluation of the Adhesive Bonding Process Used in Helicopter Manufacture — Part 4. Nondestructive Inspection of Adhesive Bonds Using Holographic Techniques*, Picatinny Arsenal Technical Report 4419, AD 765 455 (Oct. 1972).

9. Yettito, P.R., A Thermal, Infrared Inspection Technique for Bond-Flaw Inspections, *Applied Polymer Symposia No. 3*, pp. 435–454 (1966) (M.J. Bodnar, ed.).

10. Beemer, R.D., Introduction to Weld Bonding, *SAMPE Quarterly*, 5(1):37-41 (October 1973).

11. Wu, K.C. and Bowen, B.B., Northrop Corp., Aircraft Div., *Advanced Aluminum Weldbond Manufacturing Methods*, AFML-TR-76-131, Sept. 1976. Final Report, June 1975–March 1976. AF Contract F 33615-71-C-5083. AD CAB 01685. Also in Preprint Book, *22nd National SAMPE Symposium, Vol. 22*, San Diego, Cal., pp. 536–554 (April 26-28, 1972).

12. Grosko, J.J. and Kizer, J.A., Lockheed-Georgia Co., Weldbond Flight Component Design/Manufacturing Program, AFML-TR-74-179, AFFDL-TR-74-106, Dec. 1974. (AF Contract F33615-71-C-1716 AD BOO2822L. Final Report (July 15, 1971–July 15, 1974).

11

Test Methods and Practices

INTRODUCTION

Adhesive tests are used for a variety of reasons. Some of these are:[1]

- Comparison of properties (tensile, shear, peel, flexural, impact and cleavage strength; durability, fatigue, environmental resistance, conductivity, etc.).

- Quality checks for a "batch" of adhesives to determine whether the adhesives are still up to standard.

- Checking the effectiveness of surface and/or other preparation.

- Determination of parameters useful in predicting performance (cure conditions, drying conditions, bondline thickness, etc.).

Testing is important in all aspects of materials science and engineering, but it is especially so in adhesives. Such tests evaluate not only the inherent strength of the adhesive, but also the bonding technique, surface cleanliness, effectiveness of surface treatments, etchings of surfaces, application and coverage of the adhesive, and the curing cycle.

This chapter will first discuss in a general manner the various types of testing carried out in adhesive joints. Only the more important types will be covered. Following this discussion a tabulation will be listed of 56 subject areas, with all relevant ASTM methods and practices and SAE Aerospace Recommended Practices (ARP's). Chapter 12 following will provide fairly detailed discussions of the contents of each method or practice, with two exceptions.

TENSILE

Pure tensile tests are those in which the load is applied normal to the plane of the bondline and in line with the center of the bond areas (Figure 3.1b). ASTM D 897 (see below for this and all standards discussed) is one of the oldest methods still in the ASTM book on adhesives. The specimens and grips called for require considerable machining and, because of the design, tend to develop edge stresses during the test. Because of these limitations, D 897 is being replaced by D 2095 on rod and bar specimens. These specimens, prepared according to ASTM D 2094, are simpler to align and, when correctly prepared and tested, more properly measure tensile adhesion.[2]

Tensile tests are among the most common tests used for evaluating adhesives, despite the fact that, where possible to use joint designs that load the adhesive in other than a tensile mode, the experienced designer generally does so (see Chapter 3). Most structural materials have high tensile strengths when compared to the tensile strengths of structural adhesives. One of the advantages of the tensile test is that it yields fundamental and uncomplicated tensile strain, modulus, and strength data.[1]

Williams and his associates at the California Institute of Technology[3] have analyzed the stress distribution in the tensile test ("poker chip" test) and found that the stresses were not uniformly distributed throughout the specimen, except when the modulus of the adhesive matches that of the adherend. This difference in moduli results in shear stresses being transmitted across the interface.

SHEAR

Pure shear stresses are those which are imposed parallel to the bond and in its plane. Single-lap shear specimen are illustrated in Figures 3.1c, 3.3 and 3.4. These specimens do not represent pure shear, but are practical and relatively simple to prepare. They also provide reproducible, usable results. The preparation of this specimen and method of testing are described fully in ASTM D 1002. Two types of panels for preparing multiple specimens are described.[2]

Shear tests are very common (note 18 tests listed below) because samples are simple to construct and closely duplicate the geometry and service conditions for many structural adhesives. As with tensile tests, the stress distribution is not uniform (see Figure 3.3) and, while it is often conventional to give the failure shear stress as the load divided by the bonding area, the maximum stress at the bondline may be considerably higher than the average stress. The stress in the adhesive may also differ from pure shear. Depending on such factors as adhesive thickness and adherend stiffness, the failure of the adhesive "shear" joint can be dominated by either shear or tension.[1]

Methods other than ASTM D 1002 are in use. ASTM D 3163 describes an almost identical test configuration, except for thickness. This method helps alleviate the problem of adhesive extruding out from the edges of the sample. ASTM D 3165 describes how a specimen can be prepared to determine the strength properties of adhesion in shear by tension loading of laminated assemblies. The double-lap shear test (see Figure 3.2 "Double strap" joint) offers the advantage of reducing the cleavage and peel stresses found in the single-lap shear test.[1]

Compression shear tests are also commonly used. ASTM D 2182 describes sample geometry similar to the lap-shear specimen and the compression-shear-test apparatus. ASTM D 905 describes a test for determining the shear properties of wood (hard maple, etc.). ASTM E 229 determines the shear strength and shear modulus by torsional loading. With proper sample construction and alignment the adhesive in E 229 is subjected to a more homogeneous stress distribution in this configuration than with lap-shear specimens.[1]

PEEL

Peel tests, intended to be used with flexible adhesives, are designed to measure the resistance to highly localized stresses (see Figures 3.1d and 3.16). Peel forces are therefore considered as being applied to linear fronts. The more flexible the adherend and the higher the adhesive modulus, the more nearly the stressed area is reduced to linearity. The stress then approaches infinity. Since the area over which the stress is applied is dependent on the thickness and modulus of the adherend and the adhesive, and is therefore very difficult to evaluate exactly, the applied stress and failing stress are reported as linear values, i.e. pounds per linear inch (pli). Probably the most widely used peel test for thin-gauge metal adherends is the T-peel test (ASTM D 1876). In this test all of the applied load is transmitted to the bond. This type of peel thus tends to provide the lowest values of any peel test.[2]

With elastomeric adhesives peel strength is dependent on bond thickness. The elongation characteristics of these adhesives permit a greater area of the bond to absorb the applied load as the bond thickness increases. The T-peel test is probably the most widely used peel test since it uses only one thickness of metal. The Bell peel test is designed to be peeled at a constant radius around a 1-inch steel roll and, for this reason, provides more reproducible results. ASTM D 1781 uses a metal-to-metal climbing drum in an attempt to achieve this same constant peel radius by peeling around a 4-inch diameter rotating drum. While the fixtures used with the Bell and drum-peel tests help stabilize the angle of peel, the ideal of a fixed radius of peel is not achieved because the high modulus of the metal tends to resist close conformation to the steel roll or drum. In both methods considerable energy is used in deforming the metal so that they provide higher peel values for a given adhesive than the T-peel method.[2]

ASTM D 3167 is a test for determining the floating-roller peel resistance of adhesives. The specimens for this test are made by bonding a flexible material to a comparatively rigid one. The method is of particular value for acceptance and process control testing. It may be used as an alternative to ASTM D 1781 (Climbing Drum Test). This method should be considered more severe, since the angle of peel is greater.

ASTM D 903 uses a 180° peel to determine the peel or stripping strength. In this method one of the adherends must be flexible enough so that it can essentially fold back on itself, somewhat similar to Figure 1.1d.

CLEAVAGE

Cleavage is a variation of peel in which the two adherends are rigid. The load is applied normal to the bond area at one end of the specimen. Figure 1.1e illustrates the type of load found in cleavage.[2] ASTM D 3807 describes how to measure "cleavage peel" of adhesives used with engineering plastics.

CREEP

Often when a bonded structure is subject to a permanent load in service, especially in the presence of vibration, the resistance of the adhesive to creep is important. Two ASTM methods are used to measure creep, ASTM D 2293, involving compression loading, and ASTM D 2294, involving tensile loading. ASTM D 1780 is a standard practice on conducting creep tests.[2]

FATIGUE

While static strength tests are useful in screening and selecting adhesives for most bonding applications, they do not cover the rigorous conditions of intermittently applied stress, or fatigue. The test used is ASTM D 3166. Although intended for metal/metal joints, the test can be used for plastic adherends. The single-lap shear specimen of ASTM D 1002 is used. The specimen is tested on a special tensile-testing machine capable of imposing a cyclic or sinusoidal stress on it. Ordinarily the test is carried out at 1800 cycles/minute. The number of cycles to failure at a given level is recorded and a so-called S-N curve constructed.

IMPACT

Impact tests measure the ability of an adhesive to attenuate or absorb forces applied in a very short time interval. Essentially, these tests measure the rate sensitivity of an adhesive to an applied load. ASTM D 950 describes a pendulum method for applying an impact load to a shear specimen. The results are reported as foot-pounds of energy absorbed in failing the bond of a 1-square inch specimen. Some machines use gravity to accelerate the given load that strikes the test specimen. A variation of the gravity-impact methods uses a series of weights dropped on the test specimen. The failing load, in this case, is the weight multiplied by the distance dropped. Other more sophisticated apparatus uses compressed air to decrease the time of load application to as little as 10^{-5} second.[2]

DURABILITY

A number of ASTM tests and practices involve durability, but one of the most important is the Wedge Test, ASTM D 3762. In this method a wedge is forced into the bondline of a flat-bonded aluminum specimen, thereby creating a tensile stress in the region of the resultant crack tip. The stressed specimen is exposed to an aqueous environment at an elevated temperature, or to any other desired environment. The resultant crack growth with time and failure modes is then evaluated. The test is primarily qualitative, but it is discriminatory in determining variations in adherend surface preparation parameters and adhesive environmental durability.

COMPILATION OF TEST METHODS AND PRACTICES

(see Chapter 12 for detailed descriptions)

Aging

ASTM D 1183-70 (1981)—Standard Test Methods for *Resistance of Adhesives to Cyclic Aging Conditions.*

ASTM D 1581-60 (1984)—Standard Test Method for *Bonding Permanency of Water-or Solvent-Soluble Liquid Adhesives for Labeling Glass Bottles*

ASTM D 1713-65 (1981)—Standard Test Method for *Bonding Permanency of Water-or Solvent-Soluble Liquid Adhesives for Automatic Machine Sealing Top Flaps of Fiberboard Specimens*

ASTM D 1877-77 (1982)—Standard Test Method for *Permanence of Adhesive-Bonded Joints in Plywood Under Mold Conditions*

ASTM D 3632-77 (1982)—Standard Practice for *Accelerated Aging of Adhesive Joints by the Oxygen-Pressure Method*

Amylaceous Matter

ASTM D 1488-60 (1981)—Standard Test Method for *Amylaceous Matter in Adhesives*

Ash Content

Federal Test Method Std. 175B, Method 4032.1—*Ash Content of Adhesives*

Biodeterioration

ASTM D 1382-64 (1981)—Standard Test Method for *Susceptibility of Dry Adhesive Film to Attack by Roaches*

ASTM D 1383-64 (1981)—Standard Test Method for *Susceptibility of Dry Adhesive Film to Attack by Laboratory Rats*

ASTM D 1877-77—Standard Test Method for *Permanence of Adhesive-Bonded Joints in Plywood Under Mold Conditions*

ASTM D 4299-84—Standard Test Methods for *Effect of Bacterial Contamination of Adhesive Preparations and Adhesive Films*

ASTM D 4300-84—Standard Test Methods for *Effect of Mold Contamination on Permanence of Adhesive Preparation and Adhesive Films*

Blocking Point

ASTM D 1146-53 (1981)—Standard Test Method for *Blocking Point of Potentially Adhesive Layers*

Bonding Permanency

(see Aging)

Characterization

ARP 1610—*Physico-Chemical Characterization Techniques, Epoxy Adhesive and Prepreg Resin System*

Chemical Reagents

ASTM D 896-84—Standard Test Method for *Resistance of Adhesive Bonds to Chemical Reagents*

Cleavage

ASTM D 1062-78 (1983)—Standard Test Methods for *Cleavage Strength of Metal-to-Metal Adhesive Bonds*

Cleavage/Peel Strength

(See also Peel strength)

ASTM D 3807-79—Standard Test Methods for *Strength Properties of Adhesives in Cleavage Peel by Tension Loading (Engineering Plastics-to-Engineering Plastics)*

Corrosivity

ASTM D 3310-74 (1983)—Standard Recommended Practice for *Determining Corrosivity of Adhesive Materials*

Creep

ASTM D 1780-72 (1983)—Standard Recommended Practice for *Conducting Creep Tests of Metal-to-Metal Adhesives*

ASTM D 2293-69 (1980)—Standard Test Method for *Creep Properties of Adhesives in Shear by Compression Loading (Metal-to-Metal)*

ASTM D 2294-69 (1980)—Standard Test Method for *Creep Properties of Adhesives in Shear by Tension Loading*

Cryogenic Temperatures

ASTM D 2557-72 (1983)—Standard Test Method for *Strength Proper-*

ties of Adhesives in Shear by Tension Loading in the Temperature Range From −267.8 to −55°C (−450 to −67°F)

Density

ASTM D 1875-69 (1980)—Standard Test Method for *Density of Adhesives in Fluid Form*

Durability (including Weathering)

ASTM D 1151-84—Standard Test Method for *Effect of Moisture and Temperature on Adhesive Bonds*

ASTM D 1828-70 (1981)—Standard Practice for *Atmospheric Exposure of Adhesive-Bonded Joints and Structures*

ASTM D 2918-71 (1981)—Standard Practice for *Determining Durability of Adhesive Joints Stressed in Peel*

ASTM D 2919-71 (1981)—Standard Practice for *Determining Durability of Adhesive Joints Stressed in Shear by Tension Loading*

See also Wedge Test

Electrical Properties

ASTM D 1304-69 (1983)—Standard Methods of Testing *Adhesives Relative to Their Use as Electrical Insulation*

Electrolytic Corrosion

ASTM D 3482-76 (1981)—Standard Practice for *Determining Electrolytic Corrosion of Copper by Adhesives*

Fatigue

ASTM D 3166-73 (1979)—Standard Test Method for *Fatigue Properties of Adhesives in Shear by Tension Loading (Metal/Metal)*

Filler Content

ASTM D 1579-60 (1981)—Standard Test Method for *Filler Content of Phenol, Resorcinol, and Melamine Adhesives*

Flexibility

(See Flexural strength)

Flexural Strength

ASTM D 1184-69 (1980)—Standard Test Method for *Flexural Strength of Adhesive Bonded Laminated Assemblies*

ASTM D 3111-76 (1982)—Standard Practice for *Flexibility Determination of Hot Melt Adhesives by Mandrel Bend Test Method*

Federal Test Method Std. 175B, Method 1081—*Flexibility of Adhesives*

Flow Properties

ASTM D 2183-69 (1982)--Standard Test Methods for *Flow Properties of Adhesives*

Fracture Strength in Cleavage

ASTM D 3433-75 (1980)—Standard Practice for *Fracture Strength in Cleavage of Adhesives in Bonded Joints*

Gap-filling Adhesive Bonds

ASTM D 3931-80—Standard Practice for Determining *Strength of Gap-Filling Adhesive Bonds in Shear by Compression Loading*

Grit Content

Federal Test Method Std. 175B, Method 4041.1—*Grit, Lumps, or Undissolved Matter in Adhesives*

High-temperature Effects

ASTM D 2295-72 (1983)—Standard Test Method for *Strength Properties of Adhesives in Shear by Tension Loading at Elevated Temperatures (Metal-to-Metal)*

Hydrogen-ion Concentration

ASTM D 1583-61 (1981)—Standard Test Method for *Hydrogen Ion Concentration*

Impact Strength

ASTM D 950-82—Standard Test Method for *Impact Strength of Adhesive Bonds*

Light Exposure

(See Radiation Exposure)

Low and Cryogenic Temperatures

ASTM D 2557-72 (1983)—Standard Test Method for *Strength Properties of Adhesives in Shear by Tension Loading in the Temperature Range from −267.8 to −55°C (−450 to −67°F)*.

Nonvolatile Content

ASTM D 1489-69 (1981)—Standard Test Method for *Nonvolatile Content of Aqueous Adhesives*

ASTM D 1490-82—Standard Test Method for *Nonvolatile Content of Urea-Formaldehyde Resin Solutions*

ASTM D 1582-60 (1981)—Standard Test Method for *Nonvolatile Content of Phenol, Resorcinol, and Melamine Adhesives*

Odor

> ASTM D 4339-84—Standard Test Method for *Determinaton of the Odor of Adhesives*

> Federal Test Method Std. 175B, Method 4051.1—*Odor Test for Adhesives*

Peel Strength (Stripping Strength)

> ASTM D 903-49 (1983)—Standard Test Method for *Peel or Stripping Strength of Adhesive Bonds*

> ASTM D 1781-76 (1981)—Standard Method for *Climbing Drum Peel Test for Adhesives*

> ASTM D 1876-72 (1983)—Standard Test Method for *Peel Resistance of Adhesives (T-Peel Test)*

> ASTM D 2558-69 (1984)—Standard Test Method for *Evaluating Peel Strength of Shoe Sole Attaching Adhesives*

> ASTM D 2918-71 (1981)—Standard Practice for *Determining Durability of Adhesive Joints Stressed in Peel*

> ASTM D 3167-76 (1981)—Standard Test Method for *Floating Roller Peel Resistance*

Penetration

> ASTM D 1916-69 (1980)—Standard Test Method for *Penetration of Adhesives*

pH

> (See Hydrogen ion Concentration)

Radiation Exposure (including Light)

> ASTM D 904-57 (1981)—Standard Practice for *Exposure of Adhesive Specimens to Artificial (Carbon-Arc Type) and Natural Light*

> ASTM D 1879-70 (1981)—Standard Practice for *Exposure of Adhesive Specimens to High-Energy Radiation*

Rubber Cement Tests

> ASTM D 816-82—Standard Methods of Testing *Rubber Cements*

Salt Spray (Fog) Testing

> ASTM B 117-73 (1979)—Standard Method of *Salt Spray (Fog) Testing*

> ASTM G 85-84—Standard Practice for *Modified Salt Spray (Fog) Testing*

> These standards, under the jurisdiction of ASTM Subcommittee G01.05 on Laboratory Corrosion Tests, are *not* described in Chapter 12. Prac-

tice G 85 provides a more corrosive enviornoment than Method B 117, generally using acids and SO_2 to supplement the salt.

Shear Strength (Tensile Shear Strength)

ASTM E 229-70 (1981)—Standard Test Method for *Shear Strength and Shear Modulus of Structural Adhesives*

ASTM D 905-49 (1981)—Standard Test Method for *Strength Properties of Adhesive Bonds in Shear by Compression Loading*

ASTM D 906-82—Standard Test Method for *Strength Properties of Adhesives in Plywood Type Construction in Shear by Tension Loading*

ASTM D 1002-72 (1983)—Standard Test Method for *Strength Properties of Adhesives in Shear by Tension Loading (Metal-to-Metal)*

ASTM D 1144-84—Standard Practice for *Determining Strength Development of Adhesive Bonds*

ASTM D 2182-72 (1978)—Standard Test Method for *Strength Properties of Metal-to-Metal Adhesives by Compression Loading (Disk Shear)*

ASTM D 2295-72 (1983)—Standard Test Method for *Strength Properties of Adhesives in Shear by Tension Loading at Elevated Temperatures (Metal-to-Metal)*

ASTM D 2339-82—Standard Test Method for *Strength Properties of Adhesives in Two-Ply Wood Construction in Shear by Tension Loading*

ASTM D 2557-72 (1983)—Standard Test Method for *Strength Properties of Adhesives in Shear by Tension Loading in the Temperature Range from −267.8 to −55°C (−450 to −67°F)*

ASTM D 2919-71 (1981)—Standard Practice for *Determining Durability of Adhesive Joints Stressed in Shear by Tension Loading*

ASTM D 3163-73 (1979)—Standard Recommended Practice for *Determining the Strength of Adhesively Bonded Rigid Plastic Lap-Shear Joints in Shear by Tension Loading*

ASTM D 3164-73 (1979)—Standard Recommended Practice for *Determining the Strength of Adhesively Bonded Plastic Lap-Shear Sandwich Joints in Shear by Tension Loading*

ASTM D 3165-73 (1979)—Standard Test Method for *Strength Properties of Adesives in Shear by Tension Loading of Laminated Assemblies*

ASTM D 3166-73 (1979)—Standard Test Method for *Fatigue Properties of Adhesives in Shear by Tension Loading (Metal/Metal)*

ASTM D 3528-76 (1981)—Standard Test Method for *Strength Properties of Double Lap Shear Adhesive Joints by Tension Loading*

ASTM D 3931-80—Standard Practice for Determining *Strength of Gap-Filling Adhesive Bonds in Shear by Compression Loading*

ASTM D 3983-81—Standard Practice for *Measuring Strength and Shear Modulus of Nonrigid Adhesives by the Thick Adherend Tensile Lap Specimen*

ASTM D 4027-81—Standard Practice for *Measuring Shear Properties of Structural Adhesives by the Modified-Rail Test*

Specimen Preparation

(See also Surface Preparation)

ASTM D 2094-69 (1980)—Standard Practice for *Preparation of Bar and Rod Specimens for Adhesion Tests*

Spot-Adhesion Test

ASTM D 3808-79—Standard Practice for *Qualitative Determination of Adhesion of Adhesives to Substrates by Spot Adhesion Test Method*

Spread

(Coverage)

ASTM D 898-69 (1980)—Standard Test Method for *Applied Weight Per Unit Area of Dried Adhesive Solids*

ASTM D 899-51 (1984)—Standard Test Method for *Applied Weight Per Unit Area of Liquid Adhesive*

Storage Life

ASTM D 1337-56 (1984)—Standard Test Method for *Storage Life of Adhesives by Consistency and Bond Strength*

Strength Development

ASTM D 1144-84—Standard Practice for *Determining Strength Development of Adhesive Bonds*

Stress-cracking Resistance

ASTM D 3929-80—Standard Practice for *Evaluating the Stress Cracking of Plastics by Adhesives Using the Bent Beam Method*

Stripping Strength

(See Peel Strength)

Surface Preparation

ASTM D 2093-69 (1976)—Standard Recommended Practice for *Preparation of Surfaces of Plastics Prior to Adhesive Bonding*

ASTM D 2651-79 (1984)—Standard Practice for *Preparation of Metal Surfaces for Adhesive Bonding*

ASTM D 2674-72 (1984)—Standard Methods of *Analysis of Sulfochromate Etch Solution Used in Surface Preparation of Aluminum*

ASTM D 3933-80—Standard Practice for *Preparation of Aluminum Surfaces for Structural Adhesive Bonding (Phosphoric Acid Anodizing)*

ARP 1524—*Surface Preparation and Priming of Aluminum Alloy Parts For High Durability Structural Adhesive Bonding, Phosphoric Acid Anodizing*

Tack

ASTM D 2979-71 (1982)—Standard Test Method for *Pressure Sensitive Tack of Adhesives Using an Inverted Probe Machine*

ASTM D 3121-73 (1984)—Standard Test Method for *Tack of Pressure-Sensitive Adhesives by Rolling Ball*

Tensile Strength

ASTM D 897-78 (1983)—Standard Test Method for *Tensile Properties of Adhesive Bonds*

ASTM D 1144-84—Standard Practice for *Determining Strength Development of Adhesive Bonds*

ASTM D 1344-78—Standard Method of Testing *Cross-Lap Specimens for Tensile Properties of Adhesives*

ASTM D 2095-72 (1983)—Standard Test Method for *Tensile Strength of Adhesives by Means of Bar and Rod Specimens*

Torque Strength

ASTM D 3658-78 (1984)—Standard Practice for *Determining the Torque Strength of Ultraviolet (UV) Light-Cured Glass/Metal Adhesive Joints*

Viscosity

ASTM D 1084-63 (1981)—Standard Test Methods for *Viscosity of Adhesives*

ASTM D 2556-69 (1980)—Standard Test Method for *Apparent Viscosity of Adhesives Having Shear-Rate-Dependent Flow Properties*

ASTM D 3236-73 (1978)—Standard Test Method for *Viscosity of Hot Melt Adhesives and Coating Materials*

Volume Resistivity

ASTM D 2739-72 (1984)—Standard Test Method for *Volume Resistivity of Conductive Adhesives*

Water Absorptiveness

(of Paper Labels)

ASTM D 1584-60 (1984)—Standard Test Method for *Water Absorptiveness of Paper Labels*

Weathering

(See Durability)

Wedge Test

ASTM D 3762-79 (1983)—Standard Test Method for *Adhesive Bonded Surface Durability of Aluminum (Wedge Test)*

Working Life

ASTM D 1338-56 (1982)—Standard Test Method for *Working Life of Liquid or Paste Adhesive by Consistency and Bond Strength*

REFERENCES

1. Anderson, G.P. et al., *Analysis and Testing of Adhesive Bonds*, Academic Press, N.Y. (1977).
2. DeLollis, N.J., *Adhesives, Adherends, Adhesion*, Robert E. Krieger Publishing Co., Huntington, N.Y. (1980).
3. Williams, M.L. et al. (University of California) *The Triaxial Tensile Behavior of Viscoelastic Materials*, GALCIT Rept. SM 63-6 (1963) (as reported by Anderson et. al. above.)

12

Standard Test Methods, Practices and Specifications

INTRODUCTION

Chapter 11 discussed test methods of interest in the field of structural adhesive bonding from a general point of view. This chapter will cover published specifications (138) and test methods (93) used in the United States. Included are ASTM Test Methods and Recommended Practices (currently called Practices), Federal Test Method Standards, ASTM Specifications, SAE-Aerospace Materials Specifications (AMS's), SAE-Aerospace Recommended Practices (ARP's), Federal Specifications, and Military Specifications. Most of these documents are continuously being revised and updated. In the case of ASTM standards, the year of issue is given after the number designation, as ASTM D 1084-63 (1981). The year of reapproval, without substantial change except for minor editorial changes, is given in parentheses, as (1981) in the example. Unless otherwise indicated, all ASTM standards listed are the responsibility of ASTM Committee D-14 on Adhesives. In the case of Aerospace Materials Specifications, the letters A, B, C, etc. following the AMS number, as AMS 3688A, indicate revisions. AMS 3688 would be the first issue, and AMS 3688A the first revision, AMS 3688B, the second, etc. Many AMS specifications have been adopted by the Department of Defense (DOD) for inclusion in the Department of Defense Index of Specifications and Standards (DOD) for inclusion in the Department of Defense Index of Specifications and Standards (DODISS, as it is popularly called). As such, they can be used in Military Procurement. Such documents are so indicated by the words "DOD adopted." Documents not so indicated may be DOD-adopted at a later date, since such possible acceptance is constantly being reviewed. Federal Supply Classification (FSC) numbers and Preparing Activity codes are given for all DOD-adopted ASTM and AMS documents.

In the case of Military and Federal Specifications, revisions are indicated by the letters A, B, C, etc., as with AMS documents. A number in parentheses following the specification number indicates an amendment. An Interim Amendment is indicated by the designation INT AMD 1, INT AMD 2, etc. given *below* the specification number. Such entries as (GL), (AR), (OS), (MI), etc., following the specification number or Interim Amendment number indicates a Limited Coordination document. Such documents have been prepared by a specific Military or Federal agency solely for its own use, although it may be used by other agencies. In the case of (MI) designations, many of these specifications were prepared by the Army Missile Command at Redstone Arsenal to replace purchasing descriptions, such as Missile Interim Specifications (MIS's). Often such documents were written to cover a specific commercial adhesive or sealant, although the commercial product and source are not named.

A list of symbols used for such agencies is given below in Table 12-1. These symbols are also used to designate the Preparing Activity.

In all cases of documents listed, the date given is the latest date of action on the document. If there is an amendment, the date is that of the amendment. In a number of cases, the amendment has added new types and classes, or modified the requirements. In most cases the bulk of the amendments cover editorial, spelling, and similar changes.

Table 12-1: Preparing Activities of Military and Federal Specifications Listed in This Chapter and Recognizing Agencies for Industry Standards (ASTM and SAE)

AR	Armament Research and Development Center, Dover, NJ (Army)
AS	Naval Air Systems Command
COND	Construction Design Requirements
EA	Chemical Research and Development Center, Aberdeen Proving Ground, MD (Army)
GL	Natick Research and Development Laboratories, Natick, MA (Army)
GSA-FSS	General Services Administration, Federal Supply Services
MC	U.S. Marine Corps
ME	Mobility Equipment Research and Development Command, Ft. Belvoir, VA (Army) (Belvoir Research and Development Center, as of March 1984)
MI	Missile Command, Redstone Arsenal, AL (Army)
MR	Materials and Mechanics Research Center, Watertown, MA (Army)
OS	Naval Sea Systems Command (Ordnance Systems)
SH	Naval Sea Systems Command (Ship Systems)
YD	Naval Facilities Engineering Command
20	Air Force Wright Aeronautical Laboratory (AFSC), Wright-Patterson AFB, Ohio (Air Force)
84	Warner Robins Air Logistics Center (AFLC)
99	Air Force Logistics Command Cataloging and Standardization Office, HQ AFLC

In all cases the Preparing Activity code and Federal Supply Classification (FSC) is given for Federal and Military documents. Most documents listed come under FSC 8040—Adhesives. A smaller number come under FSC 8030—Sealants. Other FSC categories are relatively few in number. A description of these listings is given in Table 12-2.

When a Qualified Products List (QPL) is available for Federal and Military specifications, that fact is indicated on the last line of the entry. The latest issue of the QPL is given. The designation QPL-21016-21 means that this is the 21st revision or updating of this document, MIL-A-21016.

The designation QPL-MMM-A131-6 means that this is the 6th revision or updating of Federal Specification MMM-A-131. QPL's are documents containing lists of Federally tested and recognized commercial products known to meet the requirements of the specification for the various types, classes, and grades in the specification. The manufacturer's designation for the product, along with the source, is given. The testing agency, along with its approval document number, is also listed.

In a few cases Commercial Item Descriptions (CID's) are given. These have the form A-A-XXX for adhesives in the following tabulation. CID's are basically simplified specifications enabling the Government to purchase off-the-shelf items where standards are not critical, thereby reducing costs and simplifying procurement. These are Federal, not Military documents.

Table 12-2: Federal Supply Classes Listed in This Chapter

FSC 12 GP	Fire control equipment
FSC 1336	Guided missile warheads and explosive components
FSC 1375	Demolition materials
FSC 4940	Miscellaneous maintenance and repair shop specialized equipment
FSC 5610	Mineral construction materials, bulk
FSC 8030	Preservation and sealing compounds
FSC 8040	Adhesives
FSC MFFP	Metal finishes and finishing processes and procedures

In this chapter the author has attempted to provide enough technical information in each standardization document description to give the reader a good idea of the content of the document. With minor exceptions, no attempt has been made to cover sampling procedures, test methods called out and described within specifications, or packaging and packing. The coverage of these documents is quite complete. However, not all specifications and test methods covering wood bonding have been included. Additional information in specifications and standards and how they are used may be found in references 1, 2, and 3.

TEST METHODS AND PRACTICES

American Society for Testing and Materials (ASTM)[4]

ASTM E 229-70 (1981)—Standard Test Method for *Shear Strength and Shear Modulus of Structural Adhesives*, 7pp

This method covers the determination of the shear strength and shear modulus of structural adhesives as they occur in thin gluelines restrained by the relatively higher-modulus adherends. This method, which is tedious to perform, is intended for high-modulus adherends such as most structural metals. The specimen consists of two large loading blocks with a thin, narrow ring between them. This simulates the torque loading of a large-diameter, thin-walled tube. Between the loading blocks is an adherend ring made of the materials under test. The load block is fitted with an alignment shaft and a suitable pin/jig arrangement to apply torque to the adherend ring area of the specimen.[5]

In the method torsional shear forces are applied to the adhesive through the circular specimen, producing a peripherally uniform stress distribution. The maximum stress in the adhesive at failure represents the shear strength of the adhesive. By measuring the adhesive strain as a function of load, a stress-strain curve can be established. The test specimen should be made from the same materials as are to be used in production. Production cleaning and bonding processes should be used when applicable.

ASTM F 402-80—Standard Practice for *Safe Handling of Solvent Cements and Primers Used for Joining Thermoplastic Pipe and Fittings*, 2 pp (DOD Adopted) (ASTM Committee F 17 on Plastic Piping Systems)
FSC COND YD

This practice covers procedures for safe handling of solvent cements and primers used in joining thermoplastic pipe and fittings. The procedures are general ones and include safeguards against hazards of fire and precautions for protection of personnel from breathing of vapors and contact with skin or eyes.

ASTM D 816-82—Standard Methods of Testing *Rubber Cements*, 8 pp (ASTM Committee D-11 on Rubber)

These methods cover adhesives that may be applied in plastic or fluid form and that are manufactured from natural rubber, reclaimed rubber, synthetic elastomers, or combinations of these materials. The tests include adhesion strength, bonding range, softening point, cold flow, viscosity, cold brittleness, density, and plastic deformation (for heavy doughs or putties).

ASTM E 864-82—Standard Practice for *Surface Preparation of Aluminum Alloys to be Adhesively Bonded in Honeycomb Shelter Panels*, 4 pp (Committee E-6 on Performance of Building Constructions)

This practice provides directions for the preparation of clean uniform surfaces of aluminum alloys suitable for formation of durable adhesive bonds to nonmetallic honeycomb materials in the manufacture of sandwich panels for tactical shelters. It is the direct responsibility of Subcommittee E06.23 on Durability Sandwich Panels for Tactical Shelters. The surface preparation includes degreasing, alkaline cleaning (in a sulfuric acid-sodium dichromate solution), and drying, with rinses at several stages.

ASTM E 874-82—Standard Practice for *Adhesive Bonding of Aluminum Facings to Nonmetallic Honeycomb Core for Shelter Panels*, 7 pp (Committee E-6 on Performance of Building Constructions)

This practice describes the materials, processes, and quality controls to be used in the manufacture of adhesive-bonded, aluminum-faced, non-metallic honeycomb-core sandwich panels for tactical shelters. The practice is the direct responsibility of Subcommittee E06.23 on Durability of Honeycomb Sandwich Panels for Tactical Shelters.

ASTM D 896-84—Standard Test Method for *Resistance of Adhesive Bonds to Chemical Reagents,* 3 pp (DOD Adopted)

FSC 8040   MR

This method covers the testing of all types of adhesives for resistance to chemical reagents. It includes provisions for reporting loss in strength in accordance with ASTM methods of test for strength properties of adhesives. The method uses standard chemical reagents described in ASTM D 543 and oils and fuels from ASTM D 471. Specially formulated supplementary reagents (a hydrocarbon mixture, two jet fuel mixtures, and a silicone fluid) are also specified. The testing involves immersion of standard test specimens in the test medium for 7 days at 23°C (73.4°F). At the end of the test period aqueous reagents are rinsed off with distilled water, and other reagents with a suitable organic solvent. After drying, the specimens are immediately tested by appropriate tests. Comparisons are made with controls using air as the contact medium. (This method replaces Federal Test Method Standard No. 175a, Method 1011.1.)

ASTM D 897-78 (1983)—Standard Test Method for *Tensile Properties of Adhesive Bonds,* 5 pp (DOD Adopted)

FSC 8040   MR

This method covers the determination of the comparative "butt" tensile properties of adhesive bonds tested on standard-shape specimens under defined conditions of pretreatment, temperature, and testing machine speed. The method is known as the "butt" joint adhesion tensile test. The method is not as commonly used as the lap-shear test (ASTM D 1002). Blocks or rods of wood or metal are shaped or machined to specified dimensions of 1 13/16 in. (46 mm) diameter for wood and 1 7/8 in. (47.5 mm) for metal for the contact surface. The wood specimens are made from hard maple, and the metal specimens may be brass, copper aluminum, steel, phosphor bronze, magnesium, or nickel silver. Two of the machined circular contact surface buttons are bonded together with the adhesive under test. A tensile testing machine is used under standarized conditions, and the maximum load at failure is recorded with the force normal to the contact area. The wood specimens must be conditioned at 23°C (73.4°F) and 50% RH, but no preconditioning is required for the metal specimens. Results are reported in pounds per square inch (psi) and kilograms per square millimeter (kg/mm^2). (This method replaces Federal Test Method Standard No. 175a, Method 1011.1.)

ASTM D 898-69 (1980)—Standard Test Method for *Applied Weight Per Unit Area of Dried Adhesive Solids,* 3 pp (DOD Adopted)

FSC 8040   MR

This method covers the determination of the quantity of adhesive solids applied in a spreading or coating operation. The apparatus consists of a balance capable of weighing the material accurately to the nearest 1%, and a suitable instrument for measuring the linear dimensions of the specimens to the same degree of accuracy. Ordinarily the specimen is conditioned at 23°C and 50% RH for 48 hours for specimens of 1/8 inch or less in thickness, and 96 hours for thicker specimens. The adhesive is applied according to the manufacturer's directions. After drying, both coated and uncoated test specimens are reweighed and the weight of adhesive solids calculated as the weight of adhesive solids applied per thousand square feet or joint or surface area. (This method replaces Federal Test Method Standard No. 175a, Method 3011.)

ASTM D 899-51 (1984)—Standard Test Method for *Applied Weight Per Unit Area of Liquid Adhesive*, 2 pp (DOD Adopted)

FSC   8040   MR

This method covers the determination of the quantity of liquid adhesive applied in a spreading or coating application. It is intended to be applied only to adhesives used for the bonding of wood. The apparatus consists of a balance capable of weighing the material accurately to the nearest 1%, and a suitable instrument for measuring the linear dimensions of the specimens to the same degree of precision. The specimen is conditioned to reach equilibrium with the atmospheric conditions prevailing under actual or contemplated operational use prior to the application of the adhesive. The adhesive is applied according to the manufacturer's directions, and the specimens are reweighed immediately. The weight of liquid adhesive applied per thousand square feet of surface or joint area is then determined. (This method replaces Federal Test Method Standard No. 175a, Method 1041.1.)

ASTM D 903-49 (1983)—Standard Test Method for *Peel or Stripping Strength of Adhesive Bonds*, 4 pp (DOD Adopted)

FSC   8040   MR

This method covers the determination of the comparative peel or stripping characteristics of adhesive bonds when tested on standard-size specimens and under defined conditions of pretreatment, temperature, and testing-machine speed. The test is essentially a 180-degree peel/stripping test where only one member of the laminate couple is bent. This is in contrast to the peel tests which apply force to both laminated structure members and in which both laminated members deform—e.g. the T-peel test (ASTM D 1876). The test is particularly useful in comparative testing of adhesive materials, such as vinyl-to-metal adhesives.[5] The test specimen is shown in Figure 12-1. It consists of one piece of flexible material, 1 × 12 in. (25 × 204.8 mm), bonded for 6 in. (152.4 mm) at one end to one piece of flexible or rigid material 1 × 8 in. (25 × 203.2 mm), with the unbonded portions of each member being face to face. Test materials must be thick enough to withstand the expected tensile pull, but not over 1/8 in. (3 mm) in thickness.

The testing machine is a power-driven machine with a constant rate-of-jaw separation, or is of the inclination balance or pendulum type. The rate of travel of the power-actuated grip shall be 12 in. (305 mm)/minute, which provides a separation of 6 in. (152 mm)/minute. Specimens are conditioned for 7 days at 23°C (73°F) and 50% RH, or until equilibrium is reached.

The actual peel or stripping strength is determined by drawing the best average load line that will accommodate the recorded curve on an autographic chart. The results are reported in pounds per inch (kilograms per millimeter) of width for separation at 6 in./min (152.4 mm/min). Figure 12-1 shows how the specimen is tested.

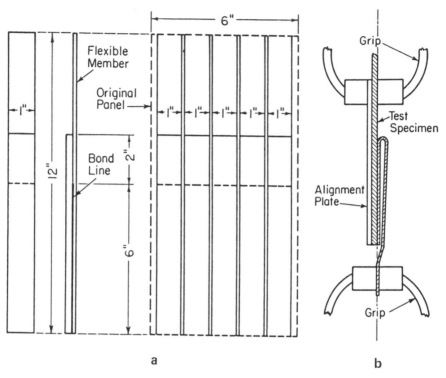

a   b

Figure 12-1: (a) Test specimen and (b) test apparatus for ASTM D 903 (reprinted, with permission, from the annual book of ASTM Standards, volume 15.06. Copyright ASTM, 1916 Race Street, Philadelphia, PA 19103)

ASTM D 904-57 (1981)—Standard Practice for *Exposure of Adhesive Specimens to Artificial (Carbon-Arc Type) and Natural Light,* **3 pp**

This practice defines conditions for the exposure of adhesive in the form of glued transparent or translucent assemblies to (1) artificial and (2) natural light sources. Where such information is of value, the same exposure conditions may be used on adhesive film or any other suitable form in which light may be a deteriorating factor. The practice is limited to the method of obtaining the exposure conditions and procedure to be followed, but *does not cover methods of test* to be used in evaluating the effects of the exposure. The artificial light is provided by a carbon-arc apparatus, and the natural light is provided by sunlight beamed at a 45-degree angle facing south.

Carbon-arc apparatus units that may be used include the Atlas Single and

Twin Arc Apparatus (AC and DC), the Atlas FadeOmeter, and the National Apparatus.

ASTM D 905-49 (1981)—Standard Test Method for *Strength Properties of Adhesive Bonds in Shear by Compression Loading*, 4 pp (DOD Adopted)   FSC   8040   MR

This method is used to determine the comparative shear strength of adhesive bonds used for bonding wood and other similar materials, when tested on a standard specimen under specified conditions of preparation, conditioning, and loading in compression. Hard maple test blocks are used as test joints. Dimensions are usually 3/4 in. × 2 1/2 in. × 12 in. (19 mm × 63.5 mm × 304 mm). The wood-block test joints are brought to the equilibrium moisture content recommended by the manufacturer, or, in the absence of such a recommendation, to 10–12% mositure. The test joints are conditioned to 23°C and 50% RH for 7 days, or until equilibrium is reached. Test specimens of 1 3/4 in. × 2 in. × 3/4 in. are cut by saw from the conditioned test joints. Figure 12-2 shows how these specimens are assembled. The width and length of the glue line is measured to determine the shear area. The test specimens are retained in the conditioning atmosphere until tested.

In testing, the test specimens are placed in a small shearing tool in such a way that the load is applied with a continuous motion at a constant rate. The shear stress at failure in pounds force per square inch (or kilopascals), based on the glue-line area, is determined and reported, together with the estimated percentage of wood failure for each specimen. (This method replaces Federal Test Method Standard No. 175a, Method 1031.)

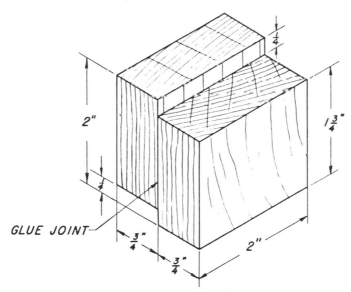

GLUE JOINT

Figure 12-2: Form and dimensions of test specimens for ASTM D 905 (reprinted, with permission, from the annual book of ASTM Standards, volume 15.06. Copyright ASTM, 1916 Race Street, Philadelphia, PA 19103)

ASTM D 906-82—Standard Test Method for *Strength Properties of Adhesives in Plywood Type Construction in Shear by Tension Loading*, 4 pp (DOD Adopted)

FSC 8040 MR

This method is used to measure comparative shear strengths of adhesives in plywood-type construction when tested on a standard specimen under specified conditions of preparation, conditioning, and testing. The method is intended to be applied only to adhesives used in bonding wood to wood. Specimens of plywood, fabricated in a prescribed way in the form of strips 1 in. (25 mm) wide and approximately 4 in. long, are cut from a conditioned test panel. Each specimen has two saw cuts (notches) two-thirds of the way through the core. These specimens are kept at the same conditioning temperature and humidity, 23°C (73°F) and 50% RH, as used for the panels, until ready for testing. A standard testing machine is used at a loading rate of 4535 to 7560 g/s (600 to 1000 lb/min), and the machine is loaded to failure. The percentage of wood failure is determined for each specimen and failing loads reported in grams per square centimeter (or pounds per square inch) of shear area. The wood used to fabricate the plywood is either sweet or yellow birch veneer.

ASTM D 950-82—Standard Test Method for *Impact Strength of Adhesive Bonds*, 6 pp (DOD Adopted)

FSC 8040 MR

This method covers the determination of the comparative impact value of adhesive bonds in shear, when tested on standard specimens under specified conditions of preparation, conditioning, and testing. The test measures pendulum-impact resistance. Different dimensions are used for specimens for metal-to-metal adhesives than are used for wood-to-wood adhesives. Preconditioning is not required for metal-to-metal bonds, but wood bonds are conditioned at 23°C (73°F) and 50% RH for 7 days or when equilibrium is reached. The specimens are mounted in special grips and placed in a standard impact machine. One adherend is struck with a pendulum hammer traveling at a prescribed speed (11 ft/s) and the energy of impact is reported in joules per square meter or foot-pounds force per square inch of bonded area. In cases of metal-to-metal adhesives, the percentages of cohesive, adhesive and contact failure are reported. In cases of wood-to-wood adhesives, the percentages of wood, glue and contact failure are reported. (This method replaces Federal Test Method Standard No. 175a, Methods 1051 and 1051.1T.)

ASTM D 1002-72 (1983)—Standard Test Method for *Strength Properties of Adhesives in Shear by Tension Loading (Metal-to-Metal)*, 5 pp (DOD Adopted) FSC 8040 MR

This method is used to determine the comparative shear strengths of adhesives for bonding metals when tested on a standard specimen under specified conditions of preparation and testing. This *lap-shear* or *tensile-shear* test is the most commonly used shear test for metal-to-metal structural adhesives. Two sections of rigid sheet material, usually 4 in. long by 1 inch wide, and 0.064 in. thick (102 × 25 × 1.65 mm) are used to fabricate the specimens, which are overlapped 1/2 in. (13 mm) and adhesively bonded together, using manufacturer's instructions, to form a one-half square inch (323 square mm) contact area. Both ends of the bonded specimen are firmly clamped into the jaws of a tensile-testing machine. Tension is applied at a prescribed speed (80 to 100 kg/cm^2) (1200 to 1400 psi) of the shear area per minute (approximately 1.3 mm or 0.05 in./min). The load at failure is reported in kiograms per square centimeter (pounds per square inch) of shear area, and the nature of the failure (cohesive, adhesive, or contact) reported. (This method replaces Federal Test Method Standard No. 175a, Methods 1032 and 1032.1T.)

ASTM D 1062-78 (1983)—Standard Test Method for *Cleavage Strength of Metal-to-Metal Adhesive Bonds*, 4 pp (DOD Adopted)

FSC   8040   MR

This method covers the determination of the cleavage properties of adhesive bonds when tested in standard-shape metal specimens under conditions of preparation and testing. It may also be used to compare adhesives used with other metallic materials having a specified surface treatment. *Cleavage strength*, as used here, means the tensile load in terms of pounds per inch of width required to cause separation of a test specimen 1 in. (25 mm) in length. Test specimens, essentially in block form, are machined to the specified dimensions, including drilled holes to accept pins for attaching the grip to the specimen. Conditioning is usually not required. A tensile load is applied at a rate of 600 to 700 lb (270 to 320 kg)/minute. The maximum load in pounds carried by the specimen at failure is determined and the cleavage strength expressed in pounds per inch. The percentage of cohesive failure is also determined. Because of the machining involved in this method, peel tests are usually preferred for cleavage.[3] This method replaces Federal Test Method Standard No. 175a, Method 1071-T.)

ASTM D 1084-63 (1981)—Standard Test Methods for *Viscosity of Adhesives*, 6 pp (DOD Adopted)

FSC   8040   MR

The four methods described cover the determination of the viscosity of free-flowing adhesives. The methods are as follows:

Method A—a cup method, applicable only to adhesives that will deliver 50 ml in a steady, uninterrupted stream from a viscosity cup (Zahn cup) designed to deliver that volume of sample in from 30 to 100 seconds at 23°C (73°F). The time required to remove the sample by flow is noted and the average consistency in seconds is recorded for each cup size.

Method B—intended for measuring the viscosity of adhesives covering a range from 50 to 200,000 cP. Method B is limited to materials that have, or approximate, Newtonian flow characteristics. The apparatus used in a Brookfield synchroelectric viscosimeter, Model RVO, RVF, MVO, or MVF, or equivalent. The Brookfield viscosimeter is an instrument with revolving spindles. A spindle suitable for the viscosity range of the material must be used. The viscosity is reported in centipoises.

Method C—intended primarily as a control method for determining that viscosity of adhesives that have, or approach, Newtonian flow characteristics.

The apparatus is a Stormer viscosimeter with double flag paddle-type rotor, as specified in ASTM D 562, a method used for determining paint consistency. The results are reported as average viscosity in seconds for 100 revolutions of the paddle, depending on the weight used in the apparatus (which is also reported).

Method D—intended primarily as a control method for determining the viscosity of materials which have an equivalent viscosity no greater than approximately 3000 cP, and is limited to materials that have or approach Newtonian flow characteristics. (Note: For non-Newtonian flow see ASTM D 2556.) In this method a set of five Zahn viscosity cups of varying orifice size are used. These are so designed as to allow a sample to flow through the calibrated orifice in approximately 1 minute or less. The best results are obtained when the flow time is between 20 and 40 seconds. The time of flow from the cup is determined and the average viscosity in seconds for the particular cup number is reported. The determination is called Zahn seconds.

ASTM D 1144-84—Standard Practice for *Determining Strength Development of Adhesive Bonds*, 3 pp

This practice covers the determination of the strength development of adhesive bonds when tested on a standard specimen under specified conditions of preparation and testing. It is applicable to adhesives in liquid or paste form requiring curing at specified conditons of time and temperature or specific surface preparation. It is intended primarily to be used with metal-to-metal adherends; however, plastics, woods, glass or combinations of these may be substituted. The recommended practice suggests test methods and provisions for reporting strength values. The test method recommended is ASTM D 1002 (lap-shear test). The manufacturer's recommended curing time and temperature are used to determine the strength values. In addition, tests are run at the same temperature and time intervals to determine fixture time (set time) at 20, 50 and 80% strength. Testing is normally carried out immediately after the specimens reach room temperature.

ASTM D 1146-53 (1981)—Standard Test Method for *Blocking Point of Potentially Adhesive Layers*, 3 pp (DOD Adopted)

FSC   8040   MR

This method covers the determination of the blocking point of a thermoplastic or hygroscopic layer or coating of potentially adhesive material. These materials are those in a substantially non-adhesive state which may be activated to an adhesive state by application of heat or solvents. Since some potentially adhesive materials are both thermoplastic and hygroscopic, this method provides means for estimating, on the same material, both thermoplastic and hygroscopic blocking. Two degrees of blocking (1st and 2nd degree) and two types of blocking (cohesive and adhesive) are covered. Thermoplastic blocking is measured in a desiccator over anhydrous calcium chloride at 38°C, and hygroscopic blocking is measured in a desiccator over 50% RH (solution of saturated sodium dichromate) at 38°C. The critical temperature (lowest temperature at which blocking of a given degree occurs) and the critical humidity (lowest humidity at which blocking of a given degree occurs) can also be determined. (This method replaces Federal Test Method Standard No. 175a, Method 2041.)

ASTM 1151-84—Standard Test Method for *Effect of Moisture and Temperature on Adhesive Bonds*, 3 pp (DOD Adopted)

FSC   8040   MR

This method defines conditions for determining the performance of adhesive bonds when subjected to continuous exposure at specified conditions of moisture and temperature. The performance is expressed as a percentage, based on the ratio of strength retained after exposure, to the original strength. The method may be used to determine the performance of adhesive bonds in terms of any desired strength property, including those measured by ASTM Methods D 897, D 903, D 906, and D 1002. Samples are conditioned for 7 days at 50% RH and 23°C. This standard prescribes 22 different conditions of temperature and moisture. The duration of the exposure is dependent upon the nature of the adhesive and type of specimens and will, therefore, be covered by material specifications. (This method replaces Federal Test Method Standard No. 175a, Methods 2052-T and 2031.)

ASTM D 1183-70 (1981)—Standard Test Methods for *Resistance of Adhesives to Cyclic Laboratory Aging Conditions*, 3 pp (DOD Adopted)

FSC   8040   MR

These methods cover the determination of the resistance of adhesives to cyclic accelerated service conditions by exposing bonded specimens to conditions of high and low temperatures and relative humidities. The extent of degradation is determined from changes in strength properties as a result of exposure to test conditions. Test panels or test specimens may be used, depending on the ASTM test method used. Controls are also required. All specimens are conditioned for 7 days at 23°C (73.4°F) and 50% RH, in one of the test environments shown in Table 12-3.

Table 12-3: Test Procedures for ASTM D 1183

Procedure Designation	Name	Period (hr)	Temperature °C	Temperature °F	RH (%)
A	Interior	24	23	73.4	85-90
		24	48.5	120	<25
		72	23	73.4	85-90
		48	48.5	120	<25
B	Interior	48	60	140	<15
		48	38.5	100	85-90
		8	-18	0	~100
		64	38.5	100	85-90
C	Exterior, land	48	71	160	<10
	and air	48	23	73.4	*
		8	-57	-70	~100
		64	38.5	100	~100
D	Exterior,	48	71	160	<10
	marine	48	23	73.4	**
		8	-57	-70	~100
		64	23	73.4	**

*Immersed in water.
**Immersed in substitute ocean water.

After such exposure, the test specimens or panels are conditioned for 7 days at 23°C (73.4°F) and 50% RH and then tested at once for their specified strength properties. The average strength value is determined and the percentage change in strength as a result of exposure to the test conditions is reported. (This method replaces Federal Test Method Standard No. 175a, Method 2051-T.)

ASTM D 1184-69 (1980)—Standard Test Method for *Flexural Strength of Adhesive Bonded Laminated Assemblies,* 4 pp (DOD Adopted)

FSC   8040   MR

This method covers the determination of the comparative properties of adhesive assemblies when subjected to flexural stresses with standard-shape specimens under specified conditions of pretreatment, temperature, relative humidity, and testing technique. The test specimen and testing technique were designed to develop a large portion of shear forces between the laminae of the test piece when the load is applied, rather than to reduce shear stress to a minimum, as is done in other ASTM test methods for flexural properties. This method is not applicable to assemblies made with nonrigid adherends.

The data are reduced to a comparable basis by means of formulas provided. For metal specimens maximum shear stress, Ss, is determined in MPa or psi. For wood specimens, flexural strength, S, is determined in MPa or psi. In testing, the bonding conditions are prescribed by the manufacturer. The test specimens are rectangular pieces 38 mm (1.5 in.) long and 19.1 mm (0.75 in.) wide. These are machined from laminated panels consisting of 8 plies of 0.3 mm (0.01 in.) thick adherend material. Each ply is coated with adhesive on both sides with an even spread and bonded. All specimens are conditioned at 23°C and 50% RH for 48 hours for metal and plastics and 7 days for wood. Testing is carried out under these same conditions. The specimens are tested as simple beams loaded at mid-span. Detailed conditions of testing are prescribed. The results, Ss for metal and S for wood, are determined from the breaking load, P. (This method replaces Federal Test Methods Standard No. 175a, Method 1021.)

ASTM D 1304-69 (1983)—Standard Methods of Testing *Adhesives Relative to Their Use as Electrical Insulation,* 3 pp

These methods cover procedures for testing adhesives in liquid, highly viscous, solid, or set states, that are intended to be cured by electronic heating, or that are intended to provide electrical insulation, or that are intended for use in electrical apparatus. The following procedures are used:

(1) *Procedures for Testing Adhesives Before Use—*
Power Factor and Dielectric Constant of Liquid Adhesives (Methods D 150)
Direct Current Conductivity (Method D257)
Extract Conductivity (Methods D 202)
Acidity and Alkalinity (Methods D 202)
pH Value (Methods D 202)

(2) *Procedures for Testing Properties of Adhesives as Used*
Power Factor and Dielectric Constant of a Dried or Cured Adhesive Film (Methods D 115 and D 150)

Dielectric Strength (Methods D 115)
Volume and Surface Resistivity (Method D 897 for preparation of areas to be cemented, and Methods D 257)
Arc Resistance (Method D 495)

In general, the ASTM methods cited above are used, with slight modifications.

ASTM D 1337-56 (1984)—Standard Test Method for *Storage Life of Adhesives by Consistency and Bond Strength*, 3 pp

This method is applicable to all adhesives having a relatively short storage life. It is intended to determine whether the storage life conforms to the minimum storage life required of an adhesive by consistency tests (Procedure A) or by bond strength tests (Procedure B), or by both. The *storage life* is defined as the time during which the adhesive can be stored under specified conditions and remain suitable for use. The storage temperature can be any of the standard temperatures specified in ASTM D 618, Conditioning Plastics and Electrical Insulating Materials for Testing. Any type of viscosimeter that provides results in fundamental units can be used to measure viscosity. In the Bond Strength Test, the lap-type shear specimens or the spool-type tension specimens, such as those described in Methods D 897, D 906, and D 1002, may be used.

ASTM D 1338-56 (1982)—Standard Test Method for *Working Life of Liquid or Paste Adhesives by Consistency and Bond Strength*, 3 pp

This method is similar to ASTM D 1337 above, except that it covers working life instead of storage life. *Working life* is defined as the time elapsed between the moment an adhesive is ready for use and the time when the adhesive is no longer usable.

ASTM D 1344-78—Standard Method of Testing *Cross-Lap Specimens for Tensile Properties of Adhesives*, 4 pp

This method covers a simplified tension test procedure for determining the comparative strength of adhesives by the use of a cross-lap assembly under specified conditions of pretreatment, temperature, and testing-machine speed. The test was especially designed for the adhesion of glass, either to itself or to other materials. The test specimens (Fig. 12-3) consist of two 25 by 38 by 2.7-mm (1 by 1 1/2 by 1/2-in.) pieces of glass, which are pressed together at right angles in such a manner that one square inch of bonded area extends with a 1/4 inch overlap on all four sides of the square bonded area. Other material (brass, copper, aluminum, steel, wood, etc.) may be varied in thickness from 1/8 to 1/2 inch, depending on its maximum strength. Preconditioning is not required for glass-to-glass or glass-to-metal bonds. Wood specimens must be conditioned at 23°C and 50% RH. A testing machine capable of maintaining a specified rate of loading is used. Several types of jigs for applying pressure during cure are recommended. The tensile strength is calculated as the breaking load, expressed in newtons per square meter (N/m^2). The percentage of adhesion, cohesion, material, or contact failure is also reported.

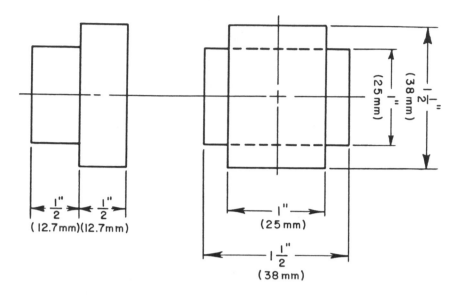

Figure 12-3: Cross-lap specimen used in ASTM D 1344 (reprinted, with per-mission, from the annual book of ASTM Standards, volume 15.06. Copyright ASTM, 1916 Race Street, Philadelphia, PA 19103)

ASTM D 1382-64 (1981)—Standard Test Method for *Susceptibility of Dry Adhesive Films to Attack by Roaches*, 2 pp

This method covers the determination of the extent to which adhesive-impregnated paper, compared with blanks, is damaged by American roaches *(Periplaneta americana)*. The method may, with appropriate changes, be adapted to other vermin. Filter paper is dipped in a freshly prepared adhesive preparation and drained, then dried for 24 hours in air. Starved American roaches, both male and female, 5 to 6 months old, are used in the testing, which is ordinarily carried out at 23°C (73.4°F) and 50% RH. The specimens, with controls, are placed in a suspension device in a glass beaker with a wooden cover. The beaker is used to contain the roaches. The percentage destruction of the impregnated filter paper is determined by weighing before and after exposure for up to 14 days.

ASTM D 1383-64 (1981)—Standard Test Method for *Susceptibility of Dry Adhesive Film to Attack By Laboratory Rats*, 2 pp

This method covers the evaluation of the susceptibility of dry adhesive films to attack by rodents by determining the comparative damage to adhesive-impreg-nated filter paper and to blanks by laboratory-bred white rats. Filter paper is dipped in a freshly prepared adhesive and drained, then dried for 24 hours in air. Both male and female starved rats are used in the testing, which is ordinarily carried out at 23°C (73.4°F) and 50% RH. The specimens, with controls, are fastened with clips to a wire-mesh cage containing with rats. Each cage contains a hard maple block for tooth conditioning. A test period of 3 days is used. The percentage destruction of the impregnated and unimpregnated filter paper is determined by weighing before and after exposure of 3 days.

ASTM D 1488-60 (1981)—Standard Test Method for *Amylaceous Matter in Adhesives*, 2 pp (DOD Adopted)

FSC   8040   MR

This method covers the determination of the presence or absence of amylaceous (starch-like) material in phenol-, resorcinol-, and melamine-resin glues. Quick spot tests specified are iodine test solution and ethyl alcohol for phenol and resorcinol adhesives, and iodine test solution and acetic acid-ethyl alcohol solution for melamine resins. With both tests, a decided darkening of the residue denotes the presence of amylaceous matter. (This method is part of Federal Test Method Standard No. 175a.)

ASTM D 1489-69 (1981)—Standard Test Method for *Nonvolatile Content of Aqueous Adhesives*, 2 pp (DOD Adopted)

FSC   8040   MR

This method covers the determination of the nonvolatile content of aqueous adhesives, such as dextrin, starch, casein, animal gelatin, etc. In the test 10-g samples of the adhesive are weighed into a small beaker to the nearest 0.01 g. The adhesive is dispersed in 50 ml of hot distilled water and transferred into a 200-ml volumetric flask. Ten-ml aliquots are then pipetted from the flask into a tared weighing bottle filled 3/4 full of silica sand dried to constant weight. The sample in the weighing bottle is dried at 105°C to constant weight, then covered and cooled in a desiccator to room temperature before weighing to the nearest 0.001 g. The percentage of nonvolatile matter is then calculated. (This method replaces Federal Test Method Standard No. 175a, Method 4021, part of Procedure B.)

ASTM D 1490-82—Standard Test Method for *Nonvolatile Content of Urea-Formaldehyde Resin Solutions*, 2 pp (The D 1490-67 (1978) version was DOD Adopted, but the current version is not.)

This method covers the determination of the apparent nonvolatile content of urea-formaldehyde solutions intended for use as wood adhesives. In the test a portion of the resin solution is placed in a tared Lunge weighing pipet and weighed to ±1 mg. An aliquot of the solution is then transferred into a tared aluminum-foil drying dish, 5 ml of water is added, and the solution is mixed. The drying dish containing the diluted specimen is then dried at 105°C (221°F) for 3 hours. The percentage of nonvolatile matter is then calculated.

ASTM D 1579-60 (1981)—Standard Test Method for *Filler Content of Phenol, Resorcinol, and Melamine Adhesives*, 2 pp (DOD Adopted)

FSC   8040   MR

This method is suitable for measuring the filler content of phenol, resorcinol, and melamine resin-base adhesives mixed with hardener or catalyst, that set at room, intermediate, and high temperatures. Results are expressed on the basis of the nonvolatile content of the mixed liquid adhesive. In the procedure, samples of the adhesive are weighed into suitable solvent solutions, depending on the resin. The solutions are then stirred and filtered. In the case of melamines, acid-washed diatomaceous earth is added. Drying is carried out at 105°C (221°F) to constant weight, except for melamine, where a 16-hour period is used. The filler content is then determined by calculation.

ASTM D 1581-60 (1984)—Standard Test Method for *Bonding Permanency of Water- or Solvent-Soluble Liquid Adhesives for Labeling Glass Bottles,* 2 pp

This method covers the determination of the bonding permanency of dextrin, casein, starch, animal gelatin, and other liquid adhesives (excluding pressure-sensitive types) used for applying printed paper labels to glass bottles. Nine exposure conditions (A through I) are provided to simulate typical storage conditions. In the test a uniform adhesive film is applied to a clean glass plate to provide a wet-film thickness of 0.025 to 0.038 mm (0.001 to 0.0015 in.). The label is applied and the adhered unit stored at room temperature and 50% RH (approximate) for at least 24 hours, or until it is no longer sticky to touch. The specimens are then stored under one of the nine test conditions for the prescribed period, usually 7 days, at the end of which they are examined for permanency of the bond. Permanency is reported as a percentage based on the amount of paper fiber remaining on the film.

ASTM D 1582-60 (1981)—Standard Test Method for *Nonvolatile Content of Phenol, Resorcinol, and Melamine Adhesives,* 2 pp

This method covers the determination of the nonvolatile content or total solids of phenol, resorcinol, and melamine adhesives, with or without hardener added, and containing high-boiling and low-boiling volatile organic solvents, or water, or both. In some cases low-molecular-weight materials in the adhesives may be lost if hardener is not used. In the test, samples are weighed out and dried at 70°C, 105°C, or 150°C with slightly different procedures, depending on the presence or absence of hardener.

ASTM D 1583-61 (1981)—Standard Test Method for *Hydrogen Ion Concentration of Dry Adhesive Films,* 2 pp (DOD Adopted)

FSC   8040   MR

This method covers determination of the hydrogen ion concentration (pH), acidity or alkalinity, of organic adhesives in the cured dry-film form. It is not designed to apply to pressure-sensitive adhesives. The pH is important because the adherends or the adhesive itself may be adversely affected in time by the acidity or alkalinity. These conditions arise from the catalysts used to cure the adhesive. A pH meter capable of measuring ±0.05 pH unit is used. Samples of the adhesive are used to form dry films on glass sheets. These films are then ground, weighed out and placed in small standard-size glass vials to which freshly boiled distilled water is added. The pH is then read at 23°C. (This method replaces Federal Test Method Standard No. 175a, Method 4011.)

ASTM D 1584-60 (1984)—Standard Test Method for *Water Absorptiveness of Paper Labels,* 2 pp

This method covers the selection of paper labels of uniform water absorptiveness for use in determining the bonding permanency of adhesives for labeling glass bottles. This method is used along with ASTM D 1581. A Cobb sizing tester is used. In the test, 10 ml of distilled water is introduced to the inside of a metal ring in contact with the unprinted surface of the paper label (Cobb tester). The water is then poured off after 110 seconds. After another 10 seconds the surface moisture is blotted from the label with a soft towel and the label reweighed. The water absorptiveness is calculated from the amount of weight gain. It is reported in g/m^2.

ASTM D 1713-65 (1981)—Standard Test Method for *Bonding Permanency of Water- or Solvent-Soluble Liquid Adhesives for Automatic Machine Sealing Top Flaps of Fiberboard Specimens,* 2 pp

This method covers the determination of the bonding permanency of dextrin, casein, starch, resin-base, and other water- or solvent-soluble liquid adhesives (excluding pressure-sensitive types) used for sealing the top flaps of fiberboard shipping cases. Six different test exposure conditions are used. Adhesive films are applied under standardized conditions, using a tube or rod-type applicator. The test specimens are 5 1/4 in. $\times$ 5 1/4 in. swatches of corrugated faced board stock, solid fiberboard stock, or the liner facing material. The glued test samples are loaded with a pressure of 0.5 psi for 5 minutes, after which the samples are stored for at least 24 hours, or until the adhesive has dried completely. The samples are then exposed to the specified test conditions for periods ranging from 1 to 7 days, depending on the condition. Bonding permanency is then determined by stripping back the top swatch under standardized conditions. The value is recorded as a percentage of failure, based on the amount of paper fiber remaining on the film.

ASTM D 1780-72 (1983)—Standard Recommended Practice for *Conducting Creep Tests of Metal-to-Metal Adhesives,* 5 pp (DOD Adopted)
FSC   8040   MR

This recommended practice covers the determination of the amount of creep of metal-to-metal adhesive bonds due to the combined effects of temperature, tensile-shear stress, and time. Test periods depend upon the reasonable life expected from the material in service. The test specimens are similar to those used for ASTM D 1002, except that the length is 5 inches. In the test, 3 fine scribe lines are made across the machined vertical edge of the prepared specimen, one across the center of the lap joint and the other two at a distance of 0.030 in. ±0.010 in. from the ends of the lap joint within the lap section. After application of a prescribed load by a direct dead weight or level procedure, the deformation is measured directly. This is accomplished by observing the displacement of the 3 scribe lines by means of a calibrated microscope. The deformation is measured at various time intervals to provide a smooth time-deformation curve. After a suitable period of time the load is removed and the specimen is permitted to recover. The recovery is measured by the calibrated microscope. The data are presented in tabular or curve form. The adhesive shear stress is reported in psi or MPa and the deformation given at specific time intervals. The creep rate in inches per hour or mm per hour is also calculated. It is often useful to plot creep as a function of time on log-log coordinates, since the plots frequently result in nearly straight-line functions.

ASTM D 1781-76 (1981)—Standard Method for *Climbing Drum Peel Test for Adhesives*, 6 pp (DOD Adopted)

FSC   8040   MR

This method covers the determination of the peel resistance of adhesive bonds between a relatively flexible adherend and a rigid adherend, and the relatively flexible facing of a sandwich structure and its core, when tested under specified conditions. The specimen consists of a laminated assembly of two adhesively bonded layers 1 inch to 3 inches wide (25 to 76 mm) and 10 inches to 12 inches long (254 to 305 mm), depending upon whether laminated assemblies, or sandwich constructions are being evaluated. In the test a flanged drum is used, onto which the end tabs of the specimens are clamped. Flexible loading straps or cables are used to guide the specimen on the drum during peeling. The outside radius of the drum is 2 inches (51 mm). One tab of the specimen is clamped to the drum, which is fastened in the lower jaw of a tensile-testing machine through the flexible metal bonds or cables. The other end is clamped to the upper joint of the testing machine through a self-clinching device.

The peel resistance over a distance of at least 6 inches (152 mm) of the bond is measured, in tension, at a crosshead speed of 1 inch/minute (25.4 mm/min). The flexible member is peeled from the test assembly at a rate of 4 inches/minute (102 mm/min) and the average load to peel the laminar specimen is recorded. This type of peel test is considered by some as superior to the T-peel test (ASTM D 1876) for laminar and sandwich construction. The force required to peel away the layer material is factored out of the measurement.[5] (This method replaces Federal Test Method Standard No. 175a, Method 1042-T.)

ASTM D 1828-70 (1981)—Standard Practice for *Atmospheric Exposure of Adhesive-Bonded Joints and Structures*, 4 pp

This practice defines the procedure for the direct exposure of adhesive bonded joints and structures to natural atmospheric environments. The procedure for sheltered atmospheric exposure, such as a Stevenson screen, of adhesive-bonded joints and specimens is similar, except for the requirements of facing south and measurement of solar radiation. This practice is limited to the procedure by which samples are exposed, and does not cover the tests that may be used to evaluate the effects of atmospheric exposure on these adhesive-bonded joints and structures. The samples may be any of several varieties, such as:

- A complete structure for test
- A section of a structure for test
- A complete structure or section with strength observations on specimens cut therefrom
- Test specimens themselves
- Any of the above, mounted under stress

The test methods to evaluate the effects of exposure may be any that are suitable, including nondestructive qualitative or quantitative observations on the same sample at prescribed intervals, or destructive tests on separate sets of specimens in accordance with such tests as ASTM D 1002. In the case of metal adherends, four types of test sites are recommended:

- Rural-pure atmosphere
- Industrial-sulfurous gases present
- Marine-seacoast sites where chlorides are deposited on the specimens
- Tropical, or southern Florida, where heat and high humidities are present (Freeport, Texas is suitable)

In cases of both metallic and nonmetallic adherends, exposure sites should be chosen to include variations in average temperature (and temperature range), relative humidity, and precipitation. The practice provides suggestions as to suitable exposure racks and climatological instruments, and gives details as to specimen preparation, exposure duration, and report preparation.

ASTM D 1875-69 (1980)—Standard Test Method for *Density of Adhesives in Fluid Form*, 2 pp (DOD Adopted)

FSC   8040   MR

This method covers the measurement of density (weight per gallon) of adhesives and adhesive components when in liquid form. It is particularly applicable where the fluid has too high a viscosity, or where a component is too volatile for a specific gravity balance determination. A weight-per-gallon cup (83.2-ml capacity at $25°C$) is used. The weight in grams of the contents, divided by 10, is the weight per gallon in pounds. (This method is part of Federal Test Method Standard No. 175a.)

ASTM D 1876-72 (1983)—Standard Test Method for *Peel Resistance of Adhesives (T-Peel Test)*, 4 pp (DOD Adopted)

FSC   8040   MR

This method is primarily intended for determining the relative peel resistance of adhesive bonds between flexible adherends by means of a T-type specimen. The T-peel strength is defined as the average load per unit width of bond line required to produce progressive separation of the bonded, flexible adherends under conditions designated in this method. To be considered "flexible," the adherends must have such dimensions and physical properties as to permit bending them through any angle up to $90°$ without breaking or cracking. The test panels are 6 inches (152 mm) wide by 12 inches (305 mm) long, and bonded only over 9 inches (241 mm) of their length. These bonded panels (Fig. 12-4) are cut into 1-inch (25 mm) wide specimens and the 3-inch (75 mm) unbonded ends of the specimen are then spread apart, perpendicular to the bond line. The bent tabs of the prepared specimen are placed in the grips of a tensile-testing machine. The load is then applied at a rate of 10 inches (254 mm)/minute. The peel resistance is measured over at least a 5-inch (127 mm) length. The results are reported in pounds per inch of width. The type of failure is also reported. (This method is part of Federal Test Method Standard No. 175a.)

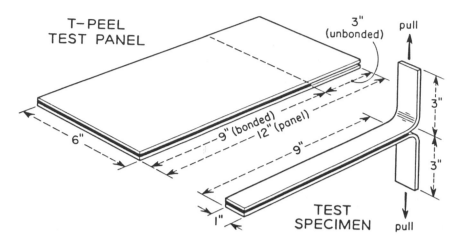

T-PEEL
TEST PANEL

3"
(unbonded)

pull

9" (bonded)
12" (panel)

9"

6"

1"

TEST
SPECIMEN   pull

3"

3"

Figure 12-4: Test panel and test specimen for ASTM D 1876 T-peel test (reprinted, with permission, from the annual book of ASTM Standards, volume 15.06. Copyright ASTM, 1916 Race Street, Philadelphia, PA 19103)

ASTM D 1879-70 (1981)$^{\epsilon 1}$—Standard Practice for *Exposure of Adhesive Specimens to High-Energy Radiation*, 5 pp

The purpose of this practice is to define conditions for the exposure of polymeric adhesives in bonded specimens to high-energy radiation prior to determination of radiation-induced changes in physical or chemical properties. This practice covers gamma or X-ray radiation, electron or beta radiation, neutrons, and mixtures of these, such as reactor radiation. It specifies only conditions of irradiation, but does not cover the preparation of test specimens, testing conditions, or the evaluation of tests. Five procedures are covered, as follows:

Procedure A—Exposure at ambient conditions

Procedure B—Exposure at controlled temperature

Procedure C—Exposure in medium other than air

Procedure D—Exposure under load

Procedure E—Exposure combining two or more of the variables listed in Procedures A to D

ASTM D 1916-69 (1980)—Standard Test Method for *Penetration of Adhesives*, 3 pp

This method covers the determination of the penetration under pressure of adhesives used in systems where at least one of the adherends is porous. It is particularly adaptable for use with starch or starch-base adhesives commonly used in the paper-converting industry. The penetration testing apparatus is so constructed as to allow a section of round steel bar stock 25 mm (1 in.) in diameter and weight 700 grams (the hammer) to fall freely through a distance of 0.8 mm (31 in.) upon a steel disk (the anvil) 6 mm (1/4 in.) thick and 51 mm

(2 in.) in diameter, in such a manner as to strike the anvil at right angles. Dimensions of the guiding tube and of the section of bar stock permit free fall of the bar stock without undue sideways movement.

The procedure involves stacking five sheets of filter paper, one on top of another, upon the anvil at the bottom of the tube. One-tenth ml of the adhesive under test is applied to the center of the bottom end of the hammer, using a 1-ml syringe. Thirty seconds after the adhesive is applied to the hammer, it is released and allowed to fall freely. The hammer is then raised, the filter papers removed and then stained with a slight excess of staining solution suitable for the adhesive being tested. KI-I_2 is appropriate for adhesive containing starch. Adhesives that are basic (high pH) may be stained with phenolphthalein or other suitable indicator solution. The stained papers are allowed to air-dry for one hour, after which is densitometer and planimeter are used to calculate the density and stained areas. The results are expressed as J_a, density times area, and maximum, minimum, and average values for penetration.

ASTM D 2093-84—Standard Recommended Practice for *Preparation of Surfaces of Plastics Prior to Adhesive Bonding*, 3 pp

This practice describes surface preparations for plastic adherends to be used prior to adhesive bonding of test specimens. It does not cover the actual preparation of test specimens, testing conditions, or evaluation of tests. *Physical treatments*, such as sanding and solvent wiping, are used to remove the glossy finish and all traces of dirt, grease, mold release, and other contaminants from the bonding surfaces. *Chemical treatments*, such as sulfuric acid-dichromate solution and sodium naphthalene complex, are used in some cases to chemically alter the surface layers of the polymer itself to improve its adhesion characteristics. Two groups of plastic adherends are considered. Group I covers plastics that need only physical treatment, and Group II covers those that need a chemical treatment after solvent wiping. The latter group includes polyolefins, chlorinated polyether, polyformaldehyde, polytrifluoromonochloroethylene, and polytetrafluoroethylene.

ASTM D 2094-69 (1980)—Standard Practice for *Preparation of Bar and Rod Specimens for Adhesion Tests*, 6 pp

This practice describes bar- and rod-type butt-joined specimens and procedures for preparing and bonding them. The specimens are intended to be used with various adherend materials in like or unlike combinations for determining the strength properties of adhesives in accordance with ASTM D 2095 following. Details are given for metal, reinforced and nonreinforced plastics (thermosetting or thermoplastic). Topics covered include geometry, machining of adherends, and surface preparation.

ASTM D 2095-72 (1983)—Standard Test Method for *Tensile Strength of Adhesives by Means of Bar and Rod Specimens*, 4 pp

This method covers the determination of the relative tensile strength of adhesives by the use of bar- and rod-shaped butt-joined specimens under defined conditions of preparation, conditioning, and testing. The method is applicable to the testing of adhesives with various adherend materials in either similar or dissimilar combinations. The directions for fabricating test specimens are given in ASTM D 2094 above. The specimens are bonded together at the machined control surface. The prepared bar or rod specimen is placed in a tensile testing machine fitted with a special self-aligning holding jig. Steel dowel pins are used to secure the specimens. The crosshead speed is adjusted to 2400–2800 psi (170–195 kg/cm^2) of bond area per minute. The maximum load carried by the specimen at failure is recorded. The percentage of the various types of failure (adhesive, cohesive, contact, or adherend) is also recorded.[5]

ASTM D 2182-72 (1978)—Standard Test Method for *Strength Properties of Metal-to-Metal Adhesives by Compression Loading (Disk Shear)*, 3 pp

This method covers the determination of the shear strength of adhesives when tested on a standard specimen under specified conditions of preparation and loading in compression. It is intended primarily as an evaluation method for adhesives for metals. The test specimen consists of a metal disk having an area of 1 square inch (6.45 cm^2) with a radius of 0.564 inch (14.3 mm) and a thickness of 1/4 inch (6.35 mm), bonded to a strip of metal 4 1/4 inches (107.95 mm) long, 1.128 inches (28.7 mm) wide, and 1/4 inch (6.35 mm) thick. The strip and disk may be of similar or dissimilar metals. The bonded surfaces must be well machined and polished before bonding. The specimen is held upright in a shearing tool or special holding jig which is then placed in the testing machine. An anvil section of the shearing tool is fitted so that the load on the disk, when a compressive load is applied, shears the disk along the plane of the plate section of the specimen. The rate of testing, in compression, is 1200–1400 psi (85–100 kgf/cm^2) per minute to failure. The shear stress at failure is determined in psi or kg/cm^2 of the shear area. The nature of the failure, including the average estimated percentage of failure (adhesive, cohesive, contact), is also reported.[5]

ASTM D 2183-69 (1982)—Standard Test Method for *Flow Properties of Adhesives*, 5 pp

This method covers the determination of the flow properties of adhesives under prescribed heating or curing conditions. The method is intended for adhesives that are used in the form of *films* and adhesives that are applied as *liquids*, but are dried to a relatively tack-free state prior to assembly of the adherends and curing or setting of the adhesives. Flow is calculated from a determination of the weight of adhesive flowing from a given area of the adhesive film during curing or setting. The method provides reasonably accurate information regarding the flow of adhesive during cure or setting under pressure. This information is useful for determination of pressure requirements for optimum bond thickness. It can also be used to determine advancement of cure during shelf aging of an adhesive. Data obtained by the method may be used for manufacturing control, specification acceptance, research, and development.

A special flow-test apparatus is used to apply pressure to the specimen. Unsupported film adhesives are tested as received. With liquid adhesives, a film is prepared by casting onto uncoated cellophane, TFE-fluorocarbon, or other suitable parting surface. The adhesive is then dried but not cured. Weight ratios are then determined by the use of circular-disk specimens cut with dies. In the test circular-disk specimens of 40.5 mm (1.6 in.) diameter are dried and placed between two slightly larger disks of uncoated cellophane or other similar parting film. The sandwich is then placed in the flow test apparatus and pressure applied according to the manufacturer's recommendations by compression of the spring. This entire assembly is then placed in an oven to flow and cure the adhesive according to the manufacturer's recommendations of time, temperature, and pressure.

Flow is determined after the required exposure period and after the test unit is removed from the oven, the cellophane is removed, and the specimens dried with a lint-free cheesecloth. A specified-size circular disk is cut from the center of the cured adhesive specimen and weighed after drying. The percentage flow is then calculated.

ASTM D 2293-69 (1980)—Standard Test Method for *Creep Properties of Adhesives in Shear By Compression Loading (Metal-to-Metal)*, 3 pp

This method covers the determination of the creep properties of adhesives for bonding metals when tested on a standard specimen and subjected to certain conditions of temperature and tensile stress in a spring-loaded testing apparatus. The specimens are similar to the ASTM D 1002-type tension lap-shear specimens, except the length of either side of the shear area is 1/4 inch (6.35 mm) rather than 3 1/2 inches (88.9 mm). Specimens have polished bond area edges scribed with three fine lines across the glueline for creep measurement. A special slotted spring is used to hold the specimen. The short-stub-overlap test specimen is placed in the slot made by compressing the spring. The specimen is confined between two washers (bushings). The correct compression load is controlled by the degree of compression of the calibrated spring. With a static load on the specimen, the creep of the bond line is measured by following the movement of the scribe lines with a traveling microscope. Creep is followed as a function of time. Static load and temperature may be varied for a particular test.[5]

ASTM D 2294-69 (1980)—Standard Test Method for *Creep Properties of Adhesives in Shear By Tension Loading (Metal-to-Metal)*, 3 pp

This method covers the determination of the creep properties of adhesives for bonding metals when tested on a standard specimen and subjected to certain conditions of temperature and tensile stress in a spring-loaded testing apparatus. The method can be used in the temperature range of −55 to 260°C (−67 to 500°F). The test specimens are similar to ASTM D 1002-type tension lap-shear specimens except that holes are drilled in the end tab about 1/2 inch (12 mm) from each end. The overlap edges are polished and have three scribelines for creep measurement. The specimen is loaded into a special tension creep test jig while it is being "compressed" at a given loading by a testing machine. The specimen is locked into place by pins placed through the holes in the specimen and tabs. When the compression is released, the lap-shear joint is under tension by the action of a spring in the test apparatus. The tensile creep of the lap joint is measured by a travelling microscope by following the movement of the scribe lines. Creep is followed as a function of time. Static load and temperature may be varied for a particular test.[5]

ASTM D 2295-72 (1983)—Standard Test Method for *Strength Properties of Adhesives in Shear By Tension Loading at Elevated Temperatures (Metal-to-Metal),* 4 pp

This method covers the determination of the comparative shear strengths of adhesives for bonding metals when tested on a standard specimen and under specified conditions of preparation and testing at elevated temperatures (315 to 850°C or 600 to 1500°F). The testing machine conforms to ASTM D 638, except that pin-type grips are used to hold the specimen. The specimens are similar to the tension lap-shear specimens described in ASTM D 1002, except that pin-type grips are used. Unless otherwise specified, the test specimens are loaded at a rate of 8.3 to 9.7 MPa (1200 to 1400 psi) per minute. The load at failure is recorded and the amount of this failure (cohesive, adhesive, substrate, etc.) reported in MPa or psi.

ASTM D 2339-82—Standard Test Method for *Strength Properties of Adhesives in Two-Ply Wood Construction in Shear By Tension Loading,* 4 pp

This method covers the determination of the comparative shear strengths of adhesives when tested on a standard specimen and under specified conditions of preparation, conditioning, and testing. It is intended to be used only for adhesives used in bonding wood to wood. By the nature of the manner in which adhesives are used in two-ply wood construction, shear strength is an important performance criterion. Sweet or yellow birch veneer is used in the method. The average load at failure and the average percentage of wood failure is reported.

ASTM D 2556-69 (1980)—Standard Test Method for *Apparent Viscosity of Adhesive Having Shear-Rate-Dependent Flow Properties,* 3 pp

This method covers the measurement of the apparent viscosity of shear-rate-dependent adhesives. The principle of measurement is based upon a reversble isothermal change in apparent viscosity with change in rate of shear. Measurement is performed with a spindle disk, T-bar, or coaxial cylinder rotational viscometer under standardized conditions with rigid control of the time intervals of measurement. Readings are obtained on the viscometer dial scale at the end of 1 minute for each rotational speed. Changes from the lowest speed to the highest speed and return to the lowest speed are made without stopping the instrument. The adhesive sample and instrument are conditioned at 23°C (73.4°F) or other temperature specified by the adhesive vendor and purchaser for at least 16 hours. The results are reported in poises or equivalent units associated with the particular test equipment at one or more selected rotational speeds. A statement is made as to whether the results were obtained while increasing or decreasing the rotational speeds. The Brookfield Synchro-Lectric Viscometers, Models LV, RV, or HV, have been found satisfactory.

ASTM D 2557-72 (1983)—Standard Test Method for *Strength Properties of Adhesives in Shear By Tension Loading in the Temperature Range From −267.8 to −55°C (−450 to −67°F),* 4 pp

This method covers the determination of the comparative shear strength of adhesives for bonding metals when tested on a standard specimen under specified conditions of preparation and testing at extreme sub-zero temperatures. The testing apparatus is similar to ASTM D 1002, except that pin-type grips are used to hold the test specimen. The cooling equipment is a cold box or cryostat filled with a gaseous or liquid refrigerant in which the standard specimen is immersed prior to and during the test. The test specimens are similar to those used in ASTM D 1002, except that doublers and pin grips are used. The selection of adherends is based on the test temperature range. Many alloys become brittle at extreme sub-zero (cryogenic) temperatures. In addition to the failing load, the nature of the failure, including the average estimated percentage of failure in the cohesion of the adhesive, contact failure, and adhesion to the metal, are reported.

ASTM D 2558-69 (1984)—Standard Test Method for *Evaluating Peel Strength of Shoe Sole-Attaching Adhesives*, 4 pp

This method covers the determination of the peel strength of the adhesive bond on shoe-soling materials. These materials are bonded to standard control materials (rubber composition or vinyl plastic) under controlled conditions. The peel-strength measurements are made under specified conditions and intervals. A tension-testing machine, a source of infrared heat to produce temperatures of 225°F (107°C) in the adhesive film within 60 seconds, a cushioned pressure device, and a constant-temperature oven to provide 105°F (41°C) or 120°F (49°C) are required. The test specimen consists of an 8 inch by 1 inch by 1/8 inch (203 mm by 25.4 mm by 3.2 mm) thick sample of shoe-soling material bonded to a control adherend of the same dimension. For vinyl plastic-soling materials, specimens are of the same thickness, but have standard specified hardness, tensile strength, specific gravity, elongation, abrasive index, and 100% modulus. In the test, the 8 inch by 1 inch strips of material are bonded together with one end of the specimen opened up 1/2 inch (12.7 mm) for clamping on the test grips of the tensile testing machine. Surface preparation procedures are prescribed prior to bonding. After bonding, various directions are given for conditioning in conditioning ovens. Vinyl plastic-soling materials are also subjected to 50,000 flex cycles on a Ross Flex Tester, followed by a second conditioning at room temperature.[5]

In testing, a load is applied at a rate of 2 inches (305 mm) per minute. The peeling force is measured during the test and the average peel strength in lb/inch (kg/cm) of width is recorded, as well as the mode of failure of the test specimen (adhesive, cohesive, or contact failure due to voids in the bond line).

ASTM D 2651-79 (1984)—Standard Practice for *Preparation of Metal Surfaces for Adhesive Bonding*, 7 pp

This practice covers procedures that have proved satisfactory for preparing various metal surfaces for adhesive bonding. Metals covered are aluminum alloy (7 methods), stainless steel (7 methods), carbon steel (2 methods), magnesium alloy (5 methods), titanium alloy (1 method), copper and copper alloys (5 methods).

ASTM D 2674-72 (1984)—Standard Methods of *Analysis of Sulfo-chromate Etch Solution Used in Surface Preparation of Aluminum*, 4pp

These methods offer a means for controlling the effectiveness of the etchant normally used for preparing the surface of aluminum alloys for subsequent adhesive bonding. As the etchant reacts with aluminum, hexavalent chromium is converted to trivalent chromium. A measure of the two and the difference can be used to determine the amount of dichromate used. The sulfochromate solution can be replenished by restoring the sodium dichromate and the sulfuric acid to the original formulation levels. Sludge can be removed from the bottom of the tank. There are three methods. Method A is intended for measuring the sulfuric acid content of a sulfochromate solution. Method B is intended for measuring the hexavalent and trivalent chromium content of a sulfochromate solution. Method C is intended as an alternate method for measuring the hexavalent and trivalent chromium content of a sulfochromate solution.

ASTM D 2739-72 (1984)—Standard Test Method for *Volume Resistivity of Conductive Adhesives*, 3 pp

This method covers the determination of the volume resistivity of resin-based conductive adhesives in the cured condition. The test is made on a thin layer as prepared in a bonded specimen. The method is used for conductive adhesives that are cured either at room temperature or at elevated temperatures. The volume resistivity of adhesive layers cured between metal adherends is measured on a resistance (Kelvin) bridge. Tensile-adhesion plugs are described for the method. Any other test specimens and materials can be used as long as similar precautions are observed regarding preparation and tolerances. Ordinarily brass tensile plugs coated with gold or silver are used. The volume resistivity is reported in ohm-cm ($\Omega \cdot$cm).

ASTM D 2855-83—Standard Practice for *Making Solvent-Cemented Joints With Poly(Vinyl Chloride) (PVC) Pipe and Fittings*, 10 pp (ASTM Committee F-17 on Plastic Piping Systems)

This practice describes a procedure for making joints with polyvinyl chloride plastic (PVC) pipes, both plain-ends and fittings, and bell ends, by means of solvent cements. These procedures are general ones for PVC piping. In nonpressure applications, simplified procedures may be used where outlined in the practice. Manufacturers may supply variations and additional specific detailed instructions for their particular products where necessary. The techniques covered are applicable only to PVC pipes, both plain and bell-end, and fittings of the same class, as described in Specification D 1784. Sections covered include General Principles, Materials, Procedure, Installation, and Safe Handling of Solvent Cement. Fourteen figures are given, along with three appendixes.

ASTM D 2918-71 (1981)—Standard Practice for *Determining Durability of Adhesive Joints Stressed in Peel*, 4 pp

The combination of stress and moisture decreases the durability of most adhesive joints. Stresses in the presence of water or water vapor may cause some adhesive joints to fail at some fraction of the stress required to break the dry joint. The time to failure for a given adhesive joint tested under moist conditions generally decreases with increasing stress, temperature, and relative humidity.

This practice provides data for assessing the durabilities of adhesive joints by means of T-peel-type specimens stressed in contact with air, air in equilibrium with certain solutions, water, aqueous solutions, or other environments at various temperatures. A total of 10 Standard Test Environments are listed, including salt fog. The practice may be used as an accelerated screening test for assessing the durability of adhesive joints. It may be used to measure the durability of adhesive joints exposed outdoors or to environmental conditions experienced by adhesive joints in service. The tests may also be used to determine the effects of various surface preparations or substrates on durabilities of adhesive joints.

The procedure, intended primarily for metal-to-metal laminated assemblies, typically uses 1/2 inch (12.7 mm) wide specimens, each 8 inches (200 mm) long, or any width desired. The 8-inch length has a 1-inch (25.4 mm) long delaminated (previously shimmed) end that can be spread apart forming a "T" until the cross bar measures 4 inches (100 mm). This requires separating the bond 1 inch beyond the delaminated end. In the procedure, one "T" tab portion of the specimen is securely clamped while the other is stressed statically by an attached weight. The procedure can be carried out under ambient conditions, as well as under water if a suitable apparatus is built. The distance peeled as a function of time is recorded. The rate of peeling is averaged over six sequential portions of the distance versus time curve.[5]

ASTM D 2919-84—Standard Test Method for *Determining Durability of Adhesive Joints Stressed in Shear By Tension Loading*, 5 pp (DOD Adopted)   FSC   8040   MR

The first paragraph of the comments for ASTM D 2918 apply to this test method too. The test method provides data for assessing the durabilities of adhesive lap-shear joints while stressed in contact with air, air in equilibrium with certain solutions, water, aqueous solutions, or other environments at various temperatures. As with ASTM D 2918, the method may be used as an accelerated screening test for assessing durability of adhesive joints.

A tensile lap-shear specimen of the ASTM D 1002-type is used. Two coupons of 5 inches (127 mm) in length by 1 inch (25.4 mm) wide by 0.064 inch (1.6 mm) thick sheet metal are overlapped at the one-inch wide end and bonded together with a 1/2 inch (12.7 mm) overlap. Holes are located 1/2 inch from the end of the tab to accommodate the gripping bolt. The test requires the construction of a special double-steel-spring jig. As in the ASTM D 2294 static-load fatigue test, the specimen is held in the spring-compressed jig, the end tabs are gripped, clamped, and the compression on the spring released, leaving the lap-shear specimen under a static tensile load. The maximum, minimum, and average length of time to failure of the stressed specimen exposed to various environments (identical to ASTM D 2918) is recorded. The test is carried out at various levels of fractional stress below the "short-term" failure stress and static stress as time-to-failure curves are plotted.[5]

ASTM D 2979-71 (1982)—Standard Test Method for *Pressure-Sensitive Tack of Adhesives Using an Inverted Probe Machine*, 4 pp

This method covers measurement of the pressure-sensitive tack of adhesives. It is applicable to those adhesives which form a bond of measurable strength rapidly upon contact with another surface, and which can be removed from that surface cleanly, without leaving a residue visible to the naked eye. For such adhesives, tack may be measured as the force required to separate an adhesive and the adherend at the interface shortly after they have been brought into contact under a defined load of known duration at a specified temperature. The method involves bringing the tip of a cleaned stainless steel probe of defined surface roughness into contact with the adhesive at a controlled rate, under a fixed pressure, for a short time, at a given temperature, and subsequently breaking the bond formed between the probe and adhesive, also at a controlled rate. Tack is measured as the maximum force required to break the adhesive bond.

ASTM D 3111-76 (1982)—Standard Practice for *Flexibility Determination of Hot Melt Adhesives By Mandrel Bend Test Method,* 4 pp

This practice covers the determination of the flexibility of a hot-melt adhesive in sheet form under specific test conditions. This is a working test and its results are useful for comparing adhesives, not for absolute characterization of adhesives. In the procedure test strips of a hot-melt adhesive properly sized and conditioned are bent 180° over a mandrel (rod). Using a fresh specimen for each test, the test is repeated with smaller-diameter mandrels until the adhesive fails on bending. The flexibility of the adhesive is the smallest-diameter mandrel over which 4 out of 5 test specimens do not break.

Rod diameters are usually 1/8 inch (3.2 mm), 1/4 inch (6.4 mm), and 1/2 inch (12.8 mm), and rods are 3 to 6 inches (75 to 150 mm) long and made from brass or stainless steel.

ASTM D 3121-73 (1984)—Standard Test Method for *Tack of Pressure-Sensitive Adhesives By Rolling Ball,* 3 pp

This method covers measurement of the comparative tack of pressure-sensitive adhesives by a rolling ball, and is most appropriate for low-tack adhesives. The method is only one of several methods available for the measurement of tack. In the method, a steel ball 3/16 inch (11.1 mm) in diameter is released at the top of an incline, allowed to accelerate down the incline and roll onto a horizontal surface covered with a pressure-sensitive adhesive. Tack is determined by measuring the distance that the ball travels across the adhesive before stoping. The ball is retarded by the adhesion between it and the adhesive (grab) and the "plowing effect" or energy required to push the adhesive out of the ball's path. The test specimen is a substrate coated with a pressure-sensitive adhesive. It is generally about 2 inches (51 mm) wide and approximately 15 inches (381 mm) long.

ASTM D 3163-73 (1984)[e1] —Standard Test Method for *Determining the Strength of Adhesively Bonded Rigid Plastic Lap-Shear Joints in Shear By Tension Loading,* 3 pp

This test method is intended to complement ASTM D 1002 and extend its application to single-lap-shear adhesive joints of rigid plastic adherends. It is useful for generating comparative shear strength data for joints made from a number of plastics. It can also provide a means by which several plastic surface treatments can be compared. The test method is limited to test temperatures below the softening point of the subject adherends, and is not intended for use on anisotropic adherends such as reinforced-plastic laminates. Where possible, test specimens should conform to the form and dimensions of ASTM D 1002. However, adherend thickness and joint overlaps must be chosen so that failure will preferably occur in the joint and not the substrate. The optimum surface preparation techniques for the plastic adherend under test should be used. Failing stress is reported in pounds per square inch (psi) or megapascals (MPa) of shear area. Both load at failure and type of failure (percentage cohesive and apparent adhesive) should be reported.

ASTM D 3164-73 (1984)[e1]—Standard Test Method for *Determining the Strength of Adhesively Bonded Plastic Lap-Shear Sandwich Joints in Shear By Tension Loading*, 4 pp

This test method is intended to complement ASTM D 1002 and extend its application to single lap-shear adhesive joints employing plastic adherends. The test method is useful for generating comparative shear strength data for joints made from a number of plastics. It can also provide a means by which several plastic surface treatments can be compared. The test method is limited to test temperatures below the softening point of the subject adherends, and is not intended for use on anisotropic adherends such as reinforced plastic laminates. Specimens must conform to the form and dimensions given in a figure, with a 1/2-inch (12.7 mm) overlap. Alloy 2024-T3 aluminum of 0.064-inch (1.62 mm) thickness is recommended for the metal specimens. The sandwich under test is metal/adhesive/plastic/adhesive/metal. It is important to use the optimum surface preparation technique for the plastic and metal adherends under test. Failing stress in pounds per square inch (psi) or megapascals (MPa) of shear area is reported, along with the types of failure, including percentages of each type.

ASTM D 3165-73 (1979)[e1]—Standard Test Method for *Strength Properties of Adhesives in Shear By Tension Loading of Laminated Assemblies*, 5 pp

This method is intended for use in determining the comparative shear strengths of adhesives in large-area joints when tested on a standard specimen and under specified conditions of preparation and testing. Adhesives are known to respond differently in large-area joints than they do in small. The method is intended for use in metal-to-metal applications, but may be specifically adapted for plastic adherends. The method is particularly useful in that the joint configuration closely simulates the actual joint configurations of many bonded assemblies and can be used to develop design parameters for such assemblies. It can also be used as an in-process quality-control test for laminated assemblies. In practice, the laminated assembly is either made oversize and test specimens removed from it, or a percentage of the assemblies are destructively tested.

In the test with sheet metals, two 0.064 inch (1.62 mm) thick layers of metal are adhesively bonded together in the form of a laminated panel. One-inch (25.4 mm) wide strips of laminate 7 inches + L inches long (77.8 + L mm) must be machined from this laminated panel. The panel is notched or cut through at opposite but displaced sides so that when the panel is tensile stressed an overlap length of L occurs. This length, then, defines the shear area in the fabricated specimen. In the test the ends of the 1-inch wide specimen are placed in the grips of a tensile-testing machine so that the long axis of the test specimen coincides with the direction of applied stress. The load is applied at a rate of 1200–1400 psi (8.3 to 9.7 MPa) of the shear area per minute. The crosshead speed is 0.05 inches/minute for a 1/2-inch^2 shear area. The load is applied continually until failure of the joint occurs. The maximum, minimum, and average values for the failing load are reported, along with comments as to the nature of the failure, including the average estimated percentages of failure in cohesion of the adhesive, contact failure, voids, and apparent adhesion to the metal.[5]

ASTM D 3166-73 (1979)—Standard Test Method for *Fatigue Properties of Adhesives in Shear By Tension Loading (Metal/Metal)*, 3 pp

This method covers the measurement of fatigue strength in shear by tension loading of adhesives on a standard specimen and under specified conditions of preparation, loading, and testing. It is intended for use in metal-to-metal applications, but may be specially adapted for plastic adherends. The single lap-shear specimen of the ASTM D 1002 test is used. The specimen is placed in the jaws of a special tensile-testing machine capable of imposing a cyclic or sinusoidal stress on it. The test is carried out at 1800 cycles/minute, unless otherwise noted. The cyclic load ranges from a maximum (to failure) to approximately 10% of the maximum, the maximum load being selected from previous "static" tests, such as ASTM D 1002. The number of cycles to failure at a given level is recorded and a so-called S–N curve is constructed to display the fatigue behavior of the adhesive material and the joint.[5]

ASTM D 3167-76 (1981)—Standard Test Method for *Floating Roller Peel Resistance of Adhesives*, 5 pp

The purpose of this test procedure is to provide for the determination of the metal-to-metal peel strength of adhesives by a method that will provide good reproducibility at low, as well as high, strength levels and yet allow for a single method of test specimen preparation and testing. The method covers the determination of the relative peel resistance of adhesive bonds between one rigid adherend and one flexible adherend when tested under specified conditions of preparation and testing. A variation in thickness will generally influence the results. For this reason, the thickness of the sheets used to make the test specimens shall be specified in the material specification. When no thickness is specified, the flexible adherend shall be 0.025 inch (0.63 mm) thick and the rigid adherend 0.064 inch (1.63 mm) thick.

The specimen is a thin laminated structural strip 0.5 inch (12.7 mm) wide with unbonded ends suitable for clamping into the test jig. The ideal specimen is composed of one rigid and one flexible adherend.

In the test a special two-roller, drum-like fixture, each roller 1 inch (25.4 mm) in diameter, is placed in the testing machine. The flexible tab of the specimen is placed between the two rollers, leaving the rigid portion resting on the top part of the rollers. A "T"-like shape is formed from the specimen-holding configuration. The flexible tab of the specimen is then clamped in the bottom jaw of the tensile machine, while the top of the roller-drum fixture is clamped in the top jaw. With the flexible tab on the specimen curled over one of the rollers, the specimen is peeled at a crosshead speed of 6 inches (153 mm) per minute. The results are obtained in terms of average peel strength in pounds force per inch (or kilonewtons per metre) of width of each combination of materials and construction under test. The type of failure is also reported.[5]

ASTM D 3236-73 (1978)—Standard Test Method for *Viscosity of Hot Melt Adhesives and Coating Materials*, 11 pp (ASTM Committee D-2 on Petroleum Products and Lubricants)

This method covers the determination of the apparent viscosity of hot-melt adhesives and coating materials compounded with additives having apparent viscosities up to 200,000 millipascal seconds (mPa·s) at temperatures up to 347°F (175°C). Higher viscosities may be practical, although their precision has not been studied. In the test a representative sample of the molten material is maintained in a thermally controlled sample chamber. Apparent viscosity is determined under temperature equilibrium conditions, using a precision rotating spindle-type viscometer (Brookfield). Data obtained at several temperatures can be plotted on appropriate semi-logarithmic graph paper, and apparent viscosity at intermediate temperatures can be estimated.

ASTM D 3310-74 (1983)—Standard *Recommended Practice for Determining Corrosivity of Adhesive Materials*, 2 pp

This recommended practice is intended to determine whether an adhesive material, cured or uncured, is corrosive to a metal. It is a general method intended to screen out those materials that give a visible sign of corrosion. The metal under test is enclosed in a glass container with the adhesive material. The containers are stored in temperature-controlled ovens at various temperatures, with and without water present. A control consisting of glass jars containing samples (strips) of the same metal, with and without water, but with no adhesive, is stored in the same oven for comparison. Oven temperatures recommended are 160°F (71°C), 200°F (93°C), and 250°F (121°C). Other temperatures may also be used. Results are reported after various exposure periods as degree of corrosion on the scale of five. Unless corrosion is rapid, when 1-hour exposure periods may be used, the exposure period is usually 1, 3, and 7 days.

ASTM D 3433-75 (1980)—Standard Practice for *Fracture Strength in Cleavage of Adhesives in Bonded Joints*, 10 pp

This procedure covers the determination of fracture strength in cleavage of adhesives when tested on standard specimens under specified conditions of preparation and testing. It is useful in developing design parameters for bonded assemblies. While intended primarily for metal-to-metal applications, it may be used for plastic adherends if consideration is given to the thickness and rigidity of the plastic adherends.

Cleavage testing is carried out in such a way that a crack is made to extend by a tensile force acting in a direction normal to the crack surface. Load versus load-displacement across the bondline is recorded autographically. The G_{Ic} and G_{Ia} values are calculated from the load by equations that have been established on the basis of elastic stress analysis. The practice will measure the fracture strength of a bonded joint which is influenced by adherend surface condition, adhesive, adhesive-adherend interactions, primers, adhesive-supporting scrims, etc., and will determine in which of these areas the crack grows. Two types of specimens are used: (1) flat adherend specimens, and (2) contoured double-cantilever beam specimens.

G_{Ic} (fracture toughness, to start) and G_{Ia} (fracture toughness from arrest load) are reported, along with the nature of failure, including estimated percentages of each type.

ASTM D 3482-76 (1981)—Standard Practice for *Determining Electrolytic Corrosion of Copper By Adhesives*, 4 pp

This practice covers the determination of whether an adhesive has any corrosive effect on copper. It is ordinarily intended to distinguish materials that might cause corrosion in electrical and electronic equipment. The procedure is a subjective test for which precision and accuracy data have not been established. It is not recommended for adhesives on backing. Although designed specifically for use with copper, other metals may be used.

In the practice, two parallel helices of fine copper wire are laid in etched grooves on a glass tube. The adhesive material is coated over the wires and the tube and then allowed to set or cure. The wired tube is exposed to high humidity with a d–c potential applied between the wires. Corrosion products are observed visually, usually after 1, 3, and up to 15 days.

ASTM D 3528-76 (1981)—Standard Test Method for *Strength Properties of Double Lap Shear Adhesive Joints By Tension Loading*, 6 pp

This method covers the determination of the tensile shear strengths of adhesives for bonding metals when tested in an essentially peel-free standard specimen that develops adhesive stress distribution representative of that developed in a typical low-peel-production type structural joint. The reproducibility of the strengths achieved is directly related to conformance with specified conditions of preparation and testing. To obtain precise comparative results in lap-shear properties, specimen configuration preparation and speed and environment of testing must be carefully controlled and reported.

The test specimens must conform to one of two alternative types, A or B, as specified in the procedure. A testing machine meeting the requirements of ASTM E 4 is used. The maximum, minimum, and average failing stresses are reported in terms of kilograms per square centimeter (kg/cm^2) or pounds-force per square inch (psi) of total shear area. The nature of the failure, including percentages of various types, is also reported.

ASTM D 3632-77 (1982)—Standard Practice for *Accelerated Aging of Adhesive Joints By the Oxygen-Pressure Method*, 12 pp

This practice describes a procedure for estimating the relative resistance to deterioration of adhesive films and adhesive-bonded joints placed in a high-pressure oxygen environment. The instructions cover both wood-to-wood and wood-to-metal joints, as well as free films of adhesive. The effects of chemicals such as fire retardants, preservatives, or wood extractives, can be evaluated by using materials containing these chemicals for adherends. The practice is intended primarily for elastomer-based *construction* adhesives, but can be used for other types that may be susceptible to oxygen degradation. This accelerated test does not correlate exactly with the natural aging of the adhesive because of the varied conditions of natural aging and the absence of factors such as moisture and stress. The results of this accelerated test are only comparative and must be evaluated against the performance of bonded joints whose natural and accelerated aging characteristics are known.

The practice entails subjecting specimens with known physical properties to a controlled aging environment for specified time periods, then observing the physical properties again and noting any changes. The controlled environment consists of elevated temperature $(158°F)(70°C)$ and oxygen at elevated pressures (300 psi) (2.07 MPa). Three types of specimens are used, as follows:

Type A—for wood-to-wood lap (tests shear strength)

Type B—for wood-to-metal lap (tests shear strength)

Type C—for unsupported film (tests flexibility)

Three different oxygen-pressure aging specimens are used. Exposure is for: (1) 500 hours, at the end of which a single physical property is tested, (2) 1,000 hours, at the end of which a single physical property is tested, and (3) up to 1,000 hours, with tests of the physical property after 200, 400, 600, 800, and 1,000 hours. This practice is useful to the adhesive manufacturer in research and development and in manufacturing control. The results are also used for specification acceptance, or as a guide in adhesive selection.

ASTM D 3658-78 (1984)[e1]—Standard Practice for *Determining the Torque Strength of Ultraviolet (UV) Light-Cured Glass/Metal Adhesive Joints*, 5 pp

This practice covers the simplistic comparison of strengths of glass/metal joints when the adhesive is cured by ultraviolet (UV) radiation and standard specimens are used and tested under specified conditions of preparation, radiation, and load. It involves torque-loading UV-bonded hexagonal metal blocks to glass plates. The practice may be used to obtain comparative torque strength-to-failure data for other bonded joint systems, whether or not radiation cured.

ASTM D 3762-79 (1983)—Standard Test Method for *Adhesive-Bonded Surface Durability of Aluminum (Wedge Test)*, 5 pp

This method simulates, in a qualitative manner, the forces and effects on an adhesive-bond joint at metal-adhesive/primer interface. It has proven to be highly reliable in determining and predicting the environmental durability of adherend surface preparations. The method correlates well with service performance in a manner that is much more reliable than conventional lap-shear or peel tests. It is used primarily in aluminum-to-aluminum applications, but it may be used for other metals and plastics, provided consideration is given to thickness and rigidity of the adherends.

In the method a wedge is forced into the bondline of a flat-bonded aluminum specimen, thereby creating a tensile stress in the region of the resulting crack tip. The stressed specimen is exposed to an aqueous environment at an elevated temperature, or to an appropriate environment relative to the use of the bonded structure. The resultant crack growth with time and failure modes is then evaluated. Variations in adherend surface quality are easily observable when the specimens are forcibly, if necessary, opened at the test conclusion.

The test is primarily qualitative, but is very discriminatory in determining variations in adherend surface preparation parameters and adhesive environmental durability. The test has been found useful in controlling surface preparations and in screening surface preparations, primer, and adhesive systems for durability. In addition to determining crack growth rate and assigning a value to it, the failure mode should be evaluated and reported. The wedges are preferably of the same composition as the adherends.

ASTM D 3807-79 (1984)$^{\epsilon 1}$—Standard Test Method for *Strength Properties of Adhesives in Cleavage Peel By Tension Loading (Engineering Plastics-to-Engineering Plastics)*, 4 pp

This method covers the determination of the comparative cleavage/peel strengths of adhesives for bonding engineering plastics when tested on a standard specimen and under specified conditions of preparation and testing. The bonded test panels are cut into 1-inch (25 mm) wide and 7 inches (177 mm) long test specimens. These specimens are bonded for only approximately 3 inches (76 mm) of their length. A tension-testing machine is used to apply a load at a constant crosshead speed of 0.5 inch (12.7 mm) per minute. The average load in kilonewtons per meter width of specimen required to separate the adherends is determined for the first 2 inches (50.8 mm) of cleavage/peel after the initial peak. This determination is best made from the curve, using a planimeter.

ASTM D 3808-79 (1984)$^{\epsilon 1}$—Standard Practice for *Qualitative Determination of Adhesion of Adhesives to Substrates By Spot Adhesion Test Method*, 2 pp

This practice provides a simple qualitative procedure for quickly screening whether an adhesive will, under recommended application conditions, bond to a given substrate. It is not necessary to make bonded assemblies. The practice can be used to determine whether or not an adhesive will continue to adhere to the substrate under specified environmental conditions. It can also be used to evaluate adhesion of a particular adhesive to a variety of substrates, or to obtain "subjective" comparative data between several adhesives on a given substrate by noting the relative ease of failure between the adhesives tested. It is particularly applicable to adhesives that cure or set when exposed to "air" (ambient, heated, etc.) and can be used for anaerobic adhesives if testing is carried out in an oxygen-free atmosphere.

In this practice, spots of adhesive, usually about 1/4 inch (6 mm) in diameter, are placed onto a substrate using the application procedure and curing conditions acceptable to both user and supplier of the adhesive. The substrate preparation and environmental exposure of the spot of adhesive after cure or setting can be varied as desired. The determination of whether or not an adhesive bonds to the substrate is made by simply trying to pry the spot of adhesive from the substrate. The mode of failure is readily evident by examining whether the bond separated adhesively or cohesively, either in the adhesive or substrate.

ASTM D 3929-80 (1984)—Standard Practice for Evaluating the *Stress Cracking of Plastics By Adhesives Using the Bent-Beam Method*, 5 pp

This is a procedure for determining the compatibility of adhesives with plastics, based on whether the adhesive causes cracking of stressed samples. It can be used for sheets, strips, injection-molded tensile specimens, or flexural bars of plastics. The practice involves the qualitative determination of the compatibility of adhesives with plastics by observing the effect of adhesives applied in the liquid state on stressed plastic specimens. Bars of plastic are bent in a three-point loading fixture to cause a predetermined tensile stress on the surface of the bar. The liquid adhesive is then applied to the area of maximum stress, which is checked periodically for crazing or cracking. Due to the stress relaxation behavior of certain plastics, initial stress only can be determined. The stress may decrease significantly during the course of the test. Suggested stress levels for testing are 1000, 2000, and 3000 psi (7, 14, and 21 MPa). For initial screening, 3000 psi (21 MPa) is recommended, since incompatible adhesives will cause rapid crazing or cracking. Lower stress levels may be used for discerning finer differences in materials.

This test should be carried out for 5 days for adhesives that harden or cure. For adhesives that remain in the liquid state on the specimens (i.e., anaerobic adhesives) a 10-day period should be used. Since stress cracking is more likely to occur immediately after application of the adhesive, it is desirable to observe the specimens at frequent intervals at the beginning of the test. A suggested observation schedule is: 1 minute, 5 minutes, 1 hour, 24 hours, and daily thereafter. Test results are reported as the presence or absence of cracks or crazes at a given level.

ASTM D 3931-80—Standard Practice for Determining *Strength of Gap-Filling Adhesive Bonds in Shear By Compression Loading*, 5 pp

This practice covers the determination of comparative shear properties of gap-filling adhesives in wood-to-wood joints at specified thicknesses of bondline in the dry condition, when tested on standard specimens under specified conditions of preparation, conditioning, and loading in compression. It is intended as an evaluation of gap-filling adhesives such as those used to bond plywood to lumber, lumber to lumber, and other similar materials in building constructions.

Hard maple blocks conditioned at 73.4°F (23°C) are used as adherends. A testing machine with a capability of about 15,000 lb (6818 kg) in compression, or of sufficient capacity to test the adhesive in use, is used. The mean shear strength is ordinarily determined at bondline thicknesses of 0.006 and 0.060 inch (0.15 and 1.52 mm). The mean percentage of wood failure is also reported.

ASTM D 3933-80—Standard Practice for *Preparation of Aluminum Surfaces for Structural Adhesives Bonding (Phosphoric Acid Anodizing)*, 5 pp

This practice describes the requirements for phosphoric acid anodizing of aluminum and its alloys for structural adhesive bonding. The steps are too lengthy and complex to outline here. A process flow chart is provided.

ASTM D 3983-81—Standard Practice for *Measuring Strength and Shear Modulus of Nonrigid Adhesives By the Thick Adherend Tensile Lap Specimen*, 17 pp

This practice describes a method of measuring the shear modulus and rupture stress in shear of adhesives in bonded joints. It employs lap-shear specimens with wood, metal, or composite adherends, with adhesives having shear moduli up to 100,000 psi (700 MPa). The practice is suitable generally for joints in which the ratio of adherend tensile modulus to adhesive shear modulus is greater than 300 to 1. It is not suitable for adhesives that have a high shear modulus in the cured state, and that also require elimination of volatile constitutents during cure.

In the procedure lap-shear speimens are prepared with the adhesive under test, using selected adherends. The load-deformation properties of the specimens are measured under specific recommended conditions to yield a first estimate of adhesive shear modulus. This estimate is used to determine the optimized joint geometry for best attainable uniformity of stress distribution in the joints. A second set of specimens is prepared having the optimized joint geometry. The final values for load-deformation properties are then measured under a variety of controlled environmental and experimental conditions. This practice is based on the theoretical analysis of Goland and Reissner, relating stress concentrations in single-lap joints to the geometry of the joint and the mechanical properties of the materials involved.

ASTM D 4027-81—Standard Practice for *Measuring Shear Properties of Structural Adhesives By the Modified-Rail Test*, 12 pp

This practice describes the equipment and procedures used to measure shear modulus and shear strength of adhesive layers between rigid adhesives. The equipment (a universal testing machine and modified-rail shear tool) may also be used for determining the shear creep compliance of the adhesive, the effects of strain history, such as cyclic loading, upon shear properties, and a failure criterion for biaxial stress conditions, such as shear plus tension and shear plus compression. High-density wood is the preferred substrate, but wood-base composites, metal, plastic, reinforced plastics, and other common construction materials may also be used as adherends.

In the practice, shear force is applied to the adhesive through the adherends by a modified-rail shear tool. The adherends are firmly clamped between two pairs of rigid rails, one of which is fixed and the other movable. The rigid rails limit undesired adherend deformaton. The pair of movable rails is fixed to two counter-movement pivot arms. These arms restrict the attached rails (and clamped adherend) to collinear motion with respect to the fixed rails (and clamped adherend). The results are nearly uniform stress and strain distribution and the reduction of normal stress in the adhesive layer under load. A known amount of uniform tensile or compressive force can be applied to the adhesive layer by the shear tool in order to develop a fracture criterion for the adhesive under combined states of stress, such as shear plus tension, or shear plus compression, which commonly occur in bonded structures.

This practice reports the bond shear strength determined at failure and permits the construction of the stress-strain diagram determined from the plot of the load on the shear tool versus the shear displacement of the bond line. Bond strength and the stress-strain diagram may be obtained for a variety of enivronmental and loading conditions.

ASTM D 4299-84—Standard Test Methods for *Effect of Bacterial Contamination on Permanence of Adhesive Preparations and Adhesive Films*, 6 pp [Supersedes ASTM D 1174 (1981)].

These methods cover the determination of the effect of bacterial contamination on the permanency of (1) *adhesive preparations*, by determining the comparative viscosities and observing the appearance of the uncontaminated and the contaminated adhesives when tested under defined conditions, and (2) *adhesive films*, by exposure to cultures of selected species of bacteria growing on agar plates. The test method for determining the resistance of adhesive preparations to bacterial contamination (Section 12) is used to determine susceptibility of the adhesive to biodegradation. This method is useful in demonstrating whether an adhesive preparation is sufficiently protected with biocide to resist attack by bacteria during its shelf life. The method depends on changes in viscosity and physical appearance of the adhesive to provide indications of inadequate protection. Laboratories not equipped to handle a more accurate microbiological determination are generally capable of using this procedure. This method is not suitable for use with mastic adhesives which are mixed with water at the time of use.

The test methods given in section 13, which cover the resistance of adhesive films to bacterial attack, provide an indication of whether the adhesive will carry into the glue line sufficient anti-microbial properties to prevent attack by bacteria.

Three test organisms are used in these methods: *Pseudomonas fluorescens*, *Bacillus subtilis*, and *Proteus vulgaris*. In both the adhesive preparation test and the adhesive film test either a mixed culture or a separate culture inoculation may be used. Controls are set up for all samples with no culture in the adhesive preparation. In the *adhesive preparation test* checks are made on appearance and viscosity at suggested time intervals. In the adhesive film test different procedures are given for *low-viscosity* and *high-viscosity* adhesives or mastics. These test methods were adapted from ASTM D 1174-55 (1981) and from existing microbiological tests currently used in industry.

ASTM D 4300-84—Standard Test Methods for *Effect of Mold Contamination on Permanence of Adhesive Preparations and Adhesive Films*, 6 pp [Supersedes ASTM D 1286-57 (1979)]

These test methods cover the determination of the effect of mold contamination on the permanency of adhesive preparations and adhesive films, as follows: adhesive preparations, by determining the comparative viscosities and observing the appearance of the uncontaminated and the contaminated adhesives when tested under defined conditions; and adhesive film, by exposure to cultures of selected species of mold (*Aspergillus niger v. Tiegh, Aspergillus flavus*, and *Penicillium lutem*), growing on agar plates. The test for the resistance of adhesive preparations to mold contamination is used to determine susceptibility of the adhesive to biodegradation. This method is useful in demonstrating whether an adhesive preparation is sufficiently protected with biocide to resist attack by molds during its shelf life. Laboratories not equipped to handle a more accurate microbiological determination are generally capable of using this procedure. The test for resistance of adhesive films to mold attack provides an indication of whether the adhesive will carry into the glue sufficient anti-mold properties to prevent attack by molds which occur naturally in wood.

ASTM D 4338-84—Standard Test Method for *Flexibility of Supported Adhesive Films By Mandrel Bend Test*, 4 pp (This method is intended to replace Federal Test Method Std. 175A, Method 1081)

This method covers the determination of the flexibility of an adhesive film bonded to a flexible substrate. This is a "working test". The results are useful for comparing flexibility of adhesives and not for absolute characterization of adhesives. In the test a substrate coated with a film of adhesive properly sized and conditioned, is bent 180° over a mandrel (rod), with the adhesive side away from the mandrel. Using a fresh specimen for each test, the test is repeated with progressively smaller diameter mandrels until the adhesive fails (cracks) on bending. The *flexibility value* of the adhesive is the smallest-diameter mandrel over which four out of five test specimens do not break. Test rods 1/8 inch (3.2 mm), 1/4 inch (6.4 mm) and 1/2 inch (12.8 mm) are suggested.

ASTM D 4339-84—Standard Test Method for *Determination of the Odor of Adhesives*, 2 pp (This method is intended to replace Federal Test Method Std. 175A, Method 4051). (Note that Federal Test Method Std. 175B, Method 4051.1 below is similar to this method).

This test provides a means of comparing the odor of an adhesive sample to a reference sample. It is not intended to give an absolute value for the odor of a sample. It can be used for wet or dry samples. A sample of the adhesive, whether liquid or solid, is placed in a 1-quart (1-liter) wide-mouth bottle fitted with a screw cap. The mouth of the jar is covered with aluminum foil and the cap screwed on. A reference sample (control) is handled in a similar manner. The bottles and contents are conditioned for at least 24 hours at ambient temperature and humidity prior to testing by an odor panel of at least three people. The panel members report whether the sample odor is less than, equal to, or greater than the reference material odor.

ASTM D 4426-84—Standard Test Method for *Determination of Percent Nonvolatile Content of Liquid Phenolic Resins Used for Wood Laminating*, 2 pp

This test method is a fast and economical procedure for the determination of the nonvolatile or total solids content of liquid pheonolic resins used for laminating wood. For greater precision and accuracy Test Method D 1582 is recommended. In the test approximately 1 gram of resin is added to a tared aluminum foil dish and the dish and sample weighed rapidly to the nearest 0.1 mg. The samples are dried at 125± 1°C for 1-3/4 hour, then removed and placed in a desiccator for 5-15 minutes, after which they are again weighed. The nonvolatile content is determined from the weight loss.

Society of Automotive Engineers (SAE)[6]

Aerospace Recommended Practices (ARP'S)—These are few in number (only 3), and are somewhat similar to ASTM Practices, which were formerly called Recommended Practices. They describe recommended methods for carrying out procedures which may be called out in the Aeronautical Materials Specifications (AMS's) following.

ARP 1524, April 15, 1978—*Surface Preparation and Priming of Aluminum Alloy Parts for High Durability Structural Adhesive Bonding, Phosphoric Acid Anodizing,* 11 pp

This recommended practice describes the processing system and techniques for the surface preparation and priming of aluminum parts for structural adhesive bonding to achieve optimum bondline durability, corrosion resistance, and manufacturing producibility. Although originally developed and validated for high-strength aluminum alloys 2024, 7075, and 7475 in the hardened condition, it is expected to be applicable to other alloys and tempers. The system has been validated for use with 180°F (82°C) service elastomer-modified epoxy adhesive and corrosion-inhibiting primer. This document outlines the recommended procedures from solvent cleaning through application and cure of the corrosion-inhibiting adhesive primer. This is a *tank procedure*, as opposed to ARP 1575 below, which is a hand-applied procedure.

ARP 1575, October 15, 1979—*Surface Preparation and Priming of Aluminum Alloy Parts for High Durability Structural Adhesive Bonding, Hand Applied Phosphoric Acid Anodizing,* 7 pp

This recommended adhesive practice describes a *hand-applied*, non-tank, phosphoric acid-anodizing process for surface preparation of aluminum alloys for structural adhesive bonding to achieve optimum bonding durability, corrosion resistance, and manufacturing producibility. Although specifically developed and validated for high-strength aluminum alloys 2024, 7075, and 7475 in the hardened condition, it is expected to be applicable to other alloys and tempers. The system has been validated for use with 180°F (82°C) service elastomer-modified adhesive and corrosion-inhibiting primer. This document outlines the recommended procedures from hand solvent cleaning, surface abrasion, and *non-tank* phosphoric acid-anodizing through application and use of the corrosion-inhibiting adhesive primer. The phosphoric acid is gelled for this procedure.

ARP 1610, October 1, 1981—*Physico-Chemical Characterization Techniques, Epoxy Adhesive and Prepreg Resin System,* 28 pp

This recommended practice describes the physical and chemical characterization techniques for *identification* of epoxy adhesive and prepreg resin systems in order to verify the chemical formulation, resin B-staging, cure reaction rates, adhesive moisture content, and resin component mix ratios, as necessary to achieve manufacturing and quality producibility and engineering performance. Although developed specifically for epoxy adhesive and prepreg resin systems, these techniques are expected to be applicable to other thermoset materials, such as phenolics, polyimides, or polyesters.

Detailed directions are given for carrying out procedures in the following analytical techniques:

Spectroscopy
    Infrared
    Ultraviolet
    Emission
Chromatography
    Thin layer
    Gel permeation
Moisture content (Karl Fischer)
Percent reinforcement
Hydrolyzable chlorine
Percent sulfur
Atomic absorption for boron analysis (in accelerators)

General Services Administration—Federal Test Methods

Federal Test Method Standard No. 175B—*Adhesive: Methods of Testing*, September 1, 1983, 25 pp

FSC   8040   MR

This document currently consists of only four test methods. The large number of test methods in earlier issues have been replaced by ASTM methods. The standard indicates the ASTM methods superseding each of the old methods. The current four methods are as follows:

Method 1081—*Flexibility of Adhesives*, 2 pp

This method provides a means for determining the flexibility (elasticity) of single films or systems of films of adhesives. A uniform coating of the adhesive is applied to the clean surface of the test panel or coating system. The adhesive is dried and the test panel placed with the adhesive side up on a mandrel of specified diameter at a point equally distant from the top and bottom edges of the panel. The panel is then bent double (180°) or 90°, as specified in the materials specification, in approximately one second. The specimen is then checked for cracks, flakes and chips.

Method 4032.1—*Ash Content of Adhesives*, 4 pp

These two procedures are intended for use in measuring the ash content of adhesives. They are not applicable to adhesives containing decomposable salts, such as zinc chloride. Procedure A is for use with a glue or starch adhesive wherein there is no danger of the nonvolatile content forming a rubbery residue when ignited. Procedure B employs nitric acid to avoid the nonvolatile residue being transformed into a viscous foam when ignited. Procedure A uses a crucible with a capacity of 30 ml or more. A muffle furnace kept at 1022-1112°F (550-600°C) is used to combust the sample. Procedure B uses an evaporating dish of 150-ml capacity and a muffle furnace kept at 1112°F (600°C).

Method 4041.1—*Grit, Lumps, or Undissolved Matter in Adhesives*, 1 p

This method is for determining whether a liquid, water-based adhesive such as starch, dextrin, casein, latex or resin-base contains grit, lumps, or undissolved matter. Unless otherwise specified, approximately 10 grams of the adhesive material are used. The adhesive components are mixed in the proper proportions and a thin layer of the resultant product is applied to a sheet of white bond paper. The sheet is placed face-down on a piece of glass and rolled with a paper or print roller to eliminate air bubbles and wrinkling. The sheet is then felt by hand to determine any unevenness caused by grit or undissolved matter.

Method 4051.1—*Odor Test for Adhesives*, 3 pp

This method is completely different from the earlier Method 4051. In the method 20 grams of adhesive sample are placed into a wide-mouth glass bottle, covered with aluminum foil, and the cap screwed on. The bottle with sample is then aged for at least 24 hours at $23°C$ and 50% RH, along with a control containing a reference adhesive known to have an acceptable odor. Heating to higher temperatures for 2 hours or longer may be used. The bottles are then opened and smelled by a panel of at least five members. The sample is rated as to whether the odor of the test adhesive is less than, equal to, or greater than the reference material.

SPECIFICATIONS

American Society for Testing and Materials (ASTM)[4]

ASTM C 557-73 (1978)—Standard Specification for *Adhesives for Fastening Gypsum Wallboard to Wood Framing*, 7 pp (DOD Adopted) (ANSI Adopted (ASTM Committee C-11 on Ceilings and Walls)

FSC   8040   YD

This specification covers minimum standards for bonding the back surfaces of gypsum wallboard to wood framing members. It also covers test requirements and test methods for the adhesive used for the application of all thicknesses of gypsum wallboard. Organic adhesives are required. Requirements cover workability, consistency, open time, wetting characteristics, shear strength, tensile strength, bridging characteristics, aging, freeze-thaw stability, vinyl-covered gypsum board compatability, and adhesive staining.

ASTM F 656-80—Standard Specification for *Primers for Use in Solvent Cement Joints of Polyvinyl Chloride (PVC) Plastic Pipe and Fittings*, 2 pp (ASTM Committee F 17 on Plastic Piping Systems)

This specification provides general requirements for primers for use with polyvinyl chloride (PVC) pipe and fittings that are to be joined by PVC solvent cements meeting the requirements of Specification D 2564. These primers are recommended for use in pressure and nonpressure applications with either plain-end pipe and socket-type fittings or bell-end pipe. A recommended procedure for using the primer with cement is given in Recommended Practice D 2855.

The primers shall be organic liquids with water-like viscosities and shall not contain any undissolved particles. The solvent system to be used is not specified. Requirements include dissolving ability and stability.

ASTM E 865-82—Standard Specification for *Structural Film Adhesives for Honeycomb Sandwich Panels*, 5 pp (Committee E-6 on Performance of Building Construction)

This specification covers film adhesives for bonding of honeycomb sandwich panels used in tactical shelters. The adhesives may be used fcr new production or depot repair. The adhesives should be suitable for forming bonds that can withstand long exposures to temperatures from -67 to $200°F$ (-55 to $93°C$) and also withstand the combinations of stress, temperature, and relative humidity expected to be encountered in service. The adhesives are for use in bonding aluminum alloy facing to nonmetallic core, inserts, edge attachments and other components of a sandwich panel. The specification is the direct responsibility of Subcommittee E06.23 on Durability of Honeycomb Sandwich Panels for Tactical Shelters. The adhesive covers thermosetting films only. Requirements include working characteristics (application, curing), storage life, film weight, normal-, high-, and low-temperature shear strength, humidity exposure, saltspray exposure, low-temperature floating-roller-peel strength, dead-load stress durability, normal- and high-temperature climbing-drum-peel strength, and flatwise-tensile strength.

ASTM E 866-82—Standard Specification for *Corrosion-Inhibiting Adhesive Primer for Aluminum Alloys to Be Adhesively Bonded in Honeycomb Shelter Panels*, 5 pp (Committee E-6 on Performance of Building Construction)

This specification covers sprayable, pigmented liquid primers for use on aluminum alloys that are to be adhesively bonded in the fabrication of honeycomb sandwich panels for tactical shelters. When applied to a properly cleaned surface of aluminum alloy, the primer imparts corrosive resistance and forms a surface suitable for structural bonding using adhesives complying with Specification E 865 and for coating with shelter paint finishes. The specification is the direct responsibility of Subcommittee E06.23 on Durability of Honeycomb Sandwich Panels for Tactical Shelters. The primer is a pigmented liquid composed of a modified epoxy resin system, compounded so that it can be spray-applied to produce a continuous uniform coating without addition of solvent. Requirements for the *uncured primer* include solids content, viscosity, color and sprayability. Requirements for the film of *cured primer* on the primed surfaces include adhesion to metal, impact resistance, pencil hardness, corrosion resistance, humidity resistance, heat resistance, and low-temperature shock. Requirements for the *bonded specimens* (using Specification E 865 adhesive) include lap-shear strength at normal temperatures of $-67°F$ ($-55°C$), $200°F$ ($93°C$), $93°C$ and 95% RH, and $95°F$ ($35°C$) with salt spray, and metal-to-metal peel strength at RT and at $-67°F$ ($-55°C$).

ASTM D 1580-60 (1984)—Standard Specification for *Liquid Adhesives for Automatic Machine Labeling of Glass Bottles*, 2 pp

This specification covers starch, dextrin, casein, animal gelatin, and other liquid adhesives (except pressure-sensitive types) used for applying paper labels to glass bottles. It includes provisions for adhesive selection based on satisfactory machinery characteristics on specific labeling equipment, and adequate adhesion under specified storage conditions. It also provides for the control of uniformity between lots of an adhesive selected on the above basis by limiting variations in nonvolatile content and viscosity.

ASTM D 1779-65 (1983)—Standard Specification for *Adhesive for Acoustical Materials*, 3 pp (DOD Approved) (ANSI Adopted)

FSC   8040   YD

This specification covers an adhesive for bonding prefabricated acoustical materials to the inside walls and ceilings of rooms in buildings. The adhesive must maintain a tensile adhesion (bond strength) of not less than 3.45×10^4 dynes/cm^2 (1/2 psi) for a long period of time under the temperature and moisture conditions likely to be encountered, and to maintain sufficient plasticity to allow for movement of parts of the building as it ages. Requirements covered are: composition (general), toxicity, workmanship, consistency, wetting, strength, shrinkage, cracking, migration or bleeding, and storage. Test methods are given for each requirement.

ASTM D 1874-62 (1981)—Standard Specification for *Water- or Solvent-Soluble Liquid Adhesives for Automatic Machine Sealing of Top Flaps of Fiberboard Shipping Cases*, 2 pp

This specification covers starch, dextrin, casein, resin base, and other liquid adhesives (except pressure-sensitive types) used for sealing the top flaps of fiberboard shipping cases. It includes provisions for adhesive selection based on: (1) satisfactory machining characteristics on specific equipment, and (2) adequate adhesion under specified storage conditions. It also provides for the control of uniformity between lots of an adhesive selected on the above basis by limiting variations in nonvolatile content, consistency, bonding permanency, and water absorptiveness.

ASTM D 2235-81—Standard Specification for *Solvent Cement for Acrylonitrile-Butadiene Styrene (ABS) Plastic Pipe and Fittings*, 5 pp (ASTM Committee F-17 on Plastic Piping Systems)

This specification provides general requirements for solvent cement for use in assembling acrylonitrile-butadience-styrene (ABS) plastic pipe and socket-type fittings. The cement is intended for application by brush or other suitable applicator. Recommended procedures for using the cement covered in this specification to obtain satisfactory joints with ABS pipe and fittings are given in an Appendix. The ABS cement shall be a solution of ABS in methyl ethyl ketone (MEK). Minimum and maximum contents of acrylonitrile, butadiene, and styrene are specified. The cement shall be free-flowing, shall not contain lumps, undissolved particles, or foreign matter, and shall show no gelation or separation that cannot be removed by stirring. Other requirements include resin content, dissolution, viscosity, and lap-shear strength.

ASTM D 2559-84—Standard Specification for *Adhesives for Structural Laminated Wood Products for Use Under Exterior (Wet Use) Exposure Conditions*, 7 pp

This specification covers adhesives suitable for bonding of wood, including treated wood, into structural laminated wood products for general construction, for marine use, or for other uses where a high-strength, waterproof adhesive bond is required. The adhesive shall be classified by the manufacturer as to general type, such as resorcinol, phenol-resorcinol, phenol, and melamine. Requirements cover resistance to shear by compression loading, resistance to delamination during accelerated exposure to wetting and drying, and resistance to deformation under static load. A pH requirement of no less than 2.5 is also given. Test methods for each requirement are prescribed.

ASTM D 2560-80—Standard Specification for *Solvent Cements for Cellulose Acetate Butyrate (CAB) Plastic Pipe, Tubing, and Fittings*, 5 pp (ASTM Committee F-17 on Plastic Piping Systems) (ANSI Adopted)

This specification provides the general requirements for cellulose acetate butyrate (CAB) solvent cements to be used in joining cellulose acetate butyrate pipe, tubing, and fittings. A recommended procedure for joining CAB pipe, tubing, and fittings with the solvent cements specified is given in an Appendix. Requirements include composition (weight percent of CAB and thinner and composition of thinner), viscosity, and lap-shear strength.

ASTM D 2564-80—Standard Specification for *Solvent Cements for Polyvinyl Chloride (PVC) Plastic Pipe and Fittings*, 5 pp (Committee F-17 on Plastic Piping Systems) (ANSI Approved) (DOD Approved; Supersedes MIL-A-22010A as of Feb. 9, 1984)

This specification provides general requirements for polyvinyl chloride (PVC) solvent cements to be used in joining polyvinyl chloride pipe and socket-type fittings. A recommended procedure for joining PVC pipe and fittings is given in Recommended Practice D 2855. General requirements are given and detailed requirements cover resin content (10% minimum PVC), dissolution, viscosity, lap shear strength and hydrostatic burst strength. An appendix on PVC solvent cement selection is included.

ASTM D 2851-70 (1981)—Standard Specification for *Liquid Optical Adhesive*, 3 pp

This specification covers liquid optical adhesive for use in bonding glass to glass or other transparent adherends. The adhesive must be liquid and free from solvent. Each component must be completely reactive (without residual volatile products). The adhesive may be heat-, catalyst-, or radiation-cured. Requirements cover: volatility, viscosity, color, cleanliness, refractive index, stability, light transmission, environmental exposure, and bond strength.

ASTM D 3024-84—Standard Specification for *Protein-Base Adhesives for Structural Laminated Wood Products for Use Under Interior (Dry Use) Exposure Conditions*, 5 pp

This specification covers adhesives suitable for the bonding of wood into structural laminates where the use environment will not produce an equilibrium moisture content in the wood in excess of 16%. The specification covers vegetable or animal protein, such as isolated soy protein and casein. The adhesive must be mold- and water-resistant. The performance requirements of the adhesive are based on tests of the bonded wood, as measured by (1) shear strength and wood failure on laminated wood, and (2) shear strength on yellow birch or Douglas fir plywood; dry, wet, and after exposure to mold growth.

ASTM D 3110-82—Standard Specification for *Adhesives Used in Nonstructural Glued Lumber Products*, 14 pp

This specification establishes acceptable performance levels for glues or adhesives to be used in nonstructural glued lumber products, including interior and exterior moldings, window and door stock, and glued lumber panels. Adhesives that meet the respective requirements of the qualification test are considered capable of providing an adequate bond for use under the conditions described in the class. The specification is used to evaluate only the adhesives, not the glued wood product made with the adhesives. Two types of joints are covered—laminate joints and finger joints. A block-shear test is required for laminate joints, and a flexure test and tension test for finger joints. Two basic types of exposure conditions are prescribed—wet use and dry use. Tests for wet use include cured (dry) tests, a boil test, and a vacuum-pressure test. Tests for dry use include cured (dry) tests, a soak test (three cycles), an elevated-temperature (165°F) (2 methods), and a temperature-humidity test.

ASTM D 3122-80—Standard Specification for *Solvent Cements for Styrene-Rubber (SR) Plastic Pipe and Fittings*, 5 pp (Committee F-17 on Plastic Piping Systems) (ANSI Adopted)

This specification provides general requirements for styrene-rubber solvent cements to be used in joining styrene-rubber (SR) plastic pipe and fittings. A recommended procedure for joining styrene-rubber pipe and fittings is given in the appendix. General requirements include composition, solvent (methyl ethyl ketone or toluene), flowability, foreign particles, lumps, gelation, etc. The rubber shall be of the polybutadiene or butadiene-styrene type. Detail requirements cover resin content, dissolution, viscosity, and lap-shear strength.

ASTM D 3138-83—Standard Specification for *Solvent Cements for Transition Joints Between Acrylonitrile-Butadiene-Styrene (ABS) and Polyvinyl Chloride (PVC) Non-Pressure Piping Components*, 5 pp (Committee F-17 on Plastic Piping Systems)

This specification provides general requirements for solvent cements used in joining acrylonitrile-butadiene-styrene (ABS) plastic pipe or fittings to polyvinyl chloride (PVC) plastic pipe or fittings. These cements are intended for use in cementing transition joints between ABS and PVC materials in non-pressure applications only (170 kPa or 25 psi or less). The intended application is for joining ABS building drain to a PVC sewer-pipe system. The solvent cements used for joining PVC pipe and fittings are specified in Specification D 2564. Solvent cements used in joining ABS pipe and fittings are specified in Specifications D 2235. A recommended procedure for joining ABS to PVC pipe and fittings for non-pressure application is given in the Appendix. General requirements cover the composition, solvent (tetrahydrofuran in combination with cyclohexanone or methyl ethyl ketone, or both), flowability, lumps, gelation, etc. Detailed requirements cover resin content, dissoluton, viscosity, lap-shear strength, and hydrostatic-burst strength.

ASTM D 3498-76 (1981)—Standard Specification for *Adhesive for Field-Gluing Plywood to Lumber Framing for Floor Systems*, 13 pp

This specification covers minimum performance standards and test requirements for *gap-filling construction adhesives* for bonding plywood to lumber framing, particularly floor joists, at the construction site. The specification provides a basis for ensuring the quality of the adhesives and is not intended as an application specification. The adhesive must be gap-filling and must set at temperatures as low as 40°F (4.4°C). It must not support mold or bacterial growth. When completely set, the adhesive must form a durable resilent bond when protected from direct exposure to the weather. The adhesive must be suitable for application by a caulking gun, or other pressurized application equipment. It must have an open-assembly time of not less than 10 minutes. It must be applicable when applied to surfaces having a temperature range from 0° to 100°F (−17.8° to 37.8°C). It must also be functional when applied to lumber framing free of standing water, ice or snow. Test methods are specified for: shear strength, gap-filling effect on strength, durability (moisture resistance), durability (oxidation resistance), mold resistance, and bacterial resistance.

ASTM D 3930-84—Standard Specification for *Adhesives for Wood-Based Materials for Construction of Mobile Homes*, 23 pp

This very lengthy specification provides the means of measuring and evaluating the performance of adhesives for structural or semi-structural bonding of wood to wood in the manufacture of mobile homes. The wood, as used here, includes lumber, plywood, particle board, gypsum board, and all materials having wood-based surfaces at the bondline. The adhesives are classified into types and groups. Classifications are based on: (1) resistance to deformation, (2) resistance to water and water vapor, and (3) gap-filling ability. Group A is structural and Group B is semi-structural. There are 3 types, based on specific applications. The classification system groups adhesives according to resistance to deformation under load, resistance to moisture, and gap-filling ability. Minimum test values are specified for certain properties of durability, while other property requirements are left open-ended. Tests include: shear strength, moisture and temperature limitations, high- and low-temperature resistance, aging and oxidation resistance, and mold resistance.

ASTM D 4317-84—Standard Specification for *Polyvinyl Acetate-Based Emulsion Adhesives*, 6 pp

This specification is a replacement for Federal Specification MMM-A-180C, Class B (see below). It has been expanded to include the more water-resistant polyvinyl acetate adhesives currently available. The adhesive-bonding properties are measured by tests performed on maple-block specimens prepared and tested in accordance with Test Methods D 905 and D 906, respectively. The classification is as follows:

Type 1 – Wet use [16–21% moisture in wood, to 71°C (160°F)]

Type 2 – Intermediate use (for interior or protected conditions)

Type 3 – Dry use [RH low and temperature to 71°C (140°F)]

Requirements for the adhesive include viscosity, density, nonvolatile content, pH, freeze-thaw stability, and storage life. Requirements for the adhesive bond include block shear strength (compression), plywood shear tests [dry shear at 24°C (75°F) and 71°C (160°F), two-cycle boil, 48-hour soak, and humidity exposures].

Society of Automotive Engineers (SAE)[6] —Aerospace Materials Specifications (AMS's)

AMS 2491C, October 1, 1981—*Surface Treatment of Polytetrafluoro-ethylene*, 5 pp (DOD Adopted)

FSC MFFP   MR

This specification, which is really a process specification, covers the engineering requirements for preparing surfaces of polytetrafluorethylene (PTFE) for bonding. It also covers the properties resulting from the treatment. The bonding preparation described may adversely affect the electrical properties of the PTFE. A solution of sodium or other alkali metal in anhydrous liquid ammonia or tetrahydrofuran-naphthalene or other suitable solvent is used under prescribed conditions. Test requirements are given for use of the preparation with a specific epoxy adhesive type on aluminum. These tests cover tensile strength and shear strength.

AMS 3106A, April 1, 1983—*Primer, Adhesive, Corrosion-Inhibiting, —67° to +200°F (—55° to +95°C)*, 1 p. (DOD Adopted)

FSC 8040   AS

The requirements of this specification are embodied in the latest issue of AMS 3107 (see below).

AMS 3107, April 1, 1983—*Primer, Adhesive, Corrosion-Inhibiting, for High Durability Structural Adhesive Bonding*, 14 pp (DOD Adopted)

FSC 8040   AS

This specification supersedes AMS 3106 A, dated March 1, 1974. The specification and its four supplementary detail specifications cover corrosion-inhibiting, modified-epoxy primers in the form of ready-to-use sprayable liquids. The adhesive primers are for use primarily in metal surfaces in preparation for high-durability structural adhesive bonding of sandwich panels and metal-to-metal attachments. They may also be used as primers in preparation for final paint-type finishing. The primer should be a sprayable liquid composed of a resin or resin mixture, pigmented and compounded to be compatible with expoxy-base film or paste adhesives. The primer must impart corrosion resistance and contribute to the adhesive properties of the primer and adhesive system.

The primer is useful over the temperature range specified in the applicable detail specification following. Requirements for the *uncured primer* include: color, solids content, inhibitor content, weight per volume, viscosity, sprayability, and pot life. Requirements for the *cured film* applied to panels include: adhesion, flexibility, impact resistance, hardness, fluid resistance, corrosion resistance, heat resistance, low-temperature shock, compatibility with sealant, and compatibility with top coat. Requirements for the cured primer used *with the adhesive* specified in the applicable detail specification include room-temperature lap shear and metal-to-metal peel.

AMS 3107/1, April 1, 1983—*Primer, Adhesive, Corrosion Inhibiting, High Durability Epoxy,* −55° to +95°C (−65° to +200°F), 4 pp (DOD Adopted)   FSC 8040   AS

This material must meet the requirements of the basic specification, AMS 3107, above. It must also be compatible with the AMS 3695/1 epoxy film adhesives (see below) and MIL-C-83286 polyurethane coating. Detailed requirements are given for a large number of properties (19, with some subdivisions).

AMS 3107/2, April 1, 1983—*Primer, Adhesive, Corrosion-Inhibiting, High Durability Epoxy,* −55° to +120°C (−65° to +250°F), 4 pp (DOD Adopted)   FSC 8040   AS

This material must meet the requirements of the basic specification, AMS 3107, above. It must also be compatible with the AMS 3695/2 epoxy film adhesive (see below) and MIL-C-83286 polyurethane coating. Detailed requirements are given for a large number of properties (19, with some subdivisions).

AMS 3107/3, April 1, 1983—*Primer, Adhesive, Corrosion-Inhibiting, High Durability Epoxy,* −55° to +175°C (−65° to +350°F), 5 pp (DOD Adopted)   FSC 8040   AS

This material must meet the requirements of the basic specification, AMS 3107, above. It must also be compatible with the AMS 3695/3 epoxy film adhesive (see below), and silicone resin-based top coat. Detailed requirements are given for a large number of properties (19, with some subdivisions).

AMS 3107/4, April 1, 1983—*Primer, Adhesive, Corrosion-Inhibiting, High Durability Epoxy,* −55° to +215°C (−65° to +420°F), 4 pp (DOD Adopted)   FSC 8040   AS

This material must meet the requirements of the basic specification, AMS 3107, above. It must also be compatible with the AMS 3695/4 epoxy film adhesive (see below) and silicone resin-based top coat. Detailed requirements are given for a large number of properties (19, with some subdivisions).

AMS 3681B, July 1, 1983—*Adhesive, Electrically Conductive, Silver-Organic Base,* 6 pp

This specification covers an electrically-conductive adhesive supplied in two components: a paste of silver-filled epoxy-base adhesive, and a separate curing agent which may be paste or liquid. The adhesive is to be used primarily to provide an electrically conductive bond between metallic, thermosetting plastic, and ceramic surfaces, and as an electrically conductive sealing compound. The silver may be in the form of powder, flakes, or balls. The curing agent shall be an amine-type material. Shelf life must be at least 6 months. Requirements are given for the mixed adhesive for pot life, corrosive or noxious vapors, and consistency. Requirements for the cured adhesive are lap-shear strength (aluminum specimens), volume resistivity, salt-spray resistance, fungus resistance, and corrosivity.

AMS 3685A, December 1, 1951—*Adhesive, Synthetic Rubber, Buna N Type*, 3 pp

This specification covers a Buna N synthetic rubber-base stock dispersed or dissolved in suitable solvents to form a homogeneous product, ready for use without additions other than solvents for thinning with no more than 7.5% by volume of solvent. Requirements are given for brushing characteristics, weathering, corrosion, skinning, package stability, and toxicity. Requirements also cover: adhesion strength in tension, adhesion strength in shear (aluminum to synthetic rubber) adhesion strength in shear (aluminum to cotton webbing), adhesion strength (stripping method), bonding range, aromatic fuel resistance, water resistance, lubricating-oil resistance, cold flow, and softening point.

AMS 3686, September 15, 1975—*Adhesive, Polyimide Resin, Film and Paste, High Temperature Resistant, 315°C or 600°F*, 8 pp (DOD Adopted)   FSC 8040   AS

This specification covers a high-temperature, electrical-grade, polyimide-resin adhesive in the form of film or paste. The adhesive is used primarily for bonding polyimide-laminate-faced sandwich structures for use as radar-transparent assemblies. It is useful over the temperature range of −55° to +315°C (−67° to +600°F). The adhesive is a polyimide-resin system supplied in sheets, rolls of film, or in paste form. It must be suitable for use in electrical applications and shall contain no metal fillers or other inorganic additives, except for anti-oxidants and thixotropic agents of not more than 35% by weight, total, based on cured resin solids. When in film-form, the carrier shall be "E" glass cloth. Storage life must be at least 6 months. Requirements for the *uncured adhesive* cover: solids content of paste adhesive, volatile content of film adhesive, and weight of film adhesive. Requirements for the *cured adhesive* include: tensile shear at six different exposure conditions, and flatwise tensile at three different exposure conditions.

AMS 3687, March 1, 1974—*Adhesive Film, Humidity-Resistant, for Sandwich Panels, −55° to +95°C (−67° to +203°F)*

This specification covers a high-humidity-resistant, modified-epoxy adhesive in the form of film. It is used primarily for bonding aluminum-faced sandwich panels in the construction of lightweight portable shelters. The adhesive is useful in the temperature range specified in the title. The adhesive is a modified-epoxy film supplied in sheets or rolls, consisting entirely of adhesive material, or of a carrier impregnated with adhesive, with a suitable nonadhering separator film on both surfaces. The adhesive must posses high-humidity resistance, be compatible with AMS 3106 corrosion-inhibiting primer (now superseded by AMS 3107 primer), and must not have a deleterious effect on the surfaces or materials being bonded. The storage life must be at least 6 months. Requirements for the *uncured adhesive* include tack, volatile content, and color.

Requirements for the *cured adhesive* (2-hour maximum cure at 175°C (347°F) and 20 psi pressure) include tensile shear (5 tests at various conditions), fatigue strength, creep rupture (2 tests at various conditions), blister detection, tensile shear, sandwich peel (3 tests at various conditions), flatwise-tensile strength (4 tests at various conditions), flexural strength (total load) (4 tests at various conditions, creep deflection in flexure (2 tests at various conditions), with a total of 21 tests. The humidity requirement is for 95% to 100% RH at 95°C (203°F).

AMS 3688A, October 1, 1981—*Adhesive, Foaming, Honeycomb Core Splice, Structural, −55° to +80°C (−67° to +180°F),* 11 pp (DOD Adopted)   FSC 8040   AS

This specification covers a foaming-type, heat-curing, resin-base adhesive in the form of a paste or sheet. It is used primarily in splicing aluminum alloy or nonmetallic honeycomb core, and for providing a shear tie between core edges and inserts or edge members in honeycomb assemblies for use over the temperature range specified in the title. It is useful for filling gaps between core faces which are inserted into channels or similar areas where bonding pressure cannot be obtained. The adhesive is a heat-curing, nominally 120°C (250°F), resin system containing fillers and foaming agents as necessary to meet the requirements specified. The adhesive must have low sagging properties to ensure complete filling of core-splice gaps when splices are cured in a vertical position. The adhesive must be suitable for splicing nonperforated honeycomb core without damage to core node bonds located adjacent to the splice. When supplied as a sheet, the sheet shall be unsupported and be provided with a suitable non-adhering separator film on both surfaces. When in paste form the paste shall be a thixotopic, single-component, or two-part system suitable for extrusion from a disposable cartridge or caulking gun. The *uncured adhesive*, when supplied as a single component, shall have a storage life of at least 3 months at −18°C (0°F). When supplied as a two-part paste system, the storage life at the time and temperature specified by the manufacturer, shall be at least 3 months. Requirements are also given for working life. Requirements for the *cured adhesive* are given for sagging, expansion ratio, peak exotherm, density, and beam shear at −55°C (−67°F), 24°C (75°F), and 82°C (180°F).

AMS 3689A, October 1, 1981—*Adhesive, Foaming, Honeycomb Core Splice, Structural, −55° to +175°C (−67° to +350°F),* 11 pp (DOD Adopted)   FSC 8040   AS

This specification is similar in all respects to AMS 3688A above, except for the curing temperature of 175°C (350°F), and the range of usefulness.

AMS 3690, January 15, 1960—*Adhesive Compound, Epoxy, Room Temperature Curing,* 3 pp

This specification calls for an epoxy-resin adhesive consisting of two components, an epoxy resin and a hardener. Fillers and modifiers may be included in either component. The adhesive is for general-purpose use for *nonstructural bonding* of aluminum, corrosion-resistant steel, brass, and many thermosetting plastics to themselves and to each other. It is primarily intended as an adhesive for electrical components and devices operating at temperatures no higher than 185°F. Requirements cover curing, pot life, corrosively, tensile-shear strength at various conditions, including temperature extremes and thermal cycling, and fluid resistance.

AMS 3691, January 15, 1960—*Adhesive Compound, Epoxy, Medium Temperature Application*, 3 pp

This specification is similar to AMS 3690 above, except for the maximum-use temperature of 121°C (250°F) and curing conditions which call for testing cures carried out at no higher than 121°C (250°F) with less than 10 psi (.07MPa) pressure.

AMS 3692, January 15, 1960—*Adhesive Compound, Epoxy, High Temperature Application*, 3 pp

This specification is similar to AMS 3690 above, except for the maximum-use temperature 260°C (500°F) and curing conditions which call for cures carried out according to the manufacturer's recommendations, with less than 10 psi (.07MPa) pressure.

AMS 3693B, July 1, 1983—*Adhesive, Modified Epoxy, Moderate Heat Resistant, 120°C (250°F) Curing, Film Type*, 7 pp (This specification is similar to Federal Specification MMM-A-132, Type 1, Class 2 and Military Specification MIL-A-25463, Type 1, Class 2, which are described below.)

This specification covers a modified-epoxy adhesive in the form of supported film supplied in rolls or sheets. It is intended primarily for structural bonding of metallic alloys and rigid nonmetallic surfaces to themselves and to each other, and for bonding of internal and external structural honeycomb components operating in the range of −55° to +80°C (−65° to +180°F). A liquid primer suitable for spray or brush application may be required for use with the adhesive. The material consists of a supported film adhesive with protective liners. Uniformly dispersed fillers may be included in the film. The resin shall conform to MMM-A-132, Type 1, Class 2, and MIL-A-25463, Type 1, Class 2. Shelf life must be at least 6 months at −18°C (0°F). *Cured product* requirements cover tensile shear strength at three different temperatures up to 80°C (180°F), and after salt-spray exposure, humidity exposure, and fuel immersion, flatwise tensile strength and flexural strength at various temperatures, T-peel strength at 24°C (75°F), sandwich peel strength at three different temperature conditions up to 80°C (180°F), creep deformation and corrosivity. The adhesive film in each roll or sheet must be protected on one or both sides by nonadhering separator film.

AMS 3695, April 1, 1983—*Adhesive Film, Epoxy-Base, for High Durability Structural Adhesive Bonding*, 14 pp

This specification and its supplementary detail specifications cover film adhesives compounded from modified-epoxy resin in the form of ready-to-use sheet, supplied in rolls, either supported by mat or by woven monofilaments, or unsupported. It is intended primarily for bonding metal to metal or aluminum honeycomb sandwich assemblies for service usage over the temperature range specified in the applicable detail specifications following. The adhesive shall be a modified-epoxy film in sheet or rolls, consisting entirely of adhesive material, or of a carrier impregnated with adhesive, with a suitable nonadhering separator film on both surfaces. The adhesive shall be used with AMS 3107 corrosion-inhibiting primer and cured in accordance with the applicable detail specification to form an adhesive/primer system. The adhesive must not have a deleterious effect on the surface of materials being bonded.

Requirements for the *uncured adhesive* cover color, solids content, weight, thickness and working life. Requirements for the *cured adhesive* for *metal-to-metal* applications cover tensile shear (dry, and after exposure to humidity, salt spray, aromatic fuel, JP-4 fuel, phosphate ester fluid, hydraulic fluid, anti-icing fluid, diester lubricating oil, and polyol ester), fatigue, creep-rupture-deformation, climbing peel (dry, after humidity exposure, and after salt-spray crack extension test (2-edge test), and sustained stress loading. Requirements for the cured adhesive for metal-to-honeycomb application cover climbing peel (dry) and flatwise-tensile strength.

AMS 3695/1, April 1, 1983—*Adhesive Film Epoxy-Base, High Durability, for 95°C (200°F) Service*, 9 pp

This adhesive is intended for use with the AMS 3107/1 corrosion-inhibiting primer (above) to form an adhesive/primer system meeting the requirements specified in the basic specification, AMS 3695 (above). Detailed requirements are given for a large number of properties (15, with subdivisions).

AMS 3695/2, April 1, 1983—*Adhesive Film, Epoxy-Base, High Durability, for 120°C (250°F) Service*, 9 pp

This adhesive is intended for use with the AMS 3107/2 corrosion-inhibiting primer (above) to form an adhesive/primer system meeting the requirements specified in the basic specification, AMS 3695 (above). Detailed requirements are given for a large number of properties (15, with subdivisions).

AMS 3695/3, April 1, 1983—*Adhesive Film, Epoxy-Base, High Durability, for 175°C (350°F) Service*, 10 pp

This adhesive is intended for use with the AMS 3107/3 corrosion-inhibiting primer (above) to form an adhesive/primer system meeting the requirements specified in the basic specification, AMS 3695 (above). Detailed requirements are given for a large number of properties (15, with many subdivisions).

AMS 3695/4, April 1, 1983—*Adhesive Film, Epoxy-Base, High Durability, for 215°C (420°F) Service*, 10 pp

This adhesive is intended for use with the AMS 3107/4 corrosion-inhibiting primer (above) to form an adhesive/primer system meeting the requirements specified in the basic specification, AMS 3695 (above). Detailed requirements are given for a large number of properties (15, with many subdivisions).

AMS 3698, October 15, 1979—*Adhesive Film, Hot-Melt, Addition-Type Polyimide, for Foam Sandwich Structure, —55° to +230°C (—67° to +450°F)*, 11 pp

This specification covers one type of hot-melt, polyimide adhesive in the form of supported film furnished in rolls or cut sheets. It is intended primarily for structural-adhesive bonding of foam sandwich and glass-fiber, honeycomb-core sandwich assemblies requiring high strength, excellent electrical properties, and heat resistance up to 230°C (450°F). The material must be a bis-maleimide, hot-melt, addition-type, polyimide adhesive compounded to meet the requirements specified, and supplied in film form with an E-glass scrim carrier, such as Style 112, coupling with AMS 3824, in rolls or as cut sheet, with a suitable non-adhering separator film on both surfaces. Requirements are given for storage life and working life of the *uncured adhesive*. The adhesive must be compatible with, and capable of being co-cured with, the hot-melt, addition-type polyimide resin-impregnated cloth conforming to AMS 3844 or AMS 3849. Requirements for the *cured adhesive* cover co-cured facings and precured facings separately. The following properties are covered for the uncured adhesive: color, solids content, weight, thickness, tack, and drape. For the co-cured and precured facings there are requirements for flatwise tensile strength at three different temperatures up to 450°F. Honeycomb climbing-peel tests are required for the precured facings at two different temperatures, and tensile-shear (lap) tests are required for aluminum alloy facings. The adhesive film in each roll or sheet must be protected on both sides by nonadhering separator film.

General Services Administration—Federal Specifications and Commercial Item Descriptions[7]

MMM-A-100D, February 7, 1978—*Adhesive, Animal Glue*, 15 pp
FSC 8040   GSA-FSS

This specification covers animal glues for use in woodworking. There are two types, as follows:

Type I – Dry form (flake, ground, or powdered, as specified)
  Grade J1 – lower viscosity and jelly strength
  Grade J2 – higher viscosity and jelly strength

Type II – Liquid form

Type I, grades J1 and J2 adhesive is intended for use for edge joints in furniture, etc. Type II liquid form is intended for use in edge joints, veneering and dowelling in furniture. It may also be used when short assembly periods or high fabricating temperatures are involved. Requirements for Type I include moisture content, viscosity, jelly strength, pH, foam, and odor and keeping quality. Requirements for Type II include hygroscopicity, viscosity, pH, shear strength, foam, and odor.

MMM-A-105, June 21, 1967—*Adhesive and Sealing Compounds, Cellulose Nitrate Base, Solvent Type*, 10 pp
FSC 8040   AR

This specification establishes the requirements for cellulose nitrate-based adhesive and sealing compounds that are dissolved or dispersed in organic solvents. There are two types, as follows:

Type I – Label adhesive

Type II – Adhesive and sealer, general purpose

Type I is for use in attaching printed paper labels to shipping containers. Application of additional compound to the top surface of the label will make it water-resistant. Type-II compound is for use in repairing and mending many materials, including glass, metals, leather, textiles, paper, china, and some types of plastics, as well as in anchoring glass, metal and some plastic laboratory equipment. Requirements include solvent toxicity (benzene and chlorinated solvent prohibited), water resistance, application to brass (Type II only), cold stability (−13°F), application to labels (Type I only), resistance to blocking (Type I only), accelerated weathering (Type I only), nonvolatile matter, viscosity, air-dry time (tack-free), peel strength, flexibility after high- and low-temperature exposure, color, cellulose nitrate polymer, and blushing. [See A-A-529 below for a commecial item description (CID) which may be used to specify this material.]

MMM-A-110B, INT AMD-2, January 12, 1976—*Adhesive, Asphalt, Cut-Back Type (for Asphalt and Vinyl Asbestos Tiles)*, 6 pp and 2-page amendment   FSC 8040   MR

This specification covers one cut-back type of asphalt adhesive suitable for the installation of asphalt and vinyl-asbestos tiles. The adhesive consists of an asphaltic base material, a volatile solvent, and an asbestos fiber or other mineral filler modified to meet the requirements of this specification. The adhesive is intended for adhering asphalt tile and vinyl-asbestos tile to primed and unprimed concrete subfloors, either suspended, on grade, or below grade. It may also be used for bonding these floor coverings to steel or other metal subfloors and suspended plywood or hardwood subfloors that have been properly primed.

The basic specification spells out authorized combinations of solvents permissable, and the amendment modifies this list for temporary use to meet emission requirements. The requirements include volatile solvent (no benzene nor halogenated compounds), (acceptable combinations of solvents, with percentages, are given), viscosity, shelf-storage life, drying time, retained volatile solvent, sag, dry-film sensitivity to pressure, accelerated aging, alkali resistance and flashpoint.

MMM-A-115C, August 6, 1979—*Adhesive, Asphalt, Water Emulsion Type (for Asphalt and Vinyl Asbestos Tile)*, 5 pp

FSC 8040   YD

This specification covers a clay-dispersed water-emulsion type of asphalt adhesive suitable for the installation of asphalt and vinyl-asbestos tiles. The adhesive is intended for adhering asphalt tile and vinyl-asbestos tile to concrete subfloors, either suspended, on grade, or below grade. It may also be used for bonding these floor coverings to steel or other metal subfloors and suspended plywood or hardwood subfloors. Requirements include condition in container, viscosity, drying time, sag, accelerated aging, alkali/resistance, freeze-thaw resistance, workability, and tensile strength before and after aging.

MMM-A-121, December 16, 1966—*Adhesive, Bonding Vulcanized Synthetic Rubber to Steel*, 10 pp

FSC 8040   SH

This specification covers nonstructural adhesives for bonding vulcanized synthetic-rubber gaskets, matting and similar items to steel. Bond strengths are low (about 5 pounds per inch of width). This is a performance specification. Material type is not specified. The requirements include viscosity, solids content, weight of filled containers, wet adhesion, initial adhesion, adhesion after salt-water immersion, adhesion at 140°F, and stability of adhesive (wet-adhesion test). A Qualified Products List is available as QPL-A-121-16, 21 Sept. 1984.

MMM-A-122C, October 30, 1978—*Adhesive, Butadiene-Acrylonitrile Base, Medium Solids, General Purpose*, 6 pp

FSC 8040   MR

This specification covers a high-strength general-purpose adhesive specifically for use where resistance to oil, gasoline, and aromatic fuel is essential. The adhesive is a butadiene-acrylonitrile base material modified to the extent that the requirements of this specification are met. Permissable solvents, including percentages, are specified to meet emission requirements. The adhesive is intended for high-strength bonding of a wide variety of materials, including metal, glass, plastics, and synthetic rubber, particularly the nitrile types. It may also be used as a *primer* for other adhesives. The requirements include solvent content (percentage of allowable solvents specified), appearance, condition in container, nonvolatile content, ash content of nonvolatile matter, viscosity, accelerated aging, low-temperature flexibility (cold brittleness), strip adhesive strength, water resistance, oil resistance, and fuel resistance.

MMM-A-125D, August 27, 1984—*Adhesive, Casein-Type, Water and Mold Resistant*, 10 pp

FSC 8040   MR

This specification covers casein types of adhesives for adhering wood surfaces. The adhesive must be in the form of a dry, uncaked powder, and in such condition that it can be mixed with water. There are two types, as follows:

Type I – Water-resistant

Type II – Water- and mold-resistant

Type I is for use primarily in the woodworking industry, while Type II is used primarily for lumber laminating. Requirements include plywood shear tests (dry shear and wet shear), block-shear test, working life, setting properties, and mold resistance.

MMM-A-130B, INT AMD-3, April 7, 1976—*Adhesive, Contact*, 8 pp plus 1-page amendment

FSC 8040   ME

This specification covers two types of flexible adhesive for contact bonding of plastic decorative laminates to clean, dry, and smooth wood and metal surfaces. The adhesives will firmly bond such materials as leather, wood, fabrics, unglazed ceramics, wallboards, and carpet to themselves and to each other. The adhesive may be used to install cove-base corners and roll-edge counter-top material. There are two types as follows:

Type I - Volatile-organic-solvent type

Type II - Water-dispersion type

This adhesive is composed of polychloroprene (neoprene) rubber and synthetic resins modified to the extent that the requirements of this specification are met. Permitted solvent formulations are given, including percentages, for Type I. Other requirements include nonvolatile content, ash content, viscosity, density, shear strength of bonded joints, edge life, bonding range, accelerated aging, shelf-storage life, and freeze-thaw resistance. [(See also MIL-A-87134 (USAF)]. A QPL is available as QPL-MMM-A-121-16, September 21, 1984.

MMM-A-131A, January 6, 1966—*Adhesive, Glass-to-Metal (for Bonding of Optical Elements)*, 9 pp

FSC 8040   AR

This specification covers solvent-type, synthetic-resin adhesives in a two-part liquid form and in dry-film classes suitable for bonding glass to metal, using heat and pressure. Applications include glass prisms and other optical glass elements for bonding to metal supports in optical fire-control instruments. The classification is as follows:

Type I - Liquid adhesive (two-part thermosetting resin)
   Class 1 - Phenolic-neoprene
   Class 2 - Phenolic-polyamide

Type II - Dry-film adhesive
   Class 1 - Fiberglass (loosely woven) carrier supporting
         phenolic neoprene fabric
   Class 2 - Polyamide (nylon) carrier supporting phenolic
         polyamide

Requirements include thread count (Type II), viscosity (Type I), solids content (Type I), curing treatment, bond strength (with varying conditions of temperature and humidity), shock resistance, and storage life.

A Qualified Products List is available as QPL-MMM-A 131-6, May 19, 1982.

MMM-A-132A (3), November 22, 1982—*Adhesive, Heat Resistant, Airframe Structural, Metal-to-Metal*, 26 pp plus 2-page amendment

FSC 8040   AS

This specification covers the requirements for heat-resistant adhesives for use in bonding primary and secondary structural and external metallic airframe parts which will be exposed to temperatures within the range of $-67°$ to $500°F$ ($-55°$ to $260°C$). The following types and classes, all for long-term exposure, except as noted, are specified:

Type I   - for $-67°$ to $180°F$ ($-55°$ to $82°C$)
   Class 1 - High T-peel and blister detection

Class 2 – Normal T-peel and blister detection

Class 3 – No T-peel and no blister detection

Type II – for $-67°$ to $300°F$ ($-55°$ to $149°C$)

Type III – for $-67°$ to $300°F$ ($-55°$ to $149°C$) *and for short-term exposures* to $300°$ to $500°F$ ($149°$ to $260°C$)

Type IV – for $-67°$ to $500°F$ ($-55°$ to $260°C$)

Form F – film

Form P – paste

The adhesive must be thermosetting, but there are no restrictions as to the chemical nature of the adhesive. Curing agents must be supplied, or must be incorporated into the adhesive. Fillers, high moisture- and corrosion-resistant, are permitted. Primers may be furnished with the adhesive.

Requirements are given for curing and post-curing conditions and storage life. Other requirements include tensile shear at various conditions of temperature and humidity, T-peel, blister detection-tensile shear, fatigue strength, and creep rupture at various temperatures. Types I and II are intended primarily for use on clad aluminum alloys. Types III and IV are designed for use with corrosion-resistant steel or titanium alloys. A Qualified Products List (QPL) is available as QPL-MMM-132-8, April 3, 1984.

MMM-A-134, August 17, 1970—*Adhesive, Epoxy Resin, Metal-to-Metal Structural Bonding*, 29 pp

FSC 8040   AS

This specification covers epoxy resin adhesives for structural bonding, such as in the fabrication and repair of airframe parts, components, and other applications requiring bonding of a similar quality. There are three types, as follows:

Type I   – (two-part) – Room-temperature setting

Type II – (two-part) – Intermediate-temperature setting

Type III – (one-part, film or wet form) – High-temperature setting

These adhesives are primarily for use in clad aluminum alloys. Requirements include application life, curing time and temperature, curing pressure, storage life, tensile-shear at various conditions of temperature, creep rupture at RT and high temperature, fatigue strength, tensile-shear after salt-water immersion, after accelerated weathering, and after immersion in various fluids, and cleavage strength. A Qualified Products List (QPL) is available as QPL-MMM-A134-8, April 3, 1984.

MMM-A-137D, August 4, 1976—*Adhesive, Resilent Flooring (Water Soluble)*, 9 pp   FSC 8040   MR

This specification covers two classes of adhesive for securing resilient flooring to above-grade floors. It is intended for use in the installation of opaque or translucent vinyl-surface floor coverings with backing, and linoleum and other resilient floorings. It is not recommended for installation of coverings on steel or other metal subfloors. There are two classes as follows:

Class 1 - High viscosity, for trowel application (80,000-250,000 cP)

Class 2 - Low viscosity, for spray application (40,000-120,000 cP)

The adhesive consists of a binder in a water-base suspension, intimately mixed with inorganic filler and other desirable additives to control odor and prohibit mold growth. Requirements include uniformity, alkalinity, viscosity, bonding strength, adhesion, and shelf-storage life.

MMM-A-138A(1), June 19, 1975—*Adhesive, Metal to Wood, Structural*, 12 pp plus 1-page amendement

FSC 8040   AS

This specification covers synthetic-resin adhesives for the structural bonding of clad aluminum alloy to wood. There are two types as follows:

Type I - Single-adhesive system

Type II - Two-adhesive system
   Condition A - Primary adhesive with RT, 75-90°F (24-32°C), setting secondary adhesive
   Condition B - Primary adhesive with intermediate-temperature, 90-180°F (32-82°C), setting secondary adhesive

The adhesive consists of a resin or resins furnished in liquid or jelly form, with or without hardener, or in solid form, as stick, powder, or film. The chemical nature of the adhesive is not specified. The Type-I adhesive is intended primarily for use in the structural bonding of aluminum alloy to wood. It is not to be used for structural bonding. The Type-II adhesive is intended primarily for use in fabricating aluminum-wood-aluminum structural sandwich panel constructions, especially where low-density wood cores are used, as well as for the structural bonding of aluminum-to-wood attachments not of sandwich-type construction. Type-II, Condition A adhesive is for applications where small areas are involved, or where heat cannot be used in the final assembly. Type-II, Condition B adhesive is for use in cases where the fabrication of large items necessitates a final assembly time exceeding that permissable with Condition A adhesive. Requirements include working life of adhesives in liquid or jelly form film form composition, accelerator or hardener or modifier, solubility, application, curing conditions, storage life, and shear strength (initial, after high-temperature exposure, immersion in salt water, immersion in fresh water, and immersion in hydrocarbon test fluid), and pH value. A Qualified Products List (QPL) is available as QPL-MMM-A138-6, January 24, 1984.

MMM-A-139A(1), January 31, 1977—*Adhesive, Natural or Synthetic-Natural Rubber*, 17 pp plus 2-page amendement

FSC 8040   AS

This specification covers natural or synthetic-natural-rubber adhesives, primarily for the manufacture and repair of articles made of materials cured with natural or synthetic natural rubber. The classification is as follows:

Class 1 - Heat cure for manufacture

Class 2 – Room-temperature cure for manufacture

Class 3 – Room-temperature cure for repair

There is currently only one form—a two-part system consisting of the base compound and separate accelerator. These adhesives are intended for adhering natural or synthetic-natural rubber, and in particular, for the manufacturing and repair of articles fabricated from natural-rubber-coated fabrics. They must not be used for bonding nylon. Class–3 adhesive is for use only in bonding natural-rubber items in the repair of life-saving equipment. Requirements include total solids, viscosity (as received and after accelerated storage), seam strength (as received and after bond aging), strip adhesion (as received, after bond aging, accelerated aging, and after water immersion), and dead load (at $60°C/140°F$), brushability, working life (two-part adhesive), and accelerated storage. A Qualified Products List (QPL) is available for this class only as QPL-MMM-A139-4, March 18, 1975.

MMM-A-00150B (GSA-FSS), April 29, 1976—*Adhesive for Acoustical Materials*, 9 pp

FSC   8040   GSA-FFS

This interim specification covers one type of adhesive for bonding prefabricated acoustical materials to the inside walls and ceilings of rooms in buildings. The adhesive is not recommended as the sole means of bonding acoustical materials weighing more than 2 1/2 pounds per square foot to ceiling surfaces. The adhesive consists of a resin in a volatile solvent modified to meet the requirements of this specification. Requirements of solvent composition are given to meet emission control requirements. Other requirements include shelf-storage life, consistency, tensile strength (before and after aging), adhesion, wetting, shrinkage, cracking, migration or bleeding, and accelerated aging.

MMM-A-179B, August 18, 1982—*Adhesive: Paper Label*, 16 pp

FSC   8040   MR

This specification supersedes MMM-A-178A, Adhesive, Paper Label, Water Resistant, as well as MMM-A-179A. It covers the requirements for water-resistant, water-emulsion adhesives suitable for adhering printed paper labels to various substrates, such as soft wood, solid fiberboard, hot-rolled low-carbon steel with a black iron oxide coating ("black iron"), glass, tin-coated steel, and enamel-painted metal. There are two types, as follows:

Type I – Water-resistant

Type II – Water-resistant, water emulsion

Latitude is permitted in the selection of raw materials for these adhesives. Requirements include condition in container, toxicity, brushing properties, odor, color, blushing, viscosity (Type- I), drying time, water resistance, transparency, adhesion (under various conditions—high and low temperature, oil, fresh water, salt water, and accelerated aging), bleeding, flexibility, accelerated weathering, flash point (Type-I), flame resistance (Type-I), application over ink, fungus resistance, and storage life.

MMM-A-180C, August 6, 1979—*Adhesive, Polyvinyl Acetate Emulsion,*
5 pp   FSC   8040   AS

This specification covers two types of polyvinyl acetate emulsion adhesives, as follows:

Class A – General-purpose adhesive

Class B – Wood adhesive (see ASTM D 4317-84 for replacement
        specification)

The Class A adhesive is for bonding leather to such materials as metal, wood, cloth, paper, etc. and for the general bookbinding in hand operations. The Class B adhesive is for assembly gluing of wood items which are placed or stored in normal indoor-temperature service conditions where the RH is not high and does not fluctuate excessively. It is intended for such joints as dowel, mortise-tenon, lock, and finger in the assembly of wood patterns and models. It is not suitable for edge gluing and laminating for furniture parts. Requirements include weight per gallon, nonvolatile content, ash, pH, viscosity (as received and after freeze-thaw), peel strength (as received and after freeze-thaw), compression strength on hardwood and plywood (as received and after freeze-thaw), tensile lap shear (at standard temperature, at $71°C$, and after moisture exposure), condition in container, working properties, alkali dispersibility (Class A only), and flexibility.

MMM-A-181D, January 23, 1980—*Adhesives, Phenol, Resorcinol, or Melamine Base,* 13 pp

FSC   8040   MR

This specification covers general-purpose, two-part adhesives, or optionally, a one-part adhesive for Type-III, for wood-assembly gluing. The adhesives are for use in the manufacture of laminated members and other wooden articles where a high-strength, durable adhesive is required. The classification is as follows:

Type-I   – Room-temperature setting (75 to 95°F) (24 to 35°C) glue-
          line temperature
    Grade A – Two years storage life
    Grade B – Six months storage life

Type-II – Intermediate-temperature setting (95 to 190°F (35 to 88°C)
          glue-line temperature
    Grade C – Six months storage life
        Class 1 – Liquid*
        Class 2 – Powder*

Type-III – High-temperature setting (190 to 300°F (88 to 149°C) glue-
          line temperature
        Class 1 – Liquid*
        Class 2 – Powder*

    *Unless otherwise specified, Class 1 will be supplied.

The material shall be furnished as a liquid with a separate hardener, or, when specified, as a powder with a separate hardener. The hardener for the

Type-III adhesive may be incorporated as a component part of the resin. Only resins based on phenol, melamine or resorcinol may be used. These three resins are formulated with formaldehyde, and the D revision contains a cautionary note on the health hazards of formaldehyde compounds. Requirements include insoluble matter, hardener, fillers, pH of set film, cleanability, liquid working life, curing cycles, shear strength by compression load (Type-I and II only), plywood shear strength (dry and wet).

MMM-A-182A, October 6, 1978—*Adhesive, Rubber*, 6 pp

FSC 8040   EA

This specification covers a cold-bonding rubber adhesive for use with natural or synthetic rubber. It is commercially known as *rubber cement* for cold patching. There are two types, as follows:

Type I - Adhesive with a minimum of 6% rubber

Type II - Adhesive with a minimum of 10–12% rubber

The rubber used in making the adhesive shall be first-quality smoked sheet, pale crepe, or fine Para. It shall be plasticized only enough to provide good brushing qualities. Modifying resins may be added to permit the requirements of the specification to be met. The solvent used is naphtha. The Type-I adhesive is for use in cold patching of tire inner tubes and similar rubber items. The Type-II adhesive is primarily for bonding of components of natural rubber in the assembly of protective masks. Requirements include type of rubber, solvent composition, total nonvolatile content, consistency, and adhesion.

MMM-A-185B, Int Amd 2, April 16, 1975—*Adhesive (for Paper Bonding)* 8 pp plus 1-page int amendment

FSC 8040   AS

This specification covers one type of rubber adhesive, commonly known as *rubber cement*, suitable for mounting photographic prints, maps, drawings and charts, and for general paper-to-paper bonding. The adhesive is a crepe-natural-rubber or synthetic-rubber-base material modified to meet the requirements of this specification. Permissable solvents, including percentages, are specified to meet emission requirements. Application to only one surface will provide only temporary joining. If more permanent joining is desired, *both* surfaces should be coated liberally with the adhesive, allowed to dry, and then brought together under firm pressure. Requirements include nonvolatile content, acetone extract of nonvolatile components, ash content, viscosity, color, discoloration of paper, adhesion strength, wrinkle, curl and shrink properties, shelf-storage life, and flash point. (See A-A-863 below for a Commercial Item Description (CID) which may be used to specify this material)

MMM-A-187B, September 11, 1974—*Adhesive, Epoxy Resin Base, Low and Intermediate Strength, General Purpose*, 7 pp

FSC 8040   AS

This specification covers two types of epoxy resin-based adhesive for general-purpose application:

Type I - Low strength, paste form

Type II - Intermediate strength, paste or liquid form

This adhesive is a two-part system consisting of a base polymer of the epoxy type and a suitable accelerator. The two parts are furnished in a kit. In Type I (paste form) the resin and curing agent are of different colors to aid in determining when mixing is complete. Equal parts of resin and curing agent are to be used. In Type II (paste form) the requirement is similar, except that equal parts are not required. The adhesive specified is intended for use in repair operations, or in the bonding of metals, porcelain, ceramic materials, leather, wood, and various porous and nonporous materials to themselves and to each other. Requirements include form, solids, setting and curing, pressure, pot life, viscosity, sagging (paste form), tensile-shear strength, and storage life.

MMM-A-188C, January 2, 1975—*Adhesive, Urea-Resin-Type (Liquid and Powder)*, 6 pp   FSC 8040   AS

This specification covers a *urea-formaldehyde* thermosetting resin adhesive for the assembly gluing of wood items and the bonding of plastic laminate sheets to plywood. The adhesive is intended for use where moderate water resistance is required. It is *not* a fully waterproof adhesive. The types are:

Type I   - Powder (with separate curing agent)

Type II - Powder (with incorporated curing agent)

Type III - Liquid (with separate curing agent)

Requirements include curing agent, mixing properties, insoluble matter, filler, pH, working life, pot life, shear strength by compression loading (block shear strength), shear strength by tension loading (plywood shear strength), and storage stability.

MMM-A-189-C, March 18, 1985—*Adhesive, Synthetic Rubber, Thermoplastic, General Purpose*, 16 pp

FSC 8040   MR

This specification covers organic-solvent-base synthetic-rubber thermoplastic adhesives for general-purpose use. The classes are as follows:

Class 1 - 20% nonvolatile content

Class 2 - 30% nonvolatile content

The adhesive is dissolved in an organic solvent or blend of solvents. It is modified to meet the requirements of this specification. Permissable solvents, including percentages, are specified to meet emission requirements. Since these adhesives are thermoplastic, they are not recommended for applications where they will be subject to either continuous high stress, or stress at an elevated temperature. The Class 2 adhesive, with its higher solids content, forms heavier films with controlled penetration and is used for porous surfaces. Requirements other than solvents include condition in container, specific gravity, viscosity, nonvolatile content, ash content of nonvolatile matter, pH of adhesive-water mixture, shear strength at various conditions, including immersion in toluene, peel strength, corrosivity, shelf-storage life, toxicity, and compatibility with explosives or propellants.

MMM-A-250C(1), August 12, 1976—*Adhesive, Water-Resistant (for Closure of Fiberboard Boxes)*, 10 pp plus 1-page amendment
FSC 8040   MR

This specification covers water-resistant adhesives intended for closure of fiberboard boxes, cartons, and cases. The adhesives are of the following types:

Type I   - For application by automatic box-closing equipment

Type II - For hand application, by brushing

Type III - For hand application, from pressurized container (aerosol)

This adhesive must not contain toxic materials, including polychlorinated biphenyls (PCB's). Other requirements include nontoxicity, bond strength, shelf-storage life, ability to be applied by automatic equipment, brushing properties, aerosol properties, dispenser (including component parts), condition in container, gross weight, total solids, and spraying properties.

MMM-A-260C, May 10, 1984—*Adhesive, Water-Resistant (for Sealing Waterproofed Paper)*, 14 pp
FSC 8040   MR

This specification covers water-resistant adhesive intended for application to the seams in the manufacture and closure of waterproofed paper bags, wrappers, and case liners. It may also cover other applications. The following types, grades, and classes are specified:

Type I   - For application by machine

Type II - For hand application, by brushing
    Grade A - For subsistence items
    Grade B - For other than subsistence items
        Class 1 - Solvent-base adhesive
        Class 2 - Water-emulsion adhesive
        Class 3 - Hot-melt adhesive

Requirements include fungus resistance (when specified), low-temperature flexibility, resistance to water penetration, adhesion after subjection to heat, initial tack (Type II), resistance to flow, stability (under high and low temperature), consistency (classes 1 and 2), toxicity, condition in container, and shelf storage life.

This adhesive supersedes MIL-A-140A (November 23, 1951) and O-A-166 (GSA-FSS) (January 18, 1962).

A-A-342A, January 8, 1980—*Adhesive, Semi-Solid, Stick Form, with Dispensers*, 1 p   FSC 8030   GSA-FSS

This is a Commercial Item Description (CID). GSA has authorized its use in lieu of Interim Federal Specification MMM-A-002015, described below.

The adhesive is a nontoxic, mold-resistant, even-spreading material housed in an impact-resistant container with a cap. It must have a pleasant odor and a smooth uniform consistency. The container must have a mechanism which allows the adhesive stick to be mechanically advanced or retracted without making contact with the adhesive. There are a few simple requirements for nonvolatile content, adhesion, freeze-thaw, and workmanship. The CID is complete on one page.

A-A-529, September 25, 1979—*Adhesive and Sealing Compound, Cellulose Nitrate Base, Solvent Type*, 1 p

FSC   8040   GSA-FSS

This is a Commercial Item Description (CID). GSA has authorized its use in lieu of Federal Specification MMM-A-105, described above.

The CID covers a cellulose nitrate base adhesive and sealing compound intended for use in attaching paper labels, waterproofing, and as a general-purpose adhesive. The material must not cause any decrease in legibility of the printed label and must be resistant to blocking. The cellulose-nitrate film must be water-resistant, brushable, and show no signs of gelation, separation, sedimentation, or any other deterioration after storage at 25°C for 6 hours. The film must remain water-white in color and dry tack-free in air within 5 minutes. The CID is complete on one page.

A-A-863, March 1, 1984—*Adhesive (Rubber, for Paper Bonding)*, 1 p
FSC   8040   GSA-FSS

This is a Commercial Item Description (CID). GSA has authorized its use in lieu of Federal Specification MMM-A-185, described above.

The adhesive shall be suitable for bonding paper, maps, drawings and charts. It shall be smooth and uniform and free from dirt, lumps, and coarse particles. It shall be colorless, transparent, or translucent. It shall remain colorless after drying when applied before or after accelerated aging (7 days at 105°F (40.5°C). There is a viscosity requirement. Thue use of recovered material is encouraged.

MMM-A-1058A, INT AMD 1, January 26, 1978—*Adhesive, Rubber Base (in Pressurized Dispensers)*, 7 pp plus 3-page amendment
FSC   8040   AS

This interim amendment covers one type of rubber-base adhesive, a regular (high-solids) type, which is relatively colorless, fast drying, and packaged in a pressurized container commonly known as an aerosol dispenser. No chlorofluorocarbons or vinyl chloride may be used in the aerosol propellant, which must be a hydrocarbon or blend of hydrocarbons, including halogenated halocarbons. The adhesive in pressurized containers is intended to provide a quick, convenient method for bonding a wide variety of adherends. It will bond such materials as paper, glass, cardboard, cloth, wood, plastic film, metal, and foam rubber to themselves, to one another, or to other materials. It is *not* intended for structural use, or for bonding certain equipment components where such critical properties as high pressure or dielectric strength may be required.

Requirements include propellant, dispenser (including component parts), valve operations, color, condition in container, storage stability, total solids, adhesion (peel strength), net contents of dispenser package, internal pressure, delivery rate, spraying properties, and nonflammability (Type II).

MMM-A-1617(2), June 18, 1981—*Adhesive, Rubber Base, General Purpose*, 14 pp plus 2-page amendment

FSC 8040   AS

This specification covers natural- and synthetic-rubber-base adhesives intended for noncritical uses in applications where the unit stress on the adhesive is not appreciable. The adhesive must not be used for structural purposes, or for life rafts, inflatable boots, radome covers, pontoons, deicer boot manufacture, or repair. These adhesives will bond duck, leather, felt, cork, and similar relativery *porous* materials to themselves or to each other, or to relatively non-porous materials, such as wood, aluminum, steel, rubber, and plastics. The following types are specified:

Type I   – Natural-rubber-base, synthetic-natural (polyisoprene), styrene styrene-butadiene (SBR), reclaim, or combinations thereof; non-oil-resistant

Type II   – Polychloroprene-rubber-base, oil-resistant

Type III – Butadiene-acrylonitrile (nitrile) rubber base; fuel-resistant

Natural rubber can best be bonded with Type I and Type II adhesive. Type III adhesive has relatively poor resistance to natural rubber. Type II is definitely superior for bonding neoprene, and Type III is superior for bonding Buna-N and vinyl compounds. Preliminary check tests should be made in case of doubt.

Types I and II may be thinned, if necessary, but type II usually cannot be thinned. The amendment permits the use of 1, 1, 1-trichloroethane (methyl chloroform), but prohibits the use of trichloroethylene. Requirements include adhesive characteristics, nature of formulation changes, strip adhesion strength (as received, after immersion, after bond aging, and after accelerated storage, all with aluminum to various substrates with the immersion being in water, oil, and fuel), dead load (Type II), brushability, accelerated storage, and shelf-storage life. A QPL is available as QPL-MMM-A1617-1, August 21, 1972, Amend 1, August 1, 1979.

MMM-A-1754, July 1, 1973—*Adhesive and Sealing Compound, Epoxy, Metal Filled*, 7 pp

FSC 8040   AS

This specification establishes the requirements for one type of metal-filled, room-temperature curing, two-part epoxy adhesive and sealing compound. The metal filler particles may be aluminum, steel, or both. Other fillers may be used as needed. The activator shall be of the *polyamide* type. The base polymer must be a paste, while the activator may be *paste* or *liquid*. The colors of the two components must be different, resulting in a neutral gray when uniformly mixed.

This adhesive can be used as an edge-sealing or molding compound for both general-purpose and structural applications, such as on fairing surfaces. The cured adhesive can be sanded, filed, or machined with conventional tools. The adhesive will bond to steel, iron, aluminum, glass (not Pyrex), concrete, ceramics, porcelain, tile, marble, wood, and thermosetting plastics such as phenolic, polyester, or epoxy. It will bond with less strength to brass, copper, and lead. It is not recommended for nickel, tin, zinc, or thermoplastics, such as polyethylene. Proper surface treatment is necessary for all bonds.

Requirements include base polymer, activator, thinner, form, color, corrosivity, solids, working characteristics (application and sag flow), application life (pot life), curing time, tensile shear (at three temperatures), tensile strength, flexural strength, flexural bond strength, and compressive yield.

MMM-A-1931, September 19, 1975—*Adhesive, Epoxy, Silver Filled, Conductive*, 7 pp

FSC 8040   AS

This specification covers the requirements for two types of silver-filled, conductive, two-part, room-temperature-curing adhesives. These materials are used primarily to install static discharger bases to exterior aircraft surfaces. They are also suited for applications such as conductive paths on circuit boards, or for grounding and bonding metal components without welding, brazing, or soldering. The adhesives may be used where hot soldering is not practical, such as to nichrome wire or conductive pastics, or in locations which cannot be subjected to high temperatures. The adhesives may be used in preparing electrodes on specimens used to measure capacitance and loss characteristics. Cost and conductivity requirements should govern choice of types. Proper surface treatments are necessary for all bonds. The types are as follows:

Type I – Volume resistivity 0.010 ohm/cm

Type II – Volume resistivity 0.003 ohm/cm

The base polymer, in paste form, contains the silver particles, but must not contain copper or other metals, and shall contain no fillers or extenders. The curing agent shall contain no metals, extenders, or fillers. The curing agent may be paste or liquid. Requirements include base polymer, curing agent, thinner, form, color, solids, corrosivity, application, application life (pot life), sag flow, curing time, tensile shear strength at three temperatures, volume resistivity, and storage stability.

MMM-A-001993(GSA-FSS), February 28, 1978—*Adhesive, Epoxy, Flexible, Filled (for Binding, Sealing, and Grouting)*, 6 pp

FSC 8040   AS

This Interim Federal Specification covers a two- or three-component, mineral-filled, flexible epoxy-resin-base adhesive to be used in binding, sealing, and grouting *concrete*. These adhesives are intended to be used to repair spalls and other defects in portland-cement pavements. In many cases, 40-mesh sand should be added to the adhesive in a proportion that achieves an optimum balance between performance characteistics, heat of mixing, and cost. The types are as follows:

Type I – A two-component system, with mineral filler pre-mixed into each component

Type II – A three-component system, including the epoxy resin, curing agent, and mineral filler

The inert filler is a finely divided quartz silica flour or feldspathic aluminum silicate flour. Requirements include kit, base resin, curing agent, mineral filler, mixing instructions, consistency, compressive single shear strength, compressive flexible strength, nonvolatile content, and pot life.

MMM-A-002015(GSA-FSS), April 6, 1977—*Adhesive, Semi-Solid, Stick Form, with Dispensers*, 7 pp

FSC 8040   GSA-FSS

This Interim Federal Specification covers a convenience-type, quick-setting office adhesive in stick form, suitable for bonding paper, photographs, and fabrics. It is intended for mounting photographic prints, maps, drawings, and charts. It is also suitable for bonding fabrics in cases where high strength is not required. Requirements include unit containers, nonvolatile content, set time, "wrinkle, shrink, and curl," mold growth discoloration of paper and photographs, stability of container, loss of weight, odor, workability, stability of adhesive, adhesion strength, shelf-storage life, and toxicity. (See A-A-342A above for a Commercial Item Description (CID) which may be used to specify this material.)

MMM-A-002408(GSA-FSS), April 1, 1978—*Adhesive, Fire-Resistant, Thermal Insulation*   FSC 8040   GSA-FSS

This Interim Federal Specification covers fire-resistant adhesive for securing cloth and tape to certain thermal insualtions and for securing certain thermal insulations to metal surfaces. There are three classes, as follows:

Class 1 – For bonding fibrous glass cloth to unfaced fibrous glass insulation; for bonding cotton brattice cloth to faced and unfaced fibrous glass insulation board; for sealing the edges of, and bonding fibrous glass tape to the joints of fibrous glass board; and for bonding lagging cloth to thermal insulation.

Class 2 – For attaching fibrous glass insulation to metal surfaces

Class 3 – For attaching cork and fibrous glass insulation to metal surfaces

Requirements include condition in container, consistency, flash point, freeze-thaw stability, drying time (class 1), adhesive strength, stripping strength, tensile strength (class 3), corrosivity, color (class 1), flexibility, scrub resistance (class 1), paintability (class 1), fire resistance, and puncture resistance (class 2).

Department of Defense (DOD)—
Military Specifications[7]

MIL-G-413B, October 9, 1963—*Glue, Marine and Aviation Marine (Waterproof)*, 5 pp

FSC 8040   SH

This specification covers material for fastening sheeting between inner and outer wood skins of floats and flying-boat hulls and for faying wooden seams of ship decks. There are two classes, as follows:

Class 1 – Aviation marine glue for fastening sheeting in built-up skins of floats, flying-boat hulls, and double-plank construction on wood boats

Class 2 – Marine glue for faying wooden seams of ship decks

The class-1 glue shall contain rosin, pine tar. denatured alcohol, and a drying oil. The drying oil shall be tung oil, rosin oil, or linseed oil. The class-2 glue shall contain no materials of coal-tar origin. Requirements for class 1 include volatile matter content, penetration, plasticity, rosin content, mineral content, specific gravity, waterproofness and tackiness. Requirements for class 2 include viscosity at 177°C (350°F), boiling point, working properties, odor, penetration at three different temperatures, flexibility, flow, and filler content.

MIL-C-2399B(2), December 10, 1975—*Cement, Liquid, Tent Patching*, 11 pp plus 3-page amendment

This specification covers a liquid, synthetic-rubber-base cement for use in patching tentage, as specified in applicable maintenance manuals. The adhesive is based on copolymers of butadiene and acrylonitrile in a solvent blend of petroleum naphtha, toluol, xylol, acetone, methyl ethyl ketone, or methyl isobutyl ketone. Solvent formulations, with allowable percentages are specified to meet emission requirements. Requirements include consistency, flexibility, adhesion, shear strength and stripping strength (4 conditions), and stability.

MIL-A-3167A(OS), January 5, 1976—*Adhesives (for Plastic Inhibitors)*, 11 pp   FSC 8040   OS

This limited-coordination Navy specification covers adhesives for use with ethyl cellulose or cellulose acetate molded plastic inhibitors. These adhesives are intended for use in adhering web and end plastic inhibitors to cruciform-shape double-base powder grains. The classification is as follows:

Type I – Cellulose nitrate
Class 1 (see tables)
Class 2 (see tables)

Type II – Cellulose acetate
Class 1 (see tables)
Class 2 (see tables)
Class 3 (see tables)
Class 4 (see tables)

Type III – Solvent adhesive
Class 1 (see tables)
Class 2 (see tables)

The Type I, classes 1 and 2 adhesives are for use in adhering ethyl cellulose inhibitors, as specified in MIL-I-3166. The Type-I, class 2 adhesives may be diluted with up to 30% by weight of butyl acetate. All the Type-II adhesive classes are for use in adhering cellulose acetate inhibitors, as specified in MIL-I-3166. Class 1 may be diluted up to 20% by weight with acetone. Type-III, class 1 adhesive may be used to adhere either ethyl cellulose or cellulose acetate inhibitors, as specified in MIL-I-3166. The Type-III, class 2 adhesive is intended for cementing ethyl cellulose plastic to propellant powder, or to other ethyl cellulose plastic.

The composition of the various types and classes is given in a table. Another table gives the requirements for non-volatile content, viscosity, and specific gravity of the adhesives.

MIL-A-3316B(7), September 4, 1983—*Adhesives, Fire-Resistant, Thermal Insulation*, 14 pp plus 6-page amendment

FSC   8040   SH

This specification covers fire-resistant adhesives for securing cloth and tape to certain thermal insulations and for securing certain thermal insulations to metal surfaces. The amendment, the seventh covering this specification, is a long one (six pages), and makes a large number of significant changes in the basic specification. The classification, as modified by the amendment, is as follows:

Class 1 - (for brush application). For bonding fibrous glass cloth to unfaced fibrous glass cloth insulation; for sealing the edges of, and bonding fibrous glass tape to the joints of fibrous glass board; and for bonding lagging cloth to thermal insulation

        Grade A - Pigmented white

        Grade B - Pigmented red (sealer coating to identify asbestos-free system)

Class 2 - For attaching fibrous glass insulation to metal surfaces

        Grade A - Pigmented white

Class 3 - For attaching cork and fibrous glass insulation board to metal surfaces

        Grade A - Pigmented white

The adhesives specified must be effective for the purpose intended without heating or the addition of other ingredients. They must be free of all ingredients which may adversely affect the serviceability, have a deleterious effect on thermal insulation and fibrous glass cloth, or cause corrosion of base steel in the adhesion tests on primed steel in service. The adhesives must not contain mercury compounds and must be asbestos-free. Requirements include storage stability, freeze-thaw stability, toxicity and irritancy, flash point, consistency, coverage and finished weight, adhesive strength before drying (class 1 only), drying time and stripping strength, tensile-adhesive strength (class 3 only), color (class 1 only), flexibility, washability (class 1 only), paintability (class 1 only), fire resistance, and puncture resistance (class 1 only). A QPL is available as QPL-3316-66, August 5, 1983.

MIL-A-3562B, October 2, 1959—*Adhesive, Sealing (for Filters)*, 4 **pp**
FSC 8040   EA

This specification covers a liquid nitrile-rubber-base adhesive for use as a sealing material in the manufacture of filters used in collective protectors. The adhesive is dissolved in acetone. Test requirements include dioctylphthalate (DOP) penetration, total solids content, viscosity, softening point, cold flow, and cold brittleness.

MIL-A-392OC, February 21, 1977—*Adhesive, Optical, Thermosetting*, 11 pp   FSC 8040   AR

This is a performance specification covering a thermosetting liquid resin adhesive for bonding optical elements for use in military instruments. There are no restrictions on the chemical type. Requirements include refractive index, viscosity, light transmission, lint and dust particles (in adhesive and activator), cure conditions, environmental exposure (in warm water, at low temperature, and at high temperature and humidity), and cold exposure ($-62°C$) ($-80°F$) of bonded doublets. A QPL is available as QPL-3920-7, October 16, 1984.

MIL-A-5540B(2), December 28, 1973—*Adhesive, Polychloroprene*, 16 pp plus 2-page amendment
FSC 8040   AS

This specification covers neoprene adhesives for joining neoprene-coated fabric to itself and to nylon. The adhesive is for use where adherence to polychloroprene-coated fabric is required, such as in the manufacture and repair of life vests, decoy targets, and pontoons. (Note: polychloroprene is sold commercially as neoprene.) The classes are as follows:

Class 1 – Heat-cure for manufacturing [coating-to-coating, heat stable to $60°C$ ($140°F$)]

Class 2 – Room-temperature cure for manufacture [coating-to-coating, heat stable to $60°C$ ($140°F$)]

Class 3 – Room-temperature cure for repair [coating-to-coating, heat stable to $60°C$ ($140°F$)]

Class 4 – Heat-cure for manufacture (coating-to-nylon)

Class 5 – Room-temperature-cure for repair (coating to nylon)

All five classes must be furnished as Form A or B:

Form A – One-part adhesive

Form B – Two-part adhesive

Classes 1, 2, and 3 adhesives are required to meet a dead-load test (must be capable of supporting a dead load of 75 pounds for 24 hours at $60°C$ ($140°F$) without separation or creep of any of the three seams tested). Other requirements include viscosity (as received and after accelerated storage), strip adhesion after various conditionings for coating-to-coating and coating-to-nylon, brushability, working life, and toxicity. A QPL is available as QPL-5540-23(1), October 4, 1979.

MIL-A-8576(2), November 13, 1984—*Adhesive, Acrylic Base, for Acrylic Plastic*, 13 pp

FSC 8040   AS

This specification covers acrylic-monomer-base adhesives intended for use in bonding acrylic plastics. There are three types, in kits, as follows:

Type I   – Solvent type

Type II   – Non-solvent, high viscosity

Type III – Non-solvent, medium viscosity

Type I is intended for use in bonding acrylic plastic conforming to MIL-P-5425 and may be used with suitable precautions in bonding other acrylics. It should not be used for bonding material conforming to MIL-P-8184. Types II and III are intended for use in bonding acrylic plastic conforming to specifications MIL-P-3184 and MIL-P-5425. They are self-polymerizing and are applicable where the cushion technique is not necessary. Requirements cover index of refraction, specific gravity and viscosity, along with tensile strength of bonded butt joints at room temperature and 158°F (70°C).

MIL-A-9117D, April 19, 1971—*Adhesive, Sealing, for Aromatic Fuel Cells and General Repair*, 4 pp

FSC 8040   84

This specification covers a one-part synthetic elastomeric adhesive for fuel cell repair and for other general repair work where resistance to aromatic fuel is required. The classification is as follows:

Class G – For use under normal circumstances

Class L – For use where Air Pollution Regulations
are enforced

Both classes must have a nonvolatile content in the range of 24–30%. The volatile component of the class-L adhesive shall be a non-photochemically reactive solvent mixture meeting the formulation requirements given, which specify percentages permitted of various solvents and solvent combinations. Other requirements include aromatic fuel resistance (with a bond strength requirement), bond strength of unaged adhesive, heat resistance, storage stability, strength after storage, and toxicity.

MIL-A-13374E, August 25, 1983—*Adhesive, Dextrin, for Use in Ammunition Containers*, 11 pp

FSC 8040   MR

This specification covers four classes of vegetable dextrin adhesives produced from starch and intended for use in the manufacture of spirally wound containers, and in the fabrication of chipboard spacers to be used in such containers, for use in packing ammunition and components. The classes are as follows:

Class 1 – A liquid adhesive in prepared form

Class 2 – A cold-water-soluble type which must be mixed with water before use

Class 3 – A prepared dry adhesive which must first be cooked with water before use

Class 4 – Dextrin base in dry form to which additional optional chemical ingredients may be added during preparation with water before use

Optional ingredients in four groups may be added as follows:

Group A – Alkaline chemicals

Group B – Fillers

Group C – Preservatives

Group D – Defoamers

Details are given for permissable ingredients for each of these four groups. Requirements include percent water, percent grit and dirt, pH of water solution, and adhesion (fiber failure), expressed as a percentage of the bonded area.

MIL-A-14064C, May 5, 1980—*Adhesive: Grinding Disk*, 7 pp
FSC   8040   MR

This specification covers one grade of grinding-disk adhesive intended for use for the bonding of abrasive disks to metal disks. Latitude is allowed in the materials to be used, since this is a performance specification. Requirements cover brushing properties, odor, viscosity, peel strength at room and accelerated temperature, total solids, and workmanship.

MIL-A-17682E, March 11, 1982—*Adhesive, Starch*, 9 pp
FSC   8040   MC

This specification covers requirements for one type of adhesive used for mounting paper or cloth targets. The adhesive may contain reclaimed materials. The base material for the adhesive must be pure wheat, in fine-powder form, conforming to Federal Specification N-F-481, containing sufficient preservatives to prevent decomposition, weevil and mold growth in the package, and fermentation after mixing with water. Requirements include pH, odor, toxicity, brushing properties, color and transparency, homogeneity, bacteria and fungus resistance, and peel strength.

MIL-A-21016E, January 26, 1967—*Adhesive, Resilient Deck Covering*,
6 pp   FSC   8040   SH

This specification covers adhesives for securing resilient coverings to decks. The adhesive is a water-base latex, free of all ingredients which may affect the serviceability, or have a deleterious effect on metal or resilient covering. The adhesive must be free of grits, lumps, and skins and must be trowelable. It must be suitable for application without heating or addition of other ingredients. Other requirements include stability, edge-adhesive strength (before and after water immersion), fire resistance, corrosivity, and viscosity. A QPL is available as QPL-21016-22, January 18, 1984.

MIL-A-21366A(SHIPS), February 16, 1966—*Adhesive, for Bonding Plastic Table Top Material*, 10 pp

FSC 8040   SH

This limited coordination Navy specification covers a flexible adhesive for bonding thermosetting-plastic table-top material conforming to MIL-T-171 to aluminum at room temperature with only nominal pressure. The adhesive is a one-part system. It may be a neoprene-resin combination or other combination of ingredients. Requirements include storage life, working consistency, solubility, adhesive strength (tensile edge lift and dead load).

MIL-A-22397, INT AMD I (SHIPS), June 12, 1964—*Adhesive, Phenol and Resorcinol Resin (for Marine Service Use)*, 12 pp plus 5-page amendment   FSC 8040   SH

This specification covers adhesives for bonding wood where an adhesive bond with high strength, resistance to saltwater, extreme shrinking and swelling resistance, and long-time durability is required. These adhesives are particularly suitable for use in fabricating wood laminates for ship and boat use and for other severe exterior service. Use of these adhesives in laminating various species of wood is covered by separate laminate specifications. There are two grades, as follows:

Grade A – One year storage life

Grade B – Two-month storage life (immediate use)

The resins used in making the adhesive must be based on phenol, resorcinol, or a combination of both. The hardener shall be provided in either powder or liquid form. When in powder form, it may be combined with part or all of the filler. The filler, when used, must be an inert, insoluble powder. It may be combined with the resin or supplied separately. Amylaceous fillers, such as flour or starch, and protein fillers are not permitted. Requirements include pH of set film, liquid working life, cleanability, adhesive spread, assembly period, cure, resistance to shear by compression loading in white oak, resistance to delamination in white oak, and storage life. A QPL is available as QPL-22797-19, March 2, 1984.

MIL-A-22434A(WEP), March 7, 1961—*Adhesive, Polyester, Thixotropic*, 13 pp   FSC 8040   OS

This limited-coordination Navy specification covers a thixotropic polyester adhesive for use in bonding glass-cloth epoxy-resin laminate rings to metal components of warhead assemblies, where special application and assembly techniques are involved. The adhesive must bridge voids up to 0.020 inches when applied to vertical and horizontal surfaces. The adhesive is a homogeneous compound prepared by combining colloidal silica with a liquid unsaturated alkylstyrene type thermosetting resin. Methyl ethyl ketone peroxide is used as the catalyst and cobalt naphthenate as the accelerator to polymerize the resin. The requirements include the nature and purity of the thixotropic agent (silicon dioxide), catalyst, and accelerator, and the composition of the adhesive, tensile strength, and viscosity. Suggested sources of the specified ingredients are given in the text of the specification (para 6.3).

MIL-S-22473E(1), February 2, 1984—*Sealing, Locking and Retaining Compounds: (Single Component)*, 22 pp plus 1-page amendment
FSC 8030   MR

This specification for anaerobic sealants covers single-component compounds and their primers suitable for sealing, locking, and retaining metal parts. The components are normally liquid and are converted to an infusible, insoluble state when confined between closely fitting metal surfaces. The compounds are intended for use in sealing threaded fasteners, plugs, and other threaded fittings against fluid pressure, for locking such threaded assemblies against working loose under shock and vibration (threaded fasteners), and for retaining existing or replacement ball bearings in worn housings, thereby obviating the need for establishing a press fit. Thirteen examples of end-use applications of these materials are given. Fifteen grades are specified, each with varying viscosity and locking torque. Different colors are used to distinguish these grades, although there is duplication of colors.

Each compound must be visible under ultraviolet light when tested as specified. Requirements include viscosity, flash point, and solubility.

MIL-A-22895(1)(SHIPS), June 25, 1962—*Adhesive, Metal Identification Plate*, 6 pp plus 2-page amendment
FSC 8040   SM

This limited-coordination Navy specification covers adhesives for use in bonding, without heat or pressure, of metal identification plates to painted or unpainted surfaces. The adhesive is used in place of mechanical attachment and must be resistant to severe use conditions of interior or exterior exposure to vibration and shock. There are two classes, as follows:

Class A – Two component liquid *polysulfide* adhesive, characteristically non-solvent, gap-filling, and completely reactive, for knife application to metal identification plates up to 0.048 inch thick. The service temperature range is up to 121°C (250°F), but must withstand 204°C (400°F) for short-time emergency conditions.

Class B – Single-component, *neoprene or nitrile rubber* adhesive in a volatile solvent for brush application to metal identification plates not exceeding 0.030 inch thickness. This adhesive requires smooth surfaces and has a service range up to 93°C (200°F). The solvent must not exceed 65% by weight.

Requirements include shock and vibration resistance, impact bending resistance, tack, working life, color, consistency, and storage life.

MIL-C-23092B(SHIPS), February 26, 1976—*Cement, Natural Rubber, Magnetic Minesweeping Cable Repair*, 6 pp

FSC 8040   SH

This limited-coordination Navy specification covers a natural-rubber cement for making vulcanized and unvulcanized bonds of natural or synthetic rubber for the repair of magnetic minesweeping cable. The cement is dissolved in a solvent. No reclaimed rubber is permitted. Requirements include solids content, viscosity, initial and aged adhesion, and stability.

MIL-A-23940(1)(AS), September 1, 1966—*Adhesive, Silicone Rubber*, 6 pp plus 2 pp amendment

FSC 1336   AS

This limited-coordination Navy specification establishes the minimum requirements for a general-purpose room-temperature-vulcanizing silicone-rubber adhesive for use in rocket-motor igniters. Requirements include curing, viscosity, pot life, shrinkage, corrosivity, hardness, tensile strength, elongation, and brittle point.

MIL-A-23941A(AS), September 1, 1966—*Adhesive, Epoxy Type, Two Part*, 7 pp   FSC 1336   AS

This limited-coordination Navy specification establishes the minimum requirements for a two-part epoxy adhesive. No other restrictions are established on the chemical or physical nature of the adhesive. Requirements cover pot life, tensile shear under room temperature use and under moderate temperature 66°C (150°F) use, when tested at -55°C, 24°C, and 74°C (-67°F, 75°F, and 165°F).

MIL-A-24179A(2)(SHIPS), May 12, 1980—*Adhesive, Flexible Unicellular-Plastic Thermal Insulation*, 11 pp plus 2-page amendment

FSC 8040   SM

This limited-coordination Navy specification covers high-initial-strength, heat-and water-resistant, contact-type adhesives for bonding flexible unicellular-plastic thermal insulation to itself and to metal surfaces. The types and classes are as follows:

Type I   – Dispersed in water

Type II – Dispersed in nonhalogenated organic solvent
Class 1 – Low flash point
Class 2 – Intermediate flash point

Type III – Dispersed in nonflammable organic solvent

The Type-I adhesive is intended for use at installation temperatures of 10°C (50°F) or higher, and in locations where flammability and explosion hazards prohibit the use of Type-II. Type-II adhesive is intended for use at lower temperatures and for prefabrication of parts for insulation in areas where adequate fire protection is provided and a limitation on the quantity of adhesive present is enforced. Class 1 adhesive is for assembly of prefabricated parts from plastic foam insulation and for joining seams in plastic pipe insulation, where very short assembly periods are desirable in these operations. Class 1 adhesive is also suitable for attaching sheets of insulation to the interior surfaces of ships hulls, if the temperature of the hulls are 7.2°C (45°F) or lower and if rapid air movement over the open surfaces coated with adhesive cannot be provided. Class 2 adhesive is recommended for use when rapid air movement is provided at low temperatures. Class 2 adhesive is also appropriate for use at higher temperatures in very humid, still air since, in this atmosphere, Type-I adhesive requires very long drying periods.

Type-III adhesive is intended for use in areas where Types I and II are not specified and where halogenated solvents can be tolerated. Type III can be used in these areas with relatively little flammability and explosion risks. Types II and III require much shorter drying time before assembly than Type I. The requirements include dispersing medium or solvent composition, including halogens, flash point, and fire point; application techniques (roller or brush, storage life, physical properties); tensional adhesion under load at 38°C (100°F) in air, 71°C (160°F) in air, −6.7°C (20°F) in air, and 38°C (100°F) in distilled water; tensile strength (Types II and III); thermal shrinkage displacement of bonded insulation at primed steel interface; viscosity; and corrosivity. A QPL is available as QPL-24179-15, November 4, 1983.

MIL-A-24456(3)(SHIPS), June 15, 1979—*Adhesive for Plastic Vibration-Damping Tiles*, 10 pp plus 2-page amendment

FSC 8040   SM

This limited-coordination Navy specification covers a two-part epoxy adhesives for bonding plastic vibration-damping tiles to metal structures on board ships. The vibration-damping tiles are of the type specified in MIL-P-23653. The material shall be supplied in the form of a two-part epoxy system. Requirements include color, viscosity, pot life, hardness, resistance to accelerated aging, fuel and water, vertical slippage, overhead sag, resistance to shock (high impact), adhesion to painted steel and to damping tile, and storage life. A QPL is available as QPL-24456-10, June 9, 1980.

MIL-A-25463B, March 31, 1982—*Adhesive, Film Form, Metallic Structural Sandwich Construction*, 15 pp

FSC 8040   AS

This specification covers the requirements for adhesives in film form for bonding metal facings to metal cores and to metal components of sandwich panels intended for use in primary and secondary structural airframe parts that may be exposed to temperatures up to 260°C (500°F). Included in this specification is the bonding of metal facings to metal components within the sandwich panels which must be cured under the same conditions as the sandwich panel. Fabrication and inspection must be in accordance with MIL-A-83377.

Adhesives covered by this specification can be used for sandwich construction other than metal-to-metal (such as plastic-to-metal), provided testing proves such use is possible. Classification is as follows:

Type I   - For long-time exposure to temperatures from $-67°$ to 180°F ($-55°$ to 82°C)

Type II - For long-time exposure to temperatures from $-67°$ to 300°F ($-55°$ to 149°C)

Type III - For long-time exposure to temperatures from $-67°$ to 300°F ($-55°$ to 149°C) and short-time exposure from 300° to 500°F (149° to 260°C)

Type IV - For long-time exposure from $-67°$ to 500°F ($-55°$ to 260°C)

> Class 1 - For bonding metal facings to metal cores only
>
> Class 2 - For bonding metal facings to metal cores, inserts, edge attachments, and other components of completed sandwich structures

Cure temperature groups are as follows.

> Group 1 - 100°F (38°C) or less
>
> Group 2 - Over 100°F (38°C) to 200°F (93°C)
>
> Group 3 - Over 200°F (93°C) to 300°F (149°C)
>
> Group 4 - Over 300°F (149°C)

The adhesives must be *thermosetting* and must not have a deleterious effect on the metal surfaces being bonded over the range of temperatures at which they will be used. There are no ingredient restrictions other than those posed by the technicial requirements. Class 2 adhesives must meet the requirements of MMM-A-132, as applicable to the type, plus the requirements specified. The adhesives shall be in *film* form and shall consist either entirely of adhesive, or of a carrier impregnated with adhesive. A liquid primer may be furnished for use with the adhesive.

The requirements include applicability at the manufacturer's instructions at 65° to 85°F (18° to 29°C) and 60% RH maximum, curing conditions, storage life, and mechanical properties such as sandwich-peel strength, flatwise-tensile strength, flexural strength, and creep deflection. Test conditions vary, depending on the adhesive type being tested. A QPL is available as QPL-25463-30, October 22, 1984.

MIL-A-43316A(1)(GL), December 29, 1969—*Adhesive, Patching for Chloroprene Coated or Chlorosulphonated Polyethylene Coated Fabrics*, 11 pp plus 2-page amendment

FSC 8040   GL

This limited-coordination Army specification covers requirements for one type of solvent adhesive for patching coated fabrics. It is intended for use in repair and patching of inflated dual-wall shelters and other items that are also made of chloroprene-base-coated, chlorosulphonated-polyethylene top-coated synthetic fabric. The adhesive is required to effect repairs in a wide temperature range, including $-40°F$ ($-40°C$), when applied in the specified manner. The adhesive is based on an oil-resistant elastomer in a solvent blend of petroleum naphtha, toluol, xylol, or ketones. Requirements include solids content, shear strength, resistance to dead load [initial, after 24 hours aging of seam, and at $-40°F$ ($-40°C$)], peel strength (dead load) [dead load at $-40°F$ ($-40°C$)], stability, storage stability, brushing consistency, and shear strength after freeze-thaw.

MIL-A-45059C, December 2, 1983—*Adhesive for Bonding Chipboard to Terneplate, Tinplate, and Zincplate*, 8 pp

FSC   8040   MR

This specification covers a one-part, ready-to-use brushable adhesive for use in bonding chipboard to terneplate, tinplate, and zincplate in the manufacture of fiber ammunition containers and similar applications where resistance to water, oil, mold, etc., is not required. The adhesive is intended primarily to provide a bond between chipboard and the metal between time of manufacture and loading of the ammunition into the fiber containers. The requirements include shelf life, shear strength after one hour and after one week, at $-40°F$, $73.5°F$, and $140°F$ ($-40°C$, $23°C$, and $60°C$).

MIL-G-46030E(MR), June 4, 1984—*Glue, Animal (Protective Colloid)*, 8 pp   FSC   8040   MR

This specification covers one type of animal glue for use as a protective colloid in the manufacture of propellants. The glue must be made from raw materials of animal origin. It is intended for use in the hardening operations in ball propellant manufacture. The glue is in granular or pelletized form and must meet a sieve requirement. Other requirements include moisture content, grease content, ash content, pH, keeping qualities, performance, and compatibility with explosives.

MIL-A-46050C(2), February 3, 1983—*Adhesives, Cyanoacrylate, Rapid Room Temperature Curing, Solventless*, 13 pp plus 1-page amendment

FSC   8040   MR

This specification covers solventless, room-temperature-curing cyanoacrylate adhesives for use with or without an activator, when speed of curing is a primary consideration. It also covers the activator, which may be used to provide still faster curing and to enable the adhesives to bond to otherwise inhibiting surfaces. These adhesives are used for nonstructural applications requiring one-component bonding of small, well-mated surfaces where heat and/or pressure cannot be applied.

Cyanoacrylate adhesives are normally used to adhere plastics, elastomers, metals, and combinations of these. Generally, Type-I is preferred for metal-to-metal applications, and Type-II for rubber, plastics, and rubber/plastics-to-metal applications. Manufacturer's instruction sheets should be consulted for specific adherends recommended for each type. The material is not to be used for adhesive bonds in aerospace vehicles for primary or secondary structures where the adhesive is expected to perform a structural function. Setting time is viscosity-related. Lower viscosities will set more rapidly than higher viscosities. Class 1 will set more rapidly than Class 2 or 3. Not all types may be commercially available in all classes.

The types and classes are as follows:

Type I   – Methyl-2-cyanoacrylate

Type II – Ethyl-2-cyanoacrylate

Type III – Isobutyl-2-cyanoacrylate

Type IV – n-butyl-2-cyanoacrylate

Type V – Allyl-2-cyanoacrylate

Class 1 – Low viscosity

Class 2 – Medium-high viscosity

Class 3 – Medium-high viscosity

Class 4 – High viscosity

The surface activators shall be of the following types, as specified:

Type IA   – for use with Type I

Type IIA – for use with Type II

Type IIIA – for use with Type III

Type IVA – for use with Type IV

(Note: Type VA is not listed)

Requirements for the adhesive include appearance, viscosity, speed of cure, and storage life. Requirements for the surface include appearance, activating ability, and storage life. In some cases, compatibility with explosives may be a requirement.

MIL-A-46091B, July 22, 1981—*Adhesive, Brake Lining to Metal,* 10 pp
FSC 8040   Mr

This specification covers one type of thermosetting, one-component adhesive for use in bonding brake linings and clutch facings to steel and aluminum. There are two forms, as follows:

Form I – Liquid

Form II – Film

The material must be a one-component, heat-curing material. There are no restrictions on the chemical type or physical nature of the materials. The film form shall consist of unsupported film, protected by a liner. The film thickness range is specified. If a solvent is required as a separate thinner or activator, it must be furnished by the adhesive manufacturer. Requirements include working characteristics (applicability, curing conditions) and physical conditions[(disk-shear strength at ambient temperature and $400°F$ $(204°C)$] and axial shear strength at ambient temperature after brake-fluid immersion, after water immersion, and at $400°F$ $(204°C)$, using different combinations of adherends.

MIL-A-46106A(2), February 11, 1974—*Adhesive-Sealants, Silicone, RTV, General-Purpose*, 24 pp plus 2-page amendment

FSC 8040   MR

This specification covers two types of one-part, room-temperature-vulcanizing (RTV), nonfuel-resistant silicone compounds which cure to durable rubber sealants and adhesives upon contact with moisture in the air. The specification also covers primers. These compounds are available as thixotropic pastes, or as self-leveling liquids, and thus they can be used for a variety of application techniques. They can be used in the automotive, marine-appliance, metal-working, aerospace, aircraft, building construction, communication, computer, electrical/electronic and other industries. The specification suggests a large number of typical uses. The compounds are not resistant to many types of fluids, such as fuel and hydraulic fluid. When cured in contact with certain metals such as copper and other sensitive metals, a slight corrosion may result. This condition should be thoroughly investigated for electrical performance. Primers are recommended for optimum results when the silicone compounds will be exposed to water, high humidity, and elevated-temperature conditions. The primer must be of the type recommended by the sealant manufacturer for the particular silicone compound. Two types of silicone compounds are specified:

Type I   – Soft spreadable thixotropic paste

Type II – Self-leveling liquid

(The primer, if required, shall also be Type I or II)

The requirements for the uncured silicone compound include solids content, extrusion rate, flow, viscosity, tack-free time, and storage life. The requirements for the cured silicone compound include brittle point, hardness, tensile strength, elongation, peel strength, resistance to heat (392°F or 200°C), hydrolytic stability at 200°F or 93°C, dielectric constant, dissipation factor, and dielectric strength. The requirements for the primer include color, specific gravity, and total solids content.

MIL-A-46146A(2), November 25, 1981—*Adhesive-Sealants, Silicone, RTV, Noncorrosive (for Use With Sensitive Metals and Equipment)*, 24 pp plus 2-page amendment

FSC 8040   MR

The specification covers three types of one-part, room-temperature-vulcanizing, non-fuel-resistant silicone compounds which cure to durable rubber sealants and adhesives upon contact with moisture in the air. It also covers primers for use with the silicone compounds.

Since these compounds are available as thixotropic pastes or as self-leveling liquids, they lend themselves to a variety of application techniques which are easily adapted to specialty uses, as well as to production-line methods. These materials are noncorrosive to copper and other sensitive metals and are therefore gaining wide acceptance as preferred adhesives and sealants where delicate electronic devices are involved. They are used in sealing electronic devices, as terminal sealants, for potting electronic components, and as high-temperature sealants. There are three types, as follows:

Type I   - Soft, spreadable thixotropic paste

Type II - Self-leveling liquid

Type III - High strength, noncorrosive

The primer, if required, shall be Type I or II, as recommended by the manufacturer of the silicone compound. The silicone compound must vulcanized at room temperature to produce a rubbery compound. The primer must be an air-drying liquid. Requirements for the uncured silicone compound include total solids content, extrusion rate, flow, viscosity, tack-free time, corrosivity, and storage life. Requirements for the cured silicone compound include brittle point, hardness, tensile strength, elongation, peel strength, resistance to heat, hydrolytic stability (93°C and 95% RH), and electrical properties (volume resistivity, dielectric constant, dissipation factor, and dielectric strength). Requirements for the primer include color, specific gravity, total solids content, and corrosivity.

MIL-A-46864A(MI), December 20, 1983–*Adhesive, Epoxy, Modified, Flexible, Two Component*, 14 pp

FSC 8040   MI

This limited-coordination Army specification establishes the requirements for a two-component modified-epoxy adhesive system capable of curing at room temperature. The adhesive is intended primarily for use as a room-temperature-curing, flexible adhesive and sealant system for heat-shrinkable sleeving and molded components. This two-part system may also be employed in areas where bonding to rubber, metals, plastic, or combinations thereof, is required. Each potential application must be properly investigated before using the adhesive. Requirements for the uncured adhesive cover flow and application time. Requirements for the cured adhesive include peel strength (as-cured, after heat exposure, after thermal cycling, and after exposure to various fluids), tensile shear, dielectric constant, dielectric strength, and corrosivity. The adhesive system must also meet requirements for accelerated storage, after which it must conform to the flow and "as-cured" peel strength values. The shelf life must be certified by the manufacturer. Curing conditions for laboratory preparation and for field use are given separately.

MIL-A-47040(1)(MI), October 4, 1976—*Adhesive-Sealant, RTV, High Temperature*, 18 pp plus 2-page amendment

FSC 8040   MI

This limited-coordination Army specification covers one type of single-package, high-strength, thixotropic vulcanizing silicone rubber for use in the −85°F (−65°C) to 600°F (315°C) range. This adhesive is for use in the aerospace industry. The adhesive must vulcanize at room temperature to produce a rubber compound to meet the physical and electrical requirements of the specification. Requirements for the uncured silicone compound include total solids content, extrusion rate, flow, specific gravity, tack-free time, and storage life. Requirements for the cured silicone compound include brittle point, hardness, tensile strength, elongation, peel strength, tear strength, resistance to heat (392°F or 200°C), and dielectric strength. A precautionary statement is made to the effect that this compound is *not* resistant to many types of fluids, such as fuel and hydraulic fluid. When cured in contact with certain metals, such as copper and other sensitive metals, a slight corrosion may occur, causing possible problems with electrical performance. Material meeting this specification utilizes atmospheric moisture and liberates acetic acid during cure, which may result in corrosion. This type of material can also cause fracture of stressed high-strength steel when applied at high relative humidity. Particular care should be taken if this material is to be used in electrical equipment, especially when in close proximity to small-gauge wire and electrical contacts.

MIL-A-47074A(MI), March 2, 1983—*Adhesive System, Epoxy, for Dissimiliar Metal Bonding*, 11 pp

FSC   8040   MI

This limited-coordination Army specification covers two types of epoxy-resin adhesive systems for bonding of dissimilar metals, or similar metals, and where dimensional stability is required. There are two types and two classes, as follows:

Type I – Flowable (Unless otherwise specified, Type II shall be furnished.)

Type II – Nonflowable

        Class 1 – High peel strength (Unless otherwise specified, class 2 shall be used.)

        Class 2 – No peel strength required

The materials for both Type I and Type II shall consist of nitrile-phenolic primer and a thixotropic epoxy-based adhesive compound. The adhesive shall be a two-part epoxy-resin-based 100%-solids system, consisting of Part A, which contains *epichlorohydrin-bisphenol A-type resin* plus iron oxide filler, and Part B, which contains a *polyamide resin* plus titanium dioxide. In the case of Type II compound, a colloidal silica filler is added to increase viscosity. Requirements include flow, shear strength at −65°F (−54°C), 77°F (25°C), 160°F (71°C), and 200°F (93°C), peel strength (class 1 only), and thermal cycling.

MIL-A-47089(MI), May 1, 1974—*Adhesive, Metal Filled, Conductive, Electrical and Thermal*, 11 pp

FSC   8040   MI

This limited-coordination Army Specification covers one type of electrically conductive metal-filled adhesive. (Note: No mention is made of thermal conductivity in the specification, although the term is in the title.) The adhesive is intended to be used in making electrical connections when hot soldering is impractical, and in filling voids where electrical conductivity must be maintained. The adhesive shall be a silver-filled epoxy-type resin and shall have a separate hardener. Cure shall be effected by catalytic action and shall not depend on evaporation. Requirements include shelf life, pot life, low-temperature storage (of cured adhesive), and volume resistivity and shear strength of cured adhesive.

MIL-A-47126(MI), May 24, 1974—*Adhesive (Viscous), Epoxy Resin, Metal-to-Metal Bonding and Sealing,* 6 pp

FSC 8040   MI

This limited-coordination Army specification covers a viscous bonding material composed of an epoxy resin and activator. It is intended for bonding metal-to-metal parts in manufacturing to produce high-strength parts. The adhesive is two-part and the ratio of activator to resin in parts by weight is 4.50 to 100. Requirements include shelf life, consistency at RT, pot life, viscosity (75–80 poises), specific gravity, bond strength at $-65°F$ ($-54°C$), $70°F$ ($21°C$), and $200°F$ ($93°C$), weight loss after aging at $300°F$ ($149°C$) for 6 months, bond strength after 1 year at $300°F$ ($149°C$), coefficient of linear expansion, solvent resistance, hardener, and color.

MIL-P-47170(1)(MI), August 17, 1976—*Primer, Silicone Rubber Sealant,* 5 pp plus 1-page amendment

This limited-coordination Army specification covers the requirements for a room-temperature-drying silicone rubber sealant primer. The primer may be used to promote adhesion of the sealant covered by MIL-S-47162 (this specification is no longer in use) and similar hard-to-bond surfaces (such as MIL-A-46106, Type I, which see). The usage of the primer should be substantiated by tests for the surfaces in question. The primer may be applied by dipping, brushing, or spraying. The primer and sealant, used as a system, shall be manufactured by the same company. Requirements include nonvaolatile solids, viscosity, appearance, and shelf life.

MIL-P-47275(1)(MI), June 15, 1977—*Primer, Silicone,* 7 pp plus 2-page amendment

FSC 8040   MI

This limited-coordination Army specification establishes the requirements for one type of silicone primer system used to insure adhesion of silicone rubber to surfaces after vulcanization. It is intended for use as the primer system to insure adhesion of a silicone-rubber gasket seal that is vulcanized to aluminum alloy. For best results bonding operations should be completed in the 24 hours following primer application. The primer must be suitable for service temperatures between $-180°$ to $400°F$ ($-100°$ to $204°C$). The primer shall be supplied as a one-component system. The diluent shall be methanol or ethanol (one part primer and up to five parts of diluent). Requirements include adhesion, solids content, specific gravity, storage life, and shelf life.

MIL-P-47276(1)(MI), August 17, 1976—*Primer, Bonding,* 11 pp plus 2-page amendment

FSC 8040   MI

This limited-coordination Army specification covers one type of liquid organic nitrile-phenolic resin primer. It is intended for use with an adhesive tape which will be pressed between two metal surfaces primed as specified. Requirements include solids content, viscosity, density, dilution stability, surface finish, film thickness, and shear strength of joints bonded with AF6 film adhesive conforming to MIL-A-5090. (Note: This specification is no longer in use.)

MIL-P-47279(1)(MI), June 17, 1977—*Primer, Silicone Adhesive*, 6 pp, plus 2-page amendement

FSC 8040   MI

This limited-coordination Army specification covers the requirements for a room-temperature-drying silicone adhesive primer. It is intended for use in obtaining maximum adhesion when adhesive conforming to MIL-A-25457 (Note: This specification is no longer in use.) is applied to metal, glass, ceramics, and plastics. The primer may be applied by dipping, brushing, or spraying. The primer and silicone adhesive used as a system should be manufactured by the same company. Requirements cover nonvolatile content, specific gravity, flash point, color, and shelf life.

MIL-A-47280(MI), August 9, 1974—*Adhesives, Epoxy*, 11 pp

FSC 8040   AR

This limited-coordination Army specification covers two types of room-temperature-curing epoxy adhesive intended for use in bonding metal-to-metal, metal to plastics, plastics-to-plastics, and electronic components to boards or metal surfaces where excellent physical and electrical properties are required. There are two types, as follows:

Type I – High viscosity

Type II – Low viscosity

The adhesive shall consist of an epoxy resin and an activator. Requirements for the uncured adhesive include viscosity, appearance, contamination, shelf life, and pot life. Requirements for the cured adhesive include specific gravity, tensile strength, compressive strength, and flexural strength. The curing time is specified for 25°C, 65°C, and 100°C.

MIL-A-47284(1)(MI), November 1, 1976—*Adhesive, Epoxy Resin Base*, 7 pp plus 1-page amendment

FSC 8040   MI

This limited-coordination Army specification covers one type of adhesive consisting of an epoxy resin base and an amine-type curing agent. The epoxy resin contains an inorganic filler. The adhesive is intended for bonding metal-to-metal. Other constructions, such as metal-to-plastic, or plastic-to-plastic, may be bonded, provided the use of the adhesive is substantiated by testing. Requirements for the resin include viscosity, specific gravity, and solids content. Requirements for the curing agent include viscosity and specific gravity. Requirements for the cured adhesive include tensile strength, elongation, shear strength, and hardness. Shelf life of the resin and activator are also specified.

MIL-A-48611A(2), April 14, 1983—*Adhesive System, Epoxy-Elasto-meric, for Glass to Metal*, 12 pp plus 1-page amendment

FSC 8040   AR

This specification covers elastomer-modified epoxy-resin adhesive bonding systems for the structural joining of optical glass prisms to metal, the sealing of glass and metal components, and other applications requiring bonding of a similar nature. The materials to be bonded include glass prisms and other optical elements which are bonded to their metal supports in optical fire-control instruments. For maximum reliability and environmental resistance, especially to humid conditions, the Type-I adhesive system is recommended. Both Types-I and-II adhesive systems may be used for general-purpose bonding of materials such as aluminum, stainless steel, brass, bronze, plastics, glass, etc., to themselves or in dissimilar combinations. Various bond-line thicknesses may be used for bonded assemblies. However, for the structural bonding of glass prisms to metal mounts, a bond-line thickness of 0.036 to 0.41 cm is recommended. The adhesive is also used for mirror assemblies.

The types are as follows:

Type I – System composed of an epoxy primer component and an epoxy-adhesive component

Type II – System composed of an epoxy-adhesive component

The adhesive must be a thermosetting resin based on epichlorohydrin-bisphenol A-type epoxy resin, modified with acrylonitrile-butadiene rubber. Various additives, modifiers, or fillers may be included. The primer is a two-component room-temperature-curing liquid epoxy. Requirements for the adhesive include tensile shear strength and tensile strength, with requirements depending on the type, low-temperature thermal stability (Types I and II), mechanical shock stability at high humidity and at low temperature (Type I), and storage life. Requirements for the adhesive component include application time, curing time, temperature, and pressure. Requirements for the primer component include viscosity, drying time, mixing, and pot life. The formulation for an adhesive system that meets the requirements of this specification is described in a report available from the U.S. Army Research and Development Center, Dover, New Jersey 07801, ATTN: DRSMC-TST-S(D). Recommended procedures such as surface preparation, component mixing and application, and work-area environment stabilization to be followed during the use of bonding systems conforming to this specification are contained in MIL-B-48612 following. The adhesive system described in this specification should be used in place of that covered by MMM-A-131 for military applications. A QPL is available as QPL-48611-4, August 1, 1984.

MIL-B-48612(MU), January 27, 1977—*Bonding With Epoxy-Elastomeric Adhesive System, Glass to Metal*, 9 pp

FSC   12GP   AR

This limited-coordination Army specification covers the bonding of glass optical components to metal mounts for use in military instruments. Bonding thicknesses of 0.014 to 0.016 inch are recommended for most bonds, especially where aluminum is the metal mounting material. Other bondline thicknesses may be used, however. The bonding of mirrors and similar components by this specification is not recommended without confirmatory tests for optical quality. There are two types, as follows:

Type I – Bond with Type-I adhesive, MIL-A-48611

Type II – Bond with Type-II adhesive, MIL-A-48611

The quality of a bonded assembly shall be of the following two grades:

Grade A – Bonds for prism assemblies or similar components on optical instruments where high quality is mandatory

Grade B – Bonds for noncritical bonded assemblies

(Grade A shall apply to all bonded prism assemblies where grade quality is not specified.)

This is a process specification. Topics covered include assembling components, cleaning, primer preparation, primer application, adhesive preparation, adhesive application, bond fixture, adhesive curing, performance requirements, and bond life requirements. Performance requirements include mechanical shock. Bond-life requirements include low-temperature thermal stability, elevated-temperature thermal stability, and high-mechanical-shock stability.

MIL-A-50926A(PA), June 10, 1975—*Adhesive MR-23 (for Use in Ammunition)*, 7 pp   FSC 1376   AR

This limited-coordination Army specification covers one type of adhesive, MR-23, for use as an adhesive in nonmetallic cartridge cases. The adhesive, MR-23, is intended for use in the 152-mm ammunition system. The adhesive is a uniform mixture of nitrocellulose and nitroglycerine in acetone and nitromethane. The composition of the adhesive shall be in accordance with drawing 9255426. Requirements include viscosity and solids content. The viscosity is to be measured by a Brookfield viscosimeter, using an LOF model or equivalent, with spindle no. 1 at 60 RPM at 25°C. An alternate "fall-ball" method is also authorized and is described in detail.

MIL-A-52194A(MR), January 16, 1967—*Adhesive, Epoxy (for Bonding Glass Reinforced Polyester)*, 10 pp

FSC 8040   MR

This limited-coordination Army specification covers the requirements for a two-part epoxy-based adhesive suitable for bonding glass-reinforced polyester where no peel or cleavage is anticipated. Both the resin and the hardener must be in paste form. This specification was prepared to cover material for bonding the two halves of plastic gunstocks. The adhesive is intended for use in applications where its paste consistency is necessary (1) to prevent adhesive drain-off during cure, or (2) to bond adherend surfaces that are not plane. The adhesive is not intended for use where service temperatures exceed 200°F (93°C). Requirements include shear strength at −65°F (−54°C), 73°F (23°C), and 160°F (71°C), impact strength, viscosity of adhesive at ambient temperature, and 50% RH after mixing resin base and hardener, and storage life of adhesive.

MIL-A-52685, May 11, 1970—*Adhesive: Bonding, Chloroprene, Air Field Membrane Surfacing*, 9 pp

FSC 5610   ME

This specification covers one type of adhesive for bonding chloroprene-coated fabrics. It is intended to be used for the joining, by bonding, of airfield surfacing membranes. The adhesive is a rubber-based adhesive of noncuring polychloroprene-type, modified with a cyclic unsaturated hydrocarbon resin, usable without heating or additives. The requirements include total solids content, specific gravity, viscosity, ash content, tack range, shear strength (initial, wet, hot, freeze-thaw, and after-thickness test), peel strength (initial, wet, hot, freeze-thaw), and shelf life of the adhesive.

MIL-A-60091(1)(AR), April 29, 1981—*Adhesive for Bonding Demolition Charges to Structural Surfaces*, 7 pp plus 5-page amendment

FSC 1375   AR

This limited-coordination Army specification covers a one-part, ready-to-use adhesive suitable for use in bonding demolition charges to dry, wet, and underwater surfaces of structural materials under all weather conditions. It may also be used in other bonding applications. The adhesive is a synthetic-elastomer solvent-type material, with required pigment (olive drab). Its consistency is such that it can be dispensed readily from its collapsible metal-tube container by being squeezed by hand. It must be spreadable by means of a wooden tongue depressor. Requirements include consistency, adhesive strength under various conditions, compatibility with explosives (when specifically required), viscosity, and storage stability.

MIL-A-81236(2)(OS), September 9, 1968—*Adhesive: Epoxy Resin With Polyamide Curing Agent*, 7 pp plus 4-page amendment

FSC 8030   OS

This limited-coordination Navy specification covers one type of adhesive consisting of an epoxy resin and a polyamide curing agent for general use under normal circumstances. It provides for an additional type of material for use under Air Pollution Regulations. The classification is as follows:

Type I – For general use under normal circumstances

Type II – For use where Air Pollution Regulations are enforced

The material is a two-part thermosetting epoxy adhesive with a polyamide curing agent. The solvent in Type I shall be of the type normally used when the finished product is for general use under normal circumstances. For Type II adhesive the solvent shall be of a nonphotochemically reactive solvent (defined, with requirements given as to types and proportions of solvents permitted). Requirements for the epoxy resin include viscosity, density, and epoxide equivalent. Requirements for the amine curing agent include viscosity, density, and amine value. Requirements for the adhesive, consisting of 3 parts by weight of resin and 2 parts by weight of curing agent, include curability and tensile-shear strength.

MIL-A-81253(1)(OS), November 16, 1966—*Adhesive, Modified Epoxy Resin With Polyamine Curing Agent*, 6 pp plus 1-page amendment

FSC 8040   OS

This limited-coordinated Navy specification covers one kind of low-viscosity, modified-epoxy resin adhesive. The material is intended for use as an adhesive in rocket-motor systems. It includes a triethylenetetramine curing agent. The requirements for the epoxy resin include viscosity, epoxy value, color, and weight per gallon. The requirements for the curing agent include bonding range and specific gravity. The requirements for the cured resin covers tensile-shear strength.

MIL-A-81270(1)(OS), September 9, 1968—*Adhesive, Synthetic Rubber*, 5 pp plus 2-page amendment

FSC 8040   OS

This limited-coordination Navy specification covers one type of general-purpose nitrile-phenolic synthetic-rubber adhesive for general use under normal circumstances. It provides for an additional type of material suitable for use under Air Pollution Regualtion. It is intended for use in rocket motors.

The classification is as follows:

Type I - For general use under normal circumstances

Type II - For use where Air Pollution Regulations are enforced

The solvent content of Type I adhesive shall be of the type normally used when the finished product is for general use under normal circumstances. For Type II adhesive a nonphotochemically reactive solvent shall be used. This is defined and types of solvents and proportions allowed are specified.

MIL-K-81786/11(AS), April 20, 1972—*Adhesive, Epoxy, Flexible*, 1 p

FSC 4940   AS

This is a Military Specification Sheet. The complete requirements for procuring the adhesive described consist of this document and the issue in effect of MIL-K-81786(AS) (described below).

The whole text is as follows:

> "The adhesive shall be a two-part flexible epoxy type which conforms to the requirements of NAVSHIPS Procurement Specifications 0967221-1010 (Appendix H) and shall be packaged in containers which separate the two components. Each container shall weigh 2.8 grams."
>
> MIL-K-81786A, Suppl. 1 (AS)

MIL-K-81786A, Suppl. 1 (AS), August 2, 1973—*Kit, Maintenance, Electrical-Electronic, Cable and Cable Harness*, 35 pp plus 2-page supplement

FSC 4940   AS

This limited-coordination Navy specification covers an electrical-electronic maintenance kit containing consumable components, tools, and equipment designed for use by shore-based and shipboard personnel to maintain, modify and repair electronic cables and cable harnesses. MIL-K-81786/11(AS) described above describes an adhesive required for this kit.

MIL-A-82484, June 13, 1967—*Adhesive and Sealing Compounds, Cellulose Nitrate Base, Solvent Type (for Ordnance Use)*, 11 pp

FSC 8040   OS

This specification establishes the requirements for cellulose nitrate-based adhesive and sealing compounds that are suitable for ordnance use. The classification is as follows:

Type I   – Adhesive, Ordnance (38 to 42% solids)

Type II   – Adhesive, Ordnance (low viscosity)

Type III – Sealing compound, Ordnance (clear or colored by dye)

Types I and II are intended for general Ordnance use in ammunition. General use does not imply that the compounds are to be used as all-purpose adhesives; they are not recommended for use with concrete, wood, rubber, ceramics, or paper. Types I and II may be used to provide good bonds for glass, leather, metals, textiles, and some types of thermoplastics. Type III compound is intended for use as a sealing compound in the manufacture of ammunition.

The compositions of Types I and II adhesive are given in a table of 5 ingredients with percentages specified. Type III material must be a uniform solution or homogeneous dispersion of cellulose nitrate of different viscosities and organic plasticizers in relatively low-boiling organic-solvent mixtures, so formulated as to meet all the applicable requirements of this specification. Color may be specified for Type III.

No benzene or chlorinated solvent shall be permitted in any type. Each type must meet requirements for adhesion to and corrosion of brass, cold stability, nonvolatile content, viscosity, air-dry time (tack-free), and air-dry time (dry-hard), shear strength, flexibility (after 4 hours at 121°C), color (of clear compounds), percentage of cellulose nitrate, and blushing.

MIL-A-82569(OS), January 12, 1969—*Adhesive, Neoprene Base, Medium Viscosity*, 6 pp   FSC 1336   OS

This limited-coordination Navy specification covers one type of medium-viscosity neoprene-base adhesive, using aromatic solvents as the vehicle. The adhesive is intended for use in rocket motors. Requirements include total solids content, viscosity, peel strength, and storage life.

MIL-A-82636(OS), December 23, 1974—*Adhesive, Butyl, Two-Component*, 5 pp   FSC 8040   OS

This limited-coordination Navy specification covers a two-component butyl adhesive intended for use in bonded-seam construction of chemical-agent protective clothing. The product shall be a solvent-dispersed material suitable for application by a paint brush of natural or synthetic fibers. A flexible film having tackiness, strength, stretch, and aging properties suitable for the intended purpose shall remain after evaporation of the solvent. The adhesive shall be capable of being cured in air without heat or pressure. Requirements cover adhesion strength and brushability. There is no peel strength requirement.

DOD-A-82720(OS)(1), November 1, 1984—*Adhesive, Modified Epoxy, Flexible, Two-Part (Metric)*, 10 pp plus 1-page amendment

FSC   8040   OS

This limited-coordination Navy specification covers one type of room-temperature-curing, two-part, flexible, modified-epoxy adhesive intended for use in assembly of the aft closure of the MK 56 DTRM used on the standard (MR) surface-to-air missile. (Note: The DOD designation does not mean that this is intended for DOD use, although it is available to all DOD agencies for adoption. The DOD designation means that it is a metric specification with all units in the metric system.) The material shall be a structural, thermosetting, flexible, modified epoxy adhesive capable of being used at 15° to 35°C. The adhesive shall consist of two liquid components—an accelerator hardener or curative designated as Component A, and a base resin designated as Component B, furnished in a kit such that the entire contents, when mixed thoroughly, will meet the requirements of this specification. The two components must be completely miscible.

Requirements include density and viscosity for the two components, viscosity and pot life for the uncured adhesive, and tensile strength, tensile modulus, and elongation for the cured adhesive.

MIL-A-83376A(USAF), June 28, 1978—*Adhesive Bonded Metal Faced Sandwich Structures, Acceptable Criteria*, 12 pp

FSC   8040   20

This limited-coordination Navy specification establishes the requirements for adhesive-bonded sandwich structures, including the metal-to-metal bonding found in these structures. The specification is intended to define acceptance criteria of bonded metal-faced sandwich assemblies used in aerospace structures. Two types are classified as follows:

Type I – Components which are fracture- or fatigue-critical, as defined in MIL-I-6870, components, the single failure of which would cause significant danger to operating personnel, or would result in an operational penalty. This includes loss of major components, loss of control, unintentional release, inability to release armanent stores, or failure of weapon-installation components.

Type II – All components not classified as Type I.

Defects covered in this specification are defined.

MIL-A-83377B, October 6, 1978—*Adhesive Bonding (Structural) for Aerospace and Other Systems, Requirements for*, 15 pp

FSC   8040   20

This document covers requirements applicable to structural adhesive bonding of metal, composite and core material in any combination. It and the earlier revisions replaced MIL-A-9062. It is intended that the requirements of this document be mandatory for use by contractors to assure the reliability of adhesive-bonded structural components used in aerospace and other systems.

The classification is as follows:

Type I – Primary structure. Components which are fracture- (or fatigue-critical, as defined in MIL-I-6870, components, the single failure of which would cause significant danger to operating personnel, or would result in an operational penalty. This includes loss of major components, loss of control, unintentional release, inability to release armament stores, or failure of weapon-installation components.

Type II – Secondary structure. All components not classified as Type I.

This is a process-type document to be used as a guide in preparing a process specification. Definitions of terms covered are given.

MIL-A-85705(AS), April 1, 1985—*Adhesive, Aircraft, for Structural Repair*, 19 pp

FSC 8040   AS

This limited coordination Navy specification establishes the requirements for aircraft-structural-repair adhesives capable of operating within a temperature range of $-67°$ to $220°F$ ($-55°$ to $104°C$). Metal-to-metal, composite-to-composite, and composite-to-metal repairs are covered. A Qualified Products List (QPL) will be prepared. Classification is as follows:

Type I – Suitable for autoclave curing

Type II – Suitable for vacuum-bag curing

Form F – Film

Form P – Paste (modified epoxy)

Requirements include toxicity, curing, conditions, visosity, mechanical properties (lap shear, peel, flatwise tensile, static long-term stress durability, creep, cyclic stress, and fatigue strength), effects of immersion in fluids, and storage life. Tests are carried out at normal, elevated and low-temperature conditions.

MIL-A-87134(USAF), June 20, 1978—*Adhesive, Contact, for Custom Fit Helmet Liners*, 10 pp

FSC 8040   20

This limited-coordination Air Force specification covers a flexible adhesive in a volatile organic solvent for contact bonding of leather to polyurethane foam custom-fit liners in pilots' helmets. It will also bond other materials, such as plastic decorative laminates, fabrics, and wood. The foam is covered by MIL-P-83379. The requirements of this specification are equivalent to those of MMM-A-130, except that the ketones and other solvents which attack polyurethane foam are prohibited. The material shall consist of polychloroprene (neoprene) rubber and synthetic resin modified to the extent that the requirements of this specification are met. The solvents must not attack polyether-based, Freon-blown, rigid polyurethane foam. Ketones, halogenated compounds, and benzene may not be used. Requirements include nonvolatile content, ash content, viscosity, density, shear strength, bonding range (1 hour minimum), accelerated aging, and shelf-storage life.

MIL-A-87135(USAF), February 17, 1979—*Adhesive, Nonconductive, for Electronics Application,* 7 pp

FSC 8040   20

This limited-coordination Air Force specification establishes the requirements for a moderately fast-curing adhesive used for bonding components to printed wiring assemblies to prevent vibration damage. The adhesive materials shall be such that they shall have no adverse effect on the materials used in the substrate or components attached thereon, and shall be formulated from resins, elastomers, plasticizers, catalysts, and other ingredients which meet the requirements of this specification. Requirements include compatibility, working life, and temperature, overlap-shear strength, reworkability, shelf life, volume resistivity, hydrolytic stability, and corrosivity.

REFERENCES

1. Landrock, A. H., Chapter 11, Commercial and Government Specifications and Standards, in *Handbook of Plastics and Elastomers,* (C. A. Harper, editor-in-chief), McGraw-Hill, NY, (1975).
2. Landrock, A. H., Standards and Specs: Are They Really Indigestible? *Plastics Design Forum,* 2(6): 81-88 (November/December 1977).
3. Landrock, A. H., Introduction, pp 2-3, *Specifications for Adhesives,* desktop data bank®, The International Plastics Selector, Inc., San Diego, CA, (1979).
4. ASTM Standards can be found in the latest issue of the *Annual Book of ASTM Standards.* Most of the standards listed, whether test methods, practices, or specifications, are included in Volume 15.06—Adhesives, which can be obtained from the American Society for Testing and Materials, 1916 Race Street, Philadelphia, PA 19103. Separate copies of each standard may also be purchased from ASTM.
5. Acknowledgment is made for some of this material taken from discussions of ASTM Tests in *Adhesives,* edition 3, desk-top data bank, The International Plastics Selector, Inc., San Diego, CA, (1980).
6. SAE Standards (ARP's and AMS's) are available from the Society of Automotive Engineers, Inc., 400 Commonwealth Avenue, Warrendale, PA 15096. A catalog listing all current standards is also available.
7. Military and Federal specifications and standards can be obtained at no charge from the Naval Publications and Forms Center, 5801 Tabor Avenue, Philadelphia, PA 19120.

13

Applications of Adhesive Bonding

INTRODUCTION

Although other chapters in this book mention certain specific applications of adhesive bonding, this brief chapter is intended to summarize most of the important applications in particular fields, such as aircraft, space, transportation, and building construction. The treatment is certainly not exhaustive, but merely serves to indicate typical examples of uses of adhesives. Although the emphasis is on structural adhesives, sealants and nonstructural types are discussed to some extent.

AUTOMOTIVE APPLICATIONS

This discussion covers automobiles, trucks, motorcycles and related vehicles. Adhesives are used in the automotive industry for reasons of economy, performance and manufacturing convenience. They perform structural, holding and sealing functions and are used on metal, plastic, fabric, glass, rubber, and painted parts. Automotive adhesives must meet a number of unique requirements, which are essentially independent of joint performance. They must be usable under conditions which include:[1]

- An essentially unskilled workforce, often with a high turnover rate
- High production rates (100 cars/hour) with short, unvarying times for each operation
- Minimal cleaning of surfaces (oil may be present)
- Low tolerance for health and safety standards
- Cure times, pressures and temperatures which are somewhat variable and which may need to be compatible with paint bake schedules, or low material heat-distortion temperatures

- Aversion to complex measuring and mixing

Use requirements for adhesives in the automotive industry are also severe. The modern automobile must perform well from $-40°F$ $(-40°C)$ to over $200°F$ $(93°C)$ and must endure exposure to changing temperatures, salt water, fuel, oil, high humidity, vibration, detergents, and dirt.[1]

Structural applications in the automotive industry include:[1]

- rear quarter reinforcement
- double shell roof
- hood inner and outer panel
- brake shoes
- clutch and transmission bonds
- window sealants
- roof bows
- disk-brake pads
- FRP body panels (Corvette and truck)
- roof to quarter panel (1971 Vega)
- hood-hem flanges
- tulip panel (bonds through oil)

Brake linings must withstand impact and high shear stresses at temperatures which may exceed $300°F$ $(149°C)$. A nitrile-phenolic film adhesive is used for this application. *Disk brake pads* must endure temperature in excess of $350°F$ $(177°C)$, requiring phenolic adhesives. Another application of structural adhesives is that of *glass-to-metal assembly*. Formerly the windshield and backlite (rear window) were sealed to the body with butyl rubber tape and mechanically secured, or sealed and held simultaneously with a crosslinking polysulfide material. Since the early 1970's polyurethane adhesives have begun to replace polysulfide resins for windshield and backlite installations.[1]

Polyvinyl butyral adhesives are used to bond rear-view mirrors to the windshields of cars. Such applications are called "holding adhesives". Holding adhesives are those whose primary function is to attach one material to another without transmitting significant structural loads. Such adhesives may be interior or exterior, depending on whether they are used inside or outside the passenger compartment. Other applications of *interior holding adhesives* include the following:[1]

- trim-panel fabric
- door-panel fabric
- carpet to floor
- weatherstripping
- sound-deadening panels

Application of *exterior holding adhesives* include (1):

- body side molding
- wood-grain decals
- vinyl roof
- various locking applications

Automobile hoods and roofs have a tendency to "oil can" or flutter at high speeds unless they are stiffened with an inner reinforcing member. The *hood inner members* are adhesively bonded to avoid unsightly weld marks. "Hershey drops" of plastisol adhesive are used in this application. The outer panel is often hem-flanged around the periphery of the inner panel to immobilize the parts until they cure in the paint oven. Modern plastisols are able to absorb thin films of oil from the metal as they cure. This ability to bond through oil is very important in the automobile industry where thorough cleaning before assembly is traditionally a problem.[1]

One of the largest volume uses of adhesives in the automotive industry is the neoprene solvent cement used to attach *vinyl roofs. Locking compounds* are coming into use where vibration loosening is a problem. Oil-pan screws and door interior hardware fasteners are typical examples.[1]

Sealing adhesives *(sealers* and *gaskets)* are used in automobiles and trucks to seal out air, dust, or water in order to reduce corrosion or improve comfort. In many cases they may also function as holding adhesives. Many sealer applications are out of sight in a modern automobile body. Low-viscosity rubbery sealers are used in virtually all metal-to-metal body joints prior to spot welding. Hot-melt sealants of the polyethylene type may be used in locations which are difficult to reach after assembly. A stick or rod of the sealant may be placed in a door or rocker panel at a convenient time during assembly. When the body is heated in the primer bake oven the sealant melts and flows by gravity into the joint. A number of other body sealers are extruded over welded joints to keep moisture out and reduce corrosion. These compounds are used extensively to seal firewall openings, where their ability to adjust to temperature changes without cracking and to resist moisture are important. Polyvinyl sealers are commonly used in joints and seams which will subsequently be painted and which are not subject to much flexing or expansion and contraction. These compounds adhere well, are viscous enough to resist sagging, and have good weatherability. They must withstand severe salt-spray exposure and numerous freeze-thaw cycles. Epoxy sealing compounds are being used as replacements for traditional lead body solder to conceal welded body joints.[1]

Twiss[2] has provided an excellent discussion of the use of adhesives in the automobile industry.

BUILDING CONSTRUCTION

Adhesives have been used extensively for manufacturing a number of structural products for building construction, such as softwood plywood, laminated timbers, laminated paper-base panel products, and similar items. More recently, adhesives have been used to replace mechanical fastenings in the assembly operations of building construction. These adhesives are considered as *construction adhesives*. These compounds must be capable of producing adequate bonds in poorly fitting joints with thick and variable thickness bondlines. There are two principal limitations to adhesive bonding in present building operations—the lack of adequate temperature control during bonding, and the lack of adequate processing equipment, so that mechanical fastenings are also required. Examples of bonded assemblies for building construction are the following:[3]

- complete wall sections with the sheathing bonded to wood frames
- floor and roof sections with sheathing bonded to joists and rafters to provide T-beam action
- various types of stressed-skin sandwich panels
- assembly of glued-wood roof trusses
- box beams and other types of composite beams for joists or rafters
- ultimately, adhesives for bonding together all of these components in the final structure

ELECTRICAL/ELECTRONIC APPLICATIONS

Adhesives are used in the electrical/electronics industry in a variety of ways, from holding microcircuits in place to bonding coils in mammoth electrical generators. Failure of adhesives in these applications could result in serious problems, such as causing computers to stop functioning, cities to black out, and missiles to misfire. In addition to *fastening*, adhesives in electrical applications are required to *conduct or isolate electricity*, provide *shock mounting, seal,* and *protect substrates*. Properties required for various applications cover the range of useful life from a few seconds to many years. Operating temperatures range from $-454°$F $(-270°$C) to $932°$F $(500°$C).[4]

Epoxies are the most widely used adhesives in electrical/electronic applications because of their versatility, excellent adhesion, compatibility, ease of applications, good electrical properties, and resistance to weathering. *Silicones* are used where flexibility, wide temperature range, high frequency, high humidity, and atmospheric contamination are encountered. *Hot melts* can be used where their lower strength and limited temperature range can be tolerated and where rapid assembly is important. *Acrylics* are used because of their excellent electrical properties, stability, good aging characteristics, and optical clarity. *Urethane* adhesives have the flexibility, toughness, and strength up to $250°$F $(121°$C). Precoated *polyvinyl butyral* produces bonds which are tough and easily fabricated.[4]

Three major uses of adhesives in *microelectronics* are: (1) die-bonding, (2) bonding circuit elements to the substrate, and (3) sealing the packages. Another major application is in *printed-circuit boards*. The necessity to resist $482°$F $(250°$C) soldering temperatures limits the adhesives that can be used to bond copper foil to laminated printed-circuit boards. Adhesives are also used for *large equipment,* such as generators, transformers, and other equipment that must operate at elevated temperatures, and other equipment that must operate at elevated temperatures for 20–40 years in hostile environments. The size of many pieces of such equipment preclude the possibility of oven curing, and the heat conduction of copper and other metals used makes local heating impractical. For this reason, room-temperature-setting adhesives are used.[4]

Adhesives are found in almost all *electronic equipment.* They perform such functions as the bonding of composite materials in radar antennas, providing thermal or electrical paths for electronic services (conductive adhesives), or the bonding of missile nose-cone rings. Typical applications include:[5]

- antenna reflector for the tracking/illuminating radar in the Ship-board air-defense system

- Cobra Dave phased-array radar system for bonding the outer di-electric window to an element body

- thermally conductive film adhesive for bonding various electronic components in the Trident MK 5 missile guidance computer

- air traffic-control radar—a curved epoxy laminate system bonded to a metallic backbone structure

Printed-Circuit Boards

Printed circuits may be rigid or flexible, and both types use adhesives. For epoxy-glass boards the adhesive may be vinyl-phenolic, nitrile-phenolic, or modified epoxy for maximum use temperature of about 302°F (150°C). For applications up to 500°F (260°C) glass-silicone laminated boards or copper/Teflon® combinations are used. Rigid printed-circuit boards assembled in stacks can be bonded together with either thermoplastic or thermosetting adhesive-coated plastic film. Complex assemblies sometimes require thin, flexible printed circuits. RTV silicone-rubber coatings are available, not only for use as conformal insulating coatings, but also to damp out vibrations, which would put an undesirable fatigue load on all solder joints. Epoxy and silicone adhesives are also applied to component loads to strengthen them mechanically and protect them against shock and vibration.[6]

AEROSPACE APPLICATIONS

Aircraft Applications

The aircraft industry spearheaded the development and application of structural-metal adhesives and is the principal user of such materials. Almost four acres (116,000m²) of adhesive film are used in bonding one Boeing 747 and one Lockheed C-5A. One Boeing 747 will use about 40,000 ft² (3700 m²) of adhesive film, about 950 lbs. (431 kg) of polysulfide rubber, and about 50 lbs. (23 kg) of silicone rubber sealant. The use of adhesives in aircraft has increased steadily with improvements in materials and processing, and will probably continue to increase, except in those areas where limitations are imposed by the high skin temperatures of supersonic aircraft.[6]

The B-58 Hustler used about 800 lbs. (300 kg), the F-11 approximately 1,000 lbs (450 kg), and the Boeing 727 about 5,000 lbs. (2270 kg) of adhesive per airplane. As airplanes become larger, adhesives more versatile, and weight saving and fatigue requirements more exacting, the percentage of bonded construction per aircraft will continue to rise.[6]

Sealants are also used in aircraft, particularly polysulfides, where they serve to seal fuel tanks. Other sealant applications are in fairings and edge sealants on honeycomb panels, and in pressurized joints. Silicone sealants are being used where heat resistance is required, as in ducts near exhaust outlets.[6]

The B-58 Hustler had about 4,500 ft^2 (420 m^2) of bonded paneling per aircraft. The wings, which doubled as aerodynamic load-bearing primary structures and as fuel tanks, were almost entirely bonded. Their structure included edge metal-to-metal bonds of nitrile-rubber phenolic adhesives with high peel, high-temperature characteristics, and aluminum skin-to-honeycomb bonds with epoxy-phenolic adhesives. Both aluminum and fiberglass-resin honeycomb were used. Shapes varied from flat panels to highly contoured leading edges.[6]

The development of the two-phase nitrile rubber-phenolic/epoxy film adhesive, and, at a later date, the nylon-epoxy and nitrile rubber-epoxy film adhesives helped overcome some of the limitations of the nitrile rubber-phenolic film by permitting the use of a single adhesive for both edge-bonding and skin-to-honeycomb bonding. Other aircraft-bonding application include glass-fabric laminations, core splicing, and core-to-edge bonding. These are usually accomplished with an epoxy-phenolic paste adhesive, since its foaming capability enables it to fill in voids and contact large areas of honeycomb edges. Commercial jet aircraft use adhesive bonding in horizontal and vertical tail surfaces, wing panels, ailerons and flags, tail-cone assemblies, and nose radomes.[6]

The PABST program demonstrated the practicality of using structural adhesives on the mid-section fuselage of short-haul jet aircraft.[7]

Helicopter Rotor Blades—Initially helicopter rotor blades were constructed of wood laminates. However, due to the high fatigue stress and weathering, these blades did not last long before they had to be replaced. Since the early 1950's rotor blades have been constructed of aluminum or fiberglass laminates using structural film adhesives. These blades are now completely bonded and include outer-skin laminations and bonds to inner, lightweight core materials and to longitudinal spars. Rather than lasting only 50–100 hours, they now last for more than 2,000 hours of flight time and then can be reworked.[6,8] Adhesives used for this application must have the following properties:[8]

- durability—must withstand salt spray, high humidity, static and cyclic fatigue stress
- creep resistance
- high metal-to-metal and honeycomb strength in shear and peel
- long shop time (working life), i.e. 30 days at 72°F (22°C) rather than cold storage
- cure-temperature versatility

European Airbus—This aircraft is a product of a consortium of various

European aerospace companies, where various sections of the aircraft are manufactured by different companies. Details of the adhesive applications, which are many, are given in Reference 8. In September of 1984 an announcement was made that Pan Am would purchase 28 of these new jets.[9]

Space Applications

Epstein[10] has presented an exhaustive discussion of space applications of adhesives in a fairly old presentation which, nevertheless, is informative and of continuing interest. Adhesives are used extensively in manned and unmanned space vehicles, both to produce component parts and to assemble or mount these parts on the vehicle. Applications are numerous and the following are representative:

- antennas (various components)
- cold plates (to maintain operating temperatures for electronics)
- containers
- covers
- detectors
- fittings and attachments
- heat shields (for reentry)
- impact/shock attenuators
- launch-vehicle components (rocket motors, engines, nozzles, nozzle extensions, shrouds, and insulations)
- mounting panels (for support of equipment)
- multilayer circuit boards
- optical components and assemblies
- platforms (spinning and despun)
- primary structures (struts, cylindrical and conical structures, frames, etc.)
- radiators (for thermal control)
- second-surface mirrors (for thermal control)
- sensors
- solar-cell modules
- thermal isolators
- wiring (location, attachment, fastening)

CIVIL ENGINEERING APPLICATIONS

In civil engineering the technique of using pre-cut concrete segments in reinforced structures is finding ever-increasing application in bridge building, stadia, underwater tunnels, foundations for lighthouses, and similar uses. In 1963 on a bridge near Paris the segments were first joined together with an epoxy adhesive instead of mortar.[7]

Hewlett and Shaw[11] have discussed the use of adhesives in concrete bonding in an excellent presentation.

BONDED AND COATED ABRASIVES

Adhesives are used in adhering abrasive particles to backing materials, flexible or rigid. A well-known example is sandpaper. In a bonded abrasive product the primary role of the bond is that of a structural adhesive. The function is to hold the abrasive particles in some desired spatial relationship and to provide mechanical strength and integrity to the composite. Grinding wheels are currently bonded primarily with phenol-formaldehyde adhesives.[12]

A one-part epoxy paste with a dicyandiamide curing agent (MIL-A-8623, Type III) is used to bond abrasive powders or granules to a solid or hollow-metal shaft. (This specification has been cancelled and superseded by MMM-A-134, which has identical types.)

RECREATION INDUSTRY

Epoxy adhesives are widely used in the manufacture of skis and archery bows. These adhesives form high-strength bonds which distribute stresses and withstand environmental exposure. Metal skis are no longer made in the U.S. All American ski manufacturers have converted to fiberglass skis. Although these fiberglass products vary widely in form and application, all use epoxy resin as the matrix. Polyester resins were tried and discarded because of their inferior interlaminar shear strength.[13]

Materials in *skis* are more highly stressed than materials in aerospace construction. Bending stresses generated during ski use are transmitted into the skin by the adhesive layer, which is subjected to extremely high shear forces during such use. The adhesive must absorb these loads and not fail, even under conditions of high temperature and humidity. Adhesives are critical to the strength, durability and performance of any ski.[13]

The average *archery bow* contains six layers of adhesive. Epoxies are ideal as bow adhesives because they will bond well to all components and will withstand the stresses of the drawn bow. They can be formulated to the ideal viscosity to be easily applied, yet not squeeze out onto the press.[13]

Epoxies developed for the ski and archery-bow industries are also being used for *tennis-racket manufacture.* In this case bulk-molding compounds are used to make racket parts. Epoxies are also being used in *skateboard* manufacture.[13] *Golf clubs* currently are made from metal, laminated wood, thermoplastics, and advanced reinforced materials. Adhesive bonding is the only practical method of joining these components. Epoxy adhesives are used because they have the necessary gap-filling properties and affinity for the different materials.[7]

MARINE APPLICATIONS

Adhesives are used in many applications in ship and boat assembly, whether fiberglass, wood, or metal construction. In addition to structural bonding of components with conventional adhesives, adhesive/sealants are used in large quantities because of the unique marine environment to which the vessels are exposed. Polysulfide liquid polymer-based sealants are used in filling construction joints, thereby eliminating the need for swelling of the wood to close the joints. These sealants are used for a number of other applications, including caulking deck seams.

PACKAGING

Application of adhesives in packaging include the following:[14]

- corrugated box manufacture (especially the manufacturer's joint)
    - —built-up pods
    - —case sealing
    - —traymaking
- folding cartons
- laminating
- tube-winding (convolute and spiral winding)
- set-up boxes
- bags
- labels

Polyvinyl acetate and *polyvinyl alcohol* are important components of adhesives used in corrugated-box manufacture. Vegetable-base adhesives are used in built-up pads and case-sealing. *Urethanes* are used in making laminates of such materials as polyester film, aluminum foil, and polyethylene. *Hot-melt* adhesives are used in food packaging for everything from carton, can, pouch, and bag to the fabrication of spiral-wound containers and the lamination of films, paper, and metal foils.[14]

Brazier[15] has discussed adhesives for use in packaging in a recent reference.

MISCELLANEOUS

Additional uses of adhesives are found in the following:

- pressure-sensitive tapes (for packaging, cellophane tape, masking tape, reinforced shipping tape, electrician's tape, surgical tape)
- gummed tapes (remoistenable, for stamps, shipping tapes)
- pastes and rubber cements (for office, school and library uses)

- handyman applications (epoxies, cyanoacrylates, vinyl emulsion wood glues, nitrile-phenolics)
- wallpaper paste
- optical applications (cementing lens elements together)
- ordnance applications

REFERENCES

1. Schneberger, G. L., Chapter 51, Adhesives in the Automotive Industry, *Handbook of Adhesives*, 2nd Edition, (I. Skeist, ed.), Van Nostrand Reinhold, NY (1977).
2. Twiss, S. B., Chapter 22, Adhesives in the Automotive Industry, *Handbook of Adhesive Bonding*, (C. V. Cagle, ed.) McGraw-Hill, NY (1973).
3. Blomquist, R. F. and Vick, C. B., Chapter 49, Adhesives for Building Construction, *Handbook of Adhesives*, 2nd Edition, (I. Skeist, ed.) Van Nostrand Reinhold, N.Y. (1977).
4. Buchoff, L. S., Chapter 50, Adhesives in the Electrical Industry, *Handbook of Adhesives*, 2nd Edition, (I. Skeist, ed.) Van Nostrand Reinhold, NY (1977).
5. Baker, T. E. and Judge, J. S., Control and Characterization of Adhesives in the Electronics Industry, *High Performance Adhesive Bonding*, (G. De Frayne, ed.), Society of Manufacturing Engineers, Dearborn, MI, pp. 240–246 (1983).
6. De Lollis, N. J., *Adhesives, Adherends, Adhesion*, Robert E. Krieger Publishing Co., Huntington, NY (1980).
7. Garnish, E. W., Chapter 13, Some Applications of Structural Adhesives, *Adhesion* 6, (K. W. Allen, ed.), Applied Science Publishers, London and New Jersey (1982).
8. Charles, W. J. and Palmer, S. J., Chapter 7, Applications of Structural Adhesives in Production, *Adhesion* 5, (K. W. Allen, ed.), Applied Science Publishers, London (1981).
9. Staff written, Europe's Airliners Raid the U.S., *Time*, 124 (13:61 (Sept. 24, 1984).
10. Epstein, G., Chapter 24, Adhesives for Space Systems, *Handbook of Adhesive Bonding*, (C. V. Cagle, ed.), McGraw-Hill, NY (1973).
11. Hewlett, P. C. and Shaw, J. D. N., Chapter 2, Structural Adesives Used in Civil Engineering, *Developments in Adhesives*—1, (W. Wake, ed.), Applied Science Publishers, London (1977).
12. Seiler, C. J. and Zimmer, W. F., Jr., Chapter 48, Bonded and Coated Adhesives, *Handbook of Adhesives*, 2nd Edition, (I. Skeist, ed.) Van Nostrand Reinhold, NY (1977).
13. Henry, R. D., The Use of Epoxy Adhesives in the Recreation Industry, *High-Performance Adhesive Bonding*, (G. DeFrayne, ed.) Society of Manufacturing Engineers, Dearborn, MI., pp. 240-246 (1983).
14. Blair, A., Chapter 26, Packaging Adhesives, *Handbook of Adhesive Bonding*, (C. V. Cagle, ed.), McGraw-Hill, NY (1973).
15. Brazier, A. D., Chapter 4, Adhesives in Packaging, *The Packaging Media*, (F. A. Payne, ed.), published under the authority of the Council of the Institute of Packaging (UK), John Wiley & Sons, NY (1977).

Appendix—Sources of Information

Considerable useful and up-to-date information on adhesive bonding is available from journals, manufacturers' bulletins, technical conferences and their published proceedings, seminars and workshops, standardization activities, trade associations, consultants and information centers, in addition to books, many of which have been cited above. These sources will be discussed briefly.

JOURNALS AND OTHER PERIODICALS

Adhesives Age

This publication is published monthly (with a 13th issue in May) by Communication Channels, Inc., 6255 Barfield Rd., Atlanta, GA 30328. The magazine is an excellent source, with the emphasis on practical, rather than theoretical aspects. It frequently publishes issues devoted largely to special subjects, such as hot-melt adhesives, pressure-sensitive products, polyurethane, etc. The extra issue in May is now the Directory issue, formerly the *Adhesives Red Book*, a well-known source. Controlled-circulation subscriptions (free) are available. The editor is Martin Kraft.

The International Journal of Adhesion and Adhesives

This quarterly publishes contributed papers and research reports, in addition to a news section covering commercial and technological developments, conference reports, book reviews, and a calendar of forthcoming events. Letters commenting on previously published material, or presenting preliminary details of current work, are also encouraged. The journal is expensive ($140/yr for 4 issues). It is published by Butterworth Scientific Limited—Journals Division, P.O. Box 63, Westbury House, Bury Street, Guildford, Surrey GU2 5BH, UK. Subscriptions in the U.S. can be obtained through Expediters of the Printed

Word Ltd., 527 Madison Ave., N.Y., N.Y. 10002, (212) 838-1304.

The Journal of Adhesion

This is an excellent quarterly that publishes papers on adhesives and adhesion contributing to the understanding of the response of a composite structure (not necessarily a composite material) to a mechanical influence. Experimental papers are required to incorporate theoretical background, and theoretical papers must relate to practice. Announcements of forthcoming conferences and meetings are included. This quarterly is very expensive ($242/yr). The editor is Louis H. Sharpe of AT&T Bell Laboratories. The publisher is Gordon and Breach Science Publishers, c/o STBS Ltd., 42 William IV Street, London WC2N, 4DE, Uk. Subscriptions are available from Gordon and Breach Science Publishers, 50 W. 23rd St., N.Y., N.Y. 10001, (212) 206-8900.

Materials Engineering

This periodical is published monthly by Penton/IPC, 1111 Chester Avenue, Cleveland, OH 44114. From time to time it publishes short articles on the state of the art or developments in adhesives. Controlled (free) subscriptions are available to qualified subscribers. The editor is John H. Bittence.

Machine Design

This magazine is published twice monthly, except for a single issue in November. In addition, one reference issue is published in each of the months of April, May, June, September, and November. The Special Reference Issue in November is the Fastening and Joining Issue, with excellent tables and illustrations, usually covering 8 or 9 pages. The reference issues may be purchased as separates for $6.00. The magazine, which includes much informative advertising, is available from Penton/IPC, 1111 Chester Avenue, Cleveland, OH 44114. Controlled (free) subscriptions are available to qualified subscribers. The editor is Ronald Kohl. From time to time excellent articles on adhesive bonding are published, in addition to announcements of new adhesive products through advertisements.

SAMPE Journal

This journal is published bimonthly by the SAMPE National Business Office, 843 West Glentana, P.O. Box 2459, Covina, CA 91722, (818) 331-0616. SAMPE is the Society for the Advancement of Material and Process Engineering, a professional society. Excellent papers on adhesive bonding are frequently published. The journal is free to SAMPE members, but is available at $31/yr to others. The editor is Stuart M. Lee.

SAMPE Quarterly

This journal is also published by SAMPE (see above). Like the SAMPE Journal, it frequently published excellent papers on adhesive bonding. It is available at a reduced price of $12/yr for members. The regular price is $25/yr.

MANUFACTURERS' BULLETINS

A number of periodicals, including *Adhesives Age, Machine Design*, and *SAMPE Journal* have announcements of new products on adhesives and related materials. By sending for more information, frequently by using tear-out postcards, the reader can obtain up-to-date detailed infomation on many new products.

TECHNICAL CONFERENCES

Organizations such as SAMPE, SPE, The Adhesion Society, and The Adhesives and Sealant Council frequently hold conferences including papers on adhesive bonding. These organizations usually supply printed conference proceedings (preprints). SAMPE publishes the proceedings of the National SAMPE Symposium (always held on the West Coast in the Spring) and the National SAMPE Technical Conference (always held in the Fall). The 29th National SAMPE Symposium was held in 1984 and the 16th National SAMPE Technical Conference was held in the same year. Each preprint volume consists of a large number of papers, including a number on adhesive bonding. This book has references from a number of these papers.

SEMINARS AND WORKSHOPS

Organizations such as The Center for Professional Development, P.O. Box H, East Brunswick, N.J. 08816-0257, McGraw-Hill, N.Y., N.Y. and many universities frequently offer seminars and hands-on workshops for 1–5 days in adhesives technology.

STANDARDIZATION ACTIVITIES

ASTM

The American Society for Testing and Materials (ASTM), 1916 Race Street, Philadelphia, PA 19103, has a Committee D-14 on Adhesives, which meets twice a year to develop and revise standard test methods, practices and specifications on adhesives. Committee D-14 has the following subcommittees:

D14.03 Research	D14.30 Wood Adhesives
D14.04 Terminology	D14.40 Adhesives for Plastics
D14.05 Editorial Review	D14.50 Special Adhesives
D14.10 Working Properties	D14.70 Construction Adhesives
D14.20 Durability	D14.80 Metal Bonding Adhesives

D14.90 Executive

The Staff Director at ASTM Headquarters is Tom Kochaba, (215) 299-5560.

Committee C-24 on Building Seals and Sealants carries out work in caulking compounds, putty, elastomeric compounds, sealing tapes, and other related materials. Committee C-24 has 20 technical subcommittees and 7 administrative subcommittees. Various other ASTM committees have written standards in both adhesive and sealant areas. The Staff Director for Committee C-24 is Martha Kirkaldy, (215) 299-5531.

Chapter 12 in this book contains a listing and description of a number of standards issued by ASTM committees. Members who are willing to work on assigned projects and participate in the balloting process are welcome to attend meetings. Nonmembers are also welcome. Occasionally ASTM committees hold seminars and workshops.

SAE (AMS's and ARP's)

The Society of Automotive Engineers (SAE), 400 Commonwealth Drive, Warrendale, PA 15096, issues Aeronautical Materials Specifications (AMS's) and Aeronautical Recommended Practices (ARP's) in the field of adhesives. A number of these standards can be found in Chapter 12 of this book. SAE does not publish test methods as such.

TRADE AND PROFESSIONAL ASSOCIATIONS

• The Adhesion Society, c/o L.S. Penn, Midwest Research Institute, 425 Volker Blvd., Kansas City, MO 64110, (816) 753-7600, has an annual three-day meeting with technical papers in January and February. Members receive an intermittent Newsletter, including abstracts of papers.

• The Adhesives and Sealant Council, Suite 1117, 1500 North Wilson Blvd., Arlington, VA 22209, (703) 841-1112, J. Rapp, Exec. Vice Pres., holds meetings three times a year, including technical conferences. Members are companies (140) manufacturing adhesives. The Council publishes a monthly newsletter and semi-annual seminar papers.

• Adhesives Manufacturers' Association, 111 E. Wacker Drive, Chicago, IL 60601, (312) 644-6610, J. Dollard Carey, Exec. Sec'y., has 30 member firms and publishers a bimonthly newsletter. AMA has a Spring meeting and two seminars annually. Associate members, usually suppliers, have only recently been admitted.

• Sealant & Waterproofers Institute, 1800 Pickwick Ave., Glenview, IL 60025, (312) 724-7700, Carl A. Wangman, Exec. Vice Pres., is an organization of sealant contractors and suppliers of sealants and related products. It develops standards for industry and promotes the exchange of ideas. SWI publishes a quarterly *Applicator* and plans to publish a monthly newsletter in 1984.

• Society of Manufacturing Engineers (SME), One SME Drive, P.O. Box 930, Dearborn, MI 48121, (313) 271-1500, has some 65,000 members, including an Adhesives Division formed in 1978. SME sponsors educational seminars at national meetings.

• Society of Plastics Engineers (SPE), 14 Fairfield Drive, Brookfield Center, CT 06085, (203) 775-0471, holds numerous conferences and courses, including adhesives and sealants. SPE is the plastics professional society.

CONSULTANTS

A number of consultants in the field of adhesives are available. One of the outstanding organizations is Adhesives & Sealants Consultants, Fred A. Keimel, President, P.O. Box 72, Berkeley Heights, NJ 07922, (201) 464-3133. Keimel and his staff cover adhesive bonding design and analysis, compounding, source information, market data analysis, and testing. Seminars are also given. Keimel publishes a monthly newsletter *(The Adhesives & Sealants Newsletter)*. This very useful eight-page publication has been published monthly since 1977. It covers new products, future markets, innovative ideas, equipment and supplies, courses and conferences, book reviews, and many other subjects. The editor is Fred A. Keimel, who served with Bell Laboratories as an adhesives consultant for 16 years, and with the Adhesives Group for 9 years before that. The July 1984 issue of the Newsletter includes a 23-page Information Resources Guide, which is an updating of a previously published paper. This guide is available only to subscribers of the Newsletter. It proved helpful in compiling this brief resources listing.

INFORMATION CENTER

The Department of Defense Plastics Technical Evaluation Center (PLASTEC) is located at the U.S. Army ARDC, Bldg. 351N, Dover, NJ 07801-5001, (201) 724-3189. The author of this book is the specialist in adhesives and has at his disposal a number of sources, including ARDC's well-known Adhesives Section. Computer and/or manual searches can be made in a number of fields, including adhesives. Except for simple problems, a fee will ordinarily be charged for these searches.

MISCELLANEOUS ITEMS

• *Adhesives*—a desk-top-data-bank®, 3rd Edition, with Supplements published by The International Plastics Selector, Inc., P.O. Box 26637, San Diego, CA 02126. This 1010-page edition was published in 1980. The 4th Edition is scheduled for publication in 1984. The volume has very helpful data on the properties of most commercial grades of adhesives. The data is taken from manufacturers' data sheets. Tables are also given for property values in ranked order. Addresses of suppliers are also given, along with useful information on commonly used test methods.
• A two-chart guide on adhesives for household and handyman use was published in *Family Circle*, June 21, 1983. One chart, "A Nitty-Gritty Guide to Glues", lists 10 basic kinds of adhesives, their gluing properties and other fac-

tors. The second chart, "Which Glue Should I Use?", offers a simple cross-reference guide to determining what glues will "make various materials stick". This material is written for the non-professional, but it should, nevertheless, be of interest to the professional, even if it does suggest *mashed potatoes* as a possible adhesive!

General Index

This index covers terms, authors and individuals, and standards mentioned in the text. It does not cover authors listed in the references unless they are mentioned in the text. The coverage is comprehensive, covering most important terms found in the text. In general, obvious terms are not listed. Page references providing especially important coverage are *italicized*. Tradenames and agency abbreviations are listed in full uppercase type, ULTEM, PLASTEC, etc. Definitions are indicated by the designation (def) following the page number. Cross references and alternate subject headings are often given in parentheses.

It is hoped that this comprehensive index of more than 2,100 entries will enable the reader to quickly locate the information he seeks.

A54 and A54A (TRW Systems) - 245
A-1100 primer - 113
Abrading (abrasion) - 56, 68, 70–73, 75, 78–80, 86–89, 91–93, 95, 97–99, 101, 102, 107, 108, 110–115, 195, 227, 234, 277
Abrasion resistance - 168, 171
Abrasive
    blast - 74, 101, 102
    cleaning - 70, 71
    planing- 117
    polishing wheel - 157, 160, 165, 177
    scouring - 95, 100, 101, 105
ABS - (see Acrylonitrile-Butadiene-Styrene)
ABS/PVC alloy - 234
ABS/PVC transition joints - 228

Accelerators - 138
Accelerated testing - 239, 247
ACCUTHANE UR-1100 - 172
Acetal copolymer (CELCON) - *84-85, 193-194*, 227
Acetal homopolymer (DELRIN) - *85-86*, 194, 227
Acetals - 137
Acetal-toughened adhesives - 164
Acetic acid - 182, 235, 264
Acetic acid-hydrochloric acid method - 81
Acetic acid release - 178
Acetone - 60, 64, 74, 77, 78, 81, 84, 85, 87–90, 93–96, 100, 101, 103, 105, 106, 110, 111, 114, 117, 118, 136, 147, 161, 165, 226, 228–230, 232
Acid cure - 147, 178

Acid etch - 55, 56, 65, 73, 74, 77,
    81, 90, 91, 98, 118, 119,
    254
Acid-permanganate etch - 96
Acid-proof cements - 157
Acid resistance (compatibility) - 92,
    99, 136, 142, 147, 150, 153,
    166, 171, 263
Acids, oxidizing - 264
Acid tank construction - 157
ACLAR - 195
Acrylates - 175, 184
Acrylic acid diester adhesives - 128,
    132, 137, 175
Acrylic adhesives and sealants - 10,
    128, 129, 131, 132, *135–136*,
    138, 142, 143, 174, 175, 191,
    192, 194, 195, 197–199, 201,
    396
Acrylic copolymer adhesives - 132
Acrylics (plastics) - 91, 166, 172,
    177, 198, 233
Acrylonitrile-butadiene-styrene
    (ABS) - 25, *86–87*, 194, 227
Activated argon plasma - 63
Activated-gas (plasma) surface
    treatment - *61*, 88, 92, 95,
    99, 100, 109–113
Activation requirements - 127
Addition cure - 178
Addition polymers (polymerization) -
    102, 138, 180
Adherence - 12 (def)
Adherend failure - 19 (def) - 182
Adherends
    adhesives for specific - 190–204
    classification by - 128
Adhesion
    definition - 12
    mechanical - 13 (def)
    molecular - 13 (def)
    specific - 13 (def)
    theories of - 4–7
(The) Adhesion Society - 406
Adhesive
    alloys - (see Alloyed adhesives)
    application methods - 207–211
    base (binder) - 134
    bonding - 1 (def)
    bonding process - Chapter 7
    carrier - 151
    cartridges - 211
    choice (selection) - 7, 8, 55

classification - 126–135
coiled cord - 211
composition - 134–135
definition - 13
failure - 8, 9, 19 (def), 55, 246
pellets - 211
preparation - 206–207
scouring - 71
storage - 205–206
support - 151
thickness - 92, 102, 166, *218*, 289
Adhesives Age (magazine) - 403
Adhesive/sealants - 2, 126, 133, 139,
    238, 395, 401
Adhesives for specific adherends -
    Chapter 6
Adhesives Manufacturers Association -
    406
Adhesives Red Book (directory) - 403
Adhesives & Sealants Consultants
    Newsletter - 407
(The) Adhesives & Sealant Council -
    406
Adhesive-starved bond areas - 152,
    276
Adhesives, stress relief - 239
Aerospace applications - 1, 76, 168,
    177, 214, *397–399*
AF 30 adhesive - 90, 268
AF 31 adhesive - 267, 268
AF 32 adhesive - 267, 268
AF 126-2 adhesive - 74
Aging - 9, 142, 145–147, 150, 159,
    162, 166, 174, 178, 196, *291*,
    396
Air blast - 105
Air bubbles - 31
Aircraft applications - 135, 182,
    238, 251, 273, *397–399*
Air Force - 256
Air plasma treatment - 109, 111–113
AJAX scouring powder - 68
Albumen adhesives (glues) - 130
ALCOA - (see Aluminum Company
    of America)
Alcohol compatibility - 92, 98, 99,
    114, 150, 165, 166
Alcohol wash - 90, 94, 118
Alcohol-phosphoric acid etch - 247,
    255
Aliphatic amine - 148
Aliphatic hydrocarbons - 52, 94,
    147, 166

Alkaline cleaning (degreasing) - 56, 65, 73, 75, 77, 78, 82, 91
Alkaline detergent solution - 69, 70, 83, 120
Alkaline etching - 76, 78, 79
Alkaline permanganate treatment - 191
Alkaline peroxide etch - 79
Alkaline (caustic) resistance - 60, 98, 136, 142, 147, 153, 157, 166, 168, 171, 263, 264
Alkaline water - 184
ALKANOX detergent - 91
Alkyl acrylates - 183
Alloyed adhesives - *130, 133, 136*, 152, 162, 164, 182, 191
Allyl diglycol carbonate (CR-39) - 87, *136*, 168
Allyl resins - 168
Alumina ceramic - 118
Alumina powder filler - 4, 192
Aluminum - 3, 4, 55, 56, 61, 62, *64-66*, 82, 93, 136, 143, 148, 156, 162, 166, 171, 177, *190-191*, 194, 195, 221, 222, 238, 241, 242, 247, 248, 250, 251, 254, 256, 257, 263, 291, 299, 398
Aluminum filler - 140
Aluminum Company of America (ALCOA) - 191, 240
Aluminum foil - 401
Aluminum oxide (alumina) - 6, 70, 78, 79, 83, 89, 98, 100, 117-119, 141
Aluminum phosphate - 157
Alvarez, R.T. - 169
AMCHEM 7 - 83
AMCHEM 17 - 83
Amide linkage - 246
Amine-cure epoxies - 190, 196, 246
Amine curing agents - 174, 206, 264
Amino resins - 161, 182
Ammonia plasma treatment - 109, 112
Ammonia solutions - 71
Ammonium hydroxide - 69, 108
Amorphous resins - 94
Amylaceous matter - 291
Anaerobic adhesives and sealants - 10, 13 (def), 132, 133, *137-138*, 181, 191, 196, 198
  resistance to radiation - 266
Analysis, sulfochromate etch solution - 298
Anatose structure - 78, 79
Angle and corner joints - 41, 42
Anhydride-cured epoxies - 246, 255, 260-262
Anhydride hardeners (curing agents) - 149, 196, 241, 246
Animal adhesives (glues) - 10, 130, 131, 133, *160*, 183-184, 201, 212, 217
Animal-origin glues - *160-162*, 269
Annealing - 233
Anodic conversion coatings - 64
Anodize (anodizing) (anodic treatment) - 64, *66*, 76, 77, 83
Anti-icing fluid - 163
Antioxidants - 194
AP-133 primer - 110, 111
Apparent viscosity - 298
Application of adhesives (quality control) - 278
Application, classification by - 128
Application methods - 8, 207-213
Application temperature - 206
Applications of adhesives - Chapter 13
Applied weight/unit area - 297
Aqueous adhesives - 294
Aqueous dispersion - 175
Aqueous phenol cement - 230
ARALDITE 100 adhesive - 93
ARALDITE AW 134 adhesive - 196
ARALDITE AV 1566-GB - 196
Archery bows - 400
ARDC - 247, 251, 407
ARDEL polyarylate - 90
Argon plasma treatment - 92
Army aircraft - 251
Aromatic amine-cured adhesives - 239
Aromatic fuels - 150, 167, 182
Aromatic hydrocarbons - 98, 99, 147, 171, 235
Aromatic polyamine hardeners - 149
Aromatic polyester - 61, 90
Aromatic polymer adhesives - 132, *138-139*, 266
Aromatic solvents - 114, 200

ARP 1524 - 297
ARP 1610 - 292
Artificial latex - 184
Arvin acoustic analysis system - 281
Asbestos adherend - 117
Ash content - 291
Asphalt adhesives - 130–132, 139,
    179, 184
Assembly - 15 (def), 278
Assembly adhesive - 13 (def)
Assembly joint - 44
Assembly time - 28 (def)
A-stage - 12 (def)
ASTM Committee C-14 - 406
ASTM Committee D-14 - 405–406
ASTM B 117 - 253, 295
ASTM D 471 - 264
ASTM D 638 - 172
ASTM D 816 - 276, 295
ASTM D 896 - 238, 264, 277, 292
ASTM D 897 - 276, 288, 297
ASTM D 898 - 276, 297
ASTM D 899 - 276, 297
ASTM D 903 - 276, 289, 295
ASTM D 904 - 277, 245
ASTM D 905 - 61, 276, 289, 296
ASTM D 906 - 276, 296
ASTM D 950 - 276, 290, 294
ASTM D 1002 - 276, 288, 290,
    296
ASTM D 1062 - 276, 297
ASTM D 1084 - 276, 297
ASTM D 1144 - 276, 296, 297
ASTM D 1146 - 277, 292
ASTM D 1151 - 238, 277, 293
ASTM D 1174 - 277
ASTM D 1183 - 277, 291
ASTM D 1184 - 276, 293
ASTM D 1286 - 277
ASTM D 1299 - 270
ASTM D 1304 - 277, 293
ASTM D 1337 - 276, 297
ASTM D 1338 - 299
ASTM D 1344 - 276, 297
ASTM D 1382 - 64, 269, 277, 291
ASTM D 1383 - 64, 269, 277, 291
ASTM D 1448 - 276
ASTM D 1488 - 291
ASTM D 1489 - 276, 294
ASTM D 1490 - 276, 294
ASTM D 1579 - 276, 293
ASTM D 1581 - 277, 291
ASTM D 1582 - 276, 294

ASTM D 1583 - 276, 294
ASTM D 1584 - 276, 299
ASTM D 1713 - 277, 291
ASTM D 1780 - 277, 290, 292
ASTM D 1781 - 276, 289, 295
ASTM D 1828 - 238, 277, 293
ASTM D 1875 - 276, 293
ASTM D 1877 - 269, 272, 291
ASTM D 1879 - 238, 260, 277, 294
ASTM D 1916 - 276, 295
ASTM D 2093 - 297
ASTM D 2094 - 288, 297
ASTM D 2095 - 276, 288, 297
ASTM D 2182 - 276, 289, 296
ASTM D 2183 - 276, 294
ASTM D 2235 - 228
ASTM D 2293 - 292
ASTM D 2294 - 277, 290, 292
ASTM D 2295 - 238, 276, 294, 296
ASTM D 2339 - 276, 296
ASTM D 2556 - 296, 297
ASTM D 2557 - 238
ASTM D 2558 - 276, 295
ASTM D 2674 - 297
ASTM D 2739 - 140, 277, 296
ASTM D 2918 - 276, 293, 295
ASTM D 2919 - 247, 293, 295
ASTM D 2979 - 276, 298
ASTM D 3111 - 276, 293
ASTM D 3121 - 267, 297
ASTM D 3138 - 228
ASTM D 3163 - 276, 288, 296
ASTM D 3164 - 276, 295
ASTM D 3165 - 276, 288
ASTM D 3166 - 290, 293, 296
ASTM D 3167 - 295
ASTM D 3236 - 276, 297
ASTM D 3310 - 277, 292
ASTM D 3433 - 294
ASTM D 3482 - 293
ASTM D 3527 - 277
ASTM D 3528 - 276, 296
ASTM D 3632 - 277, 291
ASTM D 3658 - 276, 297
ASTM D 3702 - 276
ASTM D 3762 - 291, 299
ASTM D 3807 - 276, 290, 292
ASTM D 3808 - 276, 297
ASTM D 3929 - 277, 297
ASTM D 3931 - 276, 294, 297
ASTM D 3933 - 66, 298
ASTM D 3983 - 297
ASTM D 4027 - 297

ASTM D 4299 - 238, 270, 282
ASTM D 4300 - 238, 270, 292
ASTM D 4339 - 295
ASTM E 229 - 289, 296, 297
ASTM F 493 - 236
ASTM G 85 - 295
Atomic radiation - 169
Autoclave - 15 (def)
Automobile bodies, repair of - 168
Automotive applications - 166, 170, 181, 221, *393-395*

BAB-O scouring powder - 68
Backing - 15 (def)
Backlite - 394
Bacteria - 268, 270, 292
Bag molding - 15 (def)
Baked enamel - 200
Barium sulfate (barytes) - 109
Barker, A. - 161
Bar specimens - 297, 298
Barth, B.P. - 50
Bases (compatibility with) - 99, 263
Basic solutions - 92
Basic salt cements - 134, *157*
Battelle Memorial Institute - 74, 266
Battery terminals - 141
Bayerite oxide layer - 221
Beading (of joint end) - 38, 39
Beers, M.D. - 178
Bemmels, C.W. - 175
Bending stress - 37-39, 51
Bent-beam method - 297
Benzol - 91, 230
Benzoyl peroxide catalyst - 233
Benzyl cellulose adhesives - 140
BERYLCOAT D - 67
Beryllia filler - 4, 141
Beryllium - *67*, 167, *191*
Beryllium oxide filler - 141
B-58 Hustler - 398
Bikerman, J.J. - 6
Binder - 16 (def), 153, 165, 180
Biocides - 269
Biodeterioration - 147, 158-160, 165, 166, 170, 173, 174, 176, 182, *268-270*
Biological factors - 8, *268-270*
Bite - 16 (def)
Bitumen adhesives - 192
Bituminous concrete - 119

BLACK MAX adhesives - 143
Black oxide coating - 68
Black smut - 74, 75
Blanket molding - 15 (def)
Blister - 16 (def)
Block copolymers - 116, 174, 180
Blocked curing agent - 16 (def)
Blocking - 8, 16 (def)
    point - 292
Blood adhesives (glues) - 130, 131, 133, *160*, 184, 201
Blood albumen - (see Blood adhesives)
Bluestein, C. - 181
Blushing - 16 (def), 225
Boat building - 217, 401
Bodied adhesives (cements) - 127, 225, 230-233
Body - 16 (def)
Bodying resin - 178, 194, 228, 230, 232, 233
Boehmite oxide layer - 221
Boeing 727 aircraft - 398
Boeing 747 aircraft - 397
Boeing Aircraft Company - 66, 256
Boeing-Vertol - 109
Boiling point (solvent) - 232
Bolger, J.C. - 141
Bolts - 138
Bond (verb and noun) - 16 (def)
Bonded abrasives - 400
Bond failure, mechanism of - 8, 9
Bonding agents - 153
Bonding equipment - 214
    heating - 215
    pressure - 214
Bonding permanency - 291
Bonding range - 8
Bond line - 16 (def), 127, 140, 253, 256, 284 (see also Glue line)
    spacer - 135
    temperature - 215
    thickness - 92, 127, 152
Bonds
    chemical - 6
    covalent - 6
    electrostatic - 6
    metallic - 6
    primary - 6
    secondary - 6
Bond strength - 16 (def)
Bone adhesives (glues) - 130, 160
Bone and hide glues - 133, *160*

Bookbinding - 173
Boron nitride filler - 141
Boron trifluoride hardeners - 149
Boron trifluoride monoethylamine
    (BF-MEA) - 149
BOSTIK 598-43 - 196
BOSTIK 7026 - 196
BOSTIK 7087 - 197-198
Bottle caps - 160
Bottle labels - 158, 160, 291 (see
    also Labels)
Boxes - 157, 159 (see also Corru-
    gated boxes, Set-up boxes)
BR 34 - 245
BR 89 - 196, 197
BR 92 - 197
BR 123 - 256
BR 127 - 256
Brake Linings - 165, 394
Branching - 265
Brass - *67*, 136, 148, 156, 166,
    191, 195
    wool - 100
Brazier, A.D. - 401
Brazing - 3
Brewis, D.H. - 191
Brick - 117
Bridges - 399-400
Briquettes - 157
British Aerospace - 80
Brittle adhesives (brittleness) (em-
    brittlement) - 9, 33, 145,
    161, 164, 165, 170, 171,
    195, 199, 206, 244, 265,
    267
Bromophosphate treatment - 75
Bronze - 191
Brush (brushing) - 145, 184, 207
Brushable adhesives - 127
B-stage - 15 (def), 168, 218
Buffing - 100, 107, 108
Building construction - 145, 147,
    395-396
BUNA N rubber - 112
BUNA rubber - 194
BUNA S rubber - 113
Burnham, B.H. - 138
Bush, S.M. - 178
Butadiene-acrylonitrile adhesives -
    198, 199
Butadiene-acrylonitrile rubber - *112*,
    184, 192
Butadiene rubber - 113

Butane - 226
n-Butanol - 120
Butt joint - 21 (def), 34, 37, 44
    double - 34, 50
    plain - 34, 37 42-44
    scarf - 34, 37, 42-44
Butyl acetate - 226, 228-230
Butyl alcohol - 230
Butyl cyanoacrylate adhesives - 142
Butyl methacrylate adhesives - 133
Butyl lactate - 228-230
Butyl rubber - 107, 108, *110-111*,
    147
Butyl rubber adhesive - 129, 131,
    139, 145, 179, 184, 199, 200
Butyl rubber primer - 111, 174, 192,
    201

CAB-O-SIL - 97, 109, 222
Cadmium - 68, 138, 191
Calcium chloride-ethanol (nylon-
    bodied) - 230
California Instit. of Technology -
    288
Canada Balsam - 159
Canvas - 197
Capillary action (in solvent cement-
    ing) - 135
Capsules, microscopic - 158
Carbon - 117, 118
Carbon tetrachloride - 81, 194
Carborundum - 117, 118
Carboxymethylcellulose - 184
Cardboard - 159
Carriers - 134, 151, 213
Cartons - 157
Casein adhesives (glues) - 130, 131,
    133, *160*, 183, 184, 192, 201,
    217
Case sealing - 401
Catalyst - 16 (def), 134, 135, 138,
    141, 152, 153, 158, 165, 168,
    170, 172, 206, 233, 246
Catalytic plural components-chemi-
    cal cure - 130
Catalytic reactions - 206, 213
    moisture cure - 131
Caul - 16 (def)
Caulking compounds - 406
Caulking deck seams - 401
Caulking gun - 127, 201, 207, 222
Caustic soda - 60
C-clamps - 214

CELANAR - 92
CELANEX - 92, 93
CELCON - 84–86
Cellophane - 139, 175
   tape - 401
Cellosolve - 229
Cellular plastics - 5, 106, 194, 228–
   230 (see also Plastic foams)
Cellulose acetate - 87, 194, 226,
   228, 229
   adhesives - 129, 139, 140
Cellulose acetate butyrate - 62, 87,
   91, 194, *228-229*
   adhesives - 124, 139, 140, 178,
   179
Cellulose caprate adhesives - 139
Cellulose derivative adhesives -
   131, 178
Cellulose esters - 213
   adhesives - 131, *139-140*, 175
Cellulose ethers - 213
   adhesives - 140
Cellulose nitrate - 87, 194, 226,
   229-230
   adhesives - 129, 139, 140, 178,
   179, 194
Cellulosics - 25, 87, 227
   adhesives - 132
Cellulose propionate plastics - 87,
   194, 229-230
Cell-wall (foam) collapse - 199,
   200
Cement (noun and verb) - 17 (def)
Cement-asbestos board - 136
Ceramic-based adhesives - 213
Ceramics - 118, 130, 138, 148–
   150, 161, 162, 169, 202
Ceramic seals - 157
Chain scission - 242, 265, 268
Characterization - 192
Cheese cloth - 65
Chemical composition, classification
   by - 128
Chemical etching - 246
Chemical fluids - 238
Chemically reactive adhesives - 130
Chemical resistance - 137, 138, 145,
   148-150, 168, 170, 175, 176,
   182, *263-264*, 292
Chemical treatment - 60
CHEMLOK 205 - 112, 113
CHEMLOK 220 - 112
CHEMLOK 243B - 107

CHEMLOK 305 - 196
CHEMLOK 306 - 196
CHEMLOK 507 - 72
CHEMLOK 607 - 101, 110
CHEMLOK AP 131 - 112, 113, 115
CHEMLOK AP 134 - 113
CHEMLOK Y-4310 - 115
CHEMLOK 5224 - 113
CHEMLOK Y 5254 - 115
Chilled surface (from solvent evapo-
   ration) - 73
Chipboard
   adhesives - 140
   spacers - 159
Chips (dissolved) - 236
Chlorendic anhydride - 241
Chlorinated hydrocarbons - 62, 98,
   147, 196, 233
Chlorinated natural rubber adhesives -
   192, 201
Chlorinated phenols (additives) - 160
Chlorinated polyvinyl chloride
   (CPVC) - 236
Chlorinated solvents - 95, 99, 107,
   264
Chlorination treatment - 108, *109*,
   111-115
Chlorine
   aqueous solution - 94, 115
   gas (UV light) - 94
Chlorobutyl rubber - 111
Chloroform - 228, 229
Chlorosulfonated polyethylene
   (HYPALON) - 111–112
Chlorosulfonyl polyethylene - 111–
   112
Cholesteric liquid crystals - 282-283
Chromate conversion coating - 64,
   68, 71, 82
Chromate-free etch - 65, 66
Chromic acid - 78, 90, 96, 194
   hot - 70
Chromic acid anodize - 56, 247, 248,
   253, 254-256
Chromic acid etch - 56, 66, 84, 85,
   90, 93, 94-96, 234, 239
   warm - 87
Chromic-sulfuric acid etch - 247–
   248, 255
Chromium phosphate - 157
CIAP - see Corrosion inhibiting ad-
   hesive primer
Civil engineering applications - 399-400

Clad vs bare aluminum alloys - 256–257
Clamping - 172, 175, 234–235
Cleaning (adherend surface) - 56–60
Cleanliness (of adherends) - 7
Cleavage, fracture strength in - 294
Cleavage peel - 290, 292
Cleavage strength - 25 (def)
Cleavage stress - 9, 33, 37–39, 289
Cleavage tests - 290, 292
Climbing drum peel test - 289, 295
CLOROX - 109
Cloth - 104, 135, 148, 166, 173
Coal tar - 184
Cohesion - 17 (def)
Cohesive energy density (CED) - 226
Cohesive failure - 8, 9, 19 (def), 55, 191, 246
Cohesive strength - 180
Cold curing - 181
Cold flow - 147, 162
Cold patching - 361
Cold-press adhesive - 158–160
Cold pressing - 17 (def)
Cold-setting adhesive - 13 (def)
Collagen - 17 (def)
Colophony - 17 (def)
Composites - 104–106, 149, 166, 397
Compression creep tests - 292
Compression loading - 294, 296, 297
Compression shear tests 289, 296
Concrete - 119, 148, 150, 168, 400
  bituminous - 119
Condensation - 17 (def)
  cure - 178
  polymers - 102, 180
Condensing humidity - 247, 248
Conductive adhesives - 133, 140–141, 397
Consistency - 17 (def), 297, 299
Construction'adhesives and seal-ants - 10, 144, 170, 173, 180, 201, 395–396
Consumption of adhesives and sealants - 10
Contact adhesives (cements) - 13 (def), 155, 172, 180, 191, 201
Contact-angle (test) - 17 (def), 63, 226
Contact-bond adhesives - 200

Contact bonding - 17 (def), 184, 201, 212
Contact failure - 19 (def)
Containers (adhesive), condition of - 276
Conversion coatings - 70, 82
Convolute tube winding - 401
Cook, J.P. - 170
Cooking fats and greases, resistance to - 157
Copper - 55, 68–69, 136, 148, 149, 156, 157, 166, 191
Copper filler - 140
Copper foil - 182
Copper microspheres, silver-coated - 140
Copper phosphate - 157
Coreactants - 241
Coreacting powder adhesives - 137
Corey, A.E. - 173, 174
Cork - 160–162, 173, 183
Corner blocks - 46
Corner joints - 50
Corona discharge, resistance to - 147
Corrosion - 17 (def), 68, 70, 72, 221, 234, 253, 256
Corrosion of metal adherends - 2, 4, 60, 61, 147, 206, 221, 239, 253, 254, 257
Corrosion, bond-line crevice - 256
Corrosion, electrochemical (galvanic) - 2, 64, 192
Corrosion, electrolytic - 293
Corrosion-inhibiting adhesive primer (CIAP) - 256
Corrosion-resistant adhesive primers (CRAP) - 64
Corrosivity, test method for - 292
Corrugated boxboard manufacture - 157, 401
Corundum - 72
Costs - 8, 241
Cottoning - 17 (def)
Coupling agents - 176, 177
Coverage - 8, 17 (def), 297
CR 39 - 87, 136, 168
Cracking - 91
CRAP - (see Corrosion-resistant ad-hesive primers)
Crazing (craze resistance) - 17 (def), 91, 93, 154, 170, 182, 225, 231, 232, 233
Creep - 18 (def), 128, 129, 137, 139,

Creep (cont'd.) - 144, 154, 164, 166,
    171, 173, 174, 182, 241, 277,
    290, 292, 398
Critical surface tension - 5, 63, 226–
    227
Cross laminate - 71 (def)
Cross-lap tensile test - 298
Crosslinking - 18 (def), 61, 88, 128,
    129, 134, 145, 152, 171, 175,
    176, 218, 239–241, 265, 267,
    268, 394
Cryogenic adhesives (properties) -
    128, 146, 150, 164, 168, 169,
    171, *243–245*
Cryogenic-temperature test methods -
    292–293, 294, 296
Crystalline thermoplastics - 84, 89
Crystallinity - 154, 155, 226
C-stage - 16 (def)
Cure (curing) - 18 (def), 31, 127,
    179 (def), 278–279
Cure parameters - 8, 61, 213, 240,
    393
Cure rate - 148
Curing agent - 18 (def), 144, 170,
    174, 197, 241
  aliphatic - 266–267
  amine - 264
  anhydride - 246, 255, 260–262,
    264
  aromatic - 266–267
  aromatic amine - 263
  blocked - 16 (def)
  chlorendic anhydride 241
  epoxy-phenolic - 241
  fatty polyamides - 149, 206
  phthalic anhydride - 241
  pyromellitic anhydride - 241
  radiation effects - 266
Curing temperature - 27 (def), 213
Curing time - 28 (def)
Curtain coating - 184
Cyanoacrylate adhesives - 10, 128,
    131, 133, *141–143*, 190, 191,
    193–199, 201, 402
  elastomer-modified - 143
Cyclic aging method - 291
Cycling conditions - 234
Cyclized rubber - 178, 179, 182, 183
Cyclizing (cyclization) - 107, *108–
    109*, 110–113
Cyclohexanone - 228, 229, 235, 236
N-cyclohexyl-p-toluenesulfonamide -
    144

Cylindrical joints - 39–41
  solid - 40
  tubular (hollow) - 39, 41

DACRON - 92
Dado joints - 47, 48
Damping, mechanical - 2
Dead-weight loading - 214
Dead load strength - 146
De Bruyne, N.A. - 52
Definitions - Chapter 2
Degassing - 152
Degrease-abrade-prime - 72
Degreasing - 56, 64, 65, 67, 68, 74,
    79, 80, 87, 88, 92–95, 97–99,
    113, 115, 117, 118
Degree of polymerization (DP) - 240
De Khotinsky cement - 162
Delamination - 18 (def)
De Lollis, N.J. - 72, 73, 141
Delayed-tack adhesives - 132, *143–
    144*
DELRIN - 85, 86
Density - 293
Dental cement - 157
Dental filling materials - 157
Deoxidizer - 66, 83
DEOXIDINE 526 - 248, 255
Depletion of chemicals in etch bath -
    54
Depolymerization - 179
Design criteria - *Chapter 3*, 33–37
Destructive tests - 27 (def), 279
Detergents - 87, 89–91, 95, 98, 100,
    101, 103, 109, 111, 112, 119
  resistance to - 394
Dextrin adhesives - 10, 133, *159*, 161,
    184, 212, 261
Dextrins - 18 (def), 174
Diacetone alcohol - 228, 229, 233
Diallyl phthalate - 100, 198
Diaminodiphenylmethane (MDA) -
    149
Diaminodiphenyl sulfone (DDS) -
    149
Dichlorobenzene - 235
1,2-Dichloroethylene - 234
Dichromate seal - 82
Dicyandiamide (DICY) - 149, 153,
    191, 197, 400
Dicyclohexyl phosphate (DCHP) -
    166
Dicyclohexyl phthalate - 144

Die bonding - 396
Dielectric constant - 217
Dielectric curing - 18 (def)
Dielectric properties - 147
Diethyl benzene - 232
Diethylenepropylamine (DEAPA) - 149
Diethylenetriamine (DETA) - 148–149
Diethyl ether - 226
Diffusion adhesives - 131
Diffusion theory (of adhesion) - 6
Diluent - 18 (def), 134, 148, 264, 266
Dimensional stability - 101
Dimethyl formamide - 161, 235, 264
Dimethyl sulfoxide - 264
Dioctyl phthalate (DOP) - 235, 236
1,4-Dioxane - 85, 228, 235
Diphenyl phthalate (DPP) - 144, 166
Dipping - 145
Direct heating, curing by - 215
Dirt, exposure to - 394
Disadvantages of adhesive bonding - 3
Disk brake pads - 394
Disk shear - 296
Dispensing of adhesives - 127
Dispersions - 184
Disposable tubes (syringes) - 174
Dissimilar materials - 2, 97, 136, 171, 174, 193, 231, 233
Doctor bar (blade) - 18 (def)
Domestic adhesives and sealants - 10
Dope cement - 229, 230
Double lap-shear test - 288
Double strap joint - 288
Douglas Aircraft - 265
Dovetail joints - 47
DOW 7 - 70
DOW 17 - 70
Dow Corning A-4094 - 87
Drawing - 180
Dressings, self-adhering - 175
Drip - 127
Drying temperature - 27 (def)
Drying time - 28 (def)
Dry strength - 25 (def)
Dry tack - 26 (def)
Durability - 9, 61, 65, 66, 77–79, 158–160, 164, 165, 176, 187,
190, 191, 221, *Chapter 9, 290–291*, 293, 295, 299, 398
    stressed - 238
Duranickel - 71

E-20 etchant (Marbon Co.) - 234
EA 929 - 263
EA 9614 - 195
Early strength - 179
EBONOL C - 68
EC 1469 - 267
EC 1605 - 265
EC 1639 - 267, 268
EC 2214R - 74
EC 3532 - 197
ECCOBOND 104 - 196
Edge-glued wood joints - 45
Edge sealants - 398
Eddy-sonic test method - 279, 281
Edson, D.V. - 169
EKTASOLVE EM acetate - 229, 230
Elastic deformation (of adhesives) - 128
Elasticity/inelasticity - 146, 244
Elastomer-epoxy adhesives - 132, 133, 137, 153
Elastomeric adhesives - 33, *129*, 133, *144–147*, 197–199, 238, 289
Elastomeric resins - 130, 193
Elastomer-phenolic adhesives - 245, 267, 268 (see also Nitrile-phenolic adhesives)
Elastomers - (see Rubber adherends)
Electrical applications - 183
Electrical conductivity - 4, 134, 140, 298, 397
Electrical discharge - 84, 88
Electrical/electronic applications - 396–397
Electrical heating tapes - 217
Electrical insulation - 4, 161, 167, 169, 266, 293
Electrically conductive adhesives - 133, *140–141*
Electrical properties - 293, 396
Electricians' tape - 401
Electric light bulbs - 166
Electric resistance heating - 216–217
Electrolytic corrosion - 293
Electrolytic migration - 140
Electromagnetic bonding - 97, 231
Electromagnetic radiation - 265
Electronic parts (applications) - 156, 177

Electron radiation - 265
Electroplating - 65
Electrostatic theory (of adhesion) - 6
Elevated-temperature (high-temperature) properties - 76, 92, 130, 136-139, 147, 148-150, 153, 156, 163-169, 170, 175, 177, 178, 191-193, 196, 201, 222, *240-243*, 263, 265, 268, 294, 296, 397, 398, 400
ELMER'S glue - 173
Elongation - 152, 155, 172, 202, 289
Emery cloth (paper) - 80, 81, 87, 89, 92, 95, 103, 105, 117
Emulsions - 137, 184 (def)
Encapsulated adhesives - 13 (def)
End-to-end butt joints (wood) - 45
Endurance limit - 239
Energy sink - 242
Envelopes - 174
Environmental effects - 4, 8, 9, 76, 136, 176, *Chapter 9*
EPDM - 107, 108, 110
Epichlorohydrin elastomer (HYDRIN) - 114
EPON 828 adhesive - 195
Epoxide group - 144
Epoxy adhesives (including modified epoxy) - 10, 33, 55, 88, 93, 105, 108, 128, 130-132, 134, 137, 140, 143, *144, 148-149,* 151, 160, 166, 170, 174, 178, 180, 183, 190-202, 209, 218, 221, 222, 238, 239, 241, 242, 244-246, 255, 258-262, 263, 265, 266, 395, 397, 400, 402
Epoxy adhesives, radiation effects - 267, 268
Epoxy anhydride - 257, 258, 260-262
Epoxy aromatic amine - 258
Epoxy-based system, radiation effects - 177
Epoxy concrete patch - 177
Epoxy film adhesives - 105, 164, 398
Epoxy/glass - 143
Epoxy-nitrile film adhesive - 251
Epoxy-nylon adhesives - 98, 130, 171, 191, 244, 245, 258-262, 398
Epoxy-phenolic adhesives - 130, 132,

133, 137, 139, *149-150*, 152, 153, 193, 198, 205, 239, 242, 244, 245, 258-262, 266
Epoxy-polyamide adhesives - 171, 195, 198, 201, 239, 244, 257, 258, 260-262
Epoxy-polyamine adhesives - 198
Epoxy-polysulfide adhesives - 130, 133, 137, *150*, 195, 258 (see also Polysulfide-epoxy adhesives)
Epoxy primers - 89
Epoxy resins - 101, 134, 153, 198, 206
    foam - 200
    radiation resistance - 267
Epoxy-resorcinol adhesives - 258-260
Epoxy sealing compounds - 395
Epstein, G. - 399
Ester-based polyurethanes - 246
Ester linkage - 246
Esters - 229
    compatibility - 99
Ethers (compatibility) - 92
Ethyl acetate - 226, 228-232, 235
Ethyl alcohol (ethanol) - 73, 91, 93, 101, 108, 111, 114, 136, 147, 161, 226, 235
Ethyl cellulose - 87, 194, 230
    adhesives - 133, 140, 178
Ethyl cyanoacrylate adhesives - 142
Ethylene-chlorotrifluoroethylene polymer (E-CTFE) - 61, 87, 194
Ethylene dichloride - 228-232, 273
Ethylene-ethyl acrylate adhesives - 133
Ethylene-propylene-diene terpolymer rubber (EPDM) - 107, *110*
Ethylene rubbers - 110
Ethylene tetrafluoroethylene copolymer (ETFE) - 88
Ethylene-vinyl acetate (EVA) - 88, *155*
    adhesive - 131, 133, 174, 194, 260
Ethyl lactate - 228, 229
Ethyl naphthalene - 232
European airbus - 398-399
Evaporation rates - 225, 231
Evaporative adhesives - 131
Exothermic reaction - 165, 175

Expanded polystyrene (EPS) - 199–200
Expendible tab - 273
Expulsion (in weldbonding) - 223, 285
Extenders - 19 (def), 160, 165
Extensometers - 51

Fabric - 129, 145–147, 155, 156, 163
  carrier - 210
Face-glued wood joints - 45
Failure analysis - 51
Failure mode - 8, 9, 52, 55, 63, 246, 291
Failure time - 251, 252
Fast-setting adhesives - 133
Fatigue - 2, 52, 137, 251, 253, *290*, 293, 296, 397, 398
  resistance - 33, 182
Fatigue strength - 25 (def), 144, 164, 221
Fatty polyamides - 149, 206
Faying surface - 19 (def), 64, 85, 94, 97, 104, 105
Feathering - 19 (def)
Federal Test Method Standard No. 175B - 276
  Method 1081 - 277
  Method 4032.1 - 276, 291
  Method 4041.1 - 276
  Method 4051.1 - 276
Felt - 145
Ferric chloride method - 69
  primer - 135
Ferric sulfate - 65, 69
Fiberboard - 176, 291
Fiberglass assemblies - 157
  boats - 168
  skis - 400
Filler content determination - 293
Fillers - 9, 19 (def), 134 (def), 146, 147, 149, 150, 154, 159, 165, 175, 181
  alkalies - 159
  aluminum powder - 168
  beryllium oxide - 141
  borax - 159
  boron nitride - 141
  clay - 159
  conductive carbon - 140
  conductive metal powder - 222
  copper - 140

  glass - 267
  lead oxide - 284
  metallic - 4, 140, 222
  mineral - 101, 267
  oxide - 4
  radiation resistance - 266–267
  silica (CAB-O-SIL) - 97, 109, 222
  strontium chromate - 222
Filler sheet - 19 (def)
Fillet - 20 (def)
Filleting - 183
Film - 135, 145, 146
Film adhesives - 4, 13 (def), 74, 127, 129, 130, 132, 136, 137, 139, 149, *150–153*, 156, 162, 164–166, 168, 182, 205, 218, 222, 240, 243–245, 268, 397
  application of - 209–210, 255, 278
Film-forming properties - 174
Film, plastic - 129
Film, polyester - 166
Film-supported adhesives - 14 (def)
Film-unsupported adhesives - 14 (def)
Final inspection - 279–285
First-generation acrylic adhesives - 135
Fish adhesives (glues) - 130, 133, *161*, 201
Fixture time - 181
Flame treatment (oxidation) - 84, 88, 91, 93, 94, 97
Flammability - 170, 183, 231
  limits - 60
Flashing - 61
Flash-off (moisture) - 211
Flexible adhesives - 297, 396
Flexibility - 137, 145, 147, 149, 150, 155, 162, 171, 177, 180, 235, 246, 268, 289, 293, 396
Flexibilizers - 149, 170
Flexural modulus - 171, 241
Flexural strength - 170, 293
Floating roller peel test - 289, 295
Flooring adhesives - 139, 153, 157, 168
Flour adhesives - 10, 159
Flow - 20 (def), 31, 127, 144, 233, 242, 276, 294 (see also Viscosity)
Flow brush - 207
Flow coat - 184
Flow gun - 207
Flowing, method of application - 207

Flow-out (adhesive) - 214, 218, 222
Flow properties, test method - 294
Fluidized-bed coating application
    method - 209
FLUON - 98
FLUOREL - 114
Fluorinated ethylene-propylene co-
    polymer (FEP) - 88, 194
Fluorocarbon solvents - 60
Fluoroelastomers - 107, 114, 147
Fluoroplastics (fluoropolymers) -
    6, 56, 84, 179, 194
Fluorosilicic acid - 78
Fluorosilicone elastomers - 114
FM 34 - 199, 245
FM 61 - 256
FM 123L - 256
Foam (in adhesive) - 155, 241
Foamable hot-melt adhesive - 154–
    155
Foamed adhesive - 14 (def)
Foaming adhesive - 14 (def) -154
Foam, polyurethane - 166, 171
Foams, plastic - 5, 155, 166, 171,
    (see also Cellular plastics)
Foil - 129, 130, 135, 145, 146,
    158, 164, 166, 173, 182,
    401
Fokker tester - 273, 274, 282
Folding cartons - 401
Food-packaging adhesives - 140,
    173, 401
Formaldehyde fumes - 86, 161
Formic acid - 231
Foundry
    molds - 157
    sand - 165
FPL etch (immersion) - 64, 82, 256
    paste - 65
Fracture strength in cleavage - 294
Franklin Chemical Industries - 201
Freeze-thaw cycles - 395
FREON - 67
    PLA - 90
    TF - 90, 102, 105
    TMC - 89, 105, 114
    TP - 35, 90
Free-radical reaction - 135, 138
Frisch, K.C. - 172
Frozen adhesives - 133, 174
Frozen reactive adhesives - 174
Fuel resistance - 168, 263, 264,
    394

Fuller's earth - 65
Functions of adhesives - 1
Fungi (fungal resistance) - 8, 157,
    176, 268-270
Fungicides - 269
Furane adhesives - 132, 153–154,
    198, 199, 201
Furane ring - 153
Furfural alcohol - 182
Furniture (woodworking) - 160, 165,
    201

Galvanized metals - 81, 149
Gap-filling adhesives - 14 (def), 126,
    127, 138, 154, 161, 162, 165,
    171, 173, 182, 225, 232, 294,
    297, 400
Gamma radiation - 265–268
Garnis, E.A. - 65
Gaskets - 155, 201, 395
Gasoline - 100, 263
Gas permeability - 145, 147, 174
Gear oil - 263
Gel depressant - 160
General Electric SS-4101 - 87
General Electric SS-4004 - 92
General Electric procedure for RTV
    silicone adhesives on polycar-
    bonate - 92
GENKLENE - 93
Glass - 117–118, 130, 135, 136, 138,
    148–150, 155, 157, 161, 162,
    165, 177, 181, 195, 202, 291,
    298
Glass, optical - 118
Glass cloth films or tapes - 149, 162,
    243
Glass laminates - 181
Glass-reinforced plastics - 136, 157
Glass-reinforced thermoplastics - 106
Glass-transition temperature - 240
    second-order - 170, 241
Glass-wool (fiberglass) insulation
    mats - 165
Glazing materials - 177
Glue - 20 (def)
Glue blocks - 46
Glue films - 165
Glue line - 20 (def), 142, 160, 173,
    215 (see also Bond line)
    pH - 165
    polymerization - 135
    pressure - 215

spacer - 216
temperature - 167, 216
thickness - 51, 150, 172, 207, 242, 244
    effect on radiation resistance - 267
Glycols, resistance to - 150, 182
Goland and Reissner theory - 52
Gold - 69, 191
    filler - 140
Golf clubs - 400
Goulding, T.M. - 173
Grab - 20 (def), 158
Graft copolymer - 175
Graft copolymerization - 135
Granule adhesives - 127
Graphite - 117, 140
    electric resistance heater - 216, 217
Graphite/polyimide composite - 169
Gravity impact method - 290
Grease resistance - 147, 161, 163, 164, 173, 176, 263
Green strength - 20 (def)
Grinding - 72
    wheels - 400
Grit - 72, 294
Grit blasting - 55, 56, 70, 79, 81, 87, 89, 95, 100, 101, 103, 105, 117-119, 277
Grosko, J.J. - 82
GRS rubber - 113
Grumman Aerospace Corp. - 83
Grouting - 154
Gummed adhesives - 212
Gummed tapes - 161, 212, 401
Gums - 20 (def)
Gypsum board - 144

HALAR - 87
Halo (in weldbonding) - 220
Halogenated solvents - 94
Hand-dipping application - 208
Handyman applications (adhesives) - 402, 407
Hansen, R.H. - 61
Hardboard - 136, 155
Hardening adhesives/sealants - 133, 170
Hardening agents (hardeners) - 18 (def), 20 (def), 134, 148-149, 170, 181
Hardness - 180

HCl evolution - 239
Health and safety hazards - 8, 59, 60, 207, 212, 227, 393
Heat-activated adhesives - 14 (def), 127, 131, 158, 164, 180, 209, 212
Heat-activation process - 213
Heat aging - 146, 167, 257
Heat and moisture - 9
Heat build-up (exothermic reaction) - 148
Heat capacity - 154
Heat-curing vs RT-curing durability - 239
Heat-distortion temperature - 94, 169, 170, 195, 233, 241
Heating equipment - 215
    direct heating (oven) - 215
    electric resistance heating - 216-217
    induction heating - 217
    low-voltage heating - 217
    radiation curing - 216
    ultrasonic activation - 217-218
Heat reactivation - 20 (def), 184 (def), 218
Heat resistance - 128, 134, 149, 159, 201 [see also Elevated-temperature (high-temperature) applications]
Heat seal - 20 (def)
Heat-sealing adhesive - 14 (def)
Heat sinks - 141, 215
Heat-solvent reactivation - 172
Heat source - 278
Heat stability - 154, 241
Helicopters - 247, 251
    rotor blades - 1, 76, 398
Helium plasma gas - 61, 113
Heptane - 91
HERCLOR - 114
Hershey drops - 395
Hewlett, P.C. - 400
Hexa catalyst - 153
Hexafluoroacetone sesquihydrate - 193, 227, 233
Hexafluoroisopropanol - 233
Hide glue (adhesive) - 130, 160, 201
High-frequency dielectric heating - 217
High-performance hot melts - 155
High temperature adhesives - (see Elevated-temperature applications)

High-temperature applications - (see Elevated-temperature applications)
High-temperature plastics - 90, 94, 95, 98, 102
Higgins, J.J. - 169
Hockney, M.G.D. - 239
Holding adhesives - 126, 394 (def)
Holography - 283
Honeycomb - 4, 162, 169, 183, 281
  core - 20 (def)
  facings - 162
  panels - 182, 398
  sandwich composites - 149, 150, 156
Hot-humid environments - 76, 77
Hot-melt adhesives - 10, 14 (def), 88, 129, 131, 140, *154-156*, 159, 162, 164, 166, 168, 173–175, 179, 180, 183, 195–197, 206, 218, 241, 242, 293, 298, 395, 396, 401
  application methods - 210
    melt reservoir - 210
    progressive feed - 210–211
Hot-melt adhesives, inorganic - 157
Hot-press bonding - 159, 215–216
Hot presses (platens) - 215–216
Hot-setting adhesives - 14 (def)
Hot strength - 182
Hot-water resistance - 147
Hot-water-soak test - 251, 252
Household glues (cements) - 161, 194, 407
HT 424 - 205
Hubbard, J.R.   161
Humidity, condensing - 247
Humidity (moisture) resistance - 73, 167, 182, 246–253, 257, 394, 396, 398, 400
HY-994 - 196
Hydraulic cements - 134
Hydraulic fluid, aircraft - 136
Hydraulic oils - 167, 263
Hydraulic presses - 214
HYDRIN - 114
Hydrocarbon resin adhesives - 133
Hydrocarbon solvents - 172
  resistance to - 94, 95, 145, 146, 150, 167, 263, 264
Hydrochloric acid - 73, 77, 82, 85, 109, 119
Hydrochloric acid etch - 85

Hydrofluoric acid - 77, 80, 157
Hydrofluoric-nitric acid pickle - 77
Hydrofluoric-nitric-sulfuric acid method - 80
Hydrofluorosilicic acid etch - 55, 79, 83
Hydrogen bonding - 239
Hydrogen embrittlement - 77
Hydrogen-ion concentration - 294, (see also pH)
Hydrogen peroxide - 80
Hydrolysis - 174
  inhibitors - 269
Hydrolytic degradation (reversion) (stability) - 146, 168
Hydroxyethylcellulose adhesives - 140
8-Hydroxyquinoline - 269
Hygroscopic - 91
HYPALON - 111
Hyphae, mold - 264
Hypodermic syringes - 207
HYSOL 9340 - 196
HYSOL EA 9614 - 195
HYTREL - 116

Induction heating - 217
Inhibited alkaline cleaning - 247, 248
Inhibitor - 20 (def), 159, 269
Initial set - 233
Inorganic adhesives (cements) - 156–157
Insects - 268, 269
Inspection, destructive and nondestructive - 273
Insulation materials, thermal - 157
Interfacial resistance - 140
Interlaminar shear strength - 400
Intermediate cleaning - 60
Intermediate-temperature-setting adhesives - 14 (def)
Internal stresses - 225
(The) International Journal of Adhesion and Adhesives - 403
(The) International Plastics Selector - 407
Inverted probe machine - 298
Iodo-phosphate treatment - 73, 75
Ionic polymerization - 142
Ionized inert gas treatment - 84 (see also Activated gas surface treatment)
Ionomers - 88, 89, 194

Isocyanate curing agents - 193, 194,
196, 198, 299, 201
Isocyanates - 172, 180
Isooctane - 136, 263
Isophorone - 235
Isopropyl acetate - 229
Isopropyl alcohol (isopropanol) - 72,
87, 90, 91, 93, 95, 97–100,
103, 108, 110, 113, 114, 120,
228, 234, 235, 263
Isopropylidene acetone - 229

JAN cycling - 259
Jig - 20 (def), 213
stressing - 247, 249
Joint - 21 (def)
Joint assembly methods - 211–213
curing - 213
heat activation - 213
pressure-sensitive and contact
bonding - 212
solvent activation - 212–213
wet assembly - 212
Joint design - 8, *Chapter 3*, 55, 177
Joint efficiency - 33
Joint factor - 37, 52
Joint glues - 13 (def)
Joint types - 34, 37, 38
(The) Journal of Adhesion - 404
JOY detergent - 91
JP-4 fuel - 263

Keimel, F.A. - 407
Keith, R.E. - 71
KEL-F - 92, 114, 195
KENVERT No. 14 powder - 72
Kerosene, white - 91
Ketones - 91, 98, 99, 147, 150, 162,
166, 172, 178, 229, 235, 264
Kieselguhr - 85
Knife coating - 184, 208
Kraft paper wrap - 63, 74, 77
KRATON - 116, 180
KYNAR - 99, 198

Labels - 143, 144, 155, 158, 160,
174, 291, 299, 401
Laboratory accelerated environment -
239, 257–258
Lacquers - 178
Lagging adhesive - 173
Lambeth, A.L. - 159
Laminate (laminating) - 21 (def)

Lap joint - 21 (def), 33, 37, 42, 47,
48, 51
bevelled - 34, 37
double - 34, 37, 51
double butt - 34
half - 34
joggle - 34, 37
single - 33–35, 51
step (stepped) - 34, 51
Lap-joint strength - 26 (def)
Lap-shear (overlap) strength - 62,
156, 167, 171, 177, 191, 192,
242, 251, 267, 284, 289, 290,
296, 297
LARC-13 adhesive - 199
Latex - 21 (def), 145–147, 160, 184
(def)
Latex adhesive - 14 (def), 144–147,
161, 183–184
Lavin, E. - 173
Lead - 191
Lead dioxide catalyst - 170
Leather - 129, 135, 139, 145, 147,
155, 157, 160, 161, 173, 179
Lee, H. - 143
LEFKOWELD 109 adhesive - 265
Legging - 21 (def)
LEXAN - 62, 91, 92 (see also Poly-
carbonate plastics)
Lime, resistance to - 157
Light, effects of - 8, 147, 295
Light transmittance - 181
Lignin - 183
Liquid adhesives - 126, 129, 137,
149–151, 162, 165, 172, 218,
293, 297, 299
application method - 207–208
bonding permanency - 291
Liquid bath curing - 215
Liquid crystals - 282–283
Liquid hone - 55
Liquid-hydrogen temperature - 244–
245
Liquid lock washers - 138
Liquid-nitrogen temperature - 245
Liquid pickle - 55
Litharge cements - 133, *157*
Loading - (see Stress)
Lockheed - 82
C5A - 397
Locking compounds - 395
Lock washers - 138
LOCTITE - 138, 143, 181

SUPERBONDER 306 - 196
SUPERBONDER 414 - 196
SUPERBONDER 430 - 196
SUPERBONDER 638 - 196
SUPERFLEX - 196
Longitudinal shear strength - 26 (def)
Loss factor - 217
Low-temperature adhesives (resis-
    tance) (flexibility) - 137, 150,
    164, 167, 173, 175, 177, 178,
    180, 243-245, 257, 294, 296
    (see also Cryogenic adhesives)
Low-voltage heating - 217
L/t ratio - 36, 37
Lubricants - 147
LUCITE - 95
Lumps, test method for - 294
LUSTRAN - 100, 198

M-17 solvent - 89, 105, 114
Machine Design magazine - 404
Machine (vs hand) sanding - 105,
    108
Magnesite - 157
Magnesium - 3, 56, 57, *69-70*, 149,
    162, 191-192
Magnesium cement - 157
Magnesium oxychloride cement -
    157
Magnesium phosphate - 157
Magnet wire - 177
Maleic acid esters - 183
Maleic anhydride - 183
Mandrel-bend test - 253
Manual scouring - 105
Marguglio, L.A. - 163
Marine environment (application) -
    170, 176, 254, 255, *401*
Marino, F. - 181
Markets, adhesives and sealants -
    (see Consumption of adhesives
    and sealants)
Marsden, J.G. - 178
Martin Marietta Laboratory - 80
Masking tape - 401
Masonite - 145
Mastics - 21 (def), 127, 144, 159,
    161, 201, 208-209
Mat carrier - 210
Material specifications - 276
Materials Engineering magazine - 404
Matrix - 21 (def)
Maturing temperature - 27 (def)

Maximum acceptable concentration
    (MAC) - 60
Measuring (of components) - 394
Mechanical fastening - 1
Mechanical properties test methods -
    276-277
Mechanical shock - 196
Mechanical tests - 280
Mechanical theory (of adhesion) - 5
Medical applications - 175
Melamine-formaldehyde - 101, 190
    adhesives - 128, 132, *158*, 180,
    192, 200, 201, 293, 294
Melamines - see Melamine-formalde-
    hyde
Melt-freeze - 179
Melt index - 154
Melting point - 154, 155, 179, 240
Melt-reservoir hot-melt systems - 210
Melt viscosity - 154, 172, 180
MERLON - 91
Mesityl oxide - 229
Metal ion treatment - 84
Metal joints - 50
Metals - 5, 6, 130, 138, 145-147,
    149, 150, 155, 157, 161, 162,
    164-168, 171, 172, 177-180,
    182, 183, 190, 200, 241, 268
Metal sheets - 157
Metal wool - 100
Metaphenylenediamine (MPD) - 149
Metering (of components) - 128, 135
Methacrylates - 184
Methoxyethyl acetate - 229
Methyl acetate - 120, 228-230
Methyl alcohol (methanol) - 60, 67,
    87, 88, 91-93, 95, 97-99, 101,
    108, 110, 112-115, 230, 235
Methyl Cellosolve - 228, 229
Methyl Cellosolve acetate - 228-230
Methyl cellulose - 184
    adhesives - 140
Methyl chloroform - 58, 60
Methyl cyanoacrylate adhesives - 142
Methylene chloride - 93, 94, 120,
    228, 229, 231, 232-236
Methylene chloroform - 76
Methyl ethyl ketone (MEK) - 67, 72,
    79, 83, 85, 89, 93-96, 99,
    100, 103, 105, 107, 108, 110,
    111, 114, 116, 117, 120, 212,
    228-230, 232-236
Methyl isobutyl ketone (MIBK) - 228,
    235, 236

Methyl methacrylate monomer - 231, 233
Mica - 162
Microbial susceptibility - 269
Microelectronics - 140, 396
Microencapsulated adhesive - 133, *158*
Microorganisms - 269
Migration (plasticizer) - 197, 200
MIL-A-4090 - 268
MIL-A-8623 - 400
MIL-A-9067 - 76, 77
MIL-A-13374 - 159
MIL-A-17682 - 159
MIL-A-46050 - 143
MIL-A-46146 - 178
MIL-C-3469 - 160
MIL-H-5606 - 263
MIL-M-3171 - 70
Military applications - 130, 150, 241, 402
Miller, R.S. - 45, 201
Millet, M.A. - 201
MIL-STD-304 - 239, 257-263
MIL-STD-331 - 257
Mineral acids - 98, 136
Mineral spirits - 120
Minford, J.D. - 191, 248, 255, 256
Miter joints - 47, 49
Mixer-dispenser - 206-207
Mixing of adhesive components - 128, 135, 152, 174, 206-207, 394
Mixing pouches - 207
Mixing ratios - 206
MMM-A-132 - 138
MMM-A-134 - 400
Mobile homes - 173
Mode of application of adhesives - 127
Mode of failure (see Failure mode)
Mode of setting of adhesives - 127
Modified acrylics - 135
Modified epoxies (see Epoxy adhesives)
Modified rail test - 297
Modifier - 21 (def), 165, 169, 182
Modulus of elasticity (Young's modulus) - 50, 52, 243-244, 289
Moisture - 4, 8, 238
Moisture cure - 170, 172, 178
Moisture pick-up - 155, 164

Moisture resistance (sensitivity) - 141, 145-147, 150, 158, 160, 164, 167, 172, 173, 177, 179, 201, 246-253
Moisture/stress - 4, 76
Moisture/temperature effects - 293
Mold-release agent - 100, 105, 234
Mold resistance - 174, 176, 182, 269, 270, 291, 292
Molecular weight - 155, 156, 164, 178, 235, 240, 264-265
Monel metal - 71
Monoamyl benzene - 232
Monochlorobenzene - 234
Monomer-polymer cement - 233
Morrill, J.R. - 163
Mortise-and-tenon joints - 46
Motor oil (resistance to) - 136
Mucilage - 21 (def), 126
Multiple-layer adhesive - 14 (def)
Mycelia (mold) - 269
MYLAR polyester film - 62, 92
McCurdy, P.M. - 267, 268

NACCONAL NR - 65, 70
Nameplates - 210
Naphtha - 91
Natick Research & Development Center (Army) - 269
National Bureau of Standards (NBS) - 245
Natural adhesives (glues) - 130 (def), 131, 133, *158-162*
Natural latex - 184 (def)
Natural rubber - 107, 108, 113, 145, 194
    adhesives - 129, 131, 145, *146*, 162, 174, 194, 201
NEOLITE - 166
Neoprene adhesives - 129, 131, 144, *162*, 179, 184, 192-195, 197-202, 210
Neoprene, decomposition in vacuum - 265
Neoprene-phenolic adhesives - 130, 133, 137, *162-163*, 191, 192, 200, 201, 266
Neoprene rubber adherends - 107, *108-109*, 143, 147, 184, 194
Neoprene solvent cement - 395
NEUTRA-CLEAN 7 - 98
Neutron radiation - 265
Neutron radiography - 284

Newtonian flow fluid - 22 (def)
Nickel - *70-71*, 192
  filler - 140
Niobium - 98
Nitric acid - 65, 67, 68, 70, 71, 74,
    77-79, 81-83, 157
Nitric acid bath - 81
Nitric acid etch - 70
Nitrile-epoxy adhesives - 74, 132,
    191-193, 251, 255
Nitrile-phenolic adhesives - 90, 130,
    132, 133, 137, 153, 164, 183,
    190-192, 194, 196-202, 239,
    240, 246, 248, 255, 256, 258-
    262, 266, 394, 397, 398, 402
  radiation resistance - 267
Nitrile-phenolic primer - 89, 94,
    239
Nitrile rubber (NBR) - 107, 108,
    112, 153, 201
Nitrile rubber adhesives - 129, 131,
    137, *146*, 162, 179, 193-195,
    197-201, 210, 218
Nitrocellulose (see Cellulose nitrate)
  adhesives (see Cellulose nitrate
    adhesives)
Nitrogen oxides - 65
Nitrogen plasma treatment - 61, 109,
    111-113
Nitromethane - 228, 229
Nondestructive tests - 27 (def), 273,
    279, 281-285
Nonstructural adhesives - 126
Nonvolatile content, test methods -
    294
Norland, R.E. - 161
Northrop Corp. - 82
NORYL - 89, 90
Novalak (Novalac) - 22 (def), 153,
    192
NR 150 adhesive - 199
NR 056X adhesive - 199
Nylon adhesives - 132, 155, *164*,
    239, 255
Nylon cloth carrier - 195
Nylon, decomposition in vacuum -
    265
Nylon-epoxy adhesives - 132, 133,
    137, 152, 153, 164, 191, 244,
    255, 258-261, 398
Nylon-phenolic adhesives - 164, 195,
    266
Nylon plastics - 63, 89, 194, 227,
    230-231

nylon 6 - 62, 231
nylon 6,6 - 89, 194, 230
nylon 6, 12 - 61
nylon 11 - 61, 164
nylon 12 - 61

OAKITE HD 46 - 78
OAKITE HD 126 - 77
OAKITE HD 164 - 83
Odor, in adhesives - 294
"Oil canning" - 395
Oil can application method - 208
Oils, resistance to - 98, 145-147,
    153, 161, 163, 164, 168, 173,
    182, 201, 263, 394
Oily surfaces, effect on bonding -
    136, 179, 393, 395
Oleic acid - 157
One-component adhesive - 14 (def),
    128, 144, 177, 178 (see also
    Paste adhesives)
Open assembly time - 212
Open time - 22 (def)
Open-time bonding - 164
Optical clarity (in solvent cement-
    ing) - 232, 396
Optical glass - *118*, 157, 160, 168,
    169, 181, 202, 402
Ordnance applications - 402
Organosols - 178
Orthophosphoric acid - 73
Outdoor applications - 150
Outgassing - 241
Overlap depth - 38
Overlap length - 35, 36, 53
Oxalic acid - 72, 75
Oxalic-sulfuric acid process - 75
Oxidation - 8, 60, 80, 84, 94, 140,
    147, 159, 177, 243
Oxidative degradation - 175, 179,
    193
Oxide layer (film) - 54, 66, 80, 81,
    94, 140, 221
Oxides, conductive - 140
Oxides, inorganic - 141
Oxides, metal - 6, 79, 146, 147
Oxides of nitrogen - 65
Oxidizing agents - 95, 147, 153, 170
Oxidizing flame - 94, 97
Oxygen plasma gas treatment - 61,
    92, 110
Oxygen pressure method - 291
Ozone resistance - 147, 168, 177

Pabst program - 398
Packaging adhesives and sealants -
    10, 139, 140, 143, 155,
    160, 173, 401
P4 adhesives (TRW Systems) - 245
Painted surfaces - 119–120, 145,
    200
Panek, J.R. - 170
Paper adhesives - 10, 126, 129, 130,
    139, 148, 155, 157–159, 166,
    173, 176, 179
Paperboard adhesives - 130, 155,
    157–159
Paper cartons - 158
Paper labels (see Labels)
Paper products - 145, 155, 160, 161
Paper targets - 159
Paraformaldehyde (para) - 175
Parallel laminate - 21 (def)
Para-toluene sulfonic acid - 85
PAREL 58 - 115
Particle (wood) binders - 159
Particulate radiation - 265
PASA-JELL - 76, 83, 98
Passivation - 74
Paste adhesives - 72 (def), 54, 74,
    78, 97, 127, 137, 141, 150,
    151, 178, 190, 220, 221,
    239, 240, 248, 259, 261–263,
    299, 398, 400, 401
  application of - 208–209
Patching kits - 168
PEEK polyetheretherketone - 93,
    195
Peel durability - 295
Peel-ply method - 104
Peel strength - 26 (def), 39, 129,
    136, 137, 144, 145, 148,
    150, 152, 153, 156, 162,
    163, 166, 171, 172, 177,
    182, 192, 193, 231, 241,
    242, 244, 245, 267, 268,
    295, 398
Peel stress - 9, 33, 37, 38, 51, 61,
    289, 295
Peel tests
  Bell - 289
  climbing drum - 289, 295
  floating roller - 64, 289, 295
  T-peel - 244, 289, 295
Penetrant inspection - 285
Penetration - 22 (def), 295
  by water - 239

Perchloroethylene - 58, 60, 64, 67,
    85, 86, 89, 93, 102, 107, 231,
    233 (see also Tetrachloroethyl-
    ene)
Perfluoroalkoxy resins (PFA) - 89,
    195
Perfluoroalkylene group - 193
Permanence - 27 (def), 291
Permeability - 145 (see also Gas
    permeability and Water-vapor
    permeation)
Permittivity - 217
"P" etch - 65
"P_2" etch - 65, 66
Petroleum ether - 91, 93, 99, 100,
    235
Petroleum fluids, resistance to - 173
pH - 142, 165 (see also Hydrogen
    ion concentration)
  test method for - 294
Phenol cement, aqueous - 230
Phenol-formaldehyde (phenolic)
    adhesives - 10, 128, 130–
    132, 137, 138, 158–160, *164–
    166*, 176, 179, 180, 190, 194–
    196, 198, 199, 201, 218, 239,
    242, 255
  radiation effects - 267, 268
Phenol-formaldehyde (phenolic)
    plastics - 101–102
Phenolic adhesives [see Phenol-formal-
    dehyde (phenolic) adhesives]
Phenolic composites - 166
Phenolic plastics [see Phenol formal-
    dehyde (phenolic) plastics]
Phenolic primer - 182
Phenolic-resorcinol formaldehyde
    adhesives - 201
Phenoxy adhesives - 129, 131, 133,
    166, 194
m-Phenylene diamine curing agent -
    263
Phenylene oxide-based resins
    (NORYL) - *89–90, 195*, 233–
    235
  foams - 199
Phosphate cements - 133, 157
Phosphate conversion coating - 82
Phosphate-fluoride process - 76, 77
  etch - 77
  stabilized - 77–78
Phosphor - 283
Phosphoric acid - 66, 74, 75, 83

Phosphoric acid anodizing (PAA) - 66, 247–248, 251, 256, 298
Phosphoric acid/sodium dichrom- ate anodizing (P/SD) - 221
Photoelasticity - 51
Photoengraving reagents - 161
Photographing - 22 (def), 27 (def)
Phthalic anhydride - 241
Physical form classification - 126
Physical properties test methods - 276
Picatinny Arsenal - 61, 64, 76, 80, 96, 105, 109–112, 239, 247, 251, 257, 258
Pick-up roll - 22 (def)
Pipe jointing - 166, 236
Pizzi, A. - 158, 166, 176, 182, 201
PLASKON CTFE - 92
Plasma treatment - 61, 62 (see also Activated gas surface treat- ment)
PLASTEC - 407
Plasters, self-adhering - 175
Plastic flow - 240
Plastic foams - 106, 199–201 (see aldo Cellular plastics)
thermoplastic - 199–200
thermosetting - 200–201
Plasticity - 22 (def), 180, 244
Plasticizers - 22 (def), 98, 140, 142–144, 146, 148, 158, 159, 166, 170, 178, 180, 194, 197, 200, 264, 269
Plastic laminates - 162, 182
Plastic laminations - 166
Plastic pipe - 228
Plastics - 136–142, 146, 147, 155, 156, 169, 172, 173, 176, 179
adhesives for - 135
joints for
flexible - 42
reinforced - 43
rigid - 43
surface preparation for - 84–106
Plastics Technical Evaluation Center (PLASTEC) - 407
Plastisol adhesives - 396
Plated metals - 71, 138, 192
Platen presses - 214
Platinum - 71
PLEXIGLAS - 95
Plumbing cements - 157

Plywood - 159, 160, 164–165, 176, 181, 201, 291, 296, 395
flooring - 144
Pneumatic presses - 214
Pneumatic pump - 127
Poker-chip test - 288
Polar adhesives - 196, 217
Polarity (in rubber) - 108, 110
Polar materials - 63, 226
Polar solvents - 98, 99, 171
Polyacetal adhesives - 132
Polyacrylate adhesives - 179, 200
Polyacrylic rubber - 115
Polyamide adhesives - 129, 131–133, 144, 155, 164, 178, 179, 192, 196, 197, 211
Polyamide-epoxy adhesive - 80, 199
Polyamide-epoxy primer - 112
Polyamide plastics - 194, 230–231 (see also Nylon plastics)
Polyaromatic adhesives - 138–139, 242–243
Polyarylate plastic (ARDEL) - 90
Polyaryl ether (ARYLON T) - 195
Polyaryl sulfone (ASTREL 360) - 195
Polybenzimidazole adhesives (PBI) - 128, 131, 132, 138–139, 166– 168, 191, 192, 242–243
Polybutadiene additive - 232
Polybutadiene adhesives - 191
Polycarbonate adhesives - 131, 132
Polycarbonate plastics - 25, 62, 91, 92, 106, 195, 231
foam - 200
Polychloroprene rubber (see Neoprene)
Polychlorotrifluoroethylene (PCTFE) - 92, 114, 195
Polydimethylsiloxane - 110
Polyester adhesives - 10, 128, 131– 133, 155, 160, 168, 180, 193, 196–201, 260–262
radiation resistance - 267
styrene-modified - 202
Polyester-amide hot melts - 155
Polyester film - 401
Polyester hot melts - 155
Polyester laminates - 168
Polyester plastics (thermoplastic) (linear) (saturated) - 84, 92, 93, 196, 200
Polyester plastics (thermosetting) (non-linear) (unsaturated) - 102, 198–199

radiation resistance - 267
Polyester polyols - 269
Polyetheretherketone (PEEK) - 93,
    195
Polyetherimide (ULTEM) - 94, 196,
    236
Polyether polyols - 269
Polyethersulfone (VICTREX) - 61,
    93-94, 196
Polyethylene - 6, 61-63, 84, 94,
    154, 155, 172, 191, 196, 197
    foam - 199
Polyethylene adhesives - 132, 212
Polyethyleneamine - 142
Polyethylene cartridge assemblies -
    207
Polyethylene film laminates - 401
Polyethylene sealants - 395
Polyethylene squeeze-bottle appli-
    cation method - 208
Polyethylene terephthalate film -
    62, 92 (see also Polyester
    film)
    plastic - 92
Polyhydroxyether adhesives - 166,
    192, 200
Polyimide adhesives (PI) - 128, 131,
    132, 138, 139, 152, 167,
    168-169, 191-193, 199, 222,
    242-243, 245
Polyimide resins (PI) - 102, 199,
    242-243, 245
Polyisobutylene adhesives - 129,
    133, 139, 145, 169, 174,
    178, 179, 194, 197-200
Polyisocyanate adhesives - 180,
    192, 199, 201
Polyisoprene (see Natural rubber)
Polymer - 22 (def)
Polymeric substrates - 6
Polymerization - 22 (def), 233, 264
    degree of (DP) - 240
Polymethacrylate adhesives - 179,
    184
Polymethylmethacrylate adhesives
    (PMMA) - 197-199
Polymethylmethacrylate plastics
    (PMMA) - 196
Poly-N-vinyl pyridine - 142
Polyolefin resin adhesives - 131
Polyolefin resins and hot melts -
    155
Polyolefins - 84, 95

Polyphenylene oxide - 89, 106
Polyphenylene sulfide (PPS) - 61, 96,
    196
Polyphenylquinoxaline adhesives -
    193
Polypropylene adhesives - 133
Polypropylene brushes - 65
Polypropylene oxide - 115
Polypropylene plastics - 6, 62, 84,
    96-97, 197
Polystyrene adhesives - 131-133,
    144, 169
Polystyrene-butadiene-polystyrene
    block copolymers - 116
Polystyrene foams - 199-200
Polystyrene plastics - 25, 62, 63, 97,
    169, 172, 178, 184, 197, 226,
    231-232
Polysulfide-epoxy adhesives - 195,
    199, 258 (see also Epoxy-
    polysulfide adhesives)
Polysulfide rubber - 115, 201
Polysulfide sealants - 10, 129-132,
    170-171, 177, 192, 198, 200-
    202, 265, 394, 397, 398, 401
Polysulfides, decomposition in
    vacuum - 265
Polysulfone adhesives (UDEL) - 31,
    170-171, 241-242
Polysulfone plastics (UDEL) - 98,
    197, 233
Polytetrafluoroethylene (PTFE) -
    56, 61, 98, 197
Polyurethane adhesives and sealants -
    10, 80, 93, 129-132, 146, 151,
    158, 171-172, 175, 177, 178,
    190-192, 194-201, 221, 244,
    246, 258, 260-263, 269, 394,
    396, 401
    radiation resistance - 267
Polyurethane elastomers (rubbers) -
    112-113, 116, 199
Polyurethane plastics - 91
    foam - 106, 200-201
Polyvinyl acetal adhesives - 129, 131,
    132, 137, 172-173, 182, 183
Polyvinyl acetal-phenolic adhesives -
    130
Polyvinyl acetate adhesives - 129,
    131-133, 144, 161, 173, 179,
    183, 184, 192, 194, 198-202,
    401
Polyvinyl chloride adhesives - 183,
    184, 198

Polyvinyl chloride plastics (PVC) - 98-99, 197, 235-236
  foam - 200
Polyvinylidene chloride and co-polymer adhesives - 179, 184, 199
Polyvinylidene fluoride (KYNAR) - 99, 198
Porcelain - 118, 157
Porosity - 22 (def), 264, 282
Porous bonds - 206, 243
Porous substrates - 155, 160-162, 173, 184, 197, 212
Portable hot-melt guns - 211
Portland cement concrete - 119
Postcure (postcuring) - 22 (def), 139, 149, 167, 168, 181, 263
Post vulcanization (PV) bonding - 22 (def), 106
Potassium, powdered - 97
Potassium bromide - 75
Potassium dichromate - 84, 85, 87, 90, 94-96
Potassium fluoride - 77
Potassium iodide - 74
Potassium iodide-phosphoric acid method - 73
Potassium permanganate - 96
Potassium silicate - 156
Pot life - 22 (def), 148, 175, 177, 206, 233, 244
Powder adhesives - 127, 137, 165, 166, 260-262
  application of - 209
Power factor - 217
PR 1535 - 265
PREBOND 700 - 75, 99
Prebond treatment (see Surface preparation)
Prefit - 277-278
Pre-mixed frozen adhesives - 133, 174
Prepolymers - 178
Preservatives - 159
Pressure bars - 215
Pressure equipment - 214
  autoclave - 214-215
  pneumatic (hydraulic) presses - 214
  vacuum bags - 215
Pressure-sensitive adhesives (PSA) - 14 (def), 23 (def), 127, 132, 133, 144, 147, 155, 159, 174-175, 177, 180, 197-201, 209, 242, 401
Pressure-sensitive bonding - 212
Pressure-sensitive tack - 180, 298
Primer (priming) - 23 (def), 57, 60-61, 63, 64, 68, 70-72, 85, 87-89, 92, 94, 98, 99, 101, 103, 107, 110-113, 115-118, 133, 138, 145, 149, 162, 165, 175-177, 182, 183, 191, 195, 198, 199, 239, 240, 242, 246, 256, 278
Printed circuits - 240
Process control - 54, 213, 277
  test specimens for - 273
Process specifications - 54, 285
Progressive feed systems (hot melt) - 210-211
Propyl cyanoacrylate adhesive - 142
Propylene oxide - 235
  rubber - 115
Protein-blend glues - 159
Pulse-echo test method - 282
Pulsed eddy-sonic test method - 281
PUREX - 109

Qualification test - 23 (def)
Qualified Products List (QPL) - 23 (def), 302
Quality Acceptance Tests - 276
Quality control - Chapter 10
  flow chart - 274
  raw material inspection and process control - 275-285

Radiant-energy-curable adhesives - 181, 216
Radiation resistance - 169, 170, 265-268, 295
Radio-frequency heating - 217
Radiography - 284, 285
Radioisotope inspection methods - 284
RAE etch - 80
Rats, laboratory - 269, 291
Raw material control - 275-277
Rayner (C.A.), classification by - 132
Reactive adhesives - 135, 183
  contact - 180
  fluid - 135
Rear-view mirrors, bonding of - 394

Receiving inspection requirements - 276
Reclaimed-rubber adhesives - 129, 131, 144, *145*, 179, 184, 192, 198, 200, 201
Recreation industry - 400
Refluxing - 102
Refractive index - 181
Refractory cements - 157
Reinforced plastics - 104-106, 199
    thermoplastic - 106, 199
    thermosetting - 104-106, 199
Reinforced shipping tapes - 401
Reinforcements - 134
Relative humidity - 64
Release
    agent - 23 (def)
    coat - 175
    paper - 23 (def)
Reservoir-type application system - 155
Residual strength - 251, 252
Resilience - 145
Resin - 23 (def)
Resin adhesive - 130
Resin ester adhesives - 174
Resinoid - 23 (def)
Resistance welding - 219
Resitol - 23 (def)
Resol - 23 (def), 153
Resorcinol-epoxy adhesives (see Epoxy-resorcinol adhesives)
Resorcinol-ethanol solvent cement - 230
Resorcinol-formaldehyde adhesives - 128, 131, 132, 158, 165, *175-176*, 179, 180, 194, 195, 198-201, 293, 294
Resorcinol-formaldehyde primer - 89, 182
Resorcinol-polyvinyl butyral adhesives - 198
Retarder - 170
Reversion - 246
Reynolds Metal Co. - 191
RIDOLENE 53 - 247, 248
Roaches - 269, 291
Rocket engines - 283
Rodents - 8, 268, 269, 291
Rod specimen preparation - 297
    testing - 298
Roll (roller) coat (coating) - 145, 184, 208

Rolling ball tack test method - 198
Romig, C.A. - 178
Roofing adhesives - 139
Room-temperature-setting adhesives - 15 (def)
Rosin - 23 (def)
Rosin adhesives - 133, *159-160*, 184
Rosin derivatives - 184
Roughening (see Abrading)
Royal Aircraft Establishment (RAE) - 80
RTV silicone adhesives - 92, 178, 196
Rubber, joints for - 42, 43
Rubber adherends - *106-116*, 129, 136, 142, *145-147*, 150, 163, 177, 179, 182, 183, 201
    hard rubber - 155
Rubber-base adhesives - 10, *145-147*, 178, 192-194, 197, 200, 201
    (see also specific rubber adhesive type)
Rubber cement - 401
    test methods for - 295
Rubber resin primer - 165
Run-off - 222
Russell, W.J. - 65
Rutile structure - 78, 79
RYTON - 96, 196

SAE - 406
Safety factors - 59-60, 85, 179
Sagging - 23 (def), 127
Salt fog - 253
Salts (compatibility) - 92, 98
Salts, conductive - 140
Salt spray - 4, 8, 73, 136, 138, 142, 146, 147, 150, 166-168, 182, 253-254, 295, 395, 398
Salt water immersion - 147, 253, 255-256
Salzburg, H.K. - 160
SAMPE Journal - 404
SAMPE Quarterly - 404
Sand abrasive - 87
Sandblasting - 67, 72, 73, 90, 96, 105, 279
Sanding - 80, 87, 90, 98, 100, 103, 105, 107, 193
Sandpaper - 92, 93, 95-102, 105, 107, 108, 110, 112, 114, 117, 234, 400
Sandwich construction - 150, 182, 183, 296

Sandwich panel - 23 (def)
SANTOCEL C - 109
Satinizing - 85, 86
Saturated polyester adhesives - 168, 182
Savia, M. - 158
SBR - 108 (see also Styrene-buta- diene rubber)
Scarf joint - 21 (def), 34, *42*, 43, 50, 51, 217
Schneberger, G.L. - 5
Schollenberger, C.S. - 172
Schonhorn, M. - 61
SCOTCH-CAST XR-5001 - 112
SCOTCH-GRIP 880 - 197
SCOTCH-GRIP 897 - 198
SCOTCH-GRIP 1357 - 198
SCOTCH-GRIP 2262 - 198
SCOTCH-WELD 838 - 197
SCOTCH-WELD 2214 - 74, 197
SCOTCH-WELD 2216 - 196, 197
SCOTCH-WELD 2262 - 197
Scouring powder - 99, 100 (see also AJAX, BAB-O)
Screws - 138
Seacoast weathering environment - 254-255
Sealant and Waterproofers Institute - 406
Sealants - 10, 23 (def), 68, 139, 155, 170, 171, 176, 180, 243, 245, 265, 345, 398
Sealers - 395
Sealing - 2
Sealing adhesives - 126
Sealing tapes - 406
Seal materials - 205
Seam welds - 222
Second-generation acrylics - 135, 136, 190, 194, 196, 202
Selbo, M.L. - 201
Self-vulcanizing - 24 (def)
Sell, W.D. - 190
Separate-application adhesives - 15 (def)
Serlin, V.Y. - 169
Service conditions - 24 (def)
Service environment - 238
Service-temperature range - 150, 153-155, 161, 166-168, 170, 182
Set, initial - 233
Set, *v* - 24 (def), 179

Setting - 179 (def)
    temperature - 27 (def)
    time - 28 (def)
Set-up boxes - 401
Sharpe, L.H. - 111, 141
Shavings (dissolved) - 236
Shaw, J.D.N. - 400
Shear loading (see Shear stress)
Shear modulus - 289, 296, 297
Shear strength - 26 (def), 33, 64, 136, 137, 144, 150, 158, 161, 162, 166, 171, 183, 193, 234, 241, 254, 289, 296, 297, 398
Shear stress - 244, 288, 394
Shear tests - 288-289
Shelf life - 24 (def), 128, 130, 143, 178, 205, 209, 243
Shellac adhesives (glues) - 130, 133, *161-162*, 184
Shields, J. - 32
Shims - 218
Shim stock - 222
Ship assembly - 401
Shock - 146, 244, 397
Shoe adhesives - 162, 170, 295
Shop life - 278
Shop time - 398
Shortness - 24 (def)
Shrinkage (of adhesive) - 9, 24 (def), 31, 134, 136, 143, 148, 152, 157, 178, 225, 243
Shurtronic harmonic bond tester - 281
Silane primer - 89, 101, 195
SILCOSET 153 RTV - 196
SILCOSET 2 RTV - 196
SILGRIP SR-573 - 196
Silica - 65, 68, 100, 119
Silica ceramic - 118
Silica earth, fused - 97
Silica filler - 141
Silicate soluble glue - 133
Silicon carbide - 100, 119
Silicone adhesives/sealants - 10, 92, 129, 131, 151, 175, *176-178*, 190, 191, 193, 195-199, 201, 202, 242, 243, 259-262, 396, 397
Silicone fluid resistance - 264
Silicone mold-release agents - 54
Silicone plastics - 179
Silicone primers - 92, 98, 103, 118
Silicone (resin) (polymer) - 63, 103, 147, 175, 199

Silicone rubber - 110, 144, 175, 177
    foams - 201
Silk screening - 208
Siloxanes
    polydimethylsiloxane - 176
    polyorganosiloxane - 176
Silver - 71, 72
    flake - 140
SIRA - 88
Sizing - 24 (def)
Skateboards - 400
Skinning - 24 (def)
Skis - 400
SKYDROL - 264
Slip - 24 (def)
Slippage - 24 (def)
Slip-sheet interliner - 24 (def)
S/log N curves - 253
SME classification of adhesives - 130
S-N curves - 290
Snelgrove, J.A. - 173
Snogren, R.C. - 59, 60, 115
Soap-and-water wash - 110
Society of Automotive Engineers
    (SAE) - 406
Society of Manufacturing Engineers
    (SME) - 130, 406
Society of Plastic Engineers (SPE) -
    407
Sodium carbonate - 74, 75
Sodium carboxy methyl cellulose
    adhesives - 140, 183
Sodium chloride - 71, 136
Sodium dichromate - 67, 69, 74, 82,
    83, 91, 97, 118
Sodium dichromate-sulfuric acid -
    69, 74
Sodium etch - 114
Sodium hydroxide - 65, 67, 70, 79,
    102, 108
Sodium hydroxide etch - 102
Sodium hypochlorite - 109
Sodium metasilicate - 65, 60
Sodium naphthalene etch - 87, 99,
    195
Sodium silicate adhesives - 10, 130,
    156-157, 183
Sodium sulfate - 65, 77
Softener - 24 (def), 174
Softening point - 154, 241, 243
Solar storm - 268
Soles, shoe - 170
Solidification (of adhesives) - 7, 127

Solids content - 24 (def), 185, 207
SOLPRENE - 116, 180
Solubility parameter - 226-227
Soluble nylon adhesives - 132, 137,
    153
Soluble silicates - 142, 156
Solution viscosity - 180
Solution water-based adhesives - 183-
    184
Solvent-activated adhesives - 15 (def),
    127, 209
Solvent-activation bonding - 212-213
Solvent (solvent-based) (solvent-solu-
    tion) adhesives - 15 (def), 130,
    131, 162, 164, 178-179, 183,
    201, 206
Solvent blends (mixtures) - 149, 229,
    230, 234
Solvent bonding - 25 (def)
Solvent cement - 24 (def), 84, 144,
    178, 194, 196, 199, 395
Solvent cementing - 6, 25 (def), 87,
    89-91, 94, 95, 97, 98, 100,
    106, 116, 193, 194-197, 199,
    Chapter 8
    acetal copolymer (CELCON) - 227
    acetal homopolymer (DELRIN) -
        227
    acrylonitrile-butadiene-styrene
        (ABS) - 227-228
    cellulosics - 228
        cellulose acetate - 228
        cellulose acetate butyrate
            (CAB) - 228-229
        cellulose nitrate - 229
        cellulose propionate - 229-230
        ethyl cellulose - 230
    chlorinated polyvinyl chloride
        (CPVC) - 236
    nylons (polyamides) - 290
    phenylene oxide-based resins
        (NORYL) - 233-235
    polybutylene terephthalate (PBT) -
        233
    polycarbonate - 231
    polyetherimide (ULTEM) - 236
    polymethylmethacrylate (PMMA) -
        233
    polystyrene - 231-232
    polysulfone - 233
    polyvinyl chloride (PVC) - 235-
        236
    styrene-acrylonitrile (SAN) - 232-
        233

Solvent cleaning - 57, 65, 70, 71, 99, 111, 114, 116, 119
Solvent evaporation - 129
Solvent immersion (wash) - 57, *59*, 90, 98, 100, 107, 119
Solvent mixtures - 229, 230, 234
Solvent/non-solvent mixtures - 235
Solvent pollution - 155
Solvent reactivation - 25 (def), 173, 184 (def)
Solvent resistance - 128, 142, 145, 147, 153, 158, 160, 164, 165, 168, 173, 175, 201, 227, *263–264*
Solvents, function of - 134
Solvent soak - 105
Solvent solution - 139, 162, 168, 173, 175, 178
Solvent spray - 57, *59*, 71, 184
Solvent vapor - 179
Solvent wash (see Solvent immersion)
Solvent welding - 178
Solvent wipe - 57, *59*, 64, 73, 84, 85, 90, 96, 98-101, 105, 107, 108, 110-112, 114, 115, 119, 172, 184, 234, 277
Sonic resonator - 279
Sonic tests - 279, 281, 282
Sorel cements - 134, *157*
Sources of information - Appendix (403-408)
    consultants - 407
    information center - 407
    journals and other periodicals - 402-404
    manufacturers' bulletins - 405
    miscellaneous sources - 407-408
    seminars and workshops - 405
    standardization activities - 405-406
    technical conferences - 405
    trade and professional associations - 406-407
Soviet transport aircraft - 219
Soy(a) bean glue - 159
Space applications (environment) - 138, 167, 243, 264, 268, *399*
Specimen preparation - 297
Spin welding - 231
Spiral-winding containers - 159, 401

Sponge rubber - 145, 146
Spot adhesion test - 297
Spot-weld adhesive bonding - 218–219
Spray adhesives - 127
Spraying, method of application - 145, 207-208
Spraying, solvent (see Solvent spray)
Spread - 25 (def), 297
Spring
    k factor - 247
    pressure - 214
Squeegee application method - 208
Squeeze-bottle application method - 208
Squeeze-out - 25 (def), 400
Stabilizers - 25 (def), 142, 149, 154, 180, 236, 269
Stamps - 174
Starch adhesives - 10, 130, 131, 133, *158-159*, 183-184, 201, 269
Starches - 174
    dialdehyde - 160
Starved joint - 21 (def), 142, 218
Static electricity - 157
Stationery applications - 161, 174
Stearate coating - 140
Stearic acid - 81
Steel - 72, 143, 148, 156, 166, 183, 194, 195, 221
    carbon - 7, 73, 136, 192, 242
    cold-rolled - 55, 238, 241
    mild - 72, 192
    stainless - 55, 71, 72, *74-76*, 136, 162, 167, 171, 172, 241, 245, 256, 257
Steel shot - 87, 100
Steel wool - 99-101, 103
Steger, V.Y. - 169
Steinfink, M. - 162
Stemphylium - 269
Sterman, S. - 178
Stiffeners - 171, 173, 180
Stiffening (of joints) - 39, 40
Stoneware - 157
Stops - 218
Storage life - 25 (def), 174, 297
Storage stability - 8
Storage temperature - 206
Strain gages - 51
Strap joints - 34, 38
    single - 34, 38
    double - 34, 38, 42

recessed double - 34, 38
Strength development of adhesives - 296-298
Stress - 4, 238-239, 246, 257
    analysis - 50-53
    cleavage - 32, 37
    compression - 32
    cracking - 98, 178, 195, 225, 297
    distribution - 35, 37, 288, 289
    dynamic - 4, 33
    internal - 9, 31, 233
    moisture, effects on - 246
    peel - 32, 37
    reduction - 235
    relief - 85
    residual - 51, 244
    resistance - 201
    shear - 32, 51
    static - 4, 33
    tension - 32
    types of - 31
Stress and salt-water immersion - 255
Stressed durability - 251, 252, 256
Stressed temperature/humidity test - 247, 250
Stressing jig - 247, 249
Stress-strain distribution - 51
Stringiness - 26 (def)
Stripping strength - 289, *295*
Structural adhesives - 15 (def), 126, 130, 131, 148, 150, 165, 166, 168, 182, 193, 244-246, 257, 263-265, 268, 297, 394, 398
Structural bond - 26 (def)
Structural foams - 106, 199
Stucker, N.E. - 169
Styrene-acrylonitrile (SAN) - 100, 198, 232-233
Styrene-butadiene rubber (SBR) (BUNA S) - 107, 108, 113, 146
Styrene-butadiene rubber (SBR) adhesives - 129, 131, 132, 144, 146, 159, 170, 174, 192, 200, 201
Styrene-butadiene-styrene (SB-S) adhesives - 180
Substrate - 26 (def)
Sulfochromate etch solution, analysis of - 298
Sulfur catalysts - 153

Sulfur cements - 133, *157*
Sulfuric acid - 69, 71, 72, 74, 75, 80, 82, 84, 85, 87, 90, 91, 94-97, 107-109, 118, 157
Sulfuric acid anodizing - 247, 248, 254-256, 284
Sulfuric acid-dichromate (FPL) etch - 64, 74, 251
Sulfuric acid-ferric sulfate etch - 66
Sulfuric-nitric acid pickle - 71
SUPERFLEX primer - 196
Support, adhesive - 151
Surface activators - 135, 143
Surface area (in solvent cementing) - 234
Surface exposure time (SET) - 63, 105
Surface-free energy - 63
Surface layers - 4
Surface preparation (treatment) - 4, 26 (def), *Chapter 4*, 179, 239-241, 246, 247, 256, 291, 297-298
    metals - 64-66, 298
        aluminum - 64-66, 298
        beryllium - 67
        brass - 67
        bronze - 68
        cadmium - 68
        copper - 68
        gold - 69
        magnesium - 69
        nickel - 70
        platinum - 71
        silver - 71-72
        steel - 72-74
        steel, stainless - 74-76
        tin - 76
        titanium - 76-80
        tungsten - 80
        uranium - 80
        weldbonding metals - 82-84, 221
        zinc - 81
    miscellaneous
        asbestos - 117
        brick - 117
        carbon - 117
        ceramics - 118
        concrete - 118
        control - 277
        glass - 117-118
        graphite - 117

painted surfaces - 118
plastics - 84–106, 297
    foams - 106
    reinforced/composites - 104–106, 170
    rubber - 106–116
Surface tension - 5, 6, 63, 92
    critical - 5, 63, 226–227
Surgical tape - 401
SURLYN - 88, 194
Sweep-frequency resonance test method - 282
Syneresis - 26 (def)
Syntactic foams - 200
Synthetic adhesives - 130 (def)
Synthetic latex - 184 (def)
Synthetic natural rubber - 113
Synthetic rubber adhesives - 131, 196, 198
Syringes, dispensing - 127

Tack - 26 (def), 130, 145, 146, 154, 156, 158, 161, 162, 173–175, 179, 180, 185, 201, 202
Tack (tacky) dry - 27 (def)
Tackifiers - 27 (def), 139, 144, 145, 154, 158, 159, 174
Tacking - 101, 222
Tack range (stage) - 27 (def)
Tack, test methods - 298
Tack welding - 172
Tacky adherends - 172
Tag-end portions of assembly - 277
Tape adhesives - 127, 129, 132, 136, 137, 149–153, 164
Tape, cellophane - 175
Tape, insulating - 175, 177
Tapes - 27 (def), 135, 147, 155, 174, 177, 212, 242
Target cloth - 159
Tear-ply method - 104, 105
TEDLAR - 62, 99, 198
Teeth - 27 (def)
TEFLON FEP - 88
TEFLON TFE - 98
TEFZEL - 88
Telegraphing - 27 (def)
Temperature, changing - 394
Temperature cycling - 395 (see also Thermal cycling)
Temperature environment - 238
Temperature extremes - 8, 165, 182, (see also Cryogenic adhesives

and Elevated-temperature properties)
TENITE - 92, 93
Tennis rackets - 400
Tensile shear strength - 24 (def), 151, 153, 164, 195, 221, 244, 265, 294, 296–297
Tensile strength - 26 (def), 33, 150, 155, 157, 164, 170, 172, 177, 232, 287–288, 298
Tensile stress - 51, 138, 244, 290
Terminal acetylenic groups - 193
Terne - 191
Test methods and practices - *Chapter 11*
Tetrachloroethane - 231
Tetrachloroethylene - 292 (see also Perchloroethylene)
TETRAN TFE - 98
Textiles - 157, 160
Theories of adhesion - 4–7
Thermal conductivity - 4, 134, 140, 141, 244, 397
Thermal cycling - 226, 257–263, 395
Thermal degradation - 166, 175
Thermal expansion (coefficient of) - 2, 9, 31, 50, 51, 92, 134, 136, 157, 202, 243–244
Thermal image inspection - 283
Thermal infrared inspection (TIRI) - 283–284
Thermal insulation - 4
Thermally conductive adhesives - 133, 141
Thermal oxidation - 240
Thermal "spikes" - 169
Thermal stability - 145, 166, 167, 176, 241, 242
Thermal welding - 6, 84, 90, 97, 98, 193, 196, 231
Thermoplastic adhesives - *129, 132*, 154, 240, 244, 269
Thermoplastic elastomer (rubber) adhesives - 131, 165, *180*
Thermoplastic elastomers - 115–116, 155, *180*
Thermoplastic polyester - 84, 92, 93, 96, 200
    foam - 200
Thermoplastic polyurethane - 116
Thermoplastic resins - 130, 193, 225
Thermoplastic rubber - 115–116, 139, *180*

Thermoplastics, bonding of - 84–100, 148, 193, 202
Thermosetting adhesives - 33, 128 (def), 132, 172, 180, 191, 193, 206, 218, 242, 269
Thermosetting plastics, bonding of - 100–103, 130, 138, 148, 181, 198
Thermosetting polyesters - 168
Thermosetting resins - 130, 138, 198
Thickeners - 140, 142
Thickness (adhesive) - 31 (see also Bond-line thickness and Glue-line thickness)
Thinner - 28 (def)
THIOKOL - 115
Thixotropic (thixotropy) - 28 (def), 78, 97, 141, 222, 263
Thread-locking adhesives - 138, 266
Three-component adhesives - 152
Threshold Limit Value (TLV) - 60
Tin - 75, 192
Tin stabilizer, organic - 230
Titanium - 3, 55, 76, 83, 149, 156, 167, 169, 192–193, 222, 245, 247
t/L ratio - 37
Toluene (toluol) - 73, 91–93, 99, 100, 105, 107–115, 120, 162, 178, 200, 229–232, 234, 263, 335
o/p-Toluenesulfonamide - 144
Tongue-and-groove joint - 34, 37, 44
Torque strength - 298
Torsional loading - 289
Toughness - 152, 396
Toxicity - 59–60, 65, 67, 85, 179, 183, 206, 233, 234, 236
TPX - 95
Trichloroethane - 58, 82, 94, 107, 114, 231
Trichloroethylene (TCE) - 58, 60, 67, 70, 76, 80–82, 89, 91–93, 95, 99, 100, 102, 105, 107, 115, 232, 234, 235
Trichlorotrifluoroethane/acetone azeotrope - 58
Trichlorotrifluoroethane/ethyl alcohol - 58
Trichlorotrifluoroethane/methylene chloride - 58
Trichlorotrifluoromethane - 58, 102

Tricresyl phosphate (TCP) - 166
Triethylenetriamine (TETA) - 149
Trifluoromonochloroethylene - 92
Trisodium phosphate - 74, 75, 79, 120
Trisodium pyrophosphate - 65, 69
Trowelable adhesives - 127, 209
Tube winding - 401
TUFFAK - 91
Tungsten - 80, 193
TURCO 4215-S - 82
TURCO 5578 - 78, 79
Twiss, S.B. - 395
Two-component (two-part) adhesives - 15 (def), 128, 134, 137, 144, 158, 174, 177, 178, 181, 198, 206
Two-polymer adhesives - 133, 136–137, 152

UDEL - 98, 170
    Polysulfone P-1700 - 170, 242
ULTEM - 94
Ultrasonic activation - 217–218
Ultrasonic bonding (welding) - 90, 94, 98, 100, 218, 231
Ultrasonic cleaning - 90, 91, 98, 102
    with liquid rinse - 57, 58–59, 98, 118
Ultrasonic pulse echo test method - 282
Ultrasonic scrubbing - 57
Ultrasonic test methods - 281, 282, 284
Ultrasonic vapor degreasing - 57, 58
Ultraviolet-curing adhesives - 180–181, 213, 298
Ultraviolet radiation - 257
Ultraviolet stability - 174, 177
Unbonds - 281, 282
Underwater corrosion - 253, 256
Undissolved matter, test method for - 294
UNISET A-359 - 196
Unsaturated polyester - 168
URALANE 5738 - 197
URALANE 8615 - 197
Uranium - 80–81, 193
Urea-formaldehyde adhesives - 10, 128, 132, 159, 179, 181–182, 197, 198, 200, 201, 294
    radiation effects - 267
Urea-formaldehyde resin - 103, 199

foams - 201
Urea glues - 181
Urea linkage - 246
URETHANE BOND (Dow Corning)
    adhesive - 172
Urethane primer - 98, 103
Usage, adhesives and sealants (see
    Consumption of adhesives
    and sealants)

Vacuum bag - 16 (def), 215
Vacuum blasting - 90, 106
Vacuum environment - 264-265
VALOX - 61, 92
Van Allen belt radiation - 268
Van der Waal's forces - 6
Vapor blast - 73, 81, 87, 89, 95,
    98, 99, 101, 103
Vapor degreasing - 55, 56, 57-58,
    70-75, 77, 78, 81, 82, 98,
    101, 247, 248, 254, 255
Vapor honing - 73, 83
Vapor honing/PASA JELL 107M
    procedure - 83
Vapor rinsing - 71
VAST process - 76, 78, 79, 83
Vegetable-base adhesives (glues) -
    130, 133, 158-160, 401
Vehicle - 28 (def)
Veneer glues - 13 (def)
Veneering (wood) - 201
Vermin - 158
VERSAMIDES - 149, 194, 195,
    197
VESPEL - 102
Vibration effects - 3, 33, 74, 155,
    165, 251, 293, 394, 397
Vibration loosening - 395
VICTREX PES - 93
Vinyl acetal-phenolic adhesives -
    245
Vinyl acetate copolymers - 184
Vinyl acetate-vinyl chloride co-
    polymer adhesives - 191, 192
Vinyl adhesives - 10, 129, 131, 194,
    239
Vinyl alcohol-vinyl acetate copolymer
    adhesives - 191, 192
Vinyl chloride copolymer resins -
    178, 200
Vinyl chloride-vinyl acetate co-
    polymers - 193
Vinyl emulsion adhesives - 402

Vinyl-epoxy adhesives - 131, 173,
    182, 190
Vinyl ethers - 175
Vinyl-phenolic adhesives - 55, 130-
    132, 137, 152, 153, 173, 182-
    183, 192, 194, 239, 266, 397
    primer - 87, 165, 239
Vinyl plastics - 25, 146, 163 (see
    also Polyvinyl chloride plastics)
Vinyl plastisol adhesives - 190
Vinyl primers - 98
Vinyl resin adhesives - 183
Vinyl roofs - 395
Vinyls - 25, 103, 146
Vinyl sealers - 395
Vinyl-vinylidene adhesives - 139
Viscoelastic adhesives - 218
Viscosity - 29 (def), 31, 127, 134,
    146, 154-156, 205, 211, 222,
    400
    coefficient of - 29 (def)
    test methods - 298
Visual inspection - 279, 281
VITEL - 196
VITON - 114
Voids - 155, 212, 281-284, 398
Volatile content (% volatiles) - 202,
    276
Volatiles - 31, 137, 139, 148, 152,
    153, 169, 192, 193, 212, 215,
    240, 243
Volatility - 235, 264-265
Volkersen, theory of - 52
Volume resistivity - 141, 298
Vulcanization - 29 (def), 106, 107,
    127, 145
Vulcanized rubber - 182, 183, 201

Wallpaper paste - 402
Wangsness, D.A. - 239
Warm-setting adhesive - 15 (def)
Warm-wet environment - 79
Warp (warpage) (warping) - 29 (def),
    73, 94
Water absorptiveness (of labels) -
    299
Water-activated adhesives - 212
Water-based adhesives - 131, 160,
    179, 183-185, 192, 199-201
Water-break-free test - 63, 71, 83,
    105, 106, 277
Water dispersions - 173
Water evaporation - 129

Water glass - 156
Water immersion - 142, 157, 164, *246–253*
Water penetration - 239
Water resistance - 140, 150, *157–159*, 161, 163, 164, 166, 170, 174, 176, 182, 201, 202
Water-vapor permeation - 246
Water vapor plasma treatment - 93
Wax (waxes) - 154, 212
Weak-boundary-layer theory (of adhesion) - 6, 7, 54, 246
Wear - 163
Weathering - 8, 78, 136, 142, 147, 150, 165, 166, 173, 177, 182, 201, 233, 238, 239, *257–263*, 293, 395, 396, 398
Webbing - 29 (def)
Wedge test - 66, 256, 291, 299
Wegman, R.F. - 76, 77
Weldbonding - 29 (def), 82, *218–223*
  adhesive choice for - 221–222
  advantages and limitations - 221
  configuration - 220–221
  quality control - 285
  surface preparation - 221
  techniques - 222–223
  tooling - 222
Welding - 3
Welding parameters - 220, 222
Weld nugget - 219, 220, 223, 285
Weld-thru process - 219
Wet-abrasive blasting - 95
Wet assembly - 212
Wet bonding - 184, 201
Wet strength - 26 (def)

Wet tack - 173
Wetting - 5–7, 29 (def), 31, 60, 63, 145, 156, 164, 241, 242, 246, 282
Whiskers - 60
"White" glue - 173
Williams, M.C. - 288
Windows, automobile - 181
Windshields - 394
Wire brush - 117
Wood adherends - 195, 201
Wood adhesives (glues) and sealants - 10, 129, 130, 136, 139, 145–148, 150, 155, 157–161, 164, 166, 169, 172, 173, 176, 179, 181–183, *201*, 217, 296, 395, 400, 402
Wood failure - 19 (def)
Sood, joints for - 44–50
Wood, surface preparation - 116–117
Working life - 8, 22 (def), 128, 175, 206, 299, 398

X-ray inspection methods - 284, 285
Xylene (xylol) - 73, 94, 96, 115, 230, 232, 234, 235

Y-4310 primer (Union Carbide) - 111
Y-5042 primer (Union Carbide) - 111

Zinc - *81*, 138, 193
Zinc oxide - 67
Zinc phosphate cement - 157
Zirconium phosphate - 157
Zisman, W.A. - 63
ZYTEL 101 NC-10 - 230

Index to Standards Listed in Chapter 12

This index covers the 235 standards listed in Chapter 12 on Standard Test Methods, Practices and Specifications. It includes the following standards:

ASTM Test Methods and Practices—90 standards
ASTM Specifications—19 standards
SAE ARP's—3 standards
SAE AMS's—22 standards
Federal Test Method Standard No. 175 methods—4 standards
Federal Specifications—32 standards
Commercial Item Descriptions—3 standards
Military Specifications—62 standards

Only the designations, as ASTM D 1144, are given for the standards listed. Most subject areas of ASTM Test Methods and Practices are listed in Chapter 11 and the reader will find key words on subjects listed in the General Index preceding. The following index can then be used to find the detailed description of the standards. Only ASTM Test Methods and Practices have been covered in the General Index. Keywords and subject terms for ASTM Specifications and all other standards are not covered in either index.

ASTM Test Methods and Practices
ASTM D 816 - 303
ASTM D 896 - 304
ASTM D 897 - 304
ASTM D 898 - 304
ASTM D 899 - 305
ASTM D 903 - 305
ASTM D 904 - 306
ASTM D 905 - 307
ASTM D 906 - 308
ASTM D 950 - 308

ASTM D 1002 - 308
ASTM D 1062 - 309
ASTM D 1084 - 309
ASTM D 1144 - 310
ASTM D 1146 - 310
ASTM D 1151 - 311
ASTM D 1183 - 311
ASTM D 1184 - 312
ASTM D 1304 - 312
ASTM D 1337 - 313
ASTM D 1338 - 313

ASTM Test Methods and
    Practices (cont'd)
  ASTM D 1344 - 313
  ASTM D 1382 - 314
  ASTM D 1383 - 314
  ASTM D 1488 - 315
  ASTM D 1489 - 315
  ASTM D 1490 - 315
  ASTM D 1579 - 315
  ASTM D 1581 - 316
  ASTM D 1582 - 316
  ASTM D 1583 - 316
  ASTM D 1584 - 316
  ASTM D 1713 - 317
  ASTM D 1780 - 317
  ASTM D 1781 - 318
  ASTM D 1828 - 318
  ASTM D 1875 - 319
  ASTM D 1876 - 319
  ASTM D 1879 - 320
  ASTM D 1916 - 320
  ASTM D 2093 - 321
  ASTM D 2094 - 321
  ASTM D 2095 - 321
  ASTM D 2182 - 322
  ASTM D 2183 - 322
  ASTM D 2293 - 323
  ASTM D 2294 - 323
  ASTM D 2295 - 324
  ASTM D 2339 - 324
  ASTM D 2556 - 324
  ASTM D 2557 - 324
  ASTM D 2558 - 325
  ASTM D 2651 - 325
  ASTM D 2674 - 325
  ASTM D 2739 - 326
  ASTM D 2855 - 326
  ASTM D 2918 - 326
  ASTM D 2919 - 327
  ASTM D 2979 - 327
  ASTM D 3111 - 328
  ASTM D 3121 - 328
  ASTM D 3163 - 328
  ASTM D 3164 - 329
  ASTM D 3165 - 329
  ASTM D 3166 - 330
  ASTM D 3167 - 330
  ASTM D 3236 - 331
  ASTM D 3310 - 331
  ASTM D 3433 - 331
  ASTM D 3482 - 332
  ASTM D 3528 - 332
  ASTM D 3632 - 332

  ASTM D 3658 - 333
  ASTM D 3762 - 333
  ASTM D 3807 - 334
  ASTM D 3808 - 334
  ASTM D 3929 - 335
  ASTM D 3931 - 335
  ASTM D 3933 - 335
  ASTM D 3983 - 336
  ASTM D 4027 - 336
  ASTM D 4299 - 337
  ASTM D 4300 - 337
  ASTM D 4338 - 338
  ASTM D 4339 - 338
  ASTM D 4426 - 338
  ASTM E 229 - 302
  ASTM E 864 - 303
  ASTM E 874 - 303
  ASTM F 402 - 303

ASTM Specifications
  ASTM C 557 - 341
  ASTM D 1580 - 342
  ASTM D 1779 - 343
  ASTM D 1874 - 343
  ASTM D 2235 - 343
  ASTM D 2559 - 343
  ASTM D 2560 - 344
  ASTM D 2564 - 344
  ASTM D 2851 - 344
  ASTM D 3024 - 344
  ASTM D 3110 - 345
  ASTM D 3122 - 345
  ASTM D 3138 - 345
  ASTM D 3498 - 346
  ASTM D 3930 - 346
  ASTM D 4317 - 346
  ASTM E 865 - 342
  ASTM E 866 - 342
  ASTM F 656 - 341

SAE ARP's
  ARP 1524 - 339
  ARP 1575 - 339
  ARP 1610 - 339

SAE AMS's
  AMS 2491 - 347
  AMS 3107 - 347
  AMS 3107/1 - 348
  AMS 3107/2 - 348
  AMS 3107/3 - 348
  AMS 3107/4 - 348
  AMS 3681 - 348

SAE AMS's (cont'd)
  AMS 3685 - 349
  AMS 3686 - 349
  AMS 3687 - 349
  AMS 3688 - 350
  AMS 3689 - 350
  AMS 3690 - 350
  AMS 3691 - 351
  AMS 3692 - 351
  AMS 3693 - 351
  AMS 3695 - 351
  AMS 3695/1 - 352
  AMS 3695/2 - 352
  AMS 3695/3 - 352
  AMS 3695/4 - 352
  AMS 3698 - 352

General Services Administration
    (GSA) - 340, 353
  Federal Test Methods - 340
    Federal Test Method
        Standard No. 175 - 340
      Method 1081 - 340
      Method 4032 - 340
      Method 4041 - 340
      Method 4051 - 341
  Federal Specifications and Com-
      mercial Item Descriptions -
      353
    MMM-A-100 - 353
    MMM-A-105 - 353
    MMM-A-110 - 354
    MMM-A-115 - 354
    MMM-A-121 - 355
    MMM-A-122 - 355
    MMM-A-125 - 355
    MMM-A-130 - 355
    MMM-A-131 - 356
    MMM-A-132 - 356
    MMM-A-134 - 357
    MMM-A-137 - 357
    MMM-A-138 - 358
    MMM-A-139 - 358
    MMM-A-00150 - 359
    MMM-A-179 - 359
    MMM-A-180 - 360
    MMM-A-181 - 360
    MMM-A-182 - 361
    MMM-A-185 - 361
    MMM-A-187 - 361
    MMM-A-188 - 362
    MMM-A-189 - 362
    MMM-A-250 - 363

    MMM-A-260 - 363
    A-A-342 - 363
    A-A-529 - 364
    A-A-863 - 364
    MMM-A-1058 - 364
    MMM-A-1617 - 365
    MMM-A-1754 - 365
    MMM-A-1931 - 366
    MMM-A-001993 - 366
    MMM-A-002015 - 367
    MMM-A-002408 - 367

Department of Defense (DOD) - 367
  Military Specifications - 367
    MIL-G-413 - 367
    MIL-C-2399 - 368
    MIL-A-3167 - 368
    MIL-A-3316 - 369
    MIL-A-3562 - 370
    MIL-A-3920 - 370
    MIL-A-5540 - 370
    MIL-A-8576 - 371
    MIL-A-9117 - 371
    MIL-A-13374 - 371
    MIL-A-14064 - 372
    MIL-A-21016 - 372
    MIL-A-21366 - 373
    MIL-A-22397 - 373
    MIL-A-22434 - 373
    MIL-S-22473 - 374
    MIL-A-22895 - 374
    MIL-C-23092 - 375
    MIL-A-23940 - 375
    MIL-A-23941 - 375
    MIL-A-24179 - 375
    MIL-A-24456 - 376
    MIL-A-25463 - 376
    MIL-A-43316 - 377
    MIL-A-45059 - 378
    MIL-G-46030 - 378
    MIL-A-46050 - 378
    MIL-A-46091 - 378
    MIL-A-46106 - 380
    MIL-A-46146 - 380
    MIL-A-46864 - 381
    MIL-A-47040 - 381
    MIL-A-47074 - 382
    MIL-A-47089 - 382
    MIL-A-47126 - 383
    MIL-P-47170 - 383
    MIL-P-47275 - 383
    MIL-P-47276 - 383
    MIL-P-47279 - 384

Department of Defense (DOD) (cont'd)
   Military Specifications (cont'd)
      MIL-A-47280 - 384
      MIL-A-47284 - 384
      MIL-A-48611 - 384
      MIL-B-48612 - 385
      MIL-A-50926 - 386
      MIL-A-52194 - 386
      MIL-A-52685 - 386
      MIL-A-60091 - 387
      MIL-A-81236 - 387
      MIL-A-81253 - 387

MIL-A-81270 - 388
MIL-K-81786/11 - 388
MIL-K-81786 - 388
MIL-A-82484 - 389
MIL-A-82569 - 389
MIL-A-82636 - 389
DOD-A-82720 - 390
MIL-A-83376 - 390
MIL-A-83377 - 390
MIL-A-85705 - 391
MIL-A-87134 - 391
MIL-A-87135 - 392